KINEMATICS
AND
MECHANISMS DESIGN

Board of Advisors, Engineering

Kinematics and Mechanisms Design

C. H. Suh
University of Colorado

C. W. Radcliffe
University of California

JOHN WILEY & SONS
NEW YORK   SANTA BARBARA   CHICHESTER   BRISBANE   TORONTO

Library of Congress Cataloging in Publication Data

Suh, Chung Ha.
Kinematics and mechanisms design.

Includes bibliographical references and index.
1. Machinery, Kinematics of. 2. Machinery—Design.
I. Radcliffe, Charles W., joint author. II. Title.

TJ175.S93   621.8'15   77-7102

ISBN 0-471-01461-3

Printed in the United States of America

10 9 8 7 6 5 4 3 2 1

Preface

Prior to 1950 graphical methods were used almost exclusively for kinematic analysis and synthesis of mechanisms. Graphical methods were particularly useful in the design of planar mechanisms, and methods for kinematic analysis using vector polygons continue to be included in the curriculum of mechanical engineering. Graphical methods for kinematic synthesis, although well-known in Europe, did not become a significant part of the engineering curriculum in the United States until after the series of seven "Conferences on Mechanisms" were organized at Purdue University from 1953 to 1962 through the efforts of Professor A. S. Hall, Jr. These conferences, along with the increasing availability of high-speed digital computers, led to a major reawakening of interest in what had long been considered a mature and somewhat dormant field.

At many engineering schools this "new" material was taught as an advanced undergraduate or beginning graduate-level course. Subject matter included vector methods in kinematic analysis with an emphasis on complex polar notation, analytical kinematic synthesis of planar mechanisms, graphical Burmester theory for plane linkage synthesis, and planar curvature theory with applications to synthesis.

The development of theory and methods for analysis and synthesis of spatial mechanisms proceeded rapidly after 1950. Many new and useful methods, often requiring considerable mathematical sophistication in their development, have been published. Unfortunately, it is difficult to merge these methods into a consistent and comprehensive set of principles applicable to both plane and spatial mechanisms design.

This textbook presents modern kinematics with an emphasis on numerical design methods. The mathematical background required for an understanding of the material is modest and easily within the capabilities of advanced undergraduate or beginning graduate engineering students. Throughout the text an effort has been made to introduce notation and methods that provide an orderly transition from planar kinematics to the more complex geometry of spatial mechanisms.

v

The subject-matter is definitely computer-oriented. It is recognized that graphical methods offer a degree of physical visualization and interaction sometimes lacking with numerical methods. However, the advantages of greater accuracy, extension of possible solutions to spatial problems, and speed of computation possible with high-speed digital computers appear to offer definite advantages, particularly with the development of advanced minicomputers and computer graphics capability at modest cost.

To use this textbook effectively, both instructor and student must have some background in FORTRAN programming. The assignments are often aimed at the development of a program that can solve a *class of problems* instead of one problem with specific geometry, as would be typical for a graphical solution. The student first identifies the constraints and carries out the algebraic vector or matrix operations to form an algorithm for the problem. This algorithm then can be coded in FORTRAN and tested, using numerical results published with the many examples given. Once tested, the program is available for the solution of any problem of similar class. In many cases the time required to develop the code will be comparable to that required for a complete graphical solution of one problem.

Many of the solutions given in the textbook require the use of general purpose programs for solution of linear or nonlinear algebraic equations and for constrained nonlinear minimization. Programs DESIGN, PCON, and LSTCON, as listed in the appendix, have been found useful for these purposes; it is essential that these or similar programs be made available to students.

We acknowledge the help of students and colleagues who have aided in the development of computer programs and other material for the book. We hope that as a result of their efforts the published computer codes will be useful to future students and that design of mechanisms will be a rewarding and challenging experience.

C. H. Suh
C. W. Radcliffe

Contents

CHAPTER 1 BASIC CONCEPTS   1

    1.1 Mechanisms   1
    1.2 The Rigid Body in Kinematics—The Link   2
    1.3 Rigid Body Displacement   2
    1.4 Kinematic Connections—The Kinematic
         Pair   3
    1.5 Relative Motion—The Kinematic Chain   3
    1.6 Kinematic Inversion   3
    1.7 Transmission of Motion   6

**CHAPTER 2 VECTOR METHODS IN PLANE
KINEMATICS**   8

    2.1 Introduction   8
    2.2 Complex Polar Vector Notation   8
        Differentiation of complex polar vectors   9
    2.3 Kinematic Analysis Using Complex Polar
         Notation   9
        The four-bar linkage   9
        Velocity analysis of the four-bar linkage   11
        Acceleration analysis of the four-bar
         linkage   12

2.4   Cartesian Vector Notation   14
      The scalar product of two vectors—the
          dot product   14
      The vector product—the cross product   15
      Differentiation of vectors   16
2.5   Kinematic Analysis of Plane Mechanisms
          Using Cartesian Vector Notation   17
      The four-bar linkage   17
      Position Analysis   17
          Velocity analysis   17
          Acceleration analysis   18
      The oscillating slider mechanism   19
          Position analysis   19
          Velocity analysis   21
          Acceleration analysis   21
2.6   A General Method for Computer-Aided
          Kinematic Analysis of Plane Mechanisms   21
      Rigid body motion   22
      The two-link dyad   23
      The oscillating slider   25
      The rotating guide   27
      Rigid body motion subroutines GEOM
          DISP, POS, VEL, ACC, and MOTION   29
      Two-link dyad subroutines CRANK2, PDYAD,
          VDYAD, and ADYAD   32
      Input-output subroutines RDACC and
          WRACC   34
      Oscillating slider subroutines POSC,
          VOSC, and AOSC   36
      Rotating guide subourtines PGUIDE,
          VGUIDE, and AGUIDE   40

Appendix A   The Coriolis Component—
Acceleration of a Point Relative to a Rotating
Reference System   42

CHAPTER 3   MATRIX METHODS IN KINEMATICS   45

3.1   Introduction   45
3.2   Rigid Body Rotation Matrices   45
        Rotation about Cartesian axes   45
3.3   Spatial Rotation Matrices   47
3.4   Rigid Body Displacement Matrices   51
3.5   Applications of the Plane Rotation Matrix   53
        Example 3-1   Displacement of an arbitrary
                              point moving with a rigid
                              body   53
        Example 3-2   The finite rotation pole   54
3.6   The Screw Matrix   55
3.7   Numerical Displacement Matrices by
        Direct Matrix Inversion   56
        Reduction of the number of points
          required for the Matrix Inversion
          Method   57
3.8   The Inverse Displacement Matrix   58
3.9   Derivation of Screw Motion Parameters
        from Numerical Displacement Matrix
        Elements   59
3.10 Coordinate Transformations   61
        Vector and point transformations   61
        Successive coordinate transformations   62
3.11 Hartenberg-Denavit Notation   62
3.12 Differential Rotation Matrices   65
3.13 Differential Displacement Matrices   69

CHAPTER 4   KINEMATIC ANALYSIS OF SPATIAL
MECHANISMS   70

4.1   Relative Spatial Motion   70
4.2   Relative Displacement   71
4.3   Relative Velocity   72
4.4   Relative Acceleration   73
4.5   Plane Kinematic Analysis Using Relative
        Joint Rotation Angles   74
      Displacement analysis of the four-bar
        linkage   74
      Relative velocity analysis   76
      Relative acceleration analysis   77
4.6   Kinematic Analysis of Spatial Mechanisms
        in Closed Form   79
      The RSSR mechanism   79
      Velocity analysis   79
      Acceleration analysis   80
      Example 4-1 Kinematic analysis of the
                   RSSR spatial function generation
                   mechanism in closed form   80
      The RRSS mechanism   83
      Example 4-2   Kinematic analysis of the
                     RRSS path generation
                     mechanism in closed form   86
      The RCCC mechanism   89
      Example 4-3   Acceleration analysis of the
                     RCCC mechanism   94
4.7   Spatial Kinematic Analysis by Numerical
        Solution of the Constraint Equations   94
      The RRSS path generation mechanism   95

Example 4-4   The RRSC path generation
              mechanism—iterative solution
              of the nonlinear constraint
              Equations   98
Example 4-5   Displacement analysis of the
              RCCC mechanism by
              numerical solution of the
              constraint equations   99

**CHAPTER 5 MOBILITY ANALYSIS OF
            MECHANISMS**   103

5.1 Introduction   103
5.2 Constraint Analysis   103
     Degrees of freedom   103
     Maverick mechanisms   109
5.3 Number Synthesis of Plane Linkages   111
5.4 Range of Motion Analysis   112
     Motion classification for the plane four-bar
        linkage   the Crashof criterion   112
     Displacement analysis of spatial
        mechanisms   113
5.5 Mobility Analysis   114
     Mobility of the plane four-bar linkage   114
     Mobility of the RSSR mechanism   117
     Mobility of the RRSS mechanism   121
     Mobility of the RSRC mechanism   123

CHAPTER 6 RIGID BODY GUIDANCE   128

6.1 Introduction   128
6.2 Plane Rigid Body Guidance Mechanisms   129

The constant-length equations—two joint
  cranks   129
The constant slope equations—plane
  sliders   130
6.3 Three-Position Crank Synthesis   131
6.4 Three-Position Slider Synthesis   134
6.5 Input Crank Motion Parameters   136
6.6 Example Problems: Three-Position
     Synthesis   137
     Example 6-1 Crank synthesis—three
                  finitely separated positions
                  of a moving plane   137
     Example 6-2 Slider synthesis—finite
                  displacements   138
     Example 6-3 Slider synthesis with $\theta_{12} = 0.0$ 138
     Example 6-4 Slider that moves in a
                  specified direction   139
     Example 6-5 Crank synthesis—velocity
                  specified in first position
                  plus displacement to a
                  second position   139
     Example 6-6 Crank synthesis—velocity and
                  acceleration specified in
                  one position   143
6.7 Four-Position Synthesis—Crank Constraint   143
     The Newton-Raphson method   143
     Crank constraint equations—four-position
       plane rigid body guidance   144
     Example 6-7 Center and circle point
                  curves   146
6.8 Four-Position Synthesis—Slider Constraint   146

6.9   Four-Position Combined Finite-Differential
        Synthesis   146
      Example 6-8   Combined displacement-
                    velocity-acceleration
                    rigid body guidance   147
6.10 Spherical Rigid Body Guidance
        Mechanisms   148
      Spherical displacement matrices by direct
        numerical inversion   148
      Design equations for spherical rigid
        body guidance   149
      Example 6-9   Synthesis of a spherical
                    rigid body guidance
                    mechanism   150
      Three-position spherical rigid body
        guidance   151
6.11 General Spatial Rigid Body Guidance   152
      The sphere-sphere (S-S) link   152
      The revolute-sphere (R-S) link   154
      The revolute-revolute (R-R) link   155
      The revolute-cylindrical (R-C) link   157
      The cylindrical-cylindrical (C-C) link   158
      Example 6-10   Three-position rigid body
                     guidance using R-R,
                     S-S, C-S, or R-C links   159
      Appendix A   Derivation of Eq. 6.21   162

CHAPTER 7   FUNCTION GENERATION   164
7.1   Introduction   164
7.2   Problem Formulation—Specification of
        Precision Points   164

7.3   Chebyshev Spacing of Precision Points   166
7.4   Scale Factors for Input and Output
        Motion   167
7.5   Three-Position Function Generator
        Mechanisms   167
      The four-bar linkage function generator   167
      Example 7-1   The three-position four-bar
                      function generator   169
      The slider-crank function generator   169
      Example 7-2   The three-position slider-
                      crank function generator   170
      The double-slider function generator   172
      Example 7-3   The three-position double-
                      slider function generator   173
7.6   Function Generators with Combined
        Finite and Differential Motion
        Specifications   173
      The plane four-bar function generator   173
      Example 7-4   Velocity-acceleration
                      synthesis of a plane four-
                      bar generator   174
      The slider-crank function generator   175
      Example 7-5   Velocity-acceleration
                      synthesis of a plane slider-
                      crank function generator   175
      The double-slider function generator   176
      Example 7-6   Velocity-acceleration
                      synthesis of a plane
                      double-slider function
                      generator   177
7.7   Four-Position Plane Function Generators   177

The four-bar linkage function generator   177
Example 7-7   The crossed linkage four-bar
                      function generator   179
7.8   The Spherical Four-Bar Function
         Generator   179
7.9   The RSSR Spatial Four-Bar Function
         Generator   181
Example 7-8   Synthesis of an RSSR
                      function generator with
                      specified plane of
                      rotation for both input and
                      output cranks   183
Example 7-9   Design of an RSSR function
                      generator with six
                      precision points   185
7.10 The RSSR Function Generator with
         Specified Velocity and Acceleration   187
Example 7-10 Synthesis of the RSSR
                      function generator with
                      specified velocity and
                      acceleration   188

CHAPTER 8   PATH GENERATION   190

8.1   Introduction   190
8.2   The Plane Four-Bar Path Generation
         Linkage   190
Example 8-1   Design of a plane four-bar
                      path generator with five
                      path precision points   193

Example 8-2   Design of a plane four-bar
                       path generator with
                       specified length for both
                       guiding cranks   193
8.3   The Spherical Four-Bar Path Generator   195
      Example 8-3 Design of a spherical path
                       generation mechanism
                       with four path precision
                       points   197
8.4   Displacement Matrices for Spatial Path
      Generation Mechanisms   200
8.5   The RRSS Path Generation Mechanism   201
8.6   The RSSR-SS Path Generation Mechanism   202
8.7   Displacement Analysis—Path Generation
      Mechanisms   204
      Example 8-4   Displacement analysis of a
                       plane four-bar path
                       generation mechanism   204

CHAPTER 9   OPTIMAL SYNTHESIS OF
              MECHANISMS   206
9.1   Introduction   206
9.2   Minimization of the Integrated Error   206
      Design versus precision points   206
9.3   Optimal Kinematic Synthesis as a Problem
      in Nonlinear Programming   207
9.4   The Objective Function   207
9.5   Inequality Constraints   208
9.6   Equality Constraints   209
9.7   Geometrical Representation of the
      Nonlinear Programming Problem   209

9.8   Unconstrained Minimization   211
      Mathematical properties of a minimum   211
      Local versus global minima   211
      Approximation of functions   211
      The search for a minimum   212
      Conjugate directions   213
      Nonderivative search for a minimum   214
9.9   Powell's Nonderivative Direct Search
      Method   215
9.10 The Method of Least Squares   216
9.11 Powell's Nonderivative Least Squares
      Method   217
      Numerical approximation of the partial
      derivatives   218
9.12 Constrained Minimization—Penalty
      Functions   219
      The interior penalty function   219
      The exterior penalty function   220
      Equality constraints   221
      Scaling   221
9.13 Case Studies in Optimal Design of
      Mechanisms   221
      Example 9-1   Optimal synthesis of a plane
                    four-bar function generator 221
      Example 9-2   Optimal synthesis of a plane
                    four-bar linkage for
                    combined path generation
                    and rigid body guidance   223
      Example 9-3   The auto window glass
                    guidance mechanism   226
      Example 9-4   The RRSS path generation

mechanism (least squares
solution)   229

Example 9-5   The RRSS path generation
mechanism (direct search
solution)   231

Example 9-6   Optimal synthesis of the
plane four-bar function
generation mechanism
with inequality and
equality constraints   234

Appendix A   Proof of Convergence for
Conjugate Directions Search   236

Appendix B   Generation of Conjugate
Directions in the Powell Search Method   237

**CHAPTER 10   DIFFERENTIAL GEOMETRY OF
MOTION**   238

10.1   Introduction   238
10.2   Instantaneous Screw Parameters   238
10.3   Geometry of Screw Axis Surfaces   240
Instant pitch—a unique geometrical
property of rigid body motion   241
Geometric properties of screw axis
surfaces   241
Instantaneous screw calculus   242

Example 10-1   Instantaneous screw
motion parameters for
the coupler of a spatial
double-slider mechanism 244

10.4   Differential Displacement Matrices in
Terms of Geometrical Parameters   248

The Phi matrix   248
The plane phi matrices   250
Canonical coordinate systems   251
10.5   Plane Path Curvature   251
The velocity pole, $\mathbf{p}_0$ 251
Pole velocity, $\dot{\bar{\mathbf{p}}}_0$ 252
The acceleration pole   252
The inflection circle—zero normal
    acceleration   253
The zero tangential acceleration circle   255
The cubic of stationary curvature   256
Geometric analysis of plane mechanisms   257
Example 10-2   The rolling cylinder   258
Example 10-3   The double-slider
                    mechanism   260
Example 10-4   The four-bar linkage   262
Example 10-5   Higher-order kinematic
                    synthesis from the
                    instantaneous invariants
                    of the coupler motion   265
10.6   Higher-Order Path Curvature in Spatial
        Coupler Curves   268
Parametric equations   268
Geometric properties of spatial curves   269
Path curvature   269
Example 10-6   Path curvature analysis—
                    the RRSS mechanism   270

CHAPTER 11   DYNAMICS OF MECHANISMS   276
11.1   Introduction   276
11.2   Dynamics of the Plane Four-Bar Linkage   276

Example 11-1   Dynamic analysis of a plane
four-bar linkage   279

Example 11-2   The oscillating slider
mechanism   280

11.3   Dynamic Balancing of the Four-Bar
Linkage   283

Example 11-3   Dynamic balancing of a
four-bar linkage   287

11.4   The Inverse Dynamics Problem   289
Reduced mass or inertia   289
Reduced force or torque   290
The work-energy method .   290
The nonlinear differential equation of
motion   291
The predictor-corrector equations   292

Example 11-4   The inverse dynamics
problem   293

11.5   Dynamics of Spatial Mechanisms   294
Coordinate transformations   294
Vector transformations   294

Example 11-5   Coordinate transformations 295
Matrix transformations   296

11.6   The Dynamical Equations of Motion   297
Euler's equations   298
Equations of motion in matrix form   299

Example 11-6   Dynamics of a rotating
system   300

CHAPTER 12   COMPUTER PROGRAMS   305

12.1   Introduction   305

12.2   LINKPAC, a Subroutine Package Useful
       in Plane Kinematics and Mechanisms   305
       Example 12-1   Acceleration analysis of the
                      offset slider-crank
                      mechanism   306
       Example 12-2   Acceleration and dynamic
                      force analysis for the
                      plane four-bar linkage   308
12.3   DESIGN—Solution of Sets of Simultaneous
       Nonlinear Algebraic Equations by the
       Newton-Raphson Method   312
       Example 12-3   Use of program DESIGN   313
       Example 12-4   Center and circle points
                      for four-position rigid
                      body guidance   315
       Example 12-5   Displacement analysis of
                      the RRSC mechanism
                      using SPAPAC sub-
                      routines in YCOMP   317
       Example 12-6   Spatial three-position
                      guidance   320
12.4   PCON—Constrained Minimization Using
       Powell's Direct Search Conjugate
       Directions Method and the SUMT
       Procedure   326
       Example 12-7   The Rosenbrock test
                      function with added
                      inequality constraints   328
       Example 12-8   Optimal synthesis of a
                      combined path
                      generation and plane

                                    rigid body guidance
                                    mechanism (Example
                                    9-2)   330

12.5 LSTCON—Constrained Minimization
        Using Powell's Least Squares Method   335
        Example 12-9   Solution of three nonlinear
                                algebraic equations with
                                constraints   335

12.6 NONLIN—Solution of Simultaneous
        Nonlinear Algebraic Equations with
        Random Number Initial Guesses   338
        Example 12-10   Solution of a set of non-
                                linear algebraic
                                equations with random
                                initial guesses   338

12.7 SIMEQ, a program in BASIC Language
        for the Solution of Simultaneous
        Nonlinear and Linear Equations   340

12.8 PRCR2, a Predictor-Corrector Method
        for the Solution of a Second-Order
        Nonlinear Differential Equation   341
        Example 12-11   Function DERIV2 for the
                                Inverse Dynamics
                                Problem, Example 11.4   341

APPENDICES
        Appendix   1   LINKPAC   347
        Appendix   2   SPAPAC   362
        Appendix   3   DESIGN   371
        Appendix   4   NONLIN   377

Appendix   5   PCON   381
Appendix   6   LSTCON   394
Appendix   7   SIMEQ   406
Appendix   8   PRCR2   408

References   413
Problems   417
Index   431

Chapter 1
Basic
Concepts

1.1 MECHANISMS

A mechanism has been defined by Reuleaux [1] as "a combination of rigid or resistant bodies so formed and connected that they move upon each other with definite relative motion."

Mechanisms form the basic geometrical elements of many mechanical devices including automatic packaging machinery, typewriters, mechanical toys, textile machinery, and others. A mechanism typically is designed to create a desired motion of a rigid body relative to a reference member. Kinematic design of mechanisms often is the first step in the design of a complete machine. When forces are considered, the additional problems of dynamics, bearing loads, stresses, lubrication, and the like are introduced, and the larger problem becomes one of *machine design*.

The function of a mechanism is to transmit or transform motion from one rigid body to another as part of the action of a machine. There are three types of common mechanical devices that can be used as basic elements of a mechanism.

1. *Gear systems*, in which toothed members in contact transmit motion between rotating shafts. Gears normally are used for the transmission of motion with a constant angular velocity ratio, although noncircular gears can be used for nonuniform transmission of motion. The kinematic design of gearing is primarily concerned with specification of the shape of contacting tooth surfaces in order to achieve the desired angular velocity ratio.
2. *Cam systems*, where a uniform motion of an input member is converted into a nonuniform motion of the output member. The output motion may be either shaft rotation, slider translation, or other follower motions created by direct contact between the input cam shape and the follower. The kinematic design of cams involves the analytical or graphical specification of the cam surface shape required to drive the follower with a motion that is a prescribed function of the input motion. The subject of cam design is a highly specialized topic [2, 3] and will not be covered in this textbook.

1

3. Plane and spatial linkages are also useful in creating mechanical motions for a
   point or rigid body. Linkages can be used for three basic tasks.
   (a) *Rigid body guidance.* A rigid body guidance mechanism is used to guide a
       rigid body through a series of prescribed positions in space.
   (b) *Path generation.* A path generation mechanism will guide a point on a rigid
       body through a series of points on a specified path in space.
   (c) *Function generation.* A mechanism that creates an output motion that is a
       specified function of the input motion.

Planar linkages, in which all motions are parallel to a reference plane, have been
used extensively in machinery for many years. Efficient graphical methods for
analysis and synthesis of plane linkages are well known [4, 5, 6] and have been
applied successfully to many design problems. Graphical methods, because of the
need to work with isometric projections, have not been as successful when applied
to problems of spatial kinematics. Recently there has been an acceleration of
research and publication of new analytical methods encouraged by the availability
of high-speed digital computers. Spatial kinematic theory and linkage design
techniques have been developed [7, 8, 9, 10] to the point where the machine designer in industry will have many more design options in the future.

The mathematics of spatial kinematics is often complex and highly nonlinear,
but modern numerical methods [11, 12] often will minimize mathematical manipulations and give solutions with relatively modest computing machinery.

In any actual design it is to be expected that the designer will consider and use
combinations of all three types of mechanisms, plus additional power sources,
controls, and intermediate links such as flexible belts and chains, as required.

1.2 THE RIGID BODY IN KINEMATICS—THE LINK

Kinematics is the study of the relative motion between two or more physical
bodies. Each rigid body is considered to include all points that move with a set of
coordinate axes fixed in the body. Any point in space may be considered as
momentarily attached to any one of a system of rigid bodies. The rigid body
condition may be imposed analytically by requiring that the distance between any
two arbitrary points fixed in the rigid body remains constant during and after a
change in position of the body.

1.3 RIGID BODY DISPLACEMENT

Displacement refers to the change in position of a rigid body. If the displacement
of any three noncolinear points is specified, the corresponding spatial displacement of any other point in the body is defined. Rigid body displacement can be
specified graphically, analytically or, in some cases, numerically. The description

of displacement as a matrix operation, both in analytical and numerical form, is given in Chapter 3. Velocity is the rate of change of position with time and involves a first-order differential displacement. Acceleration is expressed in terms of second-order differential displacement components.

1.4 KINEMATIC CONNECTIONS—THE KINEMATIC PAIR

Two links are connected by contact at a "pair" of contacting surfaces. A "lower pair" is a connection between two links where the relative motion can be described by a single coordinate (i.e., a connection with a single degree of freedom). Examples of lower-pair connections are:

1. The revolute pair—a simple hinge without axial sliding.
2. The prismatic pair—a slider with motion in one direction.
3. The rolling pair—pure rolling with zero slip.

At a lower pair the relative motion is usually transmitted through area contact or its equivalent.

Higher-pair connections are those with more than one degree of freedom, such as the roll-slip pair at the point of contact between cam and follower or at a spherical ball and socket joint. Higher pairs are characterized by either line or point contact or their kinematic equivalent.

The six basic kinematic pairs are shown in Figure 1.1. Mechanisms involving higher pairs may be reduced to their equivalent shown as a series of lower pairs. For example, a cylindrical pair is equivalent to a combined revolute and prismatic pair.

1.5 RELATIVE MOTION—THE KINEMATIC CHAIN

A kinematic chain is the assembly of link/pair combinations to form one or more closed loops. The chain is said to be *simple* if it involves a single vector loop equation and *complex* if it involves two or more independent vector loops.

A complex mechanism will often contain at least one *compound* link. A simple link is the binary link with two lower pairs, one at each point of connection into the chain. A compound link may be classed as ternary, quaternary, and so forth, depending on the number of separate kinematic pairs with other links of the chain.

A useful mechanism will be formed only when one link in the chain is fixed to ground or to a member with specified motion.

1.6 KINEMATIC INVERSION

Inversions of a kinematic chain are the family of different mechanisms formed depending on the choice of fixed link. A kinematic inversion, in which link lengths are unchanged, does not affect the relative motion at the joints of the chain. The

Name of Pair	Geometric Form	Schematic Representations	Degrees of Freedom
1. Revolute (R)			1
2. Cylinder (C)			2
3. Prism (P)			1
4. Sphere (S)			3
5. Helix (H)			1
6. Plane (P_L)			3

FIGURE 1.1   KINEMATIC PAIRS.

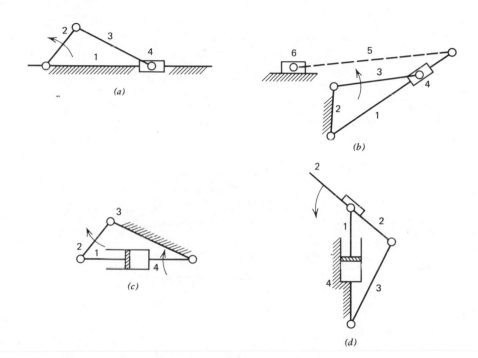

FIGURE 1.2   KINEMATIC INVERSIONS OF THE SLIDER-CRANK. (*a*) THE BASIC SLIDER CRANK MECHANISM. LINK 1 FIXED, LINK 2 INPUT, LINK 4 OUTPUT. (*b*) THE WHITWORTH QUICK RETURN. LINK 2 FIXED, LINK 3 INPUT, LINK 1 OUTPUT. (*c*) THE OSCILLATING CYLINDER. LINK 3 FIXED, LINK 2 INPUT, LINK 4 OUTPUT. (*d*) PUMP MECHANISM. LINK 4 FIXED, LINK 2 INPUT, LINK 1 OUTPUT.

absolute motion of the members or points moving with a particular member will obviously be different, considering the choice of fixed link.

The concept of kinematic inversion is also important in the motion analysis of complex mechanisms. The analysis often may be simplified by considering a kinematically simpler inversion of the mechanism, then converting the results to the actual mechanism.

Figure 1.2 shows various inversions of a four-link, four-lower-pair planar mechanism, in this case the slider-crank mechanism. Note that one of the lower pairs is a prismatic pair.

Figure 1.3 indicates the variety of mechanisms that are possible assuming a six-link, seven-lower-pair chain. There are two basic six-link chains, and each may be arranged in several ways.

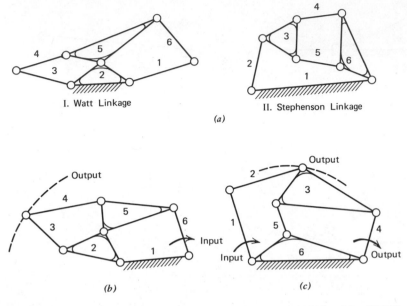

I. Watt Linkage   II. Stephenson Linkage

(a)

(b)   (c)

FIGURE 1.3 INVERSION OF BASIC SIX-LINK CHAINS. (*a*) BASIC SCHEMATIC: SIX-LINK MECHANISM. (*b*) INVERSION OF TYPE I SIX-LINK CHAIN. POSSIBLE PATH GENERATION MECHANISM. (*c*) INVERSION OF TYPE II SIX-LINK CHAIN. POSSIBLE FUNCTION GENERATOR OR PATH GENERATOR.

1.7 TRANSMISSION OF MOTION

Figure 1.4 indicates several ways in which mechanical devices may transmit motion. In each case the line n-n is the line along which there is a definite transmission of force (hence, motion). In the case of direct contact mechanisms such as cams or gear teeth, the line of transmission is along the common normal to the surfaces at the point of contact. In a linkage the line n-n is along the line of centers of the joints in the intermediate coupler link. In mechanisms with flexible tension links, such as belts, cables, and chains, the line has meaning only when the flexible link is in tension.

The transmission angle μ is the acute angle between the coupler and output link in a four-bar plane linkage. It is the complement of the pressure angle α in a cam mechanism (i.e., $\mu = 90 - \alpha$), as shown in Figure 1.5.

In order to make the force in the coupler most effective in developing torque about the output crank pin, it should have a maximum lever arm at all times (i.e., the transmission angle μ should be held as close as possible to 90° over the full

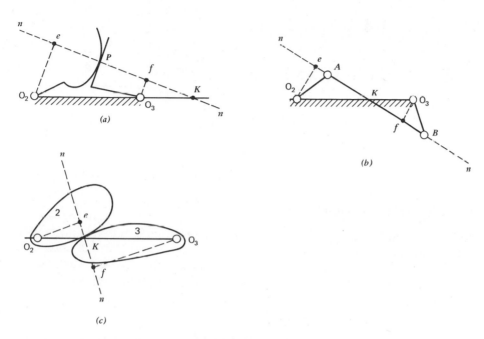

FIGURE 1.4   THE LINE OF TRANSMISSION OF MOTION. (a) DIRECT CONTACT CAM MECHANISM. (b) FOUR BAR LINKAGE. (c) ROLLING CONTACT MECHANISM.

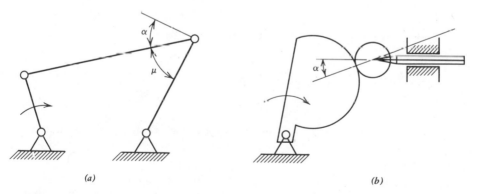

FIGURE 1.5   TRANSMISSION VERSUS PRESSURE ANGLE. (a) THE TRANSMISSION ANGLE, μ, IN LINKAGE MECHANISMS. (b) THE PRESSURE ANGLE, α, IN CAM MECHANISMS.

range of motion). In any real situation it will be necessary to approximate this condition but, in any case, a practical value for minimum allowable transmission angle will be $\geq 30°$. In designing cam systems the maximum pressure angle α should be $\leq 30°$.

Chapter 2

Vector Methods in

Plane Kinematics

2.1 INTRODUCTION

Vector notation forms a basic analytical method for the description of the motion of rigid bodies. Vectors can be described in both polar and Cartesian notation, each of which has advantages. Matrix methods, in combination with vector notation, are also useful, particularly where numerical methods are applied.

2.2 COMPLEX POLAR VECTOR NOTATION

Polar vector notation is useful in the analysis of plane mechanisms. The method is based on the polar representation of a complex number in the form

$$e^{i\theta} = \cos \theta + i \sin \theta \tag{2.1}$$

Thus, the x and y components of a vector $\mathbf{a}$ can be found from

$$\mathbf{a} = ae^{i\theta}$$
$$= a(\cos \theta + i \sin \theta) \tag{2.2}$$
$$= a_x + ia_y$$

The product of a unit vector $e^{i\phi}$ times the vector $ae^{i\theta}$ is equal to $e^{i\phi}(ae^{i\theta}) = ae^{i(\theta+\phi)}$. Thus, we note that the vector $\mathbf{a} = ae^{i\theta}$ is rotated through the angle ϕ when multiplied by the unit vector operator $e^{i\phi}$.

Multiplication of a vector $\mathbf{a} = ae^{i\theta}$ by the unit complex operator $i = \sqrt{-1}$ is equivalent to a rotation of the vector $\pi/2$ radians in the positive sense for θ; that is,

$$ie^{i\theta} = i(\cos \theta + i \sin \theta) = i \cos \theta - \sin \theta$$
$$= e^{i(\theta + \pi/2)} \tag{2.3}$$

Similarly,

$$i(ie^{i\theta}) = i^2 e^{i\theta} = -e^{i\theta} = e^{i(\theta + \pi)} \tag{2.4}$$

8

The product of a unit vector and its conjugate is equal to unity and, since

$$e^{i\theta}e^{-i\theta} = 1 \tag{2.5}$$

we see that $e^{-i\theta}$ is the complex conjugate of $e^{i\theta}$.

The following trigonometric identities will also be found useful.

$$\cos\theta = \frac{e^{i\theta} + e^{-i\theta}}{2} \tag{2.6}$$

$$\sin\theta = -i\frac{e^{i\theta} - e^{-i\theta}}{2} \tag{2.7}$$

$$\cos(\theta \pm \phi) = \cos\theta\cos\phi \mp \sin\theta\sin\phi \tag{2.8}$$

$$\sin(\theta \pm \phi) = \sin\theta\cos\phi \pm \cos\theta\sin\phi \tag{2.9}$$

Differentiation of complex polar vectors

The formation of first-order, second-order, or higher-order derivatives is straight-forward. Let the vector $\mathbf{r} = re^{i\theta}$ represent the position of a point with respect to a fixed reference origin. Taking the first derivative,

$$\frac{d}{dt}(re^{i\theta}) = \dot{r}e^{i\theta} + r(e^{i\theta}i\dot{\theta}) = \dot{r}e^{i\theta} + r\dot{\theta}ie^{i\theta} \tag{2.10}$$

It is convenient to consider each complex polar vector as being composed of two parts, a scalar magnitude such as $\dot{r}$ or $r\dot{\theta}$ in Eq. 2.10 and a unit direction vector $e^{i\theta}$ or $ie^{i\theta}$.

The second derivative becomes

$$\frac{d^2}{dt^2}(re^{i\theta}) = \ddot{r}e^{i\theta} + \dot{r}(e^{i\theta}i\dot{\theta})$$

$$+ (\dot{r}\dot{\theta} + r\ddot{\theta})ie^{i\theta} + (r\dot{\theta})(ie^{i\theta}i\dot{\theta})$$

$$= (\ddot{r} - r\dot{\theta}^2)e^{i\theta} + (r\ddot{\theta} + 2\dot{r}\dot{\theta})ie^{i\theta} \tag{2.11}$$

2.3 KINEMATIC ANALYSIS USING COMPLEX POLAR NOTATION

The four-bar linkage

Position analysis is based on a vector constraint equation formed by a closed vector loop around the basic four-link chain. The vector polygon may be in either of two forms: (1) a continuous head-to-tail chain, or (2) formed of two branches

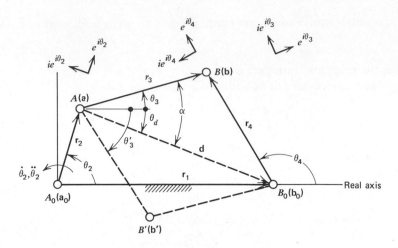

FIGURE 2.1 THE PLANE FOUR-BAR LINKAGE.

that define alternate paths to a common point of interest. The second method has been called the method of "independent position equations" by Raven [1]. Raven's notation has advantages in problems where it is desirable to have the rotation of both input and output cranks referred to fixed centers.

The position equation for the four-bar linkage is based on independent vector paths that describe the position of point B, as shown in Figure 2.1.

$$\mathbf{r}_B = r_2 e^{i\theta_2} + r_3 e^{i\theta_3} = \mathbf{r}_1 + r_4 e^{i\theta_4} \tag{2.12}$$

Assuming that the fixed vector $\mathbf{r}_1$, link lengths r_2, r_3, and r_4 plus the input angle θ_2 are known, the position analysis proceeds as follows. We first determine the length of the diagonal $\mathbf{d}$ from

$$r_2 e^{i\theta_2} + d e^{i\theta_d} = \mathbf{r}_1 \tag{2.13}$$

The angle θ_d can be eliminated by solving for $de^{i\theta_d}$ and then multiplying each side of the equation by its complex conjugate.

$$(de^{i\theta_d})(de^{-i\theta_d}) = (\mathbf{r}_1 - r_2 e^{i\theta_2})(\mathbf{r}_1 - r_2 e^{-i\theta_2}) \tag{2.14}$$

This leads to

$$d^2 = r_1^2 + r_2 - r_1 r_2 (e^{i\theta_2} + e^{-i\theta_2})$$

or   (2.15)

$$d = \sqrt{r_1^2 + r_2^2 - 2r_1 r_2 \cos\theta_2}$$

Resolve Eq. 2.13 into real and imaginary parts

$$r_2 \cos \theta_2 = r_1 - d \cos \theta_d$$

$$r_2 \sin \theta_2 = 0 - d \sin \theta_d$$

from which

$$\sin \theta_d = -\frac{r_2}{d} \sin \theta_2$$

$$\cos \theta_d = \frac{r_1 - r_2 \cos \theta_2}{d} \tag{2.16}$$

hence,

$$\tan \theta_d = \frac{-r_2 \sin \theta_2}{r_1 - r_2 \cos \theta_2}$$

The coupler angle θ_3 is found from the vector equation

$$r_3 e^{i\theta_3} = d e^{i\theta_d} + r_4 e^{i\theta_4} \tag{2.17}$$

which leads to

$$\cos(\theta_3 - \theta_d) = \frac{r_3^2 - d^2 - r_4^2}{2\, dr_3} \tag{2.18}$$

There are two possible solutions corresponding to $\pm(\theta_d - \theta_3)$. The output angle is found from the imaginary components of Eq. 2.17.

$$\sin \theta_4 = \frac{-d \sin \theta_d + r_3 \sin \theta_3}{r_4} \tag{2.19}$$

Velocity analysis of the four-bar linkage

In Eq. 2.10 the radial and transverse components of velocity are given for a point moving with respect to a fixed reference point. In the four-bar linkage all links are rigid bodies, and the first derivatives do not include the radial term. Differentiating the position equation,

$$r_2 \dot{\theta}_2 i e^{i\theta_2} + r_3 \dot{\theta}_3 i e^{i\theta_3} = r_4 \dot{\theta}_4 i e^{i\theta_4} \tag{2.20}$$

Eq. 2.20 can be interpreted graphically as a relative velocity polygon, as shown in Figure 2.2. The vector equation should be solved in three steps as follows. First,

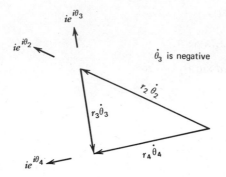

$\dot\theta_3$ is negative

FIGURE 2.2 RELATIVE VELOCITY POLY-
GON FOR THE PLANE FOUR-BAR LINKAGE
WITH DIRECTIONS INDICATED BY COM-
PLEX POLAR UNIT VECTORS.

separate the vector equation into its real and imaginary components.

$$-r_2\dot\theta_2 \sin\theta_2 - r_3\dot\theta_3 \sin\theta_3 = r_4\dot\theta_4 \sin\theta_4$$

$$r_2\dot\theta_2 \cos\theta_2 + r_3\dot\theta_3 \cos\theta_3 = r_4\dot\theta_4 \cos\theta_4 \qquad (2.21)$$

Next rearrange with the two unknowns $\dot\theta_3$ and $\dot\theta_4$ on the left side.

$$\dot\theta_3(-r_3 \sin\theta_3) + \dot\theta_4(r_4 \sin\theta_4) = r_2\dot\theta_2 \sin\theta_2$$

$$\dot\theta_3(r_3 \cos\theta_3) + \dot\theta_4(-r_4 \cos\theta_4) = -r_2\dot\theta_2 \cos\theta_2 \qquad (2.22)$$

Finally, solve Eq. 2.22 using Cramer's rule.

$$\dot\theta_3 = \frac{\begin{vmatrix} r_2\dot\theta_2 \sin\theta_2 & r_4 \sin\theta_4 \\ -r_2\dot\theta_2 \cos\theta_2 & -r_4 \cos\theta_4 \end{vmatrix}}{\begin{vmatrix} -r_3 \sin\theta_3 & r_4 \sin\theta_4 \\ r_3 \cos\theta_3 & -r_4 \cos\theta_4 \end{vmatrix}} \qquad (2.23)$$

This leads to

$$\dot\theta_3 = \dot\theta_2 \frac{r_2 r_4 \cos\theta_2 \sin\theta_4 - r_2 r_4 \sin\theta_2 \cos\theta_4}{-r_3 r_4 \cos\theta_3 \sin\theta_4 + r_3 r_4 \sin\theta_3 \cos\theta_4} \qquad (2.24)$$

$$= \dot\theta_2 \frac{r_2 \sin(\theta_4 - \theta_2)}{r_3 \sin(\theta_3 - \theta_4)} \qquad (2.25)$$

Eq. 2.25 can be verified by applying the law of sines to the velocity polygon of Figure 2.2. An expression for $\dot\theta_4$ can be found in a similar manner.

Acceleration Analysis of the Four-bar Linkage

A second differentiation of Eq. 2.12 leads to

$$r_2\ddot\theta_2(ie^{i\theta_2}) + r_2\dot\theta_2^2(-e^{i\theta_2}) + r_3\ddot\theta_3(ie^{i\theta_3}) + r_3\dot\theta_3^2(-e^{i\theta_3})$$

$$= r_4\ddot\theta_4(ie^{i\theta_4}) + r_4\dot\theta_4^2(-e^{i\theta_4}) \qquad (2.26)$$

After separation of real and imaginary components and rearranging similar to Eq. 2.22, we obtain two scalar equations with unknowns $\ddot{\theta}_3$ and $\ddot{\theta}_4$.

$$\ddot{\theta}_3(-r_3 \sin \theta_3) + \ddot{\theta}_4(r_4 \sin \theta_4) = r_2\ddot{\theta}_2 \sin \theta_2 + r_2\dot{\theta}_2^2 \cos \theta_2$$
$$+ r_3\dot{\theta}_3^2 \cos \theta_3 - r_4\dot{\theta}_4^2 \cos \theta_4 = A$$
$$\ddot{\theta}_3(r_3 \cos \theta_3) + \ddot{\theta}_4(-r_4 \cos \theta_4) = -r_2\ddot{\theta}_2 \cos \theta_2 + r_2\dot{\theta}_2^2 \sin \theta_2$$
$$+ r_3\dot{\theta}_3^2 \sin \theta_3 - r_4\dot{\theta}_4^2 \sin \theta_4 = B$$

$$(2.27)$$

from which

$$\ddot{\theta}_3 = \frac{\begin{vmatrix} A & r_4 \sin \theta_4 \\ B & -r_4 \cos \theta_4 \end{vmatrix}}{\begin{vmatrix} -r_3 \sin \theta_3 & r_4 \sin \theta_4 \\ r_3 \cos \theta_3 & -r_4 \cos \theta_4 \end{vmatrix}}$$

$$= \frac{1}{r_3} \frac{(-A \cos \theta_4 - B \sin \theta_4)}{(\sin \theta_3 \cos \theta_4 - \cos \theta_3 \sin \theta_4)} = \frac{-1}{r_3} \frac{(A \cos \theta_4 + B \sin \theta_4)}{\sin(\theta_3 - \theta_4)}$$

$$(2.28)$$

The vector equations for other common plane mechanisms are written easily. For example, position, velocity, and acceleration equations for the offset slider-crank mechanism shown in Figure 2.3 are given as

$$\mathbf{r}_B = r_2 e^{i\theta_2} + r_3 e^{i\theta_3} = x + ib$$
$$\dot{\mathbf{r}}_B = r_2(e^{i\theta_2}i\dot{\theta}_2) + r_3(e^{i\theta_3}i\dot{\theta}_3)$$
$$= r_2\dot{\theta}_2(ie^{i\theta_2}) + r_3\dot{\theta}_3(ie^{i\theta_3}) = \dot{x}$$
$$\ddot{\mathbf{r}}_B = (-r_2\dot{\theta}_2^2 e^{i\theta_2} + r_2\ddot{\theta}_2 ie^{i\theta_2}) + (-r_3\dot{\theta}_3^2 e^{i\theta_3} + r_3\ddot{\theta}_3 ie^{i\theta_3}) = \ddot{x} \qquad (2.29)$$

In this case the unknowns are $\dot{\theta}_3$, $\ddot{\theta}_3$, $\dot{x}$, and $\ddot{x}$.

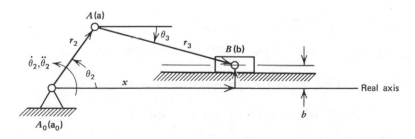

FIGURE 2.3   THE OFFSET SLIDER-CRANK.

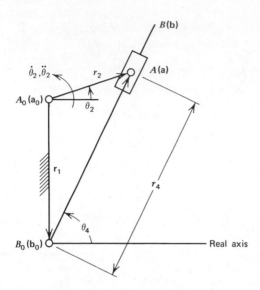

FIGURE 2.4 THE CRANK-SHAPER MECH-
ANISM.

The vector equations for the crank-shaper mechanism shown in Figure 2.4 become

$$\mathbf{r}_A = r_2 e^{i\theta_2} = -ir_1 + r_4 e^{i\theta_4}$$

$$\dot{\mathbf{r}}_A = r_2 \dot{\theta}_2(ie^{i\theta_2}) = r_4 \dot{\theta}_4(ie^{i\theta_4}) + \dot{r}_4 e^{i\theta_4}$$

$$\ddot{\mathbf{r}}_A = (-r_2 \dot{\theta}_2^2 e^{i\theta_2} + r_2 \ddot{\theta}_2 i e^{i\theta_2})$$

$$= (\ddot{r}_4 - r_4 \dot{\theta}_4^2)e^{i\theta_4} + (r_4 \ddot{\theta}_4 + 2\dot{r}_4 \dot{\theta}_4)ie^{i\theta_4} \tag{2.30}$$

with unknowns $\dot{\theta}_4$, $\ddot{\theta}_4$, $\dot{r}_4$, and $\ddot{r}_4$.

2.4 CARTESIAN VECTOR NOTATION

The method is based on the following principles of vector calculus.

The Scalar Product of Two Vectors—the Dot Product

$$\mathbf{a} \cdot \mathbf{b} = ab \cos \theta = a_x b_x + a_y b_y + a_z b_z \tag{2.31}$$

where θ is the angle between the two vectors.

For a system of orthogonal cartesian unit vectors $\hat{\imath}, \hat{\jmath}, \hat{k}$ in the x, y, z directions,

$$\hat{\imath} \cdot \hat{\imath} = \hat{\jmath} \cdot \hat{\jmath} = \hat{k} \cdot \hat{k} = 1$$

$$\hat{\imath} \cdot \hat{\jmath} = \hat{\jmath} \cdot \hat{k} = \hat{k} \cdot \hat{\imath} = 0$$

$$a \cdot \hat{\imath} = a \cos \theta = a_x$$

The dot product of any vector with a unit direction vector gives the component of the vector in the unit direction. If two vectors $\mathbf{a}$ and $\mathbf{b}$ are perpendicular, their dot product $\mathbf{a} \cdot \mathbf{b} = 0$.

The Vector Product—the Cross Product

$$\mathbf{a} \times \mathbf{b} = ab \sin \theta \tag{2.32}$$

The cross product can also be formed in terms of vector components by expansion of the determinant.

$$\mathbf{a} \times \mathbf{b} = \begin{vmatrix} \hat{\imath} & \hat{\jmath} & \hat{k} \\ a_x & a_y & a_z \\ b_x & b_y & b_z \end{vmatrix} \tag{2.33}$$

$$= \hat{\imath}(a_y b_z - b_y a_z)$$
$$+ \hat{\jmath}(a_z b_x - a_x b_z)$$
$$+ \hat{k}(a_x b_y - a_y b_x) \tag{2.34}$$

In two dimensions

$$\mathbf{a} \times \mathbf{b} - \hat{k}(a_x b_y - a_y b_x) \tag{2.35}$$

and the unit vector $\hat{k}$ is often dispensed with.
   In this case,

$$\hat{\imath} \times \hat{\imath} = \hat{\jmath} \times \hat{\jmath} = \hat{k} \times \hat{k} = 0$$

$$\hat{\imath} \times \hat{\jmath} = \hat{k}$$

$$\hat{\jmath} \times \hat{\imath} = -\hat{k}$$

If $\mathbf{a} \times \mathbf{b} = 0$, $\mathbf{a}$ and $\mathbf{b}$ are parallel.
Other useful vector identities are:

$$\mathbf{a} \times (\mathbf{b} \times \mathbf{c}) = (\mathbf{a} \cdot \mathbf{c})\mathbf{b} - (\mathbf{a} \cdot \mathbf{b})\mathbf{c}$$

$$(\mathbf{a} \times \mathbf{b}) \times \mathbf{c} = (\mathbf{c} \cdot \mathbf{a})\mathbf{b} - (\mathbf{c} \cdot \mathbf{b})\mathbf{a}$$

$$(\mathbf{a} \times \mathbf{b}) \cdot \mathbf{b} = 0$$

$$\mathbf{a} \cdot (\mathbf{b} \times \mathbf{c}) = \mathbf{b} \cdot (\mathbf{c} \times \mathbf{a}) = \mathbf{c} \cdot (\mathbf{a} \times \mathbf{b}) = -\mathbf{a} \cdot (\mathbf{c} \times \mathbf{b})$$

Differentiation of Vectors

Assume a vector $\mathbf{r} = r\hat{r}$ where $\hat{r}$ is a unit direction vector. Differentiating, we have

$$\dot{\mathbf{r}} = \dot{r}\hat{r} + r\frac{d}{dt}\hat{r} \tag{2.36}$$

The term $(d/dt)\hat{r}$ accounts for the rate of change of direction of the vector $\mathbf{r}$.

$$\frac{d}{dt}\hat{r} = \lim_{\Delta t \to 0}\frac{\Delta\hat{r}}{\Delta t} = \lim_{\Delta t \to 0}\frac{|\hat{r}|\Delta\theta}{\Delta t} = \dot{\boldsymbol{\theta}}$$

From Eq. 2.36 we see that $(d/dt)\hat{r}$ must be perpendicular to $\hat{r}$. This is specified by including the cross product with $\hat{r}$ in the form

$$\dot{\mathbf{r}} = \dot{r}\hat{r} + r(\dot{\boldsymbol{\theta}} \times \hat{r})$$

$$= \dot{r}\hat{r} + \dot{\boldsymbol{\theta}} \times \mathbf{r} \tag{2.37}$$

The second differentiation must be done carefully.

$$\ddot{\mathbf{r}} = \frac{d}{dt}(\dot{r}\hat{r} + \dot{\boldsymbol{\theta}} \times r\hat{r})$$

$$= \ddot{r}\hat{r} + \dot{r}\frac{d}{dt}\hat{r} + \ddot{\boldsymbol{\theta}} \times \mathbf{r} + \dot{\boldsymbol{\theta}} \times \frac{d}{dt}(r\hat{r})$$

where

$$\dot{r}\frac{d}{dt}\hat{r} = \dot{r}(\dot{\boldsymbol{\theta}} \times \hat{r}) = \dot{\boldsymbol{\theta}} \times \dot{r}\hat{r}$$

$$\dot{\boldsymbol{\theta}} \times \frac{d}{dt}(r\hat{r}) = \dot{\boldsymbol{\theta}} \times (\dot{r}\hat{r} + \dot{\boldsymbol{\theta}} \times \mathbf{r})$$

$$= \dot{\boldsymbol{\theta}} \times \dot{r}\hat{r} + \dot{\boldsymbol{\theta}} \times (\dot{\boldsymbol{\theta}} \times r)$$

Therefore,

$$\ddot{\mathbf{r}} = \ddot{r}\hat{r} + \ddot{\boldsymbol{\theta}} \times \mathbf{r} + \dot{\boldsymbol{\theta}} \times (\dot{\boldsymbol{\theta}} \times \mathbf{r}) + 2(\dot{\boldsymbol{\theta}} \times \dot{r}\hat{r}) \tag{2.38}$$

It should also be noted that for spatial motion,

$$\ddot{\boldsymbol{\theta}} = \frac{d}{dt}\dot{\boldsymbol{\theta}} = \frac{d}{dt}(\dot{\theta}\mathbf{u}) = \ddot{\theta}\mathbf{u} + \dot{\theta}\dot{\mathbf{u}} \tag{2.39}$$

where $\mathbf{u}$ defines the instantaneous direction of the angular velocity vector $\dot{\boldsymbol{\theta}}$.

2.5 KINEMATIC ANALYSIS OF PLANE MECHANISMS USING CARTESIAN VECTOR NOTATION

The Four-Bar Linkage

Position Analysis

Again, referring to Figure 2.1, we write the vector loop equation involving the diagonal vector $\mathbf{d}$ in the form

$$\mathbf{r}_2 + \mathbf{d} = \mathbf{r}_1$$

Solving for $\mathbf{d}$ and taking the dot product of each side with itself,

$$\mathbf{d} \cdot \mathbf{d} = (\mathbf{r}_1 - \mathbf{r}_2) \cdot (\mathbf{r}_1 - \mathbf{r}_2)$$
$$d^2 = r_1^2 + r_2^2 - 2r_1 r_2 \cos \theta_2$$

which we recognize as the law of cosines. With d known, we can proceed to find θ_d from

$$\tan \theta_d = \frac{r_{1y} - r_{2y}}{r_{1x} - r_{2x}}$$

The angle α between $\mathbf{r}_3$ and $\mathbf{d}$ is found from

$$\mathbf{r}_3 = \mathbf{d} + \mathbf{r}_4$$

or

$$\mathbf{r}_4 = \mathbf{r}_3 - \mathbf{d}$$

Again, taking the dot product on both sides,

$$\mathbf{r}_4 \cdot \mathbf{r}_4 = (\mathbf{r}_3 - \mathbf{d}) \cdot (\mathbf{r}_3 - \mathbf{d})$$
$$r_4^2 = r_3^2 + d^2 - 2r_3 d \cos \alpha$$

that is,
$$\cos \alpha = \frac{r_3^2 + d^2 - r_4^2}{2r_3 d}$$

Finally, we see that the coupler angle θ_3 becomes

$$\theta_3 = \theta_d \pm \alpha$$

after which θ_4 is easily determined.

Velocity Analysis

The position equation

$$\mathbf{r}_2 + \mathbf{r}_3 = \mathbf{r}_1 + \mathbf{r}_4$$

is differentiated as shown in Eq. 2.37, noting that for two points on a rigid body $\dot{r} = 0$; therefore, we obtain

$$\dot{\theta}_2 \times \mathbf{r}_2 + \dot{\theta}_3 \times \mathbf{r}_3 = \dot{\theta}_4 \times \mathbf{r}_4$$

where

$$\dot{\theta} = \hat{k}\dot{\theta} \qquad \text{and} \qquad \mathbf{r} = \hat{i}r_x + \hat{j}r_y$$

Expanding the cross products,

$$\hat{i}(-r_{2y}\dot{\theta}_2) + \hat{j}(r_{2x}\dot{\theta}_2) + \hat{i}(-r_{3y}\dot{\theta}_3) + \hat{j}(r_{3x}\dot{\theta}_3) = \hat{i}(-r_{4y}\dot{\theta}_4) + \hat{j}(r_{4x}\dot{\theta}_4)$$

Taking $\hat{i}$ and $\hat{j}$ components and rearranging as a set of two linear equations in $\dot{\theta}_3$ and $\dot{\theta}_4$,

$$\dot{\theta}_3(-r_{3y}) + \dot{\theta}_4(r_{4y}) = r_{2y}\dot{\theta}_2$$

$$\dot{\theta}_3(r_{3x}) + \dot{\theta}_4(-r_{4x}) = -r_{2x}\dot{\theta}_2$$

which are easily solved to yield

$$\dot{\theta}_3 = \dot{\theta}_2 \frac{r_{2x}r_{4y} - r_{2y}r_{4x}}{r_{4x}r_{3y} - r_{3x}r_{4y}}$$

$$\dot{\theta}_4 = \dot{\theta}_2 \frac{r_{2x}r_{3y} - r_{3x}r_{2y}}{r_{4x}r_{3y} - r_{3x}r_{4y}}$$

Once the angular motions are known, the velocity of any arbitrary point is easily determined. The velocity of a point $\mathbf{p}$ on the coupler is found from

$$\dot{\mathbf{p}} = \dot{\theta}_2 \times \mathbf{r}_2 + \dot{\theta}_3 \times (\mathbf{p} - \mathbf{a})$$

where $(\mathbf{p} - \mathbf{a})$ is the position of point $\mathbf{p}$ relative to the moving pivot $\mathbf{a}$.

Acceleration Analysis

The acceleration analysis proceeds from a second differentiation of the vector loop equation.

$$\ddot{\theta}_2 \times \mathbf{r}_2 + \dot{\theta}_2 \times (\dot{\theta}_2 \times \mathbf{r}_2) + \ddot{\theta}_3 \times \mathbf{r}_3 + \dot{\theta}_3 \times (\dot{\theta}_3 \times \mathbf{r}_3) = \ddot{\theta}_4 \times \mathbf{r}_4 + \dot{\theta}_4 \times (\dot{\theta}_4 \times \mathbf{r}_4)$$

Expansion of the cross products give $\hat{i}$ and $\hat{j}$ components.

$$\hat{i}(-r_{2y}\ddot{\theta}_2) + \hat{i}(-r_{2x}\dot{\theta}_2^2) + \hat{i}(-r_{3y}\ddot{\theta}_3) + \hat{i}(-r_{3x}\dot{\theta}_3^2)$$
$$= \hat{i}(-r_{4y}\ddot{\theta}_4) + \hat{i}(-r_{4x}\dot{\theta}_4^2)$$
$$\hat{j}(r_{2x}\ddot{\theta}_2) + \hat{j}(-r_{2y}\dot{\theta}_2^2) + \hat{j}(r_{3x}\ddot{\theta}_3) + \hat{j}(-r_{3y}\dot{\theta}_3^2)$$
$$= \hat{j}(r_{4x}\ddot{\theta}_4) + \hat{j}(-r_{4y}\dot{\theta}_4^2)$$

Rearranging as two linear equations with unknowns $\ddot{\theta}_3$ and $\ddot{\theta}_4$,

$$\ddot{\theta}_3(-r_{3y}) + \ddot{\theta}_4(r_{4y}) = r_{2y}\ddot{\theta}_2 + r_{2x}\dot{\theta}_2^2 + r_{3x}\dot{\theta}_3^2 \quad r_{4x}\dot{\theta}_4^2 = A$$

$$\ddot{\theta}_3(r_{3x}) + \ddot{\theta}_4(-r_{4x}) = -r_{2x}\ddot{\theta}_2 + r_{2y}\dot{\theta}_2^2 + r_{3y}\dot{\theta}_3^2 - r_{4y}\dot{\theta}_4^2 = B$$

Since all terms on the right side are known, we may solve for $\ddot{\theta}_3$ and $\ddot{\theta}_4$ using

$$\ddot{\theta}_3 = \frac{\begin{vmatrix} A & r_{4y} \\ B & -r_{4x} \end{vmatrix}}{\begin{vmatrix} -r_{3y} & r_{4y} \\ r_{3x} & -r_{4x} \end{vmatrix}} = \frac{-(Ar_{4x} + Br_{4y})}{(r_{3y}r_{4x} - r_{3x}r_{4y})}$$

$$\ddot{\theta}_4 = \frac{\begin{vmatrix} -r_{3y} & A \\ r_{3x} & B \end{vmatrix}}{\begin{vmatrix} -r_{3y} & r_{4y} \\ r_{3x} & -r_{4x} \end{vmatrix}} = \frac{-(Ar_{3x} + Br_{3y})}{(r_{3y}r_{4x} - r_{3x}r_{4y})}$$

Such calculations can be programmed for digital computation and, when combined with the previous expressions for position and velocity, they constitute a complete kinematic analysis of the basic four-bar linkage.

The Oscillating Slider Mechanism

As an example of kinematic analysis for a mechanism involving sliding contact, consider the oscillating slider mechanism shown in Figure 2.5a. In this example all motions are with respect to a fixed coordinate system; that is, the rate of change of velocity for any vector r is given by Eq. 2.38. Analysis of similar problems, expressed in terms of motion relative to a rotating reference frame, including the Coriolis Component, will be covered in Chapter 4.

Position Analysis

The position analysis is simplified by noting that the mechanism shown schematically in Figure 2.5b is kinematically equivalent to that shown in Figure 2.5a. Therefore, defining a vector $\mathbf{f} = (e_3 + r_4)\hat{r}_4$, we may describe the position of fixed point $\mathbf{b}_0$ using point $\mathbf{a}_0$ as a reference by the vector equation

$$\mathbf{r}_2 - (e_3 + r_4)\hat{r}_4 + \mathbf{r}_3 = \mathbf{r}_1$$

where $\mathbf{r}_1$, $\mathbf{r}_2$, e_3, and r_4 are known and r_3 and $\hat{r}_3$ are unknown.

To solve for $\mathbf{r}_3$, we note that the length of the diagonal vector $\mathbf{d}$ can be found from

$$d^2 = \mathbf{d} \cdot \mathbf{d} = (\mathbf{b}_0 - \mathbf{a}) \cdot (\mathbf{b}_0 - \mathbf{a})$$

$$= (e + r_4)^2 + r_3^2$$

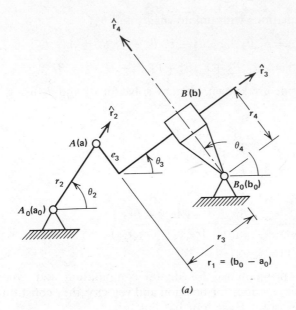

(a)

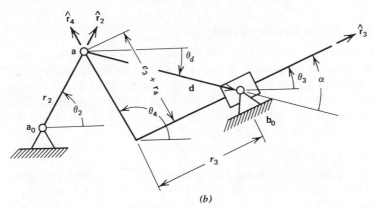

(b)

FIGURE 2.5   THE OSCILLATING SLIDER MECHANISM. (*a*) ORIGINAL
MECHANISM. (*b*) EQUIVALENT MECHANISM.

Hence

$$r_3 = \sqrt{(b_{0x} - a_x)^2 + (b_{0y} - a_y)^2 - (e_3 + r_4)^2}$$

The angles

$$\alpha = \tan^{-1}\left(\frac{e_3 + r_4}{r_3}\right)$$

and

$$\theta_d = \tan^{-1}\left(\frac{b_{0y} - a_y}{b_{0x} - a_x}\right)$$

from which

$$\theta_3 = \theta_d \pm \alpha$$

$$\hat{\mathbf{r}}_3 = (\cos\theta_3 , \sin\theta_3)$$

Velocity Analysis

Differentiating the position equation with $\mathbf{f} = (e_3 + r_4)\hat{\mathbf{r}}_4$, we obtain

$$\dot{\boldsymbol{\theta}}_3 \times (\mathbf{r}_3 - \mathbf{f}) + \dot{r}_3\hat{\mathbf{r}}_3 = -\dot{\boldsymbol{\theta}}_2 \times \mathbf{r}_2$$

which can be solved for $\dot{\theta}_3$ and $\dot{r}_3$.

Acceleration Analysis

A second differentiation leads to

$$\ddot{\boldsymbol{\theta}}_3 \times (\mathbf{r}_3 - \mathbf{f}) + \ddot{r}_3\hat{\mathbf{r}}_3 = -\ddot{\boldsymbol{\theta}}_2 \times \mathbf{r}_2 - \dot{\boldsymbol{\theta}}_2 \times (\dot{\boldsymbol{\theta}}_2 \times \mathbf{r}_2)$$

$$- \dot{\boldsymbol{\theta}}_3 \times (\dot{\boldsymbol{\theta}}_3 \times (\mathbf{r}_3 - \mathbf{f})) - 2\dot{\boldsymbol{\theta}}_3 \times (\dot{r}_3\hat{\mathbf{r}}_3)$$

with unknowns $\ddot{\theta}_3$ and $\ddot{r}_3$.

2.6   A GENERAL METHOD FOR COMPUTER-AIDED KINEMATIC ANALYSIS OF PLANE MECHANISMS

Many plane mechanisms can be shown to be assembled from one or more of three basic combinations of rigid members: the two-link dyad, the oscillating slider, and the rotating guide. Consider the plane mechanism shown in Figure 2.6. At first glance it would appear that a complete position-velocity-acceleration analysis would be a formidable task. However, with a properly prepared set of computer subprograms, the kinematic analysis is straightforward.

The input crank, whose position is defined by the specified angle θ_2, constitutes a rigid body; therefore, the motion of point **a** is known. Members 3 and 4 form a two-link dyad and can be analyzed in a manner similar to that used for the four-bar linkage. The motion of **c** and **d** can be calculated from rigid body motion relationships with points **a** and $\mathbf{b}_0$ as reference points. The relative motion at **d** is analyzed as a special case of the oscillating slider with $\mathbf{d}_0$ specified as a fixed reference point. This leads to specification of the angular motion of link 8 and, hence, the motion of point **f** is defined. Finally, with the motion of points **c** and **f** specified, we may compute the motion of point **g**, the final point of interest. The dimensions r_5 and r_7 are variable and the relative accelerations at points **d** and **f** must include the $2(\dot{\boldsymbol{\theta}} \times \dot{r}\hat{r})$ term from Eq. 2.38.

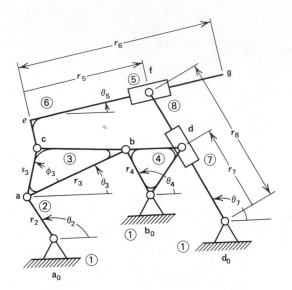

FIGURE 2.6 AN EXAMPLE OF A MULTI-ELEMENT PLANE MECHANISM THAT INVOLVES BASIC KINEMATIC ELEMENTS.

The basis for a set of computer subprograms to be listed in Chapter 12 follows. In all cases points 1 and 2 are the *reference points* and point 3 is the *point of interest*.

Rigid Body Motion

It is often convenient to describe the geometrical shape of any of the rigid bodies to be assembled into a mechanism as though it were dimensioned as an engineering drawing with one of the basic dimensions (e.g., $\mathbf{r}$, parallel to the x-axis). For example, the coupler link 3 in Figure 2.6 could be described by r_3, s_3, and ϕ_3. Once the actual position angle θ_3 is known along with the position of the reference point $\mathbf{a}$, the position of any other point such as $\mathbf{c}$ can be calculated using the rotation matrix to be described in Chapter 3.

Generally, the velocity and acceleration of any point of interest (e.g., point 3 in Figure 2.7a, are described in terms of the known motion of any reference point 1 in the form

$$\dot{\mathbf{p}}_3 = \dot{\mathbf{p}}_1 + \dot{\boldsymbol{\theta}} \times (\mathbf{p}_3 - \mathbf{p}_1) \tag{2.40}$$

$$\ddot{\mathbf{p}}_3 = \ddot{\mathbf{p}}_1 + \ddot{\boldsymbol{\theta}} \times (\mathbf{p}_3 - \mathbf{p}_1) + \dot{\boldsymbol{\theta}} \times (\dot{\boldsymbol{\theta}} \times (\mathbf{p}_3 - \mathbf{p}_1)) \tag{2.41}$$

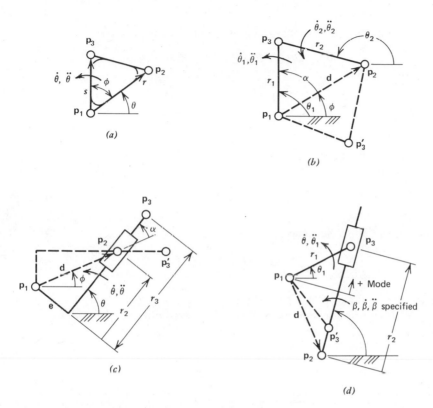

FIGURE 2.7   BASIC THREE-JOINT PLANE KINEMATIC ELEMENTS, (a) THE RIGID BODY. (b) THE TWO LINK DYAD. (c) THE OSCILLATING SLIDER. (d) THE ROTATING GUIDE.

Eq. 2.40 is programed as subroutine VEL in LINKPAC, while Eq. 2.41 is solved in subroutine ACC. These subroutines require all of the quantities on the right side as inputs. If the reference point is fixed, such as point $\mathbf{a}_0$ for the input crank in Figure 2.6, then $\dot{\mathbf{p}}_1 = \ddot{\mathbf{p}}_1 = 0$.

The Two-Link Dyad (Figure 2.7b)

In this case the analysis is similar to the four-bar linkage except that the DYAD subprograms must account for the possibility that either or both reference points 1 and 2 may be in motion.

The *position analysis* is accomplished in the following steps, where the notation is as shown in Figure 2.7b.

1. Calculate the distance **d** between reference points 1 and 2 from

$$d^2 = (p_{2x} - p_{1x})^2 + (p_{2y} - p_{1y})^2 \qquad (2.42)$$

2. Check mobility. If

$$d > (r_1 + r_2) \qquad (2.43)$$

or

$$d < |(r_1 - r_2)| \qquad (2.44)$$

the dyad cannot be assembled.
3. Calculate ϕ, the angle for vector **d**.

$$\phi = \tan^{-1}\left(\frac{p_{2y} - p_{1y}}{p_{2x} - p_{1x}}\right) \qquad (2.45)$$

4. Calculate α, the angle between $\mathbf{r}_1$ and **d**.

$$\alpha = \pm\cos^{-1}\left(\frac{r_1^2 + d^2 - r_2^2}{2r_1 d}\right) \qquad (2.46)$$

5. The sign of α is determined from the specified *mode of assembly in the first position*. With continuous mobility the mode of assembly does not change as a result of a change in position.

$$\theta_1 = \phi \pm \alpha \qquad (2.47)$$

6. The coordinates of the point of interest $\mathbf{p}_3$ becomes

$$p_{3x} = p_{1x} + r_1 \cos\theta_1$$
$$p_{3y} = p_{1y} + r_1 \sin\theta_1 \qquad (2.48)$$

7. Finally,

$$\theta_2 = \tan^{-1}\left(\frac{p_{3y} - p_{2y}}{p_{3x} - p_{2x}}\right) \qquad (2.49)$$

These calculations are programmed as subroutine PDYAD.

The *velocity analysis* is based on the equivalence of the two expressions for the velocity of the point of interest in terms of the known position and velocity of the reference points; that is,

$$\dot{\mathbf{p}}_3 = \dot{\mathbf{p}}_1 + \dot{\boldsymbol{\theta}}_1 \times (\mathbf{p}_3 - \mathbf{p}_1) = \dot{\mathbf{p}}_2 + \dot{\boldsymbol{\theta}}_2 \times (\mathbf{p}_3 - \mathbf{p}_2) \qquad (2.50)$$

This leads to

$$\dot{\theta}_1 = -\frac{(\dot{p}_{2x} - \dot{p}_{1x})(p_{3x} - p_{2x}) + (\dot{p}_{2y} - \dot{p}_{1y})(p_{3y} - p_{2y})}{(p_{3y} - p_{1y})(p_{3x} - p_{2x}) - (p_{3y} - p_{2y})(p_{3x} - p_{1x})}$$

$$\dot{\theta}_2 = -\frac{(\dot{p}_{2x} - \dot{p}_{1x})(p_{3x} - p_{1x}) + (\dot{p}_{2y} - \dot{p}_{1y})(p_{3y} - p_{1y})}{(p_{3y} - p_{1y})(p_{3x} - p_{2x}) - (p_{3y} - p_{2y})(p_{3x} - p_{1x})} \qquad (2.51)$$

$\dot{\mathbf{p}}_3$ is then calculated from

$$\dot{\mathbf{p}}_3 = \dot{\mathbf{p}}_1 + \dot{\boldsymbol{\theta}}_1 \times (\mathbf{p}_3 - \mathbf{p}_1) \tag{2.52}$$

Acceleration analysis proceeds from

$$\ddot{\mathbf{p}}_3 = \ddot{\mathbf{p}}_1 + \ddot{\boldsymbol{\theta}}_1 \times (\mathbf{p}_3 - \mathbf{p}_1) + \dot{\boldsymbol{\theta}}_1 \times (\dot{\boldsymbol{\theta}}_1 \times (\mathbf{p}_3 - \mathbf{p}_1))$$
$$= \ddot{\mathbf{p}}_2 + \ddot{\boldsymbol{\theta}}_2 \times (\mathbf{p}_3 - \mathbf{p}_2) + \dot{\boldsymbol{\theta}}_2 \times (\dot{\boldsymbol{\theta}}_2 \times (\mathbf{p}_3 - \mathbf{p}_2)) \tag{2.53}$$

which gives

$$\ddot{\theta}_1 = -\frac{E(p_{3x} - p_{2x}) + F(p_{3y} - p_{2y})}{(p_{3y} - p_{1y})(p_{3x} - p_{2x}) - (p_{3y} - p_{2y})(p_{3x} - p_{1x})} \tag{2.54}$$

$$\ddot{\theta}_2 = -\frac{F(p_{3y} - p_{1y}) + E(p_{3x} - p_{1x})}{(p_{3y} - p_{1y})(p_{3x} - p_{2x}) - (p_{3y} - p_{2y})(p_{3x} - p_{1x})} \tag{2.55}$$

where

$$E = (\ddot{p}_{2x} - \ddot{p}_{1x}) + \dot{\theta}_1^2(p_{3x} - p_{1x}) - \dot{\theta}_2^2(p_{3x} - p_{2x})$$
$$F = (\ddot{p}_{2y} - \ddot{p}_{1y}) + \dot{\theta}_1^2(p_{3y} - p_{1y}) - \dot{\theta}_2^2(p_{3y} - p_{2y})$$

and

$$\ddot{\mathbf{p}}_3 = \ddot{\mathbf{p}}_1 + \ddot{\boldsymbol{\theta}}_1 \times (\mathbf{p}_3 - \mathbf{p}_1) + \dot{\boldsymbol{\theta}}_1 \times (\dot{\boldsymbol{\theta}}_1 \times (\mathbf{p}_3 - \mathbf{p}_1)) \tag{2.56}$$

These calculations are programmed as subroutines VDYAD and ADYAD in LINKPAC.

The Oscillating Slider (Figure 2.7c)

Position analysis involves the relationships

$$d^2 = e^2 + r_2^2 = (p_{2x} - p_{1x})^2 + (p_{2y} - p_{1y})^2 \tag{2.57}$$

If $d^2 < e^2$, the oscillating slider cannot be assembled.

$$r_2 = \sqrt{(p_{2x} - p_{1x})^2 + (p_{2y} - p_{1y})^2 - e^2}$$

$$\alpha = \tan^{-1}\left(\frac{e}{r_2}\right)$$

$$\phi = \tan^{-1}\left(\frac{p_{2y} - p_{1y}}{p_{2x} - p_{1x}}\right) \tag{2.58}$$

If, in the initial position, α is positive,

$$\theta = \phi + \alpha$$

Otherwise,

$$\theta = \phi - \alpha$$

after which

$$p_{3x} = p_{1x} + r_3 \cos \theta + e \sin \theta$$
$$p_{3y} = p_{1y} + r_3 \sin \theta - e \cos \theta \qquad (2.59)$$

Velocity analysis is based on

$$\dot{\mathbf{p}}_2 = \dot{\mathbf{p}}_1 + \dot{\boldsymbol{\theta}} \times (\mathbf{p}_2 - \mathbf{p}_1) + \dot{r}_2 \hat{r}_2$$

from which

$$\dot{\theta} = \frac{(\dot{p}_{2y} - \dot{p}_{1y})\cos \theta - (\dot{p}_{2x} - \dot{p}_{1x})\sin \theta}{(p_{2x} - p_{1x})\cos \theta + (p_{2y} - p_{1y})\sin \theta} \qquad (2.60)$$

$$\dot{r}_2 = \frac{(\dot{p}_{2y} - \dot{p}_{1y})(p_{2y} - p_{1y}) + (\dot{p}_{2x} - \dot{p}_{1x})(p_{2x} - p_{1x})}{(p_{2x} - p_{1x})\cos \theta + (p_{2y} - p_{1y})\sin \theta} \qquad (2.61)$$

and

$$\dot{p}_{3x} = \dot{p}_{1x} - \dot{\theta}(r_3 \sin \theta - e \cos \theta) = \dot{p}_{1x} - \dot{\theta}(p_{3y} - p_{1y}) \qquad (2.62)$$
$$\dot{p}_{3y} = \dot{p}_{1y} + \dot{\theta}(r_3 \cos \theta + e \cos \theta) = \dot{p}_{1y} + \dot{\theta}(p_{3x} - p_{1x}) \qquad (2.63)$$

The *acceleration analysis* follows from Eq. 2.38 in the form

$$\ddot{\mathbf{p}}_2 = \ddot{\mathbf{p}}_1 + \ddot{\boldsymbol{\theta}} \times (\mathbf{p}_2 - \mathbf{p}_1) + \dot{\boldsymbol{\theta}} \times (\dot{\boldsymbol{\theta}} \times (\mathbf{p}_2 - \mathbf{p}_1)) + \ddot{r}_2 \hat{r}_2 + 2\dot{\boldsymbol{\theta}} \times \dot{r}_2 \hat{r}_2 \quad (2.64)$$

Collecting terms,

$$-\ddot{\theta}(p_{2y} - p_{1y}) + \ddot{r}_2 \cos \theta = (\ddot{p}_{2x} - \ddot{p}_{1x}) + \dot{\theta}^2(p_{2x} - p_{1x}) + 2\dot{\theta}\dot{r}_2 \sin \theta = E$$
$$\ddot{\theta}(p_{2x} - p_{1x}) + \ddot{r}_2 \sin \theta = (\ddot{p}_{2y} - \ddot{p}_{1y}) + \dot{\theta}^2(p_{2y} - p_{1y}) - 2\dot{\theta}\dot{r}_2 \cos \theta = F$$

which gives

$$\ddot{\theta} = -\frac{E \sin \theta - F \cos \theta}{(p_{2x} - p_{1x})\cos \theta + (p_{2y} - p_{1y})\sin \theta} \qquad (2.65)$$

$$\ddot{r}_2 = -\frac{E(p_{2x} - p_{1x}) + F(p_{2y} - p_{1y})}{(p_{2x} - p_{1x})\cos \theta + (p_{2y} - p_{1y})\sin \theta} \qquad (2.66)$$

and

$$\ddot{p}_{3x} = \ddot{p}_{1x} - \ddot{\theta}(r_3 \sin \theta - e \cos \theta) - \dot{\theta}^2(r_3 \cos \theta + e \sin \theta)$$
$$\ddot{p}_{3y} = \ddot{p}_{1y} + \ddot{\theta}(r_3 \cos \theta + e \sin \theta) - \dot{\theta}^2(r_3 \sin \theta - e \cos \theta) \qquad (2.67)$$

These calculations are programmed as subroutines POSC, VOSC, and AOSC and are listed in LINKPAC.

The Rotating Guide [Figure 2.7d]

In this case we note that the angular motion of the guide must be specified by β, $\dot{\beta}$, and $\ddot{\beta}$.

Position analysis proceeds as follows:

$$d = \sqrt{(p_{2x} - p_{1x})^2 + (p_{2y} - p_{1y})^2}$$

$$\phi = \tan^{-1}\left(\frac{p_{2y} - p_{1y}}{p_{2x} - p_{1x}}\right)$$

The variable r_2 is found from

$$r_1^2 = (p_{3x} - p_{1x})^2 + (p_{3y} - p_{1y})^2$$
$$= (p_{2x} + r_2 \cos \beta - p_{1x})^2 + (p_{2y} + r_2 \sin \beta - p_{1y})^2$$

which leads to a quadratic in r_2.

$$r_2^2 + r_2[2(p_{2x} - p_{1x})\cos \beta + 2(p_{2y} - p_{1y})\sin \beta] + (d^2 - r_1^2) = 0$$

or the compact form

$$r_2^2 + r_2[E] + [F] = 0$$

with two solutions

$$r_2 = \left| \frac{-E \pm \sqrt{E^2 - 4F}}{2} \right| \tag{2.68}$$

If $4F > E^2$, the rotating guide cannot be assembled.

Finally, for real values of r_2,

$$p_{3x} = p_{2x} + r_2 \cos \beta$$
$$p_{3y} = p_{2y} + r_2 \sin \beta \tag{2.69}$$

The mode-of-assembly question in the case of the rotating guide is somewhat complicated by the fact that two situations are possible as shown in Figure 2.10. In situations where point p_2 lies outside a circle of radius r_1 about p_1 there are two possible locations for point p_3. The positive mode corresponds to the larger of the two values for r_2. When p_2 lies within the circle of radius r_1, only one location for p_3 is possible, as located by vector r_2 at the specified angle β.

Velocity analysis involves the solution of

$$\dot{\mathbf{p}}_3 = \dot{\mathbf{p}}_1 + \dot{\boldsymbol{\theta}}_1 \times (\mathbf{p}_3 - \mathbf{p}_1)$$
$$= \dot{\mathbf{p}}_2 + \dot{\boldsymbol{\beta}} \times (\mathbf{p}_3 - \mathbf{p}_2) + \dot{r}_2 \hat{\mathbf{r}}_2 \tag{2.70}$$

which leads to the two scalar equations

$$-\dot{\theta}_1(p_{3y} - p_{1y}) + \dot{r}_2(-\cos \beta) = (\dot{p}_{2x} - \dot{p}_{1x}) - r_2\dot{\beta} \sin \beta = E$$

$$\dot{\theta}_1(p_{3x} - p_{1x}) + \dot{r}_2(-\sin \beta) = (\dot{p}_{2y} - \dot{p}_{1y}) + r_2\dot{\beta} \cos \beta = F$$

Therefore, we obtain

$$\dot{\theta}_1 = \frac{-E \sin \beta + F \cos \beta}{(p_{3y} - p_{1y})\sin \beta + (p_{3x} - p_{1x})\cos \beta} \qquad (2.71)$$

$$\dot{r}_2 = \frac{E(p_{3x} - p_{1x}) + F(p_{3y} - p_{1y})}{(p_{3y} - p_{1y})\sin \beta + (p_{3x} - p_{1x})\cos \beta} \qquad (2.72)$$

and

$$\dot{p}_{3x} = \dot{p}_{1x} - \dot{\theta}_1(p_{3y} - p_{1y})$$

$$\dot{p}_{3y} = \dot{p}_{1y} + \dot{\theta}_1(p_{3x} - p_{1x}) \qquad (2.73)$$

Acceleration analysis follows from

$$\ddot{\mathbf{p}}_3 = \ddot{\mathbf{p}}_1 + \ddot{\boldsymbol{\theta}}_1 \times (\mathbf{p}_3 - \mathbf{p}_1) + \dot{\boldsymbol{\theta}}_1 \times (\dot{\boldsymbol{\theta}}_1 \times (\mathbf{p}_3 - \mathbf{p}_1))$$

$$= \ddot{\mathbf{p}}_2 + \ddot{\boldsymbol{\beta}} \times (\mathbf{p}_3 - \mathbf{p}_2) + \dot{\boldsymbol{\beta}} \times \dot{\boldsymbol{\beta}} \times (\mathbf{p}_3 - \mathbf{p}_2))$$

$$+ \ddot{r}_2\hat{r}_2 + 2\dot{\boldsymbol{\beta}} \times \dot{r}_2\hat{r}_2 \qquad (2.74)$$

$$-\ddot{\theta}_1(p_{3y} - p_{1y}) + \ddot{r}_2(-\cos \beta) = (\ddot{p}_{2x} - \ddot{p}_{1x}) + \dot{\theta}_1^2(p_{3x} - p_{1x})$$

$$- \dot{\beta}^2(r_2 \cos \beta) - 2\dot{\beta}\dot{r}_2 \sin \beta$$

$$- \ddot{\beta}(p_{3y} - p_{2y}) = E$$

$$\ddot{\theta}_1(p_{3x} - p_{1x}) + \ddot{r}_2(-\sin \beta) = (\ddot{p}_{2y} - \ddot{p}_{1y}) + \dot{\theta}_1^2(p_{3y} - p_{1y})$$

$$- \dot{\beta}^2(r_2 \sin \beta) + 2\dot{\beta}\dot{r}_2 \cos \beta$$

$$- \ddot{\beta}(p_{3x} - p_{2x}) = F$$

from which

$$\ddot{\theta}_1 = \frac{-E \sin \beta + F \cos \beta}{(p_{3y} - p_{1y})\sin \beta + (p_{3x} - p_{1x})\cos \beta} \qquad (2.75)$$

$$\ddot{r}_2 = -\frac{E(p_{3x} - p_{1x}) + F(p_{3y} - p_{1y})}{(p_{3y} - p_{1y})\sin \beta + (p_{3x} - p_{1x})\cos \beta} \qquad (2.76)$$

and

$$\ddot{p}_{3x} = \ddot{p}_{1x} - \ddot{\theta}_1(p_{3y} - p_{1y}) - \dot{\theta}_1^2(p_{3x} - p_{1x})$$

$$\ddot{p}_{3y} = \ddot{p}_{1y} + \ddot{\theta}_1(p_{3x} - p_{1x}) - \dot{\theta}_1^2(p_{3y} - p_{1y}) \qquad (2.77)$$

These expressions for position, velocity, and acceleration analysis for the rotating guide element have been programmed and are listed in LINKPAC as subroutines PGUIDE, VGUIDE, and AGUIDE.

All of the plane kinematic analysis subroutines are written using a special notation in which all points in the mechanism are specified by an integer index in a two-dimensional *points* array P(30, 2). The coordinates of up to 30 points can be stored with x and y coordinates indicated by values of 1 or 2 for the column index.

Examination of the subroutine listings will disclose that acceleration subroutines call velocity subroutines and velocity subroutines call position subroutines as required. In this manner the user need only call the subroutine involving the highest-order derivatives desired. Lower-order derivatives are computed and stored as required. Velocity and acceleration components of all points are stored in arrays VP(30,2) and AP(30,2).

Rigid Body Motion Subroutines GEOM, DISP, POS, VEL, ACC, and MOTION

As a first example of the use of a LINKPAC subroutine, consider the problem of calculating the new position of a point of interest **b** on a rigid body after a displacement of the body described by two positions of a reference point **a** and the angular displacement of the body, θ.

Using LINKPAC notation, we first relabel point **a** as a member of the "points" array P(30,2). Point **a** is labeled as point N1, and its current coordinates will be stored in P(N1,1) and P(N1,2). The initial position of point N1 is stored in the PI(30,2) array in locations PI(N1,1) and PI(N1,2). Point **b** becomes point N2 stored in P(N2,1) and P(N2,2), and so forth.

Subroutine DISP, based on Eq. 3.16, can be called to calculate the new position of **b** using the FORTRAN statement

---

CALL DISP (N1,N2, THETA, P, PI)

The first argument is the index in the points array, which identifies the reference point with initial position previously defined and stored in the PI array as elements PI(N1,1) and PI(N1,2). The final position of the reference point must also be specified and has been stored in P(N1,1) and P(N1,2).

The second argument identifies the point of interest, in this case point N2. The initial position of the point of interest has been stored in PI(N2,1) and PI(N2,2).

The third argument is the angular displacement in radians of the rigid body containing points N1 and N2. Note that indices N1 and N2 could identify any two points on the same rigid body.

The fourth and fifth arguments are the two dimensional arrays P and PI where position data for up to 30 points can be stored and is available to all subprograms.

Similarly, to calculate the velocity of a point of interest, we first define and store the position and velocity components for a reference point in locations P(N1,1), P(N1,2), VP(N1,1), and VP(N1,2) where VP can be interpreted as "velocity of point." The velocity of the point of interest with index N2 is calculated using subroutine VEL and the FORTRAN statement

---

CALL VEL(N1, N2, W, P, VP)

In this case arguments N1 and N2 again identify a reference point N1 and a point of interest N2 on the same rigid body with position and velocity components stored in arrays P and VP. The angular velocity of the rigid body in radians per second must be defined as W. The velocity components for the point of interest are returned as VP(N2,1) and VP(N2,2).

The acceleration of a point of interest N2 can be found in terms of the known acceleration of a reference point N1, plus the angular velocity W, and angular acceleration A, using a FORTRAN statement

---

CALL ACC(N1, N2, W, A, P, VP, AP)

The acceleration components for the point of interest with index N2 would now be available as AP(N2,1) and AP(N2,2).

Examination of the listings of subroutines DISP, VEL, and ACC as given in LINKPAC will reveal that a single call to ACC will also call VEL to provide the necessary preliminary velocity calculations.

A complete program for analysis of rigid body motion becomes:

```
      PROGRAM MOTIONS(INPUT,OUTPUT,TAPE5=INPUT,TAPE6=OUTPUT)
      DIMENSION PI(30,2),P(30,2),VP(30,2),AP(30,2)
C
C CALCULATION OF RIGID BODY MOTION COMPONENTS FOR A POINT N2
C IN TERMS OF THE MOTION OF A REFERENCE POINT N1 GIVEN THE
C ANGULAR VELOCITY W AND ANGULAR ACCELERATION A FOR THE
C RIGID BODY
C
      DATA PI(4,1),PI(4,2),PI(7,1),PI(7,2),P(4,1),P(4,2),
     $ VP(4,1),VP(4,2),AP(4,1),AP(4,2),THETA,W,A,N1,N2 /
     $ 1.,1.,3.,2.,3.,1.,30.,20.,10.,-30.,45.,10.,100.,4,7 /
C
C CALCULATE NEW POSITION FOR POINT 7 GIVEN ITS INITIAL POSITION
C PLUS THE INITIAL AND FINAL POSITION FOR REFERENCE POINT 4.
C CONVERT ANGULAR DISPLACEMENT FROM DEGREES TO RADIANS FIRST.
C
```

```
         CON=ATAN(1.)/45.
         TH=THETA·CON
         CALL  DISP(N1,N2,TH,P,PI)
C
C CALCULATE VELOCITY AND ACCELERATION OF POINT 7.
C
         CALL  ACC(N1,N2,W,A,P,VP,AP)
C
C PRINT RESULTS
C
         WRITE(6,100)  N2,P(N2,1),P(N2,2),N2,VP(N2,1),VP(N2,2),
                       N2,AP(N2,1),AP(N2,2)
    100 FORMAT(1H1/10X,3H  P(,I1,3H)=(,F8.2,3H  ,  ,F8.2,2H )/
       $               10X,3HVP(,I1,3H)=(,F8.2,3H  ,  ,F8.2,2H )/
       $               10X,3HAP(,I1,3H)-(,F8.2,3H  ,  ,F8.2,2H )/)
         END
```

With results printed as

```
            P(7)=(    3.71 ,    3.12 )
           VP(7)=(    8.79 ,   27.07 )
           AP(7)=( -272.84 , -171.42 )
```

It is sometimes convenient to describe the geometry of an individual link in terms of a coordinate system fixed in the link. LINK PAC assumes that each link is described by a vector **r** between joints N1 and N2 in the link plus a vector **s** between joints N1 and N3. An angle ϕ defines the rotation required to rotate **r** into **s**.

Subroutine GEOM can be called to calculate initial Cartesian coordinates from polar coordinates assuming the vector **r** lies along the x-axis with point N1 at the origin and point N2 having coordinates $PI(N2,1) = |r|$ and $PI(N2,2) = 0.0$. Point N3 has coordinates $PI(N3,1) = s \cos \theta$ and $PI(N3,2) = s \sin \theta$.

As an example, consider a link described geometrically by r, s, and ϕ. We wish to "assemble" this link into a mechanism in a new position with vector **r** described by an angular position θ.

The final coordinates of points N2 = 4 and N3 = 5 can be calculated in terms of r, s, and 0, the two positions of reference point N1 = 3, and the angular position θ for vector **r**, as follows.

```
         CALL GEOM (3, 4, 5, R, S, PHI, PI)
         CALL DISP (3, 4, TH, P, PI)
         CALL DISP (3, 5, TH, P, PI)
```

In this sequence GEOM calculates initial coordinates for points 3, 4, and 5 in an initial position with point 3 at the origin, and $\theta = 0.0$.

DISP is then called for points 4 and 5 individually to calculate their final coordinates. Point 3 is the reference point with specified positions.

The same task can be accomplished using subroutine POS without storage of initial coordinate data in the PI array. In this case the final coordinates for points 4 and 5 would be calculated and stored in the points array using the statement

CALL POS (3, 4, 5, R, S, PHI, TH, P)

where PHI and TH are given in radians.

Two-Link Dyad Subroutines CRANK2, PDYAD, VDYAD, and ADYAD

Algorithms for position, velocity, and acceleration analysis of the two-link dyad were given earlier. These algorithms have been translated into FORTRAN subroutines and are available to the LINKPAC user as PDYAD for position analysis and mobility checking, VDYAD for velocity analysis, and ADYAD for acceleration analysis. A call to a higher-order subroutine will automatically call lower-order subroutines. For example, a call to ADYAD will call VDYAD internally and VDYAD will call PDYAD as required.

As an example, consider the following subroutine that could be called by another program to return the complete acceleration analysis of a four-bar linkage. Points N1 and N4 identify the fixed pivots for the linkage. Points N2 and N3 identify the two moving joints for a continuous loop around joints N1-N2-N3-N4. Points N5, N6, and N7 locate the centers of gravity for links R2, R3, and R4, as shown in Figure 2.8.

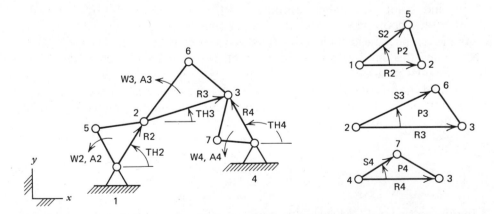

FIGURE 2.8 FOUR-BAR LINKAGE WITH LINKPAC NOTATION.

```
          SUBROUTINE FOUR1 (M, N1, N2, N3, N4, N5, N6, N7, R2, S2, PH2,
     $ R3, S3, PH3, R4, S4, PH4, TH2, TH3, TH4, W2, W3, W4, A2, A3, A4,
     $ P, VP, AP)
          DIMENSION P(30,2) VP(30,2), AP(30,2)
C
C POSITION, VELOCITY AND ACCELERATION ANALYSIS
C OF THE FOUR-BAR LINKAGE.
C
C COMPUTE MOTION OF TWO POINTS ON THE INPUT CRANK
          CALL CRANK2 (N1, N2, N5, R2, S2, PH2, TH2, W2, A2, P, VP, AP)
C
C POSITION, VELOCITY AND ACCELERATION OF BASIC
C TWO-LINK DYAD WITH MODE OF ASSEMBLY INDICATED
C BY THE SIGN OF M.
C         CALL ADYAD (M, N2, N4, N3, R3, R4, TH3, TH4, P, W3, W4, VP, A3, A4, AP)
C
C COMPUTE MOTION OF POINTS N5, N6, N7
C
          CALL POS (N2, N3, N6, R3, S3, PH3, TH3, P)
          CALL POS (N4, N3, N7, R4, S4, PH4, TH4, P)
          CALL ACC (N2, N6, W3, A3, P, VP, AP)
          CALL ACC (N4, N7, W4, A4, P, VP, AP)
C
C RESULTS MUST BE PRINTED BY THE CALLING PROGRAM
          RETURN
          END
```

An even more compact subroutine (in this case without comments) would make use of LINKPAC subroutine MOTION with array PI added to the argument list.

```
          SUBROUTINE FOUR2 (M, N1, N2, N3, N4, N5, N6, N7, R2, S2, PH2,
     $ R3, S3, PH3, R4, S4, PH4, TH2, TH3, TH4, W2, W3, W4,
     $ A2, A3, A4, PI, P, VP, AP)

          DIMENSION (PI(30,2), P(30,2), VP(30,2), AP(30,2)

          CALL CRANK2 (N1, N2, N5, R2, S2, PH2, TH2, W2, A2, P, VP, AP)
          CALL ADYAD (M, N2, N4, N3, R3, R4, TH3, TH4, P, W3, W4, VP, A3, A4, AP)
          CALL MOTION (N2, N3, N6, R3, S3, PH3, TH3, W3, A3, PI, P, VP, AP)
          CALL MOTION (N4, N3, N7, R4, S4, PH4, TH4, W4, A4, PI, P, VP, AP)
          RETURN
          END
```

The mode factor M determines whether links R3 and R4 will be assembled in either the open or crossed configuration. The open configuration corresponds to a positive mode of assembly with a positive angle α, the angle required to rotate the diagonal vector $\mathbf{d}$ into vector $\mathbf{r}_3$. The crossed configuration, with negative α, is specified by setting M negative. Once set, the mode of assembly will not change over the full range of continuous motion for the linkage.

Input-Output Subroutines RDACC and WRACC

The writing of input-output subroutines can often be time consuming; LINKPAC contains subroutines that may be helpful to the user in some cases. Consider the use of the subroutine FOUR2 for the acceleration analysis of the four-bar linkage. A main program must be prepared to read data, call FOUR2, and write results. LINKPAC subroutines RDGEOM, RDFIXED, and RDINPUT are called by subroutine RDACC to provide the information required by our new subroutine FOUR2.

```
            RDGEOM contains the statements:
            READ (5,100) R2, S2, P2, R3, S3, P3, R4, S4, P4
       100 FORMAT (3F10.0)
            READ (5,200) M
       200 FORMAT (I2)
```

In this case the angles $P2 = \phi_2$, and so on, are read in degrees, then converted to $PH2 = \phi_2$ in radians within RDGEOM. Therefore RDGEOM expects four data cards, three with link geometry data and a fourth with the mode of assembly $\pm M$.

Subroutine RDFIXED reads the coordinates of fixed pivots with indices $N1 = 1$ and $N4 = 4$ in FORMAT (4F10.0) and sets $VP(1) = VP(4) = AP(1) = AP(4) = 0.0$.

Subroutine RDINPUT reads input crank motion parameters including the number of crank displacements NTH, the initial position of the input crank $TH20 = \theta_{20}$, the crank displacement increment $DEL = \Delta\theta_2$ in degrees, and the input crank angular velocity and acceleration $W2 = \omega_2 = \dot{\theta}_2$ and $A2 = \alpha_2 = \ddot{\theta}_2$ in FORMAT (I5,4F10.0). When using the standard input subprogram RDACC, the main program must contain the labeled COMMON blocks /FRBR/ and /RD/. These COMMON blocks provide an alternate means of communication between main programs and subprograms. To take advantage of this, we first rewrite our previous subroutine FOUR2, give it a new name FOURAN, and add comments.

```
      SUBROUTINE FOURAN
      COMMON/FRBR/M, R2, S2, PH2, R3, S3, PH3, R4, S4, PH4, P(30,2),
     $ VP(30,2), AP(30,2), PI(30,2), TH2, TH3, TH4, W2, W3, W4, A2, A3, A4
C
C ACCELERATION ANALYSIS OF THE FOUR-BAR LINKAGE WITH FIXED PIVOTS
C LOCATED AT POINTS 1 AND 4. INPUT CRANK IS LINK 2.
C
C COMPUTE MOTION OF POINTS ON THE INPUT CRANK.
C
      CALL CRANK2(1, 2, 5, R2, S2, PH2, TH2, W2, A2, R, VP, AP)
C
C ACCELERATION ANALYSIS OF THE BASIC TWO-LINK DYAD
C
```

```
        CALL ADYAD(M, 2, 4, 3, R3, R4, TH3, TH4, P, W3, W4, VP, A3, A4, AP)
C
C COMPUTE MOTION OF POINTS 6 AND 7 ON LINKS 3 AND 4.
C
        CALL MOTION(2, 3, 6, R3, S3, PH3, TH3, W3, A3, PI, P, VP, AP)
        CALL MOTION(4, 3, 7, R4, S4, PH4, TH4, W4, A4, PI, P, VP, AP)
        RETURN
        END
C
C
```

The main program is written in the following form.

```
        PROGRAM FOURBAR(INPUT, OUTPUT, TAPE5=INPUT, TAPE6=OUTPUT)
        COMMON/FRBR/M, R2, S2, PH2, R3, S3, PH3, R4, S4, PH4, P(30,2)
      $ VP(30,2), AP(30,2), PI(30,2), TH2, TH3, TH4, W2, W3, W4, A2, A3, A4
        COMMON/RD/NTH, NPTS, TH20, DELTH
C
C READ LINK GEOMETRY, FIXED PIVOT COORDINATES AND INPUT CRANK MOTION
C PARAMETERS.
C
        CALL RDACC
C
C SET UP THE DO LOOP FOR ACCELERATION ANALYSIS AND PRINTING OF
C RESULTS IN STANDARD FORMAT BY SUBROUTINE WRACC.
C
        NPTS=7
        DO 10 J= 1, NTH
        TH2= TH20+(J-1)*DELTH
        CALL FOURAN
     10 CALL WRACC
        STOP
        END
```

A sample printout of results would be as follows. The data are printed after reading in RDACC. The results for NPTS = 7 are printed by WRACC for each new value of TH2.

```
        * * * * * * * * * * * * * * * * * * * * * * * *
          KINEMATIC ANALYSIS FOUR-LINK MECHANISMS
        * * * * * * * * * * * * * * * * * * * * * * * *


        LINK GEOMETRY DATA
R2=   1.000   S2=   2.000   PHI2=   30.000 DEG
R3=   2.828   S3=   3.000   PHI3=   45.000
R4=   2.000   S4=   4.000   PHI4=   60.000
MODE FACTOR M=   1


        FIXED PIVOT COORDINATES
P(1,1)=   0.   P(1,2)=   0.   P(4,1)=   3.000   P(4.2)=   0.
```

```
        INPUT CRANK MOTION PARAMETERS
NTH=  3  TH20=  0.  DELTH=   30.000 DEGREES
W2=628.000 RAD/SEC  A2=  0.  RAD/SEC/SEC

        ACCELERATION RESULTS FOR THETA2=  0.  DEG

    ANGULAR POSITION
    TH2=  0.           TH3=    .4501E+02        TH4=    .9002E+02 DEGREES

    ANGULAR VELOCITY
    W2=  .6280E+03  W3=  -.3140E+03          W4=  -.3140E+03 RAD/SEC

    ANGULAR ACCELERATION
    A2=  0.           A3=  -.8933E+02         A4=   .2957E+06 RAD/SEC/SEC
```

POINT	POSITION		VELOCITY		ACCELERATION	
NUMBER	X	Y	X	Y	X	Y
1	0.	0.	0.	0.	0.	0.
2	.100E+01	0.	-0.	.628E+03	-.394E+06	0.
3	.300E+01	.200E+01	.628E+03	.190E+00	-.591E+06	-.197E+06
4	.300E+01	0.	0.	0.	0.	0.
5	.173E+01	.100E+01	-.628E+03	.109E+04	-.683E+06	-.394E+06
6	.100E+01	.300E+01	.942E+03	.628E+03	-.394E+06	-.296E+06
7	-.465E+00	.200E+01	.628E+03	.109E+04	-.249E+06	-.122E+07

```
        ACCELERATION RESULTS FOR THETA2=  30.00 DEG

    ANGULAR POSITION
    TH2=  .3000E+02   TH3=    .3158E+02    TH4=    .8209E+02 DEGREES

    ANGULAR VELOCITY
    W2=  .6280E+03    W3=  -.2270E+03    W4=  -.1122E+02 RAD/SEC

    ANGULAR ACCELERATION
    A2=  0.           A3=   .1534E+06    A4=   .3497E+06 RAD/SEC/SEC
```

POINT	POSITION		VELOCITY		ACCELERATION	
NUMBER	X	Y	X	Y	X	Y
1	0.	0.	0.	0.	0.	0.
2	.866E+00	.500E+00	-.314E+03	.544E+03	-.342E+06	-.197E+06
3	.328E+01	.198E+01	.222E+02	-.309E+01	-.693E+06	.960E+05
4	.300E+01	0.	0.	0.	0.	0.
5	.100E+01	.173E+01	-.109E+04	.628E+03	-.394E+06	-.683E+06
6	.156E+01	.342E+01	.348E+03	.386E+03	-.825E+06	-.241E+06
7	-.156E+00	.246E+01	.276E+02	.354E+02	-.859E+06	-.110E+07

Oscillating Slider Subroutines POSC, VOSC, and AOSC

The oscillating slider element shown in Figure 2.7c forms the second basic kinematic pairing between two links. Subroutines POSC for position and mobility analysis and subroutines VOSC and AOSC for velocity and acceleration analysis

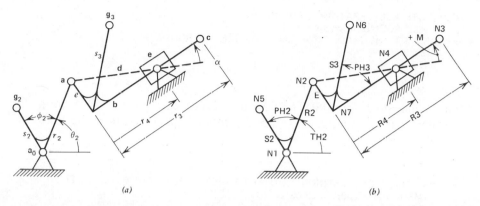

(a)   (b)

FIGURE 2.9   NOTATION FOR PROGRAM OSCSLDR, ACCELERATION ANALYSIS OF THE OSCILLATING SLIDER MECHANISM. (*a*) THE OSCILLATING SLIDER. (*b*) LINKPAC NOTATION.

are available as LINKPAC subroutines. In this case the mode of assembly is positive when the angle α that rotates vector **d** into a position parallel to r_3, as shown in Figure 2.7c, is positive.

Consider the acceleration analysis of the oscillating-slider mechanism shown in Figure 2.9a with points and dimensions relabeled as FORTRAN variables in Figure 2.9b.

A subroutine OSCSLDR can be written using LINKPAC subroutines to carry out the position, velocity, and acceleration analysis of the mechanism as follows.

```
      SUBROUTINE OSCSLDR(N1,N2,N3,N4,N5,N6,N7,M,R2,S2,PH2,R3,S3,PH3,
     $ R4,P,VP,AP,PI,TH2,TH3,W2,W3,VR4,A2,A3,AR4,E)
      DIMENSION P(30,2),PI(30,2),VP(30,2),AP(30,2)
C
C POSITION,VELOCITY,AND ACCELERATION ANALYSIS OF THE OSCILLATING-SLIDER
C
C COMPUTE MOTION OF POINTS N2 AND N5 ON THE INPUT CRANK.
C POINT N1 IS THE FIXED CRANK PIVOT. POINT N4 IS THE SLIDER PIVOT
      CALL CRANK2(N1,N2,N5,R2,S2,PH2,TH2,W2,A2,P,VP,AP)
C
C CALL AOSC FOR ACCELERATION ANALYSIS OF THE BASIC MECHANISM.
      E= SIGN(E,FLOAT(M))
      CALL AOSC(M,N2,N4,N3,E,R4,R3,TH3,P,W3,VR4,VP,A3,AR4,AP)
C
C CALCULATE THE INITIAL POSITION PI OF POINTS N6 AND N2 ON LINK 3
C AND THE FINAL POSITION FOR POINT N6.
      CALL GEOM(N7,N3,N6,R3,S3,PH3,PI)
      PI(N2,1)= 0.0 $ PI(N2,2)= E
      CALL DISP(N2,N6,TH3,P,PI)
```

```
C COMPUTE ACCELERATION OF POINT N6
      CALL ACC(N2,N6,W3,A3,P,VP,AP)
C RETURN TO THE MAIN PROGRAM TO PRINT RESULTS
      RETURN
      END
```

The main program OSCTEST reads data, sets up the DO loop for incrementing the input angle TH2, and prints results as given in the following material.

```
      PROGRAM OSCTEST(INPUT,OUTPUT,TAPE5=INPUT, TAPE6=OUTPUT)
      DIMENSION PI(30,2),P(30,2),VP(30,2),AP(30,2)
C ACCELERATION ANALYSIS OF THE OSCILLATING-SLIDER MECHANISM
C READ LINKAGE DIMENSIONS
      WRITE(6,10)
  10 FORMAT(1H1)
      READ(5,100) R2,S2,P2,R3,S3,P3,E
 100 FORMAT(3F10.0)
      CON= ATAN(1.)/45.
      PH2=P2*CON $ PH3= P3*CON
C IDENTIFY POINTS N1,N2,N3,N4,N5,N6,N7 AND MODE OF ASSEMBLY M
      READ(5,105) N1,N2,N3,N4,N5,N6,N7,M
 105 FORMAT(8I2)
C READ INPUT CRANK MOTION PARAMETERS NTH,TH20,DEL,W2,A2
      READ(5,108) NTH,TH20,DEL,W2,A2
 108 FORMAT(I5,4F10.0)
C READ FIXED PIVOT COORDINATES
      READ(5,110) P(N1,1),P(N1,2),P(N4,1),P(N4,2)
 110 FORMAT(4F10.0)
      VP(N1,1)= VP(N1,2)= VP(N4,1)= VP(N4,2)= 0.0
      AP(N1,1)= AP(N1,2)= AP(N4,1)= AP(N4,2)= 0.0
C PRINT HEADINGS FOR THE RESULTS
      WRITE(6,200) R2,S2,P2,R3,S3,P3,E,N1,N2,N3,N4,N5,N6,N7,M,
     $ TH20,DEL,W2,A2,P(N1,1),P(N1,2),P(N4,1),P(N4,2)
 200 FORMAT(20X,29H ACCELERATION ANALYSIS OF THE /
     $ 20X,29H OSCILLATING-SLIDER MECHANISM //
     $ 5X,4H R2=,F8.2,2X,4H S2=,F8.2,2X,5H PH2=,F8.2/
     $ 5X,4H R3=,F8.2,2X,4H S3=,F8.2,2X,5H PH3=,F8.2/
     $ 5X,17H ECCENTRICITY, E= ,F8.2//
     $ 5X,24H N1,N2,N3,N4,N5,N6,N7,M=,8I3//
     $ 6H TH20=,F8.2,2X,5H DEL=,F8.2,2X,4H W2=,F8.2,2X,4H A2=,F8.2//
     $ 9H P(N1,1)=,F6.2,2X,9H P(N1,2)=,F6.2,2X,9H P(N4,1)=,F6.2,2x,
     $ 9H P(N4,2)=,F6.2)
C SET UP DO LOOP FOR DESIRED RANGE OF MOTION
      DELTH= DEL*CON
      TH2ZERO= TH20*CON
      DO 1000 J= 1,NTH
      TH2= TH2ZERO +(J-1)*DELTH
      CALL OSCSLDR(N1,N2,N3,N4,N5,N6,N7,M,R2,S2,PH2,R3,S3,PH3,
     $ R4,P,VP,AP,PI,TH2,TH3,W2,W3,VR4,A2,A3,AR4,E)
      TH2= TH2/CON
      TH3= TH3/CON
```

```
1000 WRITE(6,205) J,TH2,TH3,W2,W3,A2,A3
     $ N2,P(N2,1),P(N2,2),VP(N2,1),VP(N2,2),AP(N2,1),AP(N2,2),
     $ N3,P(N3,1),P(N3,2),VP(N3,1),VP(N3,2),AP(N3,1),AP(N3,2),
     $ N5,P(N5,1),P(N5,2),VP(N5,1),VP(N5,2),AP(N5,1),AP(N5,2),
     $ N6,P(N6,1),P(N6,2),VP(N6,1),VP(N6,2),AP(N6,1),AP(N6,2),R4,VR4,AR4
 205 FORMAT(////,20X,16H POSITION  NUMBER,I3//
     $ 5X,5H  TH2=,F6.2,10X,5H  TH3=,F6.2,5X,4H  DEG/
     $ 6X,4H  W2=,F11.4,6X,4H  W3=,E11.4,8H  RAD/SEC /
     $ 6X,4H  A2=,F11.4,6X,4H  A3=,E11.4,12H  RAD/SEC/SEC//
     $ 1X,6H  POINT,7X,8HPOSITION  ,13X,8HVELOCITY,11X,12HACCELERATION /
     $ 4X,1HN,3X,6HP(N,1),5X,6HP(N,2),4X,7HVP(N,1),4X,7HVP(N,2),
     $ 4X,7HAP(N,1),4X,7HAP(N,2),/ 4(I4,1X,2E10.3,2X,2F10.3,2X,2E10.3/),
     $ 5X,4H  R4=,F11.4,5X,5H  VR4=,E11.4,5X,5H  AR4=,E11.4)
     STOP
     END
```

A sample output of results would take the form

ACCELERATION ANALYSIS OF THE
OSCILLATING-SLIDER MECHANISM

```
R2=   1.00   S2=   1.50   PH2=   30.00
R3=   3.00   S3=   2.00   PH3=   45.00
ECCENTRICITY,          E.       .50

 N1,N2,N3,N4,N5,N6,N7,M=  1  2  3  4  5  6  7  1

TH20=  60.00   DEL=  10.00   W2=  100.00   A2=  0.

P(N1,1)=  0.   P(N1,2)= 0.   P(N4,1)=  2.00   P(N4,2)= 0.
```

POSITION NUMBER 1

```
TH2= 60.00            TH3=-13.22          DEG
W2=   .1000E+03       W3=  -.1741E+02  RAD/SEC
A2= 0.               A3=    .7875E+04  RAD/SEC/SEC
```

POINT N	POSITION		VELOCITY		ACCELERATION	
	P(N,1)	P(N,2)	VP(N,1)	VP(N,2)	AP(N,1)	AP(N,2)
2	.500E+00	.866E+00	-.866E+02	.500E+02	-.500E+04	-.866E+04
3	.331E+01	-.307E+00	.107E+03	.115E+01	.339E+04	.138E+05
5	.809E-14	.150E+01	-.150E+03	.809E-12	-.809E-10	-.150E+05
6	.209E+01	.143E+01	-.767E+02	.224E+02	-.994E+04	.366E+04
	R4= .1658E+01		VR4= .1044E+03		AR4=-.5482E+03	

POSITION NUMBER 2

```
TH2= 70.00            TH3=-14.33          DEG
W2=   .1000+03        W3=  -.5370E+01  RAD/SEC
A2=  0.              A3=    .6000E+04  RAD/SEC/SEC
```

POINT	POSITION		VELOCITY		ACCELERATION	
N	P(N,1)	P(N,2)	VP(N,1)	VP(N,2)	AP(N,1)	AP(N,2)
2	.342E+00	.940E+00	−.940E+02	.342E+02	−.342E+04	−.940E+04
3	.312E+01	−.287E+00	−.101E+03	.193E+02	.386E+04	.734E+04
5	−.260E+00	.148E+01	−.148E+03	−.260E+02	.260E+04	−.148E+05
6	.194E+01	.148E+01	−.911E+02	.256E+02	−.668E+04	.167E+03
	R4= .1839E+01		VR4= .1022E+03		AR4= −.1960E+04	

Rotating Guide Subroutines PGUIDE, VGUIDE, and AGUIDE

A third basic linkage element is the rotating guide as shown in Figure 2.7d. In this case the motion of reference points $\mathbf{p}_1$ and $\mathbf{p}_2$, identified in LINKPAC subroutines by the dummy point indicates N1 and N2, are specified. In addition, the rotation of member R2 is specified by β, $\dot{\beta}$, and $\ddot{\beta}$.

The position analysis in this case requires careful consideration. As shown in Figure 2.10, several possibilities exist for the assembly of the rotating guide element. For given values of R1 and β it is possible that the ray from point N2 at angle β does not intersect the circle of radius R1 about point N1. The subroutine must test for this possibility.

Assuming that the ray at angle β intersects the circle, there are two situations depending on whether point N2 is within or outside of the circle. If point N2 is outside the circle, as shown in Figure 2.10a, we will arbitrarily assume the positive mode to correspond to the larger value for R2 calculated from Eq. 2.68. With point N2 inside the crank circle, only one value of R2 is possible for a specified

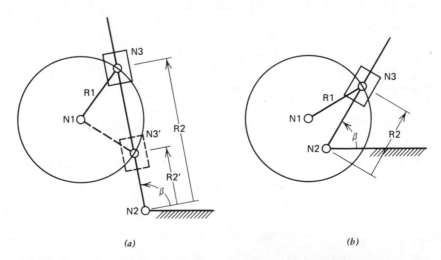

(a)   (b)

FIGURE 2.10 MODES OF ASSEMBLY FOR THE ROTATING GUIDE. (a) POINT N2 OUTSIDE R1 CIRCLE. (b) POINT N2 INSIDE R1 CIRCLE.

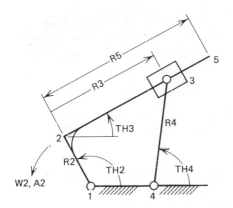

FIGURE 2.11 EXAMPLE USING ROTAT-
ING GUIDE SUBROUTINES.

value of β. These possibilities are considered in subroutine PGUIDE by first
comparing the calculated distance between points N1 and N2 and the crank
radius R1, and then branching to the appropriate section of the subroutine,
depending on the specified sign of the mode factor M.

As an example of the use of the rotating guide subroutines, consider the
mechanism shown in Figure 2.11. A minimum complexity program to carry out
the position, velocity, and acceleration analysis might be as listed below.

```
      PROGRAM FIG211(INPUT,OUTPUT,TAPE6=OUTPUT)
C ACCELERATION ANALYSIS — MECHANISM SHOWN IN FIG. 2.11
      DIMENSION P(30,2),VP(30,2),AP(30,2)
      DATA R1,R2,R4,R5,E,THET2,W2,A2,M/
     $ 1.5,1.0,2.0,4.0,0.0,120.,100.,0.0, +1 /
      WRITE(6,100) R1,R2,R4,R5,E,THET2,W2,A2,M
 100 FORMAT (1H1,5/,8E9.2,I3)
C CONVERT ANGLES FROM DEGREES TO RADIANS
      CON= ATAN(1.)/45.
      THET3= THET2 -90.
      TH2= THET2*CON $ TH3= THET3*CON
C INITIALIZE MOTION OF THE FIXED PIVOTS
      P(1,1)= 0.6 $ P(1,2)= 0. $ VP(1,1)=VP(1,2)=AP(1,1)=AP(1,2)=0.0
      P(4,1)= R1 $ P(4,2)= 0. $ VP(4,1)=VP(4,2)=AP(4,1)=AP(4,2)=0.0
C COMPUTE MOTION OF POINT 2
      CALL CRANK(1,2,R2,TH2,W2,A2,P,VP,AP)
C NOTE THAT POINTS 4,2,3 FORM A ROTATING GUIDE ELEMENT
      CALL AGUIDE(M,4,2,3,R4,R3,TH4,TH3,P,W4,W2,VR3,VP,
     $ A4,A2,AR3,AP)
C WRITE NUMERICAL RESULTS IN A SIMPLE FORMAT — IDENTIFY ON OUTPUT SHEET
      THET4= TH4/CON
      WRITE(6,200) VR3,AR3,THET4,W4,A4,P(3,1),P(3,2),VP(3,1),VP(3,2),
     $ AP(3,1),AP(3,2)
 200 FORMAT(5/,5E12.4,5/,6E12.4)
      STOP
      END
```

The program is self-explanatory with data provided by a FORTRAN DATA statement and numerical results only. The numerical results have been identified on the computer print out by comparing the WRITE statements with the printed output.

				DATA				
R1	R2	R4	R5	E	THET2	W2	A2	M
.15E+01	.10E+01	.20E+01	.40E+01	0.	.12E+03	.10E+03	0.	1

		RESULTS		
VR3	AR3	THET4	W4	A4
−.3098E+03	−.7380E+05	.9104E+02	.2342E+03	.2479E+05

P(3,1)	P(3,2)	VP(3,1)	VP(3,2)	AP(3,1)	AP(3,2)
.1464E+01	.2000E+01	−.4683E+03	−.8541E+01	−.4757E+05	−.1106E+06

APPENDIX A
THE CORIOLIS COMPONENT—ACCELERATION OF A POINT RELATIVE TO A ROTATING REFERENCE SYSTEM (SEE FIGURE 2.12)

The vector $\mathbf{a}$ locates a reference point A of the moving system in fixed system coordinates (i.e., $\mathbf{a} = a_x \hat{x} + a_y \hat{y}$). The vector $\boldsymbol{\rho} = u\hat{\imath} + v\hat{\jmath}$ locates P in the moving system, where $\hat{\imath}$ and $\hat{\jmath}$ are unit vectors in the moving system. $\dot{\boldsymbol{\theta}}$ is the absolute angular velocity of the *moving system*.

The position of point $\mathbf{P}$ in the fixed system is given by

$$\mathbf{p} = \mathbf{a} + (u\hat{\imath} + v\hat{\jmath})$$

Differentiating, noting that $\hat{\imath}$ and $\hat{\jmath}$ are unit vectors of constant length that rotate with the moving reference system,

$$\dot{\mathbf{p}} = \dot{\mathbf{a}} + (\dot{u}\hat{\imath} + u(\dot{\boldsymbol{\theta}} \times \hat{\imath}) + \dot{v}\hat{\jmath} + v(\dot{\boldsymbol{\theta}} \times \hat{\jmath}))$$

$$= \dot{\mathbf{a}} + \dot{\boldsymbol{\theta}} \times (u\hat{\imath} + v\hat{\jmath}) + (\dot{u}\hat{\imath} + \dot{v}\hat{\jmath})$$

$$= \dot{\mathbf{a}} + \dot{\boldsymbol{\theta}} \times \boldsymbol{\rho} + (\dot{u}\hat{\imath} + \dot{v}\hat{\jmath})$$

Examining the terms on the right side of the velocity vector equation, the first two terms are seen to give the velocity of a point Q on the rotating reference member, which is instantaneously coincident with the point of interest P. The second term accounts for the rotation of the reference member. The third term accounts for the relative velocity of point P with respect to the reference member. Note that $\dot{u}\hat{\imath} + \dot{v}\hat{\jmath} \neq \dot{\rho}\hat{\rho}$!

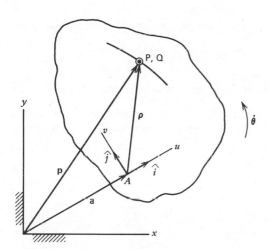

FIGURE 2.12   THE CORIOLIS COMPONENT.

Differentiating a second time,

$$\ddot{\mathbf{p}} = \ddot{\mathbf{a}} + \ddot{\boldsymbol{\theta}} \times (u\hat{\imath} + v\hat{\jmath}) + \dot{\boldsymbol{\theta}} \times (\dot{\boldsymbol{\theta}} \times (u\hat{\imath} + v\hat{\jmath})) + \dot{\boldsymbol{\theta}} \times (\dot{u}\hat{\imath} + \dot{v}\hat{\jmath})$$
$$+ \ddot{u}\hat{\imath} + \dot{u}(\dot{\boldsymbol{\theta}} \times \hat{\imath}) + \ddot{v}\hat{\jmath} + \dot{v}(\dot{\boldsymbol{\theta}} \times \hat{\jmath})$$

Collecting terms,

$$\ddot{\mathbf{p}} = \ddot{\mathbf{a}} + \ddot{\boldsymbol{\theta}} \times \boldsymbol{\rho} + \dot{\boldsymbol{\theta}} \times (\dot{\boldsymbol{\theta}} \times \boldsymbol{\rho})$$
$$+ (\ddot{u}\hat{\imath} + \ddot{v}\hat{\jmath}) + 2\dot{\boldsymbol{\theta}} \times (\dot{u}\hat{\imath} + \dot{v}\hat{\jmath})$$

which can be written in the form

$$\ddot{\mathbf{p}} = \ddot{\mathbf{q}} + \mathbf{a}_{\text{rel}} + \mathbf{a}_{\text{cor}}$$

where

$$\ddot{\mathbf{q}} = \ddot{\mathbf{a}} + \ddot{\boldsymbol{\theta}} \times \boldsymbol{\rho} + \dot{\boldsymbol{\theta}} \times (\dot{\boldsymbol{\theta}} \times \boldsymbol{\rho})$$

= acceleration of a point Q in the rotating reference system, which is instantaneously coincident with the point of interest P.

$$\mathbf{a}_{\text{rel}} = \ddot{u}\hat{\imath} + \ddot{v}\hat{\jmath}$$

= relative acceleration of point P with respect to a nonrotating reference system $\hat{\imath}, \hat{\jmath}$.

$$\mathbf{a}_{\text{cor}} = 2\dot{\boldsymbol{\theta}} \times (\dot{u}\hat{\imath} + \dot{v}\hat{\jmath})$$
$$= 2\dot{\boldsymbol{\theta}} \times \mathbf{v}_{\text{rel}}$$

= the supplemental Coriolis component, which accounts for the rotation of system $\hat{\imath}, \hat{\jmath}$.

$$\mathbf{v}_{rel} = (\dot{u}\hat{\imath} + \dot{v}\hat{\jmath})$$

= the relative velocity of point P with respect to the reference member.

The direction of the Coriolis component can be found by either the right-hand rule applied to the cross product $\dot{\boldsymbol{\theta}} \times v_{rel}$ or from components found by expansion of the determinant. The above analysis can be carried out in three dimensions using $\boldsymbol{\rho} = u\hat{\imath} + v\hat{\jmath} + \omega\hat{k}$ with the same final result.

Chapter 3
Matrix Methods in Kinematics

3.1 INTRODUCTION

In Chapter 2 it was shown that vector methods can be applied to a wide variety of problems in planar kinematic analysis. In Chapter 3 we continue our use of vector methods, but introduce matrix notation as a convenience in numerical calculations, particularly in those problems involving a description of spatial motions.

3.2 RIGID BODY ROTATION MATRICES

The analytical description of the displacement of a rigid body is based on the notion that all points in a rigid body must retain their original relative positions regardless of the new orientation of the body (i.e., the scalar distance between any two arbitrary points fixed in the body remains constant during the displacement). The total rigid body displacement can always be considered as the sum of its two basic components: the angular rotation of the body plus the linear displacement of any arbitrary reference point fixed in the rigid body. The angular motion can be described in one of several ways, the most popular being (1) a set of rotations about a right-handed set of Cartesian axes, (2) Euler's angles, and (3) angular rotation about an arbitrary axis in space.

Rotation about Cartesian Axes

First let us consider the angular displacement as consisting of a sequence of rotations γ, β, and α about a set of mutually perpendicular fixed axes x-y-z. The order in which rotations are prescribed also influences the final position of the rigid body (i.e., the rotations are not commutative). Figure 3.1 illustrates the final rigid body displacement for a rectangular body originally lying in the x-y plane after a sequence of three $90°$ rotations. In one case the rotations are taken in the α, β, γ sequence; in the second case a γ, β, α sequence is followed.

Figure 3.2 illustrates the rotation about the z-axis for a vector of constant length fixed in a rotating rigid body. All components of the vector, $\mathbf{v}$, before and after the displacement, are measured with respect to a fixed set of reference axes x-y. Figure 3.2 applies equally well to rotations parallel to the x-y plane or to the α-component of three-dimensional rotation.

45

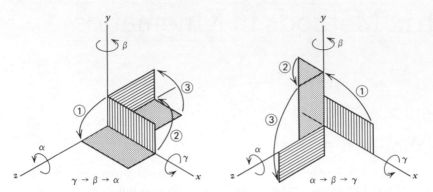

FIGURE 3.1 FINITE ROTATION ABOUT *X-Y-Z* AXES SHOWING DEPENDENCE ON ORDER OF ROTATION COMPONENTS.

The components of **v** after rotation can be found from inspection of Figure 3.2 as

$$v_{2x} = v_{1x} \cos \alpha - v_{1y} \sin \alpha$$

$$v_{2y} = v_{1x} \sin \alpha + v_{1y} \cos \alpha$$

$$v_{2z} = v_{1z} \tag{3.1}$$

Eq. 3.1 can be conveniently arranged as a matrix equation in one of the two forms

$$\begin{bmatrix} v_{2x} \\ v_{2y} \end{bmatrix} = \begin{bmatrix} \cos \alpha & -\sin \alpha \\ \sin \alpha & \cos \alpha \end{bmatrix} \begin{bmatrix} v_{1x} \\ v_{1y} \end{bmatrix} \tag{3.2}$$

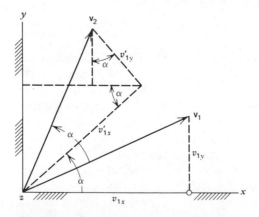

FIGURE 3.2 ROTATION OF A VECTOR WITH RESPECT TO A FIXED *Z*-AXIS.

or

$$\begin{bmatrix} v_{2x} \\ v_{2y} \\ v_{2z} \end{bmatrix} = \begin{bmatrix} \cos\alpha & -\sin\alpha & 0 \\ \sin\alpha & \cos\alpha & 0 \\ 0 & 0 & 1 \end{bmatrix} \begin{bmatrix} v_{1x} \\ v_{1y} \\ v_{1z} \end{bmatrix} \tag{3.3}$$

Eq. 3.2 is useful in describing two-dimensional plane rotation of a rigid body, whereas Eq. 3.3 forms one component of three-dimensional rigid body rotation. A similar consideration of a rotation β about the y-axis leads to

$$\begin{bmatrix} v'_{2x} \\ v'_{2y} \\ v'_{2z} \end{bmatrix} = \begin{bmatrix} \cos\beta & 0 & \sin\beta \\ 0 & 1 & 0 \\ -\sin\beta & 0 & \cos\beta \end{bmatrix} \begin{bmatrix} v_{1x} \\ v_{1y} \\ v_{1z} \end{bmatrix} \tag{3.4}$$

and a γ rotation about the x-axis gives

$$\begin{bmatrix} v''_{2x} \\ v''_{2y} \\ v''_{2z} \end{bmatrix} = \begin{bmatrix} 1 & 0 & 0 \\ 0 & \cos\gamma & -\sin\gamma \\ 0 & \sin\gamma & \cos\gamma \end{bmatrix} \begin{bmatrix} v_{1x} \\ v_{1y} \\ v_{1z} \end{bmatrix} \tag{3.5}$$

3.3   SPATIAL ROTATION MATRICES

In forming the complete spatial rotation matrix we note that a vector fixed in the rigid body may be rotated through a series of positions by a successive multiplication of the vector by one of the 3×3 matrices of Eqs. 3.3 to 3.5. These matrices are the *basic rotation matrices* from which all other finite rotation matrices can be formed. Eq. 3.2 provides a complete description of finite rotation in plane motion. Eq. 3.2 often will be written in the compact form

$$\mathbf{v}_2 = [R_\alpha]\mathbf{v}_1 \tag{3.6}$$

where $[R_\alpha]$ is the *plane rotation matrix*.

As noted in Section 3.2, the final orientation of the rigid body will depend on the order of the α, β, γ rotations. We will adopt a sequence of three rotations (1) α about the z-axis, (2) β about the y-axis, and (3) γ about the x-axis. The final position of the vector, $\mathbf{v}$ is then described by successive matrix-vector multiplications from right to left as described by

$$\mathbf{v}_2 = [R_{\gamma,\,x}][R_{\beta,\,y}][R_{\alpha,\,z}]\mathbf{v}_1 \tag{3.7}$$

$$= [R_{\alpha\beta\gamma}]\mathbf{v}_1 \tag{3.8}$$

The $[R_{\alpha\beta\gamma}]$ matrix takes the form

$$[R_{\alpha\beta\gamma}] = \begin{bmatrix} C\alpha C\beta & -S\alpha C\beta & S\beta \\ S\alpha C\gamma + C\alpha S\beta S\gamma & C\alpha C\gamma + S\alpha S\beta S\gamma & -C\beta S\gamma \\ S\alpha S\gamma - C\alpha S\beta C\gamma & C\alpha S\gamma + S\alpha S\beta C\gamma & C\beta C\gamma \end{bmatrix} \tag{3.9}$$

where $C\alpha = \cos\alpha$, $S\alpha = \sin\alpha$, and so forth.

The second method for describing rigid body rotation will be in terms of a rotation ϕ about an axis $\mathbf{u}$, a unit vector whose components are the direction cosines u_x, u_y, and u_z. In forming the complete rotation matrix in this form, we note that so far we are able to describe only rotations about either the x-, y-, or z-axes. We therefore seek some way in which the rotation ϕ can be described relative to one of the fixed coordinate axes. This is arranged by first rotating the rigid body to bring the axis $\mathbf{u}$ parallel to the z-axis, allowing the rotation ϕ to occur around this temporary location for $\mathbf{u}$, and then returning the axis $\mathbf{u}$ to its original position. The complete sequence can be described by the matrix equation

$$\mathbf{v}_2 = [R_{+\beta,\,y}][R_{-\gamma,\,x}][R_{\phi,\,z}][R_{+\gamma,\,x}][R_{-\beta,\,y}]\mathbf{v}_1 \tag{3.10}$$

which is shown graphically in Figure 3.3.

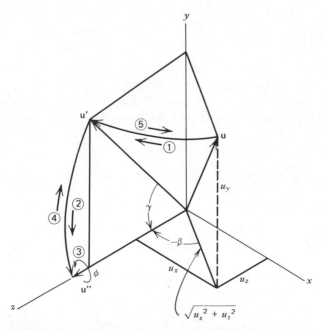

FIGURE 3.3   SEQUENCE OF FINITE ROTATIONS EQUIV-
ALENT TO A ROTATION ϕ ABOUT AN AXIS U.

In forming the rotation matrices it should be noted that the following substitutions are possible.

$$\sin \gamma = u_y \qquad\qquad \sin \beta = \frac{u_x}{\sqrt{u_x^2 + u_z^2}}$$

$$\cos \gamma = \sqrt{u_x^2 + u_z^2} \qquad \cos \beta = \frac{u_z}{\sqrt{u_x^2 + u_z^2}} \qquad (3.11)$$

Therefore,

$$\cos \gamma \sin \beta = u_x$$

$$\cos \gamma \cos \beta = u_z$$

Substituting into Eq. 3.10, we obtain

$$[R_{\phi, \mathbf{u}}] = \begin{bmatrix} u_x^2 V\phi + C\phi & u_x u_y V\phi - u_z S\phi & u_x u_z V\phi + u_y S\phi \\ u_x u_y V\phi + u_z S\phi & u_y^2 V\phi + C\phi & u_y u_z V\phi - u_x S\phi \\ u_x u_z V\phi - u_y S\phi & u_y u_z V\phi + u_x S\phi & u_z^2 V\phi + C\phi \end{bmatrix} \qquad (3.12)$$

where

$$V\phi = \text{vers } \phi = 1. - \cos \phi$$

$$S\phi = \sin \phi$$

$$C\phi = \cos \phi$$

The *axis rotation matrix* $[R_{\phi, \mathbf{u}}]$ will be found to be one of the most useful forms for description of spatial rigid body finite rotation.

Euler angles are often introduced in a discussion of the dynamics of spinning bodies (e.g., the dynamics of spinning tops or gyroscopes). They form a unique description of the displacement of a rigid body in terms of three *relative* displacement angles ψ, θ, and ϕ, as shown in Figure 3.4. For purposes of physical visualization four sets of Cartesian axes that were initially coincident are shown. The set x-y-z is fixed in the reference member 1. Axes x'-y'-z', fixed in member 2, are rotated an angle ψ about fixed axis z. Axes x''-y''-z'', fixed in member 3, have reached their final position by a sequence of two relative rotations, a rotation ψ about fixed axis z followed by a rotation θ about the new position x'. These first two rotation angle components, the precession angle ψ, and the nutation angle θ have located the z'' or spin axis as shown. A final spin ϕ for member 4, the rigid body of interest, completes the angular displacement of the rigid body. The corresponding angular velocity vectors are shown in Figure 3.5.

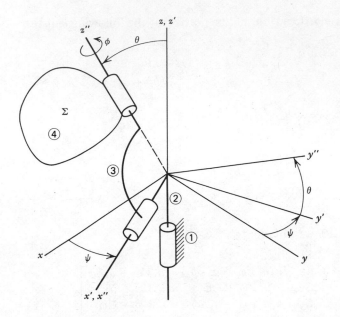

FIGURE 3.4 EULER'S ANGLES FOR FINITE ROTATION
OF A RIGID BODY.

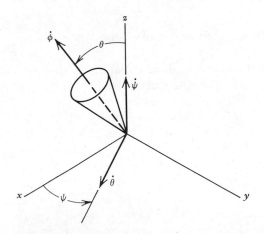

FIGURE 3.5 ANGULAR VELOCITY VEC-
TORS—EULER ANGLES.

This sequence of relative angular displacements can be described by a sequence of rotation matrices

$$[R_{\psi, \theta, \phi}] = [R_{\phi, z''}][R_{\theta, x'}][R_{\psi, z}]$$

where

$$x' = [R_{\psi, z}]x$$

$$z'' = [R_{\theta, x'}][R_{\psi, z}]z$$

A second and more direct derivation would account for the spin ϕ first followed by the relative angles θ and ψ in sequence. Since, in this derivation, all coordinate systems are coincident at the start, all rotations take place about axes parallel to the fixed x-y-z system; that is,

$$[R_{\psi, \theta, \phi}] = [R_{\psi, z}][R_{\theta, x}][R_{\phi, z}]$$

Either form leads to the *Euler rotation matrix*.

$$[R_{\psi, \theta, \phi}] = \begin{bmatrix} C\psi C\phi - S\psi C\theta S\phi & -C\psi S\phi - S\psi C\theta C\phi & S\psi S\theta \\ S\psi C\phi + C\psi C\theta S\phi & -S\psi S\phi + C\psi C\theta C\phi & -C\psi S\theta \\ S\theta S\phi & S\theta C\phi & C\theta \end{bmatrix} \tag{3.13}$$

3.4   RIGID BODY DISPLACEMENT MATRICES

The basic rotation matrix equations (Eqs. 3.3 to 3.5) describe the rotation of any vector fixed in a rigid body. The vector is conveniently described in terms of two points fixed in the body, a reference point $\mathbf{p}$ at the tail of the vector, and a point of interest $\mathbf{q}$ at the head of the vector. For plane rigid body motion (Figure 3.6), we may rewrite Eq. 3.2 as

$$\begin{bmatrix} q_x - p_x \\ q_y - p_y \end{bmatrix} = \begin{bmatrix} \cos \theta & -\sin \theta \\ \sin \theta & \cos \theta \end{bmatrix} \begin{bmatrix} q_{1x} - p_{1x} \\ q_{1y} - p_{1y} \end{bmatrix} \tag{3.14}$$

where θ is the rotation of the rigid body with respect to a fixed set of x-y axes.
Eq. 3.14 may be written in the compact form

$$(\mathbf{q} - \mathbf{p}) = [R_\theta](\mathbf{q}_1 - \mathbf{p}_1) \tag{3.15}$$

Typically, the original position $\mathbf{p}_1$ and the final position $\mathbf{p}$ for the reference point are given along with the rotation angle θ. Eq. 3.15 can then be rearranged in a form suitable for calculation of the coordinates of the new position of point $\mathbf{q}$ when its first position $\mathbf{q}_1$ is specified. Solving Eq. 3.15 for $\mathbf{q}$, we obtain

$$\mathbf{q} = [R_\theta](\mathbf{q}_1 - \mathbf{p}_1) + \mathbf{p} \tag{3.16}$$

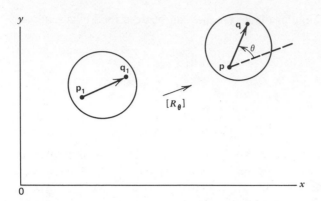

FIGURE 3.6 PLANE RIGID BODY DISPLACEMENT.

Eqs. 3.15 and 3.16 are in a convenient form for carrying out algebraic manipulations and may also serve as the basis for computer subroutines of general usefulness. For plane motion the rotation matrix is a 2×2 matrix.

An alternate form of Eq. 3.16 can be found by rearrangement of terms as

$$\mathbf{q} = [R_\theta]\mathbf{q}_1 - [R_\theta]\mathbf{p}_1 + \mathbf{p}$$

which can be written as a 3×3 matrix equation for plane displacement as

$$
\begin{bmatrix} q_x \\ q_y \\ q_z = 1 \end{bmatrix}
=
\begin{bmatrix}
\cos\theta & -\sin\theta & (p_x - p_{1x}\cos\theta + p_{1y}\sin\theta) \\
\sin\theta & \cos\theta & (p_y - p_{1x}\sin\theta - p_{1y}\cos\theta) \\
0 & 0 & 1
\end{bmatrix}
\begin{bmatrix} q_{1x} \\ q_{1y} \\ q_{1z} = 1 \end{bmatrix}
$$

$$(3.17)$$

or either of the compact forms

$$
\begin{bmatrix} q_x \\ q_y \\ 1 \end{bmatrix}
=
\begin{bmatrix}
[R_\theta] & (p - [R_\theta]p_1) \\
0\ 0 & 1
\end{bmatrix}
\begin{bmatrix} q_{1x} \\ q_{1y} \\ 1 \end{bmatrix}
\qquad (3.18)
$$

$$
\begin{bmatrix} \mathbf{q} \\ 1 \end{bmatrix}
= [D]
\begin{bmatrix} \mathbf{q}_1 \\ 1 \end{bmatrix}
\qquad (3.19)
$$

The 3×3 matrix $[D]$ is the *plane displacement matrix*. The displacement matrix equation has advantages in repetitive numerical calculations where all matrix elements are defined in terms of the specified displacement of the reference point **p** and the angular displacement of the rigid body.

Spatial rigid body displacement can be described in a manner completely analogous to Eqs. 3.15, 3.16, or 3.19. Figure 3.6 could apply equally well to spatial motion with the plane rotation matrix $[R_\theta]$ replaced by its three-dimensional equivalent $[R_{\alpha, \beta, \gamma}]$, $[R_{\phi, u}]$ or $[R_{\psi, \theta, \phi}]$.

The equivalent expressions, using the $[R_{\phi, u}]$ matrix for convenience, become

$$(\mathbf{q} - \mathbf{p}) = [R_{\phi, u}](\mathbf{q}_1 - \mathbf{p}_1) \tag{3.20}$$

$$\mathbf{q} = [R_{\phi, u}](\mathbf{q}_1 - \mathbf{p}_1) + \mathbf{p} \tag{3.21}$$

$$\begin{bmatrix} \mathbf{q} \\ 1 \end{bmatrix} = [D_{\phi, u}]\begin{bmatrix} \mathbf{q}_1 \\ 1 \end{bmatrix} = \begin{bmatrix} [R_{\phi, u}] & (p - [R_{\phi, u}]p_1) \\ 0\ 0\ 0 & 1 \end{bmatrix}\begin{bmatrix} \mathbf{q}_1 \\ 1 \end{bmatrix} \tag{3.22}$$

where $[D_{\phi, u}]$ is a 4×4 displacement matrix.

3.5   APPLICATIONS OF THE PLANE ROTATION MATRIX

Example 3-1 Displacement of an arbitrary point moving with a rigid body. Figure 3.7 illustrates the displacement of a plane rigid body. The motion is described by the displacement of a reference point $\mathbf{p}$ from position $\mathbf{p}_1$ to position $\mathbf{p}_2$ accompanied by a rotation θ_{12} for the rigid body defined by the triangular shape shown. A point of interest is arbitrarily selected in the first position with coordinates $\mathbf{q}_1$ (3,1). The second position $\mathbf{q}_2$ is found from Eq. 3.16, written in the form

$$\begin{bmatrix} q_{2x} \\ q_{2y} \end{bmatrix} = \begin{bmatrix} \cos 60^\circ & -\sin 60^\circ \\ \sin 60^\circ & \cos 60^\circ \end{bmatrix}\begin{bmatrix} (3. - 1.) \\ (1. - 1.) \end{bmatrix} + \begin{bmatrix} 3. \\ 2. \end{bmatrix} \tag{3.23}$$

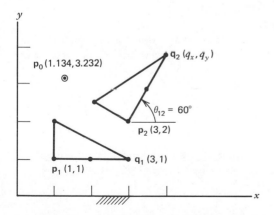

FIGURE 3.7   A PLANE DISPLACEMENT AND THE FINITE ROTATION POLE.

From which

$$q_{2x} = (2.)(0.5) - (0)(0.866) + 3. = 4.000$$

$$q_{2y} = (2.)(0.866) - (0)(0.5) + 2. = 3.732$$

In using the 3×3 displacement matrix equation we would first form the matrix $[D]$ as

$$[D_{12}] = \begin{bmatrix} 0.500 & -0.866 & [3. - 0.500(1.) + 0.866(1.)] \\ 0.866 & 0.500 & [2. - 0.866(1.) - 0.500(1.)] \\ 0 & 0 & 1 \end{bmatrix}$$

$$= \begin{bmatrix} 0.500 & -0.866 & 3.366 \\ 0.866 & 0.500 & 0.634 \\ 0 & 0 & 1 \end{bmatrix}$$

The $[D_{12}]$ matrix is then stored in the computer memory and is available to compute the coordinates of $\mathbf{q}_2$ from

$$\begin{bmatrix} q_{2x} \\ q_{2y} \\ 1 \end{bmatrix} = \begin{bmatrix} 0.500 & -0.866 & 3.366 \\ 0.866 & 0.500 & 0.634 \\ 0 & 0 & 1 \end{bmatrix} \begin{bmatrix} 3. \\ 1. \\ 1 \end{bmatrix} = \begin{bmatrix} 4.000 \\ 3.732 \\ 1 \end{bmatrix}$$

Example 3-2 The Finite Rotation Pole. As a plane is displaced from position 1 to position 2 with a specified rotation angle, there is always one point of the plane that does not change its position. The plane can be considered as having rotated about this point in the fixed plane, which is known as the *finite rotation pole*. This point should not be confused with a point of zero velocity (i.e., a *velocity pole*).

The displacement matrix can be used to locate such a point $\mathbf{p}_0(x_0, y_0)$. We first write the displacement matrix in analytical form using point $\mathbf{p}_0$ as a reference point.

$$[D_{12}] = \begin{bmatrix} \cos \theta_{12} & -\sin \theta_{12} & (p_{0x} - p_{0x} \cos \theta_{12} + p_{0y} \sin \theta_{12}) \\ \sin \theta_{12} & \cos \theta_{12} & (p_{0y} - p_{0x} \sin \theta_{12} - p_{0y} \cos \theta_{12}) \\ 0 & 0 & 1 \end{bmatrix} \quad (3.24)$$

Elements of the above matrix are then equated to corresponding elements of the numerical matrix in Example 3-1, which described the rigid body displacement. Equating elements of the third column,

$$p_{0x}(1. - \cos \theta_{12}) + p_{0y} \sin \theta_{12} = 3.366$$

$$p_{0y}(1. - \cos \theta_{12}) - p_{0x} \sin \theta_{12} = 0.634$$

Making the substitutions for $\theta_{12} = 60°$,

$$p_{0x} = \frac{\begin{vmatrix} 3.366 & 0.866 \\ 0.634 & 0.500 \end{vmatrix}}{\begin{vmatrix} 0.500 & 0.866 \\ -0.866 & 0.500 \end{vmatrix}} = 1.134$$

$$p_{0y} = \frac{\begin{vmatrix} 0.500 & 3.366 \\ -0.866 & 0.634 \end{vmatrix}}{1.0} = -3.232$$

This demonstrates a fundamental concept in numerical kinematics. Regardless of the analytical form of the elements of either the rotation or displacement matrices, the numerical values associated with a particular rigid body displacement must be the same. This allows analytical expressions for corresponding elements to be equated and solved for unknown parameters.

3.6   THE SCREW MATRIX

Eq. 3.20 is the basic matrix rotation equation for a rigid body. It is sometimes of interest to use a special reference point $\mathbf{p}$, which has its two positions on a fixed axis $\mathbf{u}$ along which the rigid body is *screwed* (Figure 3.8). The body simultaneously

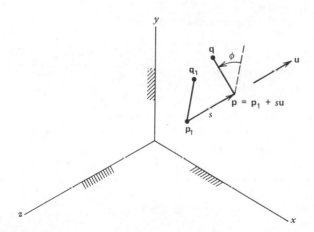

FIGURE 3.8   SCREW DISPLACEMENT OF A RIGID BODY.

slides along and rotates about the axis $\mathbf{u}$ with a linear displacement $\mathbf{s} = s\mathbf{u}$ and angular displacement ϕ. Eq. 3.20 becomes

$$(\mathbf{q} - (\mathbf{p}_1 + s\mathbf{u})) = [R_{\phi, \mathbf{u}}](\mathbf{q}_1 - \mathbf{p}_1) \tag{3.25}$$

which, when rewritten in the form of Eq. 3.22, becomes

$$\begin{bmatrix} q_x \\ q_y \\ q_z \\ 1 \end{bmatrix} = \begin{bmatrix} [R_{\phi, \mathbf{u}}] & (\mathbf{p}_1 + s\mathbf{u} - [R_{\phi, \mathbf{u}}]\mathbf{p}_1)) \\ 0\ 0\ 0 & 1 \end{bmatrix} \begin{bmatrix} q_{1x} \\ q_{1y} \\ q_{1z} \\ 1 \end{bmatrix} \tag{3.26}$$

The 4×4 matrix in Eq. 3.26 is the finite *screw displacement matrix* $[S]$.

3.7 NUMERICAL DISPLACEMENT MATRICES BY DIRECT MATRIX INVERSION

Eq. 3.26 or any other displacement matrix-vector equation leads to a linear transformation between vectors $\mathbf{q}_1$ and $\mathbf{q}$ in the form

$$\begin{bmatrix} q_x \\ q_y \\ q_z \\ 1 \end{bmatrix} = \begin{bmatrix} a_{11} & a_{12} & a_{13} & a_{14} \\ a_{21} & a_{22} & a_{23} & a_{24} \\ a_{31} & a_{32} & a_{33} & a_{34} \\ 0 & 0 & 0 & 1 \end{bmatrix} \begin{bmatrix} q_{1x} \\ q_{1y} \\ q_{1z} \\ 1 \end{bmatrix} \tag{3.27}$$

For plane displacement we have the 3×3 matrix equation

$$\begin{bmatrix} q_x \\ q_y \\ 1 \end{bmatrix} = \begin{bmatrix} a_{11} & a_{12} & a_{13} \\ a_{21} & a_{22} & a_{23} \\ 0 & 0 & 1 \end{bmatrix} \begin{bmatrix} q_{1x} \\ q_{1y} \\ 1 \end{bmatrix} \tag{3.28}$$

It is sometimes convenient to describe rigid body motion in terms of the known displacement of specified points fixed in the rigid body. For example, the displacement of a plane rigid body could be specified completely by the displacement of three arbitrary noncolinear points $\mathbf{a}$, $\mathbf{b}$, and $\mathbf{c}$ fixed in the body. Assume that the positions of the points are given as:

$$A_1(\mathbf{a}_1) = A_1(2, 4, 1) \qquad A_2(\mathbf{a}_2) = A_2(5, 1, 1)$$
$$B_1(\mathbf{b}_1) = B_1(2, 6, 1) \qquad B_2(\mathbf{b}_2) = B_2(7, 1, 1)$$
$$C_1(\mathbf{c}_1) = C_1(1, 5, 1) \qquad C_2(\mathbf{c}_2) = C_2(6, 2, 1)$$

Note that a constant z coordinate $= 1$ has been specified for all points. The three displacement equations can be combined in the form

$$[D] \begin{bmatrix} 2 & 2 & 1 \\ 4 & 6 & 5 \\ 1 & 1 & 1 \end{bmatrix} = \begin{bmatrix} 5 & 7 & 6 \\ 1 & 1 & 2 \\ 1 & 1 & 1 \end{bmatrix} \tag{3.29}$$

from which

$$[D] = \begin{bmatrix} 5 & 7 & 6 \\ 1 & 1 & 2 \\ 1 & 1 & 1 \end{bmatrix} \begin{bmatrix} 2 & 2 & 1 \\ 4 & 6 & 5 \\ 1 & 1 & 1 \end{bmatrix}^{-1} = \begin{bmatrix} 0 & 1 & 1 \\ -1 & 0 & 3 \\ 0 & 0 & 1 \end{bmatrix} \tag{3.30}$$

Comparison of the elements of the $[D]$ matrix with the analytical equivalents given in Eq. 3.17 reveals that

$$\cos \theta = 0$$

$$\sin \theta = -1$$

Hence the rigid body rotation angle $\theta = -90°$.

The displacement matrix for spatial motion can be found in a similar fashion by specification of the displacement of any four noncoplanar points fixed in the rigid body.

Reduction of the Number of Points Required for the Matrix Inversion Method

It is possible to form the displacement matrix numerically from the displacement of two arbitrary points fixed in a plane rigid body.

In this case a third point is calculated, and used as point $\mathbf{c}$, by rotating $\mathbf{b}$ $90°$ about $\mathbf{a}$. That is,

$$\mathbf{c}_1 = [R_{\theta=90°}](\mathbf{b}_1 - \mathbf{a}_1) + \mathbf{a}_1 = \begin{bmatrix} 0 & -1 \\ 1 & 0 \end{bmatrix} (\mathbf{b}_1 - \mathbf{a}_1) + \mathbf{a}_1$$

$$\mathbf{c}_2 - [R_{\theta=90°}](\mathbf{b}_2 \quad \mathbf{a}_2) \mid \mathbf{a}_2 - \begin{bmatrix} 0 & -1 \\ 1 & 0 \end{bmatrix} (\mathbf{b}_2 - \mathbf{a}_2) + \mathbf{a}_2$$

which gives

$$c_{1x} = -(b_{1y} - a_{1y}) + a_{1x}$$

$$c_{1y} = (b_{1x} - a_{1x}) + a_{1y}$$

$$c_{2x} = -(b_{2y} - a_{2y}) + a_{2x}$$

$$c_{2y} = (b_{2x} - a_{2x}) + a_{2y}$$

The displacement matrix can then be calculated in a manner similar to Eq. 3.30 from

$$[D] = \begin{bmatrix} a_{2x} & b_{2x} & (a_{2x} + a_{2y} - b_{2y}) \\ a_{2y} & b_{2y} & (a_{2y} - a_{2x} + b_{2x}) \\ 1 & 1 & 1 \end{bmatrix} \begin{bmatrix} a_{1x} & b_{1x} & (a_{1x} + a_{1y} - b_{1y}) \\ a_{1y} & b_{1y} & (a_{1y} - a_{1x} + b_{1x}) \\ 1 & 1 & 1 \end{bmatrix}^{-1}$$

3.8 THE INVERSE DISPLACEMENT MATRIX

The inverse of the plane rotation matrix can be formed in terms of the reverse or inverse displacement, formed by replacing the angle θ by its negative $-\theta$. The inverse matrix so formed can be shown to satisfy the relation

$$[R_\theta][R_{-\theta}] = [I] \tag{3.31}$$

where

$$[I] = \begin{bmatrix} 1 & 0 \\ 0 & 1 \end{bmatrix} = \text{the unit matrix}$$

Thus, we see that

$$[R_\theta]^{-1} = [R_{-\theta}] = \begin{bmatrix} \cos\theta & \sin\theta \\ -\sin\theta & \cos\theta \end{bmatrix} \tag{3.32}$$

We also note that the inverse of $[R_\theta]$ is equal to the transpose of $[R_\theta]$; that is,

$$[R_\theta]^{-1} = [R_\theta]^T \tag{3.33}$$

Hence, the $[R]$ matrix is orthogonal. It is left as an exercise for the reader to show that Eqs. 3.32 and 3.33 hold equally well for spatial rotation matrices.

It is often desirable to be able to form the inverse displacement matrix $[D]^{-1}$ quickly in terms of the elements a_{ij} of the original displacement matrix $[D]$. We first resolve the total displacement into a rotation plus translation as described by successive displacements $[D_T][D_R]$.

$$[D] = [D_T][D_R] = \begin{bmatrix} 1 & 0 & a_{13} \\ 0 & 1 & a_{23} \\ 0 & 0 & 1 \end{bmatrix} \begin{bmatrix} a_{11} & a_{12} & 0 \\ a_{21} & a_{22} & 0 \\ 0 & 0 & 1 \end{bmatrix}$$

Recalling the properties of matrix inversion,

$$[[M_1][M_2]]^{-1} = [M_2]^{-1}[M_1]^{-1}$$

we have

$$[D]^{-1} = [D_R]^{-1}[D_T]^{-1} = \begin{bmatrix} a_{11} & a_{21} & 0 \\ a_{12} & a_{22} & 0 \\ 0 & 0 & 1 \end{bmatrix} \begin{bmatrix} 1 & 0 & -a_{13} \\ 0 & 1 & -a_{23} \\ 0 & 0 & 1 \end{bmatrix}$$

That is, for a plane displacement matrix $[D]$,

$$[D]^{-1} = \begin{bmatrix} a_{11} & a_{21} & -(a_{11}a_{13} + a_{21}a_{23}) \\ a_{12} & a_{22} & -(a_{12}a_{13} + a_{22}a_{23}) \\ 0 & 0 & 1 \end{bmatrix} \qquad (3.34)$$

A similar procedure for the 4×4 spatial displacement matrix leads to

$$[D]^{-1} = \begin{bmatrix} a_{11} & a_{21} & a_{31} & -(a_{11}a_{14} + a_{21}a_{24} + a_{31}a_{34}) \\ a_{12} & a_{22} & a_{32} & -(a_{12}a_{14} + a_{22}a_{24} + a_{32}a_{34}) \\ a_{13} & a_{23} & a_{33} & -(a_{13}a_{14} + a_{23}a_{24} + a_{33}a_{34}) \\ 0 & 0 & 0 & 1 \end{bmatrix} \qquad (3.35)$$

Using Eqs. 3.34 or 3.35, it is simple to compute the elements of the inverse displacement matrix $[D]^{-1}$ from the elements a_{ij} of the displacement matrix $[D]$. This saves time in numerical calculations when repeated calculations of the inverse matrix are required (e.g., when points on the moving polode are calculated from the corresponding point on the fixed polode in plane motion or when establishing the moving axode from the fixed axode in spatial motion).

3.9 DERIVATION OF SCREW MOTION PARAMETERS FROM NUMERICAL DISPLACEMENT MATRIX ELEMENTS

The numerical displacement matrix elements can be equated to the corresponding analytical form given in matrix 3.26.

The screw rotation angle ϕ can be determined first from the sum of the diagonal elements of the rotation submatrix; that is,

$$a_{11} + a_{22} + a_{33} = (u_x^2 \text{ vers } \phi + \cos \phi) + (u_y^2 \text{ vers } \phi + \cos \phi)$$
$$+ (u_z^2 \text{ vers } \phi + \cos \phi)$$
$$= 2 \cos \phi + 1$$

Therefore,

$$\phi = \cos^{-1}\left(\frac{a_{11} + a_{22} + a_{33} - 1}{2}\right) \qquad (3.36)$$

The direction cosines of the screw axis are given by

$$u_x = \frac{a_{32} - a_{23}}{2 \sin \phi}$$

$$u_y = \frac{a_{13} - a_{31}}{2 \sin \phi}$$

$$u_z = \frac{a_{21} - a_{12}}{2 \sin \phi} \qquad (3.37)$$

The coordinates of one point **p** on the screw axis are found as follows. Assuming $u_x \neq 0$, we may assume $p_x = 0$, then solve the following three equations for p_y, p_z, and the linear displacement s.

$$a_{14} = su_x - p_y a_{12} - p_z a_{13}$$

$$a_{24} = su_y + p_y - p_y a_{22} - p_z a_{23}$$

$$a_{34} = su_z + p_z - p_y a_{32} - p_z a_{33}$$

Rearranging in matrix form, assuming $p_x = 0$, we have

$$\begin{bmatrix} s \\ p_y \\ p_z \end{bmatrix} = \begin{bmatrix} u_x & -a_{12} & -a_{13} \\ u_y & (1 - a_{22}) & -a_{23} \\ u_z & -a_{32} & (1 - a_{33}) \end{bmatrix}^{-1} \begin{bmatrix} a_{14} \\ a_{24} \\ a_{34} \end{bmatrix} \tag{3.38}$$

There are two special cases. If $a_{11} = a_{22} = a_{33} = 1$, then $\phi = 0$ and the displacement is a pure translation s along the direction **u**. The translation s is given by

$$s = \sqrt{a_{14}^2 + a_{24}^2 + a_{34}^2} \tag{3.39}$$

and the direction **u** is defined by

$$u_x = \frac{a_{14}}{s}$$

$$u_y = \frac{a_{24}}{s}$$

$$u_z = \frac{a_{34}}{s} \tag{3.40}$$

The point **p** is of no concern in this case, since no unique screw axis exists. A second special case occurs when $\phi = 180°$. In this case,

$$|u_x| = \sqrt{\tfrac{1}{2}(a_{11} + 1)}$$
$$|u_y| = \sqrt{\tfrac{1}{2}(a_{22} + 1)}$$
$$|u_z| = \sqrt{\tfrac{1}{2}(a_{33} + 1)} \tag{3.41}$$

and the proper signs are determined by comparing signs of various elements of the matrix. s and p are calculated from Eq. 3.38.

3.10   COORDINATE TRANSFORMATIONS

Vector and Point Transformations

In certain situations it is desirable to express rigid body motion components in terms of a set of axes fixed in the rigid body. This is particularly important in coordinate transformations useful in dynamical applications where inertial properties of the body are referred to principal axes fixed in the body (see Chapter 11).

Consider the plane motion of a rigid body as shown in Figure 3.9. In this example the x-y system constitutes a fixed reference system and the x'-y' system moves with the body. The components of a vector $\mathbf{v}$ can be measured in either system. The components are said to be transformed from the fixed system to the moving system by the transformation equation.

$$\begin{bmatrix} v'_x \\ v'_y \end{bmatrix} = [T] \begin{bmatrix} v_x \\ v_y \end{bmatrix}$$

(3.42)

Eq. 3.42 is related to Eq. 3.15. We see that if

$$[R] = \begin{bmatrix} \cos \phi & -\sin \phi \\ \sin \phi & \cos \phi \end{bmatrix}$$

$$[T] = [R]^{-1} = \begin{bmatrix} \cos \phi & \sin \phi \\ -\sin \phi & \cos \phi \end{bmatrix}$$

We may also wish to transform position coordinates for a point $\mathbf{p}$ from one coordinate system to the other. The coordinates $\mathbf{p}(p_x, p_y)$ and $\mathbf{p}'$ (p'_x, p'_y) are related by

$$\begin{bmatrix} p'_x \\ p'_y \end{bmatrix} = [T] \begin{bmatrix} p_x \\ p_y \end{bmatrix} - \begin{bmatrix} s_x \\ s_y \end{bmatrix}$$

(3.43)

or in terms of the displacement of a rigid body with axes x'-y' originally coincident with x-y.

$$\begin{bmatrix} p_x \\ p_y \end{bmatrix} = [R] \begin{bmatrix} p'_x \\ p'_y \end{bmatrix} + \begin{bmatrix} s_x \\ s_y \end{bmatrix}$$

(3.44)

Thus, we see that coordinate transformation matrices are the inverse of displacement matrices.

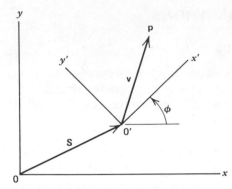

FIGURE   3.9   PLANE   COORDINATE TRANSFORMATION.

Successive Coordinate Transformations

Assume that the rotation angle ϕ in Figure 3.9 is increased by an amount ϕ_1. The rotation transformation matrix becomes

$$[T_1] = \begin{bmatrix} \cos(\phi + \phi_1) & \sin(\phi + \phi_1) \\ -\sin(\phi + \phi_1) & \cos(\phi + \phi_1) \end{bmatrix}$$

$$= \begin{bmatrix} \cos\phi_1 & \sin\phi_1 \\ -\sin\phi_1 & \cos\phi_1 \end{bmatrix}[T] = [T_1][T] \tag{3.45}$$

The rotation transformation matrix $[T_1]$ is associated with the transformation angle ϕ_1. If successive transformations $\phi_1, \phi_2, \phi_3, \ldots, \phi_n$ total 2π, then

$$[T_n] \cdots [T_3][T_2][T_1] = [T_{2\pi}] = [I] \tag{3.46}$$

where $[I]$ is the identity matrix.

3.11   HARTENBERG-DENAVIT NOTATION

Hartenberg and Denavit [1] have demonstrated the usefulness of a special coordinate transformation between a set of axes x_1-y_1-z_1 fixed in one rigid body and a second set x_2-y_2-z_2 fixed in a second adjacent rigid body in a kinematic chain.

The coordinates of point **p** in system 1 as shown in Figure 3.10 are given by the vector equation

$$\overline{O_1P} = \overline{O_2P} + \overline{O_1O_2} \tag{3.47}$$

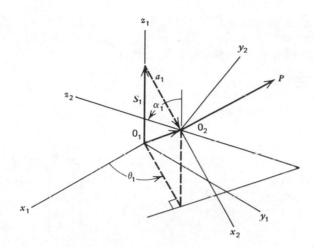

FIGURE 3.10   HARTENBERG-DENAVIT COORDINATE
TRANSFORMATION.

Eq. 3.47 can be expressed in terms of the cosines of the angles between axes in
the two coordinate systems in the following form.

$$x_1 = x_2 \cos(x_2, x_1) + y_2 \cos(y_2, x_1) + z_2 \cos(z_2, x_1) + a_1$$
$$y_1 = x_2 \cos(x_2, y_1) + y_2 \cos(y_2, y_1) + z_2 \cos(z_2, y_1) + b_1$$
$$z_1 = x_2 \cos(x_2, z_1) + y_2 \cos(y_2, z_1) + z_2 \cos(z_2, z_1) + c_1$$
$$1 = x_2(0) \qquad\quad + y_2(0) \qquad\quad + z_2(0) \qquad\quad + 1 \qquad (3.48)$$

where

$$\overline{O_1 O_2} = (a_1, b_1, c_1)$$
$$\overline{O_1 P} = (x_1, y_1, z_1)$$
$$\overline{O_2 P} = (x_2, y_2, z_2)$$

The last row is added such that the transformation equation will be in the
homogeneous matrix form

$$\begin{bmatrix} x_1 \\ y_1 \\ z_1 \\ 1 \end{bmatrix} = \begin{bmatrix} \cos(x_2, x_1) & \cos(y_2, x_1) & \cos(z_2, x_1) & a_1 \\ \cos(x_2, y_1) & \cos(y_2, y_1) & \cos(z_2, y_1) & b_1 \\ \cos(x_2, z_1) & \cos(y_2, z_1) & \cos(z_2, z_1) & c_1 \\ 0 & 0 & 0 & 1 \end{bmatrix} \begin{bmatrix} x_2 \\ y_2 \\ z_2 \\ 1 \end{bmatrix} \qquad (3.49)$$

In Figure 3.10 the axis x_2 has been selected such that it lies along the shortest common perpendicular between axes z_1 and z_2. The axis y_2 completes a right-handed set of coordinate axes. Other parameters are defined as follows.

1. a_1 is the distance between axes z_1 and z_2.
2. α_1 is the twist angle that screws axis z_1 into axis z_2 along a_1.
3. θ_1 is the angle that screws x_1 into x_2 along S_1.
4. S_1 is the distance from axis x_1 to x_2.

With these definitions, Eq. 3.49 becomes:

$$
\begin{bmatrix} x_1 \\ y_1 \\ z_1 \\ 1 \end{bmatrix} = \begin{bmatrix} \cos\theta_1 & -\sin\theta_1\cos\alpha_1 & -\sin\theta_1\sin\alpha_1 & a_1\cos\theta_1 \\ \sin\theta_1 & \cos\theta_1\cos\alpha_1 & \cos\theta_1\sin\alpha_1 & a_1\sin\theta_1 \\ 0 & -\sin\alpha_1 & \cos\alpha_1 & S_1 \\ 0 & 0 & 0 & 1 \end{bmatrix} \begin{bmatrix} x_2 \\ y_2 \\ z_2 \\ 1 \end{bmatrix}
$$

$$(3.50)$$

$$
= [T(a_1, \alpha_1, \theta_1, S_1)] \begin{bmatrix} x_2 \\ y_2 \\ z_2 \\ 1 \end{bmatrix} = [T_{21}] \begin{bmatrix} x_2 \\ y_2 \\ z_2 \\ 1 \end{bmatrix}
$$

$$(3.51)$$

The Hartenberg-Denavit matrix has been applied to the kinematic analysis of both plane and spatial mechanisms. The method involves writing a series of coordinate transformations around the complete loop of the kinematic chain of the mechanism, then making use of Eq. 3.46.

Consider the offset slider-crank of Figure 3.11. In this case:

$$[T_{21}] = [T(a_1, 90., 180., S_1)]$$

$$[T_{32}] = [T(a_2, 0.0, \theta_2, 0.0)]$$

$$[T_{43}] = [T(a_3, 0.0, \theta_3, 0.0)]$$

$$[T_{14}] = [T(0.0, 90., \theta_4, 0.0)] \tag{3.52}$$

This leads to

$$[T] = [T_{21}][T_{32}][T_{43}][T_{14}] = [I] \tag{3.53}$$

The matrix $[T]$ contains elements that are functions of $a_1, S_1, a_2, \theta_2, a_3, \theta_3$, and θ_4. Assuming that θ_4 is specified, we may solve for other unknown motion parameters θ_2, θ_3, and S_1 by equating corresponding elements of $[T]$ and $[I]$ in Eq. 3.53. Refer to reference 1 for additional detail and specific examples of spatial kinematic analysis using the matrix coordinate transformation method.

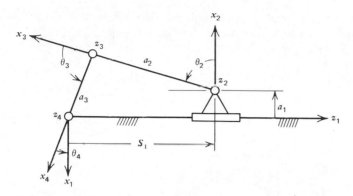

FIGURE 3.11 COORDINATE SYSTEM FOR KINEMATIC ANALYSIS OF THE OFFSET SLIDER-CRANK MECHANISM USING HARTENBERG-DENAVIT MATRIX NOTATION.

3.12 DIFFERENTIAL ROTATION MATRICES

We have seen that the rotation of a vector **v** can be described by the matrix equation

$$\mathbf{v} = [R]\mathbf{v}_1 \qquad (3.54)$$

where the $[R]$ matrix is either the plane or spatial rotation matrix in one of several possible forms.

The rate of change of position of the vector **v** is found by differentiating Eq. 3.54.

$$\dot{\mathbf{v}} = [\dot{R}]\mathbf{v}_1 + [R]\dot{\mathbf{v}}_1$$

From Eq. 3.54,

$$\mathbf{v}_1 = [R]^{-1}\mathbf{v} = [R]^{T}\mathbf{v}$$

Hence, noting that $\dot{\mathbf{v}}_1 = 0$,

$$\dot{\mathbf{v}} = [\dot{R}][R]^{T}\mathbf{v} = [W]\mathbf{v} \qquad (3.55)$$

The matrix $[W]$ will be designated as the *angular velocity matrix*. The *plane angular velocity matrix* becomes

$$[W] = \begin{bmatrix} -\sin\theta\dot{\theta} & -\cos\theta\dot{\theta} \\ \cos\theta\dot{\theta} & -\sin\theta\dot{\theta} \end{bmatrix} \begin{bmatrix} \cos\theta & \sin\theta \\ -\sin\theta & \cos\theta \end{bmatrix} = \begin{bmatrix} 0 & -\dot{\theta} \\ \dot{\theta} & 0 \end{bmatrix} \qquad (3.56)$$

The *spatial angular velocity matrix* and higher-order differential rotation matrices are conveniently described in terms of the rotation matrix $[R_\phi, \mathbf{u}]$. This matrix can be written in the compact form

$$[R_\phi, \mathbf{u}] = -[P_\mathbf{u}][P_\mathbf{u}]\cos\phi + [P_\mathbf{u}]\sin\phi + [Q_\mathbf{u}] \qquad (3.57)$$

where

$$[P_\mathbf{u}] = \begin{bmatrix} 0 & -u_z & u_y \\ u_z & 0 & -u_x \\ -u_y & u_x & 0 \end{bmatrix}$$

$$[Q_\mathbf{u}] = \begin{bmatrix} u_x^2 & u_x u_y & u_x u_z \\ u_x u_y & u_y^2 & u_y u_z \\ u_x u_z & u_y u_z & u_z^2 \end{bmatrix}$$

and

$$-[P_\mathbf{u}][P_\mathbf{u}] = [I - Q_\mathbf{u}]$$

Before proceeding with the differentiation of the spatial rotation matrix, it will be expedient to introduce the notion of vector product in matrix notation.

The vector dot product can be expressed as the scalar product of two vectors in the form

$$\mathbf{a} \cdot \mathbf{b} = (a)^T(b) = (b)^T(a)$$
$$= a_x b_x + a_y b_y + a_z b_z \qquad (3.58)$$

where

$$(a) = \begin{pmatrix} a_x \\ a_y \\ a_z \end{pmatrix} \quad \text{and} \quad (a)^T = (a_x a_y a_z)$$

The *vector cross product* $\mathbf{c} = \mathbf{a} \times \mathbf{b}$ is formed as the product of two matrices $[A]$ and (b).

$$\mathbf{a} \times \mathbf{b} = \begin{bmatrix} 0 & -a_z & a_y \\ a_z & 0 & -a_x \\ -a_y & a_x & 0 \end{bmatrix}(b) = [A](b) \qquad (3.59)$$

which can be shown to be equivalent to the usual expansion in the form of a determinant

$$\mathbf{a} \times \mathbf{b} = \begin{vmatrix} i & j & k \\ a_x & a_y & a_z \\ b_x & b_y & b_z \end{vmatrix}$$

Therefore, we see that Eq. 3.55 could be derived from

$$\dot{\mathbf{v}} = \dot{\boldsymbol{\theta}} \times \mathbf{v} = [W]\mathbf{v} \qquad (3.60)$$

where

$$[W] = \begin{bmatrix} 0 & -\dot{\theta}_z & \dot{\theta}_y \\ \dot{\theta}_z & 0 & -\dot{\theta}_x \\ -\dot{\theta}_y & \dot{\theta}_x & 0 \end{bmatrix} = \dot{\theta} \begin{bmatrix} 0 & -u_z & u_y \\ u_z & 0 & -u_x \\ -u_y & u_x & 0 \end{bmatrix} = \dot{\theta}[P_u]$$

The $[W]$ matrix of Eq. 3.60 can also be determined from Eq. 3.57 as

$$[W] - \frac{d}{dt}[R_{\phi,\,u}] - \dot{\phi}[P_u]$$
$$\phi \rightarrow 0 \qquad (3.61)$$

The *spatial angular acceleration matrix* $[\dot{W}]$ is defined as

$$[\dot{W}] = \frac{d^2}{dt^2}[R_{\phi,\,u}] = \ddot{\phi}[P_u] + \dot{\phi}[\dot{P}_u] + \dot{\phi}^2[P_u][P_u]$$
$$\phi \rightarrow 0 \qquad (3.62)$$

Eq. 3.62 is conveniently derived in the familiar vector form then converted to the matrix form of Eq. 3.57 using the matrix equivalent of the vector cross produce given by Eq. 3.59.

Eq. 3.55 can be written in vector form as

$$\dot{\mathbf{v}} = \dot{\boldsymbol{\phi}} \times \mathbf{v} \qquad (3.63)$$

Differentiating Eq. 3.63 gives

$$\ddot{\mathbf{v}} = \frac{d}{dt}(\dot{\boldsymbol{\phi}}) \times \mathbf{v} + \dot{\boldsymbol{\phi}} \times \frac{d}{dt}\mathbf{v}$$

and, since

$$\boldsymbol{\phi} = \dot{\phi}\mathbf{u} \qquad \frac{d}{dt}\mathbf{v} = \dot{\boldsymbol{\phi}} \times \mathbf{v}$$

$$\ddot{\boldsymbol{\phi}} = \frac{d}{dt}(\dot{\phi}\mathbf{u}) = \ddot{\phi}\mathbf{u} + \dot{\phi}\dot{\mathbf{u}} \qquad (3.64)$$

we have

$$\ddot{\mathbf{v}} = \ddot{\boldsymbol{\phi}} \times \mathbf{v} + \dot{\boldsymbol{\phi}} \times (\dot{\boldsymbol{\phi}} \times \mathbf{v})$$
$$= (\ddot{\phi}\mathbf{u} + \dot{\phi}\dot{\mathbf{u}}) \times \mathbf{v} + \dot{\phi}^2(\mathbf{u} \times (\mathbf{u} \times \mathbf{v}))$$

which can be written in the matrix form

$$(\ddot{\mathbf{v}}) = [\ddot{\phi}[P_{\mathbf{u}}] + \dot{\phi}[\dot{P}_{\mathbf{u}}] + \dot{\phi}^2[P_{\mathbf{u}}][P_{\mathbf{u}}]](\mathbf{v}) = [\dot{W}](\mathbf{v}) \tag{3.65}$$

The plane angular acceleration matrix is a special case of Eq. 3.68 with $[\dot{P}_{\mathbf{u}}] = 0$. This leads to

$$[\dot{W}] = \begin{bmatrix} -\dot{\theta}^2 & -\ddot{\theta} \\ \ddot{\theta} & -\dot{\theta}^2 \end{bmatrix} \tag{3.66}$$

The spatial angular acceleration matrix as defined by Eq. 3.65 can be expanded in the form

$$[\dot{W}] = \begin{bmatrix} (u_x^2 - 1)\dot{\phi}^2 & (u_x u_y \dot{\phi}^2 - \dot{u}_z \dot{\phi} - u_z \ddot{\phi}) & (u_x u_z \dot{\phi}^2 + \dot{u}_y \dot{\phi} + u_y \ddot{\phi}) \\ (u_x u_y \dot{\phi}^2 + \dot{u}_z \dot{\phi} + u_z \ddot{\phi}) & (u_y^2 - 1)\dot{\phi}^2 & (u_y u_z \dot{\phi}^2 - \dot{u}_x \dot{\phi} - u_x \ddot{\phi}) \\ (u_x u_z \dot{\phi}^2 - \dot{u}_y \dot{\phi} - u_y \ddot{\phi}) & (u_y u_z \dot{\phi}^2 - \dot{u}_x \dot{\phi} + u_x \ddot{\phi}) & (u_z^2 - 1)\dot{\phi}^2 \end{bmatrix} \tag{3.67}$$

The *spatial angular second-acceleration matrix* $[\ddot{W}]$ follows from differentiation of either the vector equation (Eq. 3.64) or the matrix equation (Eq. 3.65), which gives

$$(\dddot{\mathbf{v}}) = [\ddot{W}](\mathbf{v}) = [\dddot{\phi}[P_{\mathbf{u}}] + 2\ddot{\phi}[\dot{P}_{\mathbf{u}}] + \dot{\phi}[\ddot{P}_{\mathbf{u}}] + 3\dot{\phi}\ddot{\phi}[P_{\mathbf{u}}][P_{\mathbf{u}}]$$
$$+ 2\dot{\phi}^2[\dot{P}_{\mathbf{u}}][P_{\mathbf{u}}] + \dot{\phi}^2[P_{\mathbf{u}}][\dot{P}_{\mathbf{u}}] + \dot{\phi}^3[P_{\mathbf{u}}][P_{\mathbf{u}}][P_{\mathbf{u}}]](\mathbf{v}) \tag{3.68}$$

The *plane angular second-acceleration matrix* is again a special case where $[\dot{P}_{\mathbf{u}}] = [\ddot{P}_{\mathbf{u}}] = 0$. For plane motion specified by $\dot{\theta}$, $\ddot{\theta}$, $\dddot{\theta}$, we obtain

$$[\ddot{W}] = \begin{bmatrix} -3\dot{\theta}\ddot{\theta} & -(\dddot{\theta} - \dot{\theta}^3) \\ (\dddot{\theta} - \dot{\theta}^3) & -3\dot{\theta}\ddot{\theta} \end{bmatrix} \tag{3.69}$$

The expansion of $[\ddot{W}]$ for spatial motion from Eq. 3.68 leads to the following expressions for the elements of the *spatial angular second-acceleration matrix* $[\ddot{W}]$.

$$\ddot{W}_{11} = 3u_x \dot{u}_x \dot{\phi}^2 + 3(u_x^2 - 1)\dot{\phi}\ddot{\phi}$$

$$\ddot{W}_{12} = (2u_x \dot{u}_y + \dot{u}_x u_y)\dot{\phi}^2 + 3u_x u_y \dot{\phi}\ddot{\phi} + (\ddot{u}_z \dot{\phi} + 2\dot{u}_z \ddot{\phi} + u_z \dddot{\phi} - u_z \dot{\phi}^3)$$

$$\ddot{W}_{13} = (2u_x \dot{u}_z + \dot{u}_x u_z)\dot{\phi}^2 + 3u_x u_z \dot{\phi}\ddot{\phi} + (\ddot{u}_y \dot{\phi} + 2\dot{u}_y \ddot{\phi} + u_y \dddot{\phi} - u_y \dot{\phi}^3)$$

$$\ddot{W}_{21} = (2u_y \dot{u}_x + \dot{u}_y u_x)\dot{\phi}^2 + 3u_y u_x \dot{\phi}\ddot{\phi} + (\ddot{u}_z \dot{\phi} + 2\dot{u}_z \ddot{\phi} + u_z \dddot{\phi} - u_z \dot{\phi}^3)$$

$$\ddot{W}_{22} = 3u_y \dot{u}_y \dot{\phi}^2 + 3(u_y^2 - 1)\dot{\phi}\ddot{\phi} \tag{3.70}$$

$$\ddot{W}_{23} = (2u_y \dot{u}_z + \dot{u}_y u_z)\dot{\phi}^2 + 3u_y u_z \dot{\phi}\ddot{\phi} - (\ddot{u}_x \dot{\phi} + 2\dot{u}_x \ddot{\phi} + u_x \dddot{\phi} - u_x \dot{\phi}^3)$$

$$\ddot{W}_{31} = (2u_z \dot{u}_x + \dot{u}_z u_x)\dot{\phi}^2 + 3u_z u_x \dot{\phi}\ddot{\phi} - (\ddot{u}_y \dot{\phi} + 2\dot{u}_y \ddot{\phi} + u_y \dddot{\phi} - u_y \dot{\phi}^3)$$

$$\ddot{W}_{32} = (2u_z \dot{u}_y + \dot{u}_z u_y)\dot{\phi}^2 + 3u_z u_y \dot{\phi}\ddot{\phi} + (\ddot{u}_x \dot{\phi} + 2\dot{u}_x \ddot{\phi} + u_x \dddot{\phi} - u_x \dot{\phi}^3)$$

$$\ddot{W}_{33} = 3u_z \dot{u}_z \dot{\phi}^2 + 3(u_z^2 - 1)\dot{\phi}\ddot{\phi}$$

3.13   DIFFERENTIAL DISPLACEMENT MATRICES

The velocity matrix $[V]$ is easily formed from Eq. 3.60 with the vector $\mathbf{v}$ defined by two points fixed in the rigid body: a reference point $\mathbf{p}$ and a point of interest $\mathbf{q}$.

$$(\dot{\mathbf{q}} - \dot{\mathbf{p}}) = [W](\mathbf{q}_1 - p_1) \tag{3.71}$$

from which

$$\begin{bmatrix} \dot{\mathbf{q}} \\ 0 \end{bmatrix} = \begin{bmatrix} [W] & (\dot{\mathbf{p}} - [W]\mathbf{p}) \\ 0 & 0 \end{bmatrix} \begin{bmatrix} \mathbf{q} \\ 1 \end{bmatrix} = [V] \begin{bmatrix} \mathbf{q} \\ 1 \end{bmatrix} \tag{3.72}$$

and $[V]$ is the velocity matrix.

The $[V]$ matrix is a 3×3 matrix when used to describe the velocity of a point in a moving plane. A 4×4 velocity matrix is required for spatial rigid body motion.

Similarly, we obtain the *spatial acceleration* equation

$$\begin{bmatrix} \ddot{\mathbf{q}} \\ 0 \end{bmatrix} = \begin{bmatrix} [\dot{W}] & (\ddot{p} - [\dot{W}]\mathbf{p}) \\ 0 & 0 \end{bmatrix} \begin{bmatrix} \mathbf{q} \\ 1 \end{bmatrix} = [A] \begin{bmatrix} \mathbf{q} \\ 1 \end{bmatrix} \tag{3.73}$$

where $[A]$ is the 4×4 spatial acceleration matrix or the 3×3 plane acceleration matrix.

The *second acceleration matrix equation* becomes

$$\begin{bmatrix} \dddot{\mathbf{q}} \\ 0 \end{bmatrix} = \begin{bmatrix} [\ddot{W}] & (\dddot{p} - [\ddot{W}]\mathbf{p}) \\ 0 & 0 \end{bmatrix} \begin{bmatrix} \mathbf{q} \\ 1 \end{bmatrix} = [J] \begin{bmatrix} \mathbf{q} \\ 1 \end{bmatrix} \tag{3.74}$$

and $[J]$ is the second acceleration or "jerk" matrix.

We see that the rigid body motion matrices that describe the velocity, acceleration, or second acceleration of an arbitrary point fixed in the body are easily formed from the appropriate rotation matrices. This leads to the general relationships

$$\overset{(n)}{[D]} = \begin{bmatrix} \overset{(n)}{[R]} & \overset{(n)}{(\mathbf{p}} - \overset{(n)}{[R]}\mathbf{p}) \\ 0 & 0 \end{bmatrix} \tag{3.75}$$

$$\overset{(n)}{\mathbf{q}} = \overset{(n)}{[D]}\mathbf{q} \tag{3.76}$$

where n indicates the order of differentiation.

Chapter 4

Kinematic Analysis of Spatial Mechanisms

4.1 RELATIVE SPATIAL MOTION

Before proceeding to the kinematic analysis of spatial mechanisms, we must consider the case where two separate but joined rigid bodies are in motion with a definite and constrained relative motion occurring at the kinematic joint.

To describe the absolute motion of a point on a moving rigid body, which itself is moving relative to a moving reference member, we adopt the notation of Figure 4.1. A point $\mathbf{q}$ on the link j of a kinematic chain is constrained to move relative to the previous link $j-1$ by a cylindrical pair, a general type of joint in a spatial kinematic chain. A reference point $\mathbf{p}'_{j-1}$ on the axis of relative motion $\mathbf{u}$ is assumed to move with link $j-1$. The absolute angular displacements of the links are given by θ_j and θ_{j-1}. The relative angular displacement about axis $\mathbf{u}$ is designated as ϕ_j with a concurrent sliding displacement s along axis $\mathbf{u}$. Each of the absolute angular displacements θ_j and θ_{j-1} have an associated finite rotation axis $\mathbf{u}_j$ and $\mathbf{u}_{j-1}$. These axes have limited usefulness in kinematic analysis but may have some application in spatial kinematic synthesis.

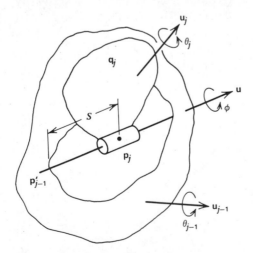

FIGURE 4.1 RELATIVE SPATIAL MOTION AT A CYLINDRICAL JOINT.

70

4.2   RELATIVE DISPLACEMENT

The absolute displacement of a point $\mathbf{q}_j$ on link j as shown in Figure 4.1 can be described in terms of the displacement of a point $\mathbf{q}'_{j-1}$ on reference link $j-1$, which is initially coincident with $\mathbf{q}_j$ plus the motion of $\mathbf{q}_j$ relative to link $j-1$. The relative motion can be described in terms of the rotation matrix (3.25) or the screw matrix (3.26). The point $\mathbf{p}'_{j-1}$ is defined by the absolute motion of link $j-1$, which itself may have motion relative to a link $j-2$ in the kinematic chain.

As an example, consider the two-link combination of Figure 4.2. We first calculate the position of a point $\mathbf{q}'_1$, assuming that link 2 rotates an angle θ about a fixed axis $\mathbf{u}_0$.

$$(\mathbf{q}'_1 - \mathbf{p}_0) = [R_{\theta,\,\mathbf{u}_0}](\mathbf{q}_1 - \mathbf{p}_0) \tag{4.1}$$

from which

$$(\mathbf{q}'_1) = [R_{\theta,\,\mathbf{u}_0}](\mathbf{q}_1 - \mathbf{p}_0) + (\mathbf{p}_0) \tag{4.2}$$

We next locate point $\mathbf{p}'_1$ from

$$(\mathbf{p}'_1 - \mathbf{p}_0) = [R_{\theta,\,u_0}](\mathbf{p}_1 - \mathbf{p}_0)$$
$$(\mathbf{p}'_1) = [R_{\theta,\,\mathbf{u}_0}](\mathbf{p}_1 - \mathbf{p}_0) + (\mathbf{p}_0) \tag{4.3}$$

The relative motion at the cylindrical joint can be accounted for in three steps.

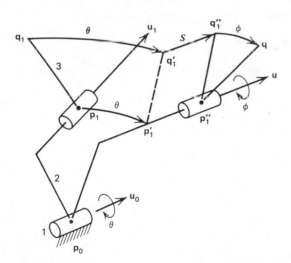

FIGURE 4.2   RELATIVE DISPLACEMENTS AT A
CYLINDRICAL JOINT.

We first find the new location of the axis $\mathbf{u}$ from

$$(\mathbf{u}) = [R_{\theta,\,\mathbf{u}_0}](\mathbf{u}_1) \tag{4.4}$$

We next locate points $\mathbf{p}_1''$ along axis $\mathbf{u}$ and $(\mathbf{q}_1'')$ from

$$(\mathbf{p}_1'') = (\mathbf{p}_1') + s(\mathbf{u}) \qquad \text{and} \qquad (\mathbf{q}_1'') = (\mathbf{q}_1') + s(\mathbf{u}) \tag{4.5}$$

Finally, from

$$(\mathbf{q} - \mathbf{p}_1'') = [R_{\phi,\,\mathbf{u}}](\mathbf{q}_1'' - \mathbf{p}_1'') \tag{4.6}$$

we obtain

$$(\mathbf{q}) = [R_{\phi,\,\mathbf{u}}](\mathbf{q}_1'' - \mathbf{p}_1'') + (\mathbf{p}_1'') \tag{4.7}$$

Substitution of Eq. 4.5 into Eq. 4.7 leads to

$$\begin{bmatrix} \mathbf{q} \\ 1 \end{bmatrix} = \begin{bmatrix} [R_{\phi,\,\mathbf{u}}] & (\mathbf{p}_1' + s\mathbf{u} - [R_{\phi,\,\mathbf{u}}]\mathbf{p}_1') \\ 0 & 1 \end{bmatrix} \begin{bmatrix} \mathbf{q}_1' \\ 1 \end{bmatrix} \tag{4.8}$$

which displays the total motion in the form of a screw matrix equation. Note that $\mathbf{p}_1'$ must be calculated in terms of $[R_{\theta,\,\mathbf{u}_0}]$, $\mathbf{p}_1$, and $\mathbf{p}_0$.

4.3 RELATIVE VELOCITY

The velocity of point $\mathbf{q}$ in Figure 4.3 is developed in terms of the relative velocity components at the cylindrical joint as follows. We first find the velocity of the point $\mathbf{q}'$ on the reference link 2, which is coincident with point $\mathbf{q}$ on link 3, from

$$(\dot{\mathbf{q}}' - \dot{\mathbf{p}}_0) = [W_{\dot{\theta},\,\mathbf{u}_0}](\mathbf{q} - \mathbf{p}_0) \tag{4.9}$$

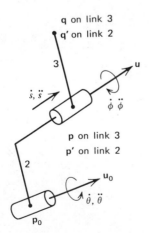

q on link 3
q' on link 2

3   u

$\dot{s},\ddot{s}$ $\dot{\phi}\,\ddot{\phi}$

p on link 3
p' on link 2

2          $\mathbf{u}_0$

$\dot{\theta},\ddot{\theta}$

$\mathbf{p}_0$

FIGURE 4.3 RELATIVE VELOCITY AND ACCELERATION COMPONENTS AT A CYLINDRICAL JOINT

and the velocity of the reference point $\mathbf{p'}$ on link 2

$$(\dot{\mathbf{p}}' - \dot{\mathbf{p}}_0) = [W_{\dot{\theta},\, \mathbf{u}_0}](\mathbf{p} - \mathbf{p}_0) \qquad (4.10)$$

The relative velocity at the cylindrical joint (assuming link 2 temporarily fixed) with point $\mathbf{p}$ on link 3 becomes

$$(\dot{\mathbf{q}}_r - \dot{\mathbf{p}}_r) = [W_{\dot{\phi},\, \mathbf{u}}](\mathbf{q} - \mathbf{p}) \qquad (4.11)$$

where $\dot{q}_r$ is the relative velocity and $\dot{p}_r - \dot{s}\mathbf{u}$, which leads to

$$(\dot{\mathbf{q}}_r) = [W_{\dot{\phi},\, \mathbf{u}}](\mathbf{q} - \mathbf{p}) + \dot{s}\mathbf{u} \qquad (4.12)$$

and the absolute velocity of point $\mathbf{q}$ on link 3 becomes

$$(\dot{q}) = [W_{\dot{\theta},\, \mathbf{u}_0}](\mathbf{q} - \mathbf{p}_0) + [W_{\dot{\phi},\, \mathbf{u}}](\mathbf{q} - \mathbf{p}) + \dot{s}\mathbf{u} \qquad (4.13)$$

4.4   RELATIVE ACCELERATION

The acceleration of point $\mathbf{q}$ can be described by the vector equation

$$\ddot{\mathbf{q}} = \ddot{\mathbf{q}}' + \mathbf{a}_{rel} + \mathbf{a}_{cor} \qquad (4.14)$$

where

$\ddot{\mathbf{q}}'$ is acceleration of a point $\mathbf{q}'$ on the reference link 2 coincident with $\mathbf{q}$ on link 3
$\mathbf{a}_{rel}$ = relative acceleration of point $\mathbf{q}$ with respect to the reference link with relative motion parameters $\dot{s}$, $\ddot{s}$, $\dot{\phi}$, and $\ddot{\phi}$
$\mathbf{a}_{cor}$ = supplemental Coriolis component due to the rotation $\dot{\theta}$ of the reference member

The terms of Eq. 4.14, for the case where the input or reference link rotates about a fixed reference axis $\mathbf{u}_0$, are given by

$$\ddot{\mathbf{q}}' = [\dot{W}_{\dot{\theta},\, \ddot{\theta},\, \mathbf{u}_0}](\mathbf{q} - \mathbf{p}_0) \qquad (4.15)$$

$$\mathbf{a}_{rel} = [\dot{W}_{\dot{\phi},\, \ddot{\phi},\, \mathbf{u}}](\mathbf{q} - \mathbf{p}) + \ddot{s}\mathbf{u} \qquad (4.16)$$

$$\mathbf{a}_{cor} = 2[W_{\dot{\theta},\, \mathbf{u}_0}]\mathbf{v}_{rel} \qquad (4.17)$$

$$= 2[W_{\dot{\theta},\, \mathbf{u}_0}](\dot{s}\mathbf{u} + [W_{\dot{\phi},\, \mathbf{u}}](\mathbf{q} - \mathbf{p})) \qquad (4.18)$$

from which

$$\ddot{\mathbf{q}} = [\dot{W}_{\dot{\theta},\, \ddot{\theta},\, \mathbf{u}_0}](\mathbf{q} - \mathbf{p}_0) + [\dot{W}_{\dot{\phi},\, \ddot{\phi},\, \mathbf{u}}](\mathbf{q} - \mathbf{p}) + \ddot{s}\mathbf{u}$$
$$+ 2[W_{\dot{\theta},\, \mathbf{u}_0}]\dot{s}\mathbf{u} + 2[W_{\dot{\theta},\, \mathbf{u}_0}][W_{\dot{\phi},\, \mathbf{u}}](\mathbf{q} - \mathbf{p}) \qquad (4.19)$$

4.5 PLANE KINEMATIC ANALYSIS USING RELATIVE JOINT ROTATION ANGLES

As an introduction to the method to be employed in spatial kinematic analysis, let us first reconsider the displacement, velocity, and acceleration analysis of the plane four-bar linkage in terms of *relative* rotation angles. The angle θ in Figure 4.4 is the *displacement angle* for the input crank, link 2. The angle α describes the *relative angular displacement* about the moving joint **a**. All displacements are measured from the first position of the linkage, as shown in Figure 4.4.

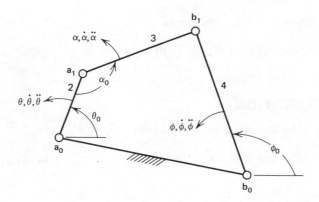

FIGURE 4.4 KINEMATIC ANALYSIS OF THE PLANE FOUR-BAR LINKAGE IN TERMS OF RELATIVE JOINT ROTATION ANGLES.

Displacement Analysis of the Four-Bar Linkage

The displaced position of point **b** can be described either in terms of the rotation θ about $\mathbf{a}_0$ plus the rotation α about **a** or by the rotation ϕ about $\mathbf{b}_0$.

In either case the output link 4 must satisfy the condition that its length remains constant during any rigid body displacement. This constraint can be imposed analytically as

$$(\mathbf{b} - \mathbf{b}_0)^T(\mathbf{b} - \mathbf{b}_0) = (\mathbf{b}_1 - \mathbf{b}_0)^T(\mathbf{b}_1 - \mathbf{b}_0) \tag{4.20}$$

We first calculate the coordinates of **b** in terms of $\mathbf{b}_1'$ and the relative rotation angle α

$$(\mathbf{b}) = [R_\alpha](\mathbf{b}_1' - \mathbf{a}) + (\mathbf{a}) \tag{4.21}$$

where b'_1 and a are both specified in terms of the input rotation angle θ from

$$(b'_1) = [R_\theta](b_1 - a_0) + (a_0) \tag{4.22}$$

$$(a) = [R_\theta](a_1 - a_0) + (a_0) \tag{4.23}$$

Expansion of Eq. 4.20, remembering that $[R_a][R_a]^{-1} = [I]$, leads to

$$2(a - b_0)^T([R_a](b'_1 - a)) + (b'_1 - a)^T(b'_1 - a)$$
$$+ (a - b_0)^T(a - b_0) = (b_1 - b_0)^T(b_1 - b_0) \tag{4.24}$$

and, since

$$[R_a] = \begin{bmatrix} \cos \alpha & -\sin \alpha \\ \sin \alpha & \cos \alpha \end{bmatrix} = \begin{bmatrix} 1 & 0 \\ 0 & 1 \end{bmatrix}\cos \alpha + \begin{bmatrix} 0 & -1 \\ 1 & 0 \end{bmatrix}\sin \alpha \tag{4.25}$$

$$[R_a](b'_1 - a) = \begin{Bmatrix} [\cos \alpha(b'_{1x} - a_x) - \sin \alpha(b'_{1y} - a_y)] \\ [\sin \alpha(b'_{1x} - a_x) + \cos \alpha(b'_{1y} - a_y)] \end{Bmatrix} \tag{4.26}$$

and

$$(a - b_0)^T[R_a](b'_1 - a) = (a_x - b_{0x})[\cos \alpha(b'_{1x} - a_x) - \sin \alpha(b'_{1y} - a_y)]$$
$$+ (a_y - b_{0y})[\sin \alpha(b'_{1x} - a_x) + \cos \alpha(b'_{1y} - a_y)] \tag{4.27}$$

Substituting into Eq. 4.24 and collecting terms, we obtain the following equation with unknown α.

$$E \cos \alpha + F \sin \alpha + G = 0 \tag{4.28}$$

where

$$E = (b'_{1x} - a_x)(a_x - b_{0x}) + (b'_{1y} - a_y)(a_y - b_{0y})$$
$$F = (b'_{1x} - a_x)(a_y - b_{0y}) - (b'_{1y} - a_y)(a_x - b_{0x})$$
$$G = -\tfrac{1}{2}[(b_1 - b_0)^T(b_1 - b_0) - (b'_1 - a)^T(b'_1 - a) - (a - b_0)^T(a - b_0)]$$

Eq. 4.28 may be converted to a quadratic form and solved for α as follows.

Let $t = \tan\left(\dfrac{\alpha}{2}\right)$

From the trigonometric identity $\tan \alpha = \dfrac{2 \tan(\alpha/2)}{1 - \tan^2(\alpha/2)}$

we have $\tan \alpha = \dfrac{2t}{1 - t^2}$

from which $\cos \alpha = \dfrac{1 - t^2}{1 + t^2}$ and $\sin \alpha = \dfrac{2t}{1 + t^2}$

Substituting into Eq. 4.28, we obtain

$$t^2(G - E) + t(2F) + (G + E) = 0 \tag{4.29}$$

with two solutions

$$t_{1,2} = \frac{-F \pm \sqrt{E^2 + F^2 - G^2}}{G - E} \tag{4.30}$$

and corresponding values of α_1 and α_2 from

$$\alpha_{1,2} = 2 \tan^{-1}(t_{1,2}) \tag{4.31}$$

The new displacement angles $\alpha_{1,2}$ are compared with the previous displacement α after which $\alpha_{1,2}$ is selected such that if the difference

$$|(\alpha_1 - \alpha)| < |(\alpha_2 - \alpha)| \quad \text{then} \quad \alpha = \alpha_1$$

$$\text{Otherwise} \quad \alpha = \alpha_2$$

The mobility of the linkage is checked by noting that when the term $(E^2 + F^2 - G^2)$ in Eq. 4.30 is negative, the solution for $\alpha_{1,2}$ does not exist and the linkage cannot be assembled.

Once a new position for $\mathbf{b}$ is known, we may determine ϕ from the displacement equation

$$(\mathbf{b} - \mathbf{b}_0) = [R_\phi](\mathbf{b}_1 - \mathbf{b}_0) \tag{4.32}$$

Taking the y-components, we obtain one scalar equation with unknown angle ϕ in the form of Eq. 4.28.

$$(b_y - b_{0y}) = \sin \phi(b_{1x} - b_{0x}) + \cos \phi(b_{1y} - b_{0y}) \tag{4.33}$$

The position angle $(\phi_0 + \phi)$ could also be calculated from

$$\tan(\phi_0 + \phi) = \frac{b_y - b_{0y}}{b_x - b_{0x}} \tag{4.34}$$

Relative Velocity Analysis

Differentiating the displacement constraint Eq. 4.20, we obtain the *velocity constraint equation*

$$(\dot{\mathbf{b}})^T(\mathbf{b} - \mathbf{b}_0) = 0 \tag{4.35}$$

The velocity of **b** is described in terms of the velocity of point **b′** on the input crank plus the relative velocity of **b** with respect to **b′**.

$$(\dot{\mathbf{b}}') = \begin{bmatrix} 0 & -\dot{\theta} \\ \dot{\theta} & 0 \end{bmatrix}(\mathbf{b} - \mathbf{a}_0) \tag{4.36}$$

$$(\dot{\mathbf{b}}) = (\dot{\mathbf{b}}') + \begin{bmatrix} 0 & -\dot{\alpha} \\ \dot{\alpha} & 0 \end{bmatrix}(\mathbf{b} - \mathbf{a}) \tag{4.37}$$

Expanding Eq. 4.37 with $\dot{\mathbf{b}}'$ known in terms of $\dot{\theta}$,

$$\dot{b}_x = \dot{b}'_x - \dot{\alpha}(b_y - a_y)$$
$$\dot{b}_y = \dot{b}'_y + \dot{\alpha}(b_x - a_x) \tag{4.38}$$

Substituting in Eq. 4.35,

$$[\dot{b}'_x - \dot{\alpha}(b_y - a_y)](b_x - b_{0x}) + [\dot{b}'_y + \dot{\alpha}(b_x - a_x)](b_y - b_{0y}) = 0$$
$$\dot{\alpha}[(b_x - a_x)(b_y - b_{0y}) - (b_y - a_y)(b_x - b_{0x})]$$
$$= -\dot{b}'_x(b_x - b_{0x}) - \dot{b}'_y(b_y - b_{0y})$$

Solving for $\dot{\alpha}$,

$$\dot{\alpha} = \frac{\dot{b}'_x(b_x - b_{0x}) + \dot{b}'_y(b_y - b_{0y})}{(b_y - a_y)(b_x - b_{0x}) - (b_x - a_x)(b_y - b_{0y})} \tag{4.39}$$

$$= \dot{\theta}\frac{(b_x - a_{0x})(b_x - b_{0x}) - (b_y - a_{0y})(b_y - b_{0y})}{(b_y - a_y)(b_x - b_{0x}) - (b_x - a_x)(b_y - b_{0y})} \tag{4.40}$$

where $\dot{\alpha}$ is the *relative angular velocity* at joint **a**.

With $\dot{\alpha}$ known, $\dot{\mathbf{b}}$ can be calculated from Eq. 4.37. Then, noting that

$$(\dot{\mathbf{b}}) = \begin{bmatrix} 0 & -\dot{\phi} \\ \dot{\phi} & 0 \end{bmatrix}(\mathbf{b} - \mathbf{b}_0) \tag{4.41}$$

we obtain

$$\dot{\phi} = \frac{\dot{b}_y}{(b_x - b_{0x})} \tag{4.42}$$

Relative Acceleration Analysis

Differentiating the velocity constraint equation, we have the acceleration constraint equation

$$(\ddot{\mathbf{b}})^T(\mathbf{b} - \mathbf{b}_0) + (\dot{\mathbf{b}})^T(\dot{\mathbf{b}}) = 0 \tag{4.43}$$

The acceleration $\ddot{\mathbf{b}}$ is given by

$$\ddot{\mathbf{b}} = \ddot{\mathbf{b}}' + \mathbf{a}_{\text{rel}} + \mathbf{a}_{\text{cor}} \tag{4.44}$$

where

$$\ddot{\mathbf{b}}' = \begin{bmatrix} -\dot{\theta}^2 & -\ddot{\theta} \\ \ddot{\theta} & -\dot{\theta}^2 \end{bmatrix} (\mathbf{b} - \mathbf{a}_0)$$

$$\mathbf{a}_{\text{rel}} = \begin{bmatrix} -\dot{\alpha}^2 & -\ddot{\alpha} \\ \ddot{\alpha} & -\dot{\alpha}^2 \end{bmatrix} (\mathbf{b} - \mathbf{a})$$

$$\mathbf{a}_{\text{cor}} = 2 \begin{bmatrix} 0 & -\dot{\theta} \\ \dot{\theta} & 0 \end{bmatrix} v_{\text{rel}}$$

$$= 2 \begin{bmatrix} 0 & -\dot{\theta} \\ \dot{\theta} & 0 \end{bmatrix} \begin{bmatrix} 0 & -\dot{\alpha} \\ \dot{\alpha} & 0 \end{bmatrix} (\mathbf{b} - \mathbf{a})$$

Eqs. 4.43 and 4.44 could be solved for $\ddot{\alpha}$ in the same way as Eqs. 4.35 and 4.37 were solved for $\dot{\alpha}$. As an alternative, it is often advantageous to solve the set of three scalar equations directly for $\ddot{b}_x$, $\ddot{b}_y$, and $\ddot{\alpha}$.

Expanding Eqs. 4.43 and 4.44, we obtain

$$\ddot{b}_x(b_x - b_{0x}) + \ddot{b}_y(b_y - b_{0y}) + \dot{b}_x^2 + \dot{b}_y^2 = 0$$

$$\ddot{b}_x = \ddot{b}_x' - \dot{\alpha}^2(b_x - a_x) - \ddot{\alpha}(b_y - a_y) - 2\dot{\theta}\dot{\alpha}(b_x - a_x)$$

$$\ddot{b}_y = \ddot{b}_y' + \ddot{\alpha}(b_x - a_x) - \dot{\alpha}^2(b_y - a_y) - 2\dot{\theta}\dot{\alpha}(b_y - a_y) \tag{4.45}$$

which can be rearranged in matrix form as

$$\begin{bmatrix} (b_x - b_{0x}) & (b_y - b_{0y}) & 0 \\ 1 & 0 & (b_y - a_y) \\ 0 & 1 & -(b_x - a_x) \end{bmatrix} \begin{bmatrix} \ddot{b}_x \\ \ddot{b}_y \\ \ddot{\alpha} \end{bmatrix} = \begin{bmatrix} -(\dot{b}_x^2 + \dot{b}_y^2) \\ \ddot{b}_x' - (\dot{\alpha}^2 + 2\dot{\theta}\dot{\alpha})(b_x - a_x) \\ \ddot{b}_y' - (\dot{\alpha}^2 + 2\dot{\theta}\dot{\alpha})(b_y - a_y) \end{bmatrix}$$

$$\tag{4.46}$$

Eqs. 4.46 can be solved by Cramer's rule for $\ddot{b}_x$, $\ddot{b}_y$, and $\ddot{\alpha}$, where $\ddot{\alpha}$ is the *relative angular acceleration* at joint $\mathbf{a}$.

The previous example of relative acceleration analysis offers no advantage in the analysis of the four-bar linkage as compared to the vector methods of Chapter 2, where the coupler motion was described in terms of the absolute angular velocity and acceleration of the coupler, link 3. The relative acceleration analysis involving coincident points $\mathbf{b}$ and $\mathbf{b}'$ and the Coriolis component has been demonstrated because of the direct analogy to the kinematic analysis of spatial mechanisms, which follows.

4.6   KINEMATIC ANALYSIS OF SPATIAL MECHANISMS IN CLOSED FORM [1]

The RSSR Mechanism

The displacement constraint equation specifies constant length for the coupler link 3 in Figure 4.5.

$$(\mathbf{a} - \mathbf{b})^T(\mathbf{a} - \mathbf{b}) = (\mathbf{a}_1 - \mathbf{b}_1)^T(\mathbf{a}_1 - \mathbf{b}_1) \tag{4.47}$$

where $\mathbf{a}$ is given in terms of the specified input angle α from

$$(\mathbf{a}) = [R_{\alpha,\,\mathbf{u}_a}](\mathbf{a}_1 - \mathbf{a}_0) + (\mathbf{a}_0) \tag{4.48}$$

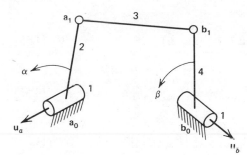

FIGURE 4.5   THE RSSR MECHANISM.

and $\mathbf{b}$ is a function of the unknown output angle β.

$$(\mathbf{b}) = [R_{\beta,\,\mathbf{u}_b}](\mathbf{b}_1 - \mathbf{b}_0) + (\mathbf{b}_0) \tag{4.49}$$

which leads to

$$E \cos \beta + F \sin \beta + G = 0 \tag{4.50}$$

where

$$E = (\mathbf{a} - \mathbf{b}_0)^T[I - Q_{\mathbf{u}_b}](\mathbf{b}_1 - \mathbf{b}_0)$$

$$F = (\mathbf{a} - \mathbf{b}_0)^T[P_{\mathbf{u}_b}](\mathbf{b}_1 - \mathbf{b}_0)$$

$$G = (\mathbf{a} - \mathbf{b}_0)^T[Q_{\mathbf{u}_b}](\mathbf{b}_1 - \mathbf{b}_0)$$

$$+ \tfrac{1}{2}[(\mathbf{a}_1 - \mathbf{b}_1)^T(\mathbf{a}_1 - \mathbf{b}_1) - (\mathbf{a} - \mathbf{b}_0)^T(\mathbf{a} - \mathbf{b}_0) - (\mathbf{b}_1 - \mathbf{b}_0)^T(\mathbf{b}_1 - \mathbf{b}_0)]$$

Solution of Eq. 4.50 using Eqs. 4.30 and 4.31 gives two possible values for β.

Velocity Analysis

The velocity constraint equation becomes

$$(\dot{\mathbf{a}} - \dot{\mathbf{b}})^T(\mathbf{a} - \mathbf{b}) = 0 \tag{4.51}$$

where $\dot{\mathbf{a}}$ is specified from

$$(\dot{\mathbf{a}}) = [W_{\dot{\alpha},\,\mathbf{u}_a}](\mathbf{a} - \mathbf{a}_0) = \dot{\alpha}[P_{\mathbf{u}_a}](\mathbf{a} - \mathbf{a}_0) \tag{4.52}$$

and $\dot{\mathbf{b}}$ involves the unknown $\dot{\beta}$ in

$$(\dot{\mathbf{b}}) = [W_{\dot{\beta},\,\mathbf{u}_b}](\mathbf{b} - \mathbf{b}_0) = \dot{\beta}[P_{\mathbf{u}_b}](\mathbf{b} - \mathbf{b}_0) \tag{4.53}$$

Substituting into Eq. 4.51, we obtain

$$\dot{\beta} = \frac{(\dot{\mathbf{a}})^T(\mathbf{a} - \mathbf{b})}{(\mathbf{a} - \mathbf{b})^T[P_{\mathbf{u}_b}](\mathbf{b} - \mathbf{b}_0)} \tag{4.54}$$

With $\dot{\beta}$ known, we can find $\dot{\mathbf{b}}$ from Eq. 4.53.

Acceleration Analysis

The acceleration constraint equation is

$$(\ddot{\mathbf{a}} - \ddot{\mathbf{b}})^T(\mathbf{a} - \mathbf{b}) + (\dot{\mathbf{a}} - \dot{\mathbf{b}})^T(\dot{\mathbf{a}} - \dot{\mathbf{b}}) = 0 \tag{4.55}$$

where $\ddot{\mathbf{a}}$ is specified by

$$(\ddot{\mathbf{a}}) = [\dot{W}_{\dot{\alpha},\,\ddot{\alpha},\,\mathbf{u}_a}](\mathbf{a} - \mathbf{a}_0) = [\ddot{\alpha}[P_{\mathbf{u}_a}] + \dot{\alpha}^2[P_{\mathbf{u}_a}][P_{\mathbf{u}_a}]](\mathbf{a} - \mathbf{a}_0) \tag{4.56}$$

and

$$(\ddot{\mathbf{b}}) = [\dot{W}_{\dot{\beta},\,\ddot{\beta},\,\mathbf{u}_b}](\mathbf{b} - \mathbf{b}_0) = [\ddot{\beta}[P_{\mathbf{u}_b}] + \dot{\beta}^2[P_{\mathbf{u}_b}][P_{\mathbf{u}_b}]](\mathbf{b} - \mathbf{b}_0) \tag{4.57}$$

Substitution into Eq. 4.55 gives

$$\ddot{\beta} = \frac{(\mathbf{a} - \mathbf{b})^T\{\ddot{\mathbf{a}} - \dot{\beta}^2[P_{\mathbf{u}_b}][P_{\mathbf{u}_b}](\mathbf{b} - \mathbf{b}_0)\} + (\dot{\mathbf{a}} - \dot{\mathbf{b}})^T(\dot{\mathbf{a}} - \dot{\mathbf{b}})}{(\mathbf{a} - \mathbf{b})^T[P_{\mathbf{u}_b}](\mathbf{b} - \mathbf{b}_0)} \tag{4.58}$$

Example 4-1 Kinematic Analysis of the RSSR Spatial Function Generation Mechanism in Closed Form. The steps outlined in Eqs. 4.47 to 4.58 have been coded as program RSSRAN as listed in Figure 4.6. Two solutions are given below. In the first example the data used with program FOURBAR in Chapter 2 is rerun as a test problem. The results for point 3 in FOURBAR are seen to compare closely with the results for point B in RSSRAN.

The second example is a spatial problem with nonparallel input and output axes $\mathbf{u}_a$, $\mathbf{u}_b$. In this example assembly of the linkage is not possible after an input displacement angle greater than 20 degrees. The mobility test is made in logical function EFG, a SPAPAC subprogram useful in solving Eq. 4.50. Both values of the relative displacement angle α_1 and α_2 are returned as arguments along with the value closest to the previous value of α. If EFG = .TRUE. assembly of the mechanism is possible.

ACCELERATION ANALYSIS OF THE RSSR MECHANISM

```
UA=    0.      0.  1.000     UB-   0.     0.     1.000
A0=    0.      0.  0.        B0=   3.000  0.     0.
A1=    1.000   0.  0.        B1=   3.000  2.000  0.
W2=  628.000 RAD/SEC
A2-    0.    RAD/SEC/SEC
```

	ALPHA	BETA		POINT B	
POSITION	0.	0.	.300E+01	.200E+01	0.
VELOCITY	.628E+03	−.314E+03	.628E+03	0.	0.
ACCELERATION	0.	.296E I 06	.592E I 06	−.197E+06	0.

	ALPHA	BETA		POINT B	
POSITION	30.00	−7.93	.328E+01	.198E+01	0.
VELOCITY	.628E+03	−.112E+02	.221E+02	−.308E+01	0.
ACCELERATION	0.	.350E+06	−.693E+06	.962E+05	0.

	ALPHA	BETA		POINT B	
POSITION	60.00	−2.64	.309E+01	.200E+01	0.
VELOCITY	.628E+03	.208E+03	−.415E+03	.191E+02	0.
ACCELERATION	0.	.176E+06	−.356E+06	−.700E+05	0.

	ALPHA	BETA		POINT B	
POSITION	90.00	9.88	.266E+01	.197E+01	0.
VELOCITY	.628E+03	.300E+03	−.590E+03	−.103E+03	0.
ACCELERATION	0.	.548E+05	−.771E+05	−.196E+06	0.

ACCELERATION ANALYSIS OF THE RSSR MECHANISM

```
UA=    0.      0.    1.000        UB=   1.000  0.     0.
A0=    0.      0.    0.           B0=   2.000  0.     0.
A1=    0.      1.000 0.           B1=   2.000  0.     1.000
W2=  102.000 RAD/SEC
A2= −10.000 RAD/SEC/SEC
```

	ALPHA	BETA		POINT B	
POSITION	0.	.0	.200E+01	−.711E−04	.100E+01
VELOCITY	.100E+03	−.200E+03	0.	.200E+03	.142E−01
ACCELERATION	−.100E+02	.236E+02	0.	−.207E+02	−.400E+05

	ALPHA	BETA		POINT B	
POSITION	10.00	−20.65	.200E+01	.353E+00	.936E+00
VELOCITY	.100E+03	−.220E+03	0.	.206E+03	−.777E+02
ACCELERATION	−.100E+02	−.261E+05	0.	.725E+04	−.546E+05

	ALPHA	BETA		POINT B	
POSITION	20.00	−46.71	.200E+01	.728E+00	.686E+00
VELOCITY	.100E+03	−.330E+03	0.	.226E+03	−.240E+03
ACCELERATION	−.100E+02	−.140E+06	0.	.165E+05	−.177E+06

NO SOLUTION IN FUNCTION EFG
NO SOLUTION IN FUNCTION EFG

```
      PROGRAM RSSRAN(INPUT,OUTPUT,TAPE5=INPUT,TAPE6=OUTPUT)
C ACCELERATION ANALYSIS OF THE RSSR MECHANISM
      LOGICAL EOF,EFG
      DIMENSION UA(3),A0(3),A1(3),AJ(3),UB(3),B0(3),B1(3),BJ(3),
     $ VAJ(3),VBJ(3),AAJ(3),ABJ(3),B1P(3),PM(3,3),QM(3,3),QIM(3,3),
     $ RM(3,3,2),WM(3,3,1),WD(3,3,1),T1(3),T2(3),T3(3),T4(3),T5(3),T6(3)
     $ ,T7(3),T8(3),T9(3),T10(3),T11(3),T12(3),VUA(3),VUB(3)
    5 READ(5,10) UA,A0,A1,UB,B0,B1,W2,A2,ALPHA,DELALPH
   10 FORMAT(6F10.0)
      IF(EOF(5)) STOP
      CON=ATAN(1.)/45.
      DEL= DELALPH*CON
      NPT= INT(ALPHA/DELALPH) + 1
      BETA= 0.0
C WRITE HEADINGS FOR THE RESULTS
      WRITE(6,20) UA,UB,A0,B0,A1,B1,W2,A2
   20 FORMAT(1H1,10X,44H ACCELERATION ANALYSIS OF THE RSSR MECHANISM
     $ // 4X,4H UA=,3F8.3,9X,4H UB=,3F8.3
     $ / 5X,3HA0=,3F8.3,10X,3HB0=,3F8.3,/ 5X,3HA1=,3F8.3,10X,3HB1=,3F8.3
     $ / 5X,3HW2=,F8.3,8H RAD/SEC,/ 5X,3HA2=,F8.3,12H RAD/SEC/SEC// )
C DISPLACEMENT ANALYSIS
      DO 100 J= 1,NPT
      ALPH= (J-1)*DEL
      DO 25 I= 1,3
      T1(I)= A1(I)-A0(I)
      T4(I)= A1(I)-B1(I)
   25 T2(I)= B1(I)-B0(I)
      CALL RMAXIS(UA,ALPH,RM,2)
      CALL ROTATE(AJ,A0,RM,A1,A0,2)
      DO 26 I= 1,3
      T5(I)= AJ(I)-A0(I)
   26 T3(I)= AJ(I)-B0(I)
      CALL PMTX(UB,PM)
      CALL QMTX(UB,QM)
      CALL QIMTX(UB,QM,QIM)
      CALL MTXVEC(T6,QIM,T2)
      E= DOT(T3,T6)
      CALL MTXVEC(T7,PM,T2)
      F= DOT(T3,T7)
      CALL MTXVEC(T8,QM,T2)
      G= DOT(T3,T8) + .5*(DOT(T4,T4) - DOT(T3,T3) - DOT(T2,T2) )
      IF(.NOT.EFG(E,F,G,BETA1,BETA2,BETA)) GO TO 100
   40 CALL RMAXIS(UB,BETA,RM,2)
      CALL ROTATE(BJ,B0,RM,B1,B0,2)
      BET= BETA/CON
      ALP= ALPH/CON
```

```
C VELOCITY ANALYSIS
      CALL  WMTX(UA,W2,WM,1)
      CALL  ROTVEC(VAJ,WM,T5,1)
      DO  45  I=  1,3
      VUA(I)=  VUB(I)=  0.0
      T11(I)=  AJ(I)-BJ(I)
   45 T9(I)=  BJ(I)-B0(I)
      CALL  MTXVEC(T10,PM,T9)
      VBETA=  DOT(VAJ,T11)/DOT(T11,T10)
      CALL  WMTX(UB,VBETA,WM,1)
      CALL  ROTVEC(VBJ,WM,R9,1)
C ACCELERATION  ANALYSIS
      CALL  WDOT(UA,VUA,W2,A2,WD,1)
      CALL  ROTVEC(AAJ,WD,T5,1)
      CALL  MTXVEC(T6,QIM,T9)
      CALL  MTXVEC(T7,PM,T9)
      DO  46  1=  1,3
   46 T12(I)=  VAJ(I)-VBJ(I)
      TEMP1=  DOT(AAJ,T11)  +  VBETA*VBETA*DOT(T6,T11)  +  DOT(T12,T12)
      TEMP2=  DOT(T11,T7)
      ABETA=  TEMP1/TEMP2
      CALL  WDOT(UB,VUB,VBETA,ABETA,WD,1)
      CALL  ROTVEC(ABJ,WD,T9,1)
C PRINT  RESULTS
      WRITE(6.200)  ALP,BET,BJ,W2,VBETA,VBJ,A2,ABETA,ABJ
  100 CONTINUE
  200 FORMAT(//20X,5HALPHA,8X,4HBETA,15X,7HPOINT  B
     $  /  5X,8HPOSITION,2X,F6.2,6X,F6.2,6X,3E10.3
     $  /  5X,8HVELOCITY,2X,E10.3,2X,E10.3,2X,3E10.3
     $  /  1X,12HACCELERATION,2X,E10.3,2X,E10.3,2X,3E10.3)
      GO  TO  5
      END
```

FIGURE 4.6   EXAMPLE 4-2. ACCELERATION ANALYSIS OF THE RSSR MECHANISM IN CLOSED FORM USING SPAPAC SUBROUTINES. PROGRAM RSSRAN.

The RRSS Mechanism

The *displacement analysis* is based on the constant length condition for the output link 4 in Figure 4.7.

$$(\mathbf{b} - \mathbf{b}_0)^T(\mathbf{b} - \mathbf{b}_0) = (\mathbf{b}_1 - \mathbf{b}_0)^T(\mathbf{b}_1 - \mathbf{b}_0) \qquad (4.59)$$

Then, since $\mathbf{a}$ and $\mathbf{b}_1'$ in terms of a specified θ are given by

$$(\mathbf{a}) = [R_{\theta,\,\mathbf{u}_0}](\mathbf{a}_1 - \mathbf{a}_0) + (\mathbf{a}_0)$$

$$(\mathbf{b}_1') = [R_{\theta,\,\mathbf{u}_0}](\mathbf{b}_1 - \mathbf{a}_0) + (\mathbf{a}_0)$$

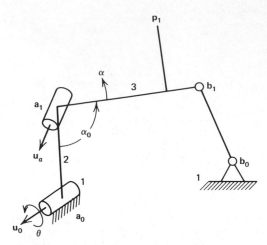

FIGURE 4.7 THE RRSS MECHANISM.

and

$$(\mathbf{u}_a) = [R_{\theta,\,\mathbf{u}_0}](\mathbf{u}_{a1})$$

We can express **b** in terms of the unknown relative rotation angle α from

$$(\mathbf{b}) = [R_{\alpha,\,\mathbf{u}_a}](b'_1 - \mathbf{a}) + (\mathbf{a}) \qquad (4.60)$$

Substitution into Eq. 4.59 again leads to

$$E \cos \alpha + F \sin \alpha + G = 0$$

where

$$E = (\mathbf{a} - \mathbf{b}_0)^T\{[I - Q_{\mathbf{u}_a}](b'_1 - \mathbf{a})\}$$

$$F = (\mathbf{a} - \mathbf{b}_0)^T\{[P_{u_a}](b'_1 - \mathbf{a})\}$$

$$G = (\mathbf{a} - \mathbf{b}_0)^T\{[Q_{u_a}](b'_1 - \mathbf{a})\}$$

$$\quad + \tfrac{1}{2}\{(b'_1 - \mathbf{a})^T(b'_1 - \mathbf{a}) + (\mathbf{a} - \mathbf{b}_0)^T(\mathbf{a} - \mathbf{b}_0) - (\mathbf{b}_1 - \mathbf{b}_0)^T(\mathbf{b}_1 - \mathbf{b}_0)\}$$

with two solutions

$$\alpha_{1,2} = 2 \tan^{-1} \frac{-F \pm \sqrt{E^2 + F^2 - G^2}}{G - E}$$

The displacement of an arbitrary point $\mathbf{p}_1$ on the coupler link 3 is calculated from

$$(\mathbf{p}) = [R_{\alpha,\,\mathbf{u}_a}](\mathbf{p}_1' - \mathbf{a}) + (\mathbf{a}) \tag{4.61}$$

where

$$(\mathbf{p}_1') = [R_{\theta,\,\mathbf{u}_0}](\mathbf{p}_1 - \mathbf{a}_0) + (\mathbf{a}_0) \tag{4.62}$$

Velocity analysis follows from

$$(\dot{\mathbf{b}})^T(\mathbf{b} - \mathbf{b}_0) = 0 \tag{4.63}$$

and, since,

$$(\dot{\mathbf{b}}') = [W_{\dot{\theta},\,\mathbf{u}_0}](\mathbf{b} - \mathbf{a}_0)$$

$$(\dot{\mathbf{b}}) = [W_{\dot{\alpha},\,\mathbf{u}_a}](\mathbf{b} - \mathbf{a}) + (\dot{\mathbf{b}}')$$

we have

$$\dot{\alpha} = -\dot{\theta}\,\frac{(\mathbf{b} - \mathbf{b}_0)^T\{[P_{\mathbf{u}_0}](\mathbf{b} - \mathbf{a}_0)\}}{(\mathbf{b} - \mathbf{b}_0)^T\{[P_{\mathbf{u}_a}](\mathbf{b} - \mathbf{a})\}}$$

The velocity of an arbitrary coupler point $\mathbf{p}$ is given by

$$(\dot{\mathbf{p}}') = \dot{\theta}[P_{\mathbf{u}_0}](\mathbf{p} - \mathbf{a}_0)$$

therefore,

$$(\dot{\mathbf{p}}) = \dot{\alpha}[P_{\mathbf{u}_a}](\mathbf{p} - \mathbf{a}) + (\dot{\mathbf{p}}') \tag{4.64}$$

Acceleration analysis is based on

$$(\ddot{\mathbf{b}})^T(\mathbf{b} - \mathbf{b}_0) + (\dot{\mathbf{b}})^T(\dot{\mathbf{b}}) = 0 \tag{4.65}$$

where

$$(\ddot{\mathbf{b}}) = (\ddot{\mathbf{b}}') + \mathbf{a}_{\text{rel}} + \mathbf{a}_{\text{cor}}$$

$$(\ddot{\mathbf{b}}') = [\dot{W}_{\dot{\theta},\,\ddot{\theta},\,\mathbf{u}_0}](\mathbf{b} - \mathbf{a}_0)$$

$$\mathbf{a}_{\text{rel}} = [\dot{W}_{\dot{\alpha},\,\ddot{\alpha},\,\mathbf{u}_a}](\mathbf{b} - \mathbf{a})$$

$$\mathbf{a}_{\text{cor}} = 2[W_{\dot{\theta},\,\mathbf{u}_0}]\{[W_{\dot{\alpha},\,\mathbf{u}_a}](\mathbf{b} - \mathbf{a})\}$$

Recalling from Eq. 3.65, with $\dot{\mathbf{u}}_a$ assumed equal to zero in $\mathbf{a}_{\text{rel}}$

$$[\dot{W}_{\dot{\alpha},\,\ddot{\alpha},\,\mathbf{u}_a}] = \ddot{\alpha}[P_{\mathbf{u}_a}] + \dot{\alpha}\overset{0}{[\dot{P}_{\mathbf{u}_a}]} + \dot{\alpha}^2[P_{\mathbf{u}_a}][P_{\mathbf{u}_a}]$$

substitution into the constraint equation (Eq. 4.65) leads to

$$\ddot{\alpha} = -\,\frac{(\mathbf{b} - \mathbf{b}_0)^T(\ddot{\mathbf{b}}' + \{\dot{\alpha}^2[P_{\mathbf{u}_a}][P_{\mathbf{u}_a}]\}(\mathbf{b} - \mathbf{a}) + \mathbf{a}_{\text{cor}}) + (\dot{\mathbf{b}})^T(\dot{\mathbf{b}})}{(\mathbf{b} - \mathbf{b}_0)^T\{[P_{\mathbf{u}_a}](\mathbf{b} - \mathbf{a})\}}$$

$$\tag{4.66}$$

Example 4-2 Kinematic Analysis of the RRSS Path Generation Mechanism in Closed Form. Eqs. 4.59 to 4.66 are solved in program RRSSAN as listed in Figure 4.8.

```
      PROGRAM RRSSAN(INPUT,OUTPUT,TAPE5=INPUT,TAPE6=OUTPUT)
C DISPLACEMENT VELOCITY AND ACCELERATION ANALYSIS OF THE RRSS MECH
C LOAD SPAPAC SUBROUTINES BEFORE EXECUTING RRSSAN
      DIMENSION U0(3),UA1(3),UA(3),VUA(3),AH(3),A0(3),A1(3),A(3),VA(3),
     1 AA(3),B0(3),B1(3),B1P(3),B(3),VBP(3),ABP(3),VB(3),AB(3),P1(3),
     2 P1P(3),P(3),VPP(3),VP(3),APP(3),AP(3),PM(3,3),QM(3,3),QIM(3,3),
     3 PM0(3,3),RM(3,3,2),WM(3,3,2),WMA(3,3,2),WD(3,3,2),WDA(3,3,2),
     4 T1(3),T2(3),T3(3),T4(3),T5(3),T6(3),T7(3)
     5 ,VU0(3),VA0(3),VB0(3),AA0(3),AB0(3)
      COMMON/PRNTR/ PRNT
      LOGICAL EOF,EFG,TEST,PRNT
      PRNT= .FALSE.
    5 READ(5,10) U0,UA1,A0,A1,B0,B1,P1,W2,A2,THETA,DELTH
      IF(EOF(5)) STOP
   10 FORMAT(3(6F10.5/),3F10.5/4F10.5 )
      CON= ATAN(1.)/45.
      DEL= DELTH*CON
      NPT= INT(THETA/DELTH) + 1
      ALPHA= 0.0
C WRITE HEADINGS FOR THE RESULTS
      WRITE(6,100) U0,UA1,A0,B0,A1,B1,P1,W2,A2,THETA,DELTH
  100 FORMAT(1H1,25X,44H ACCELERATION ANALYSIS OF THE RRSS MECHANISM //
     1 //5X,4H U0=,3F8.3,/5X,4HUA1=,3F8.3/
     2 5X,4H A0=,3F8.3,5X,4H B0=,3F8.3/5X,4H A1=,3F8.3,5X,4H B1=,3F8.3/
     3 5X,4H P1=,3F8.3/2X,3HW2=,F8.3,8H RAD/SEC,5X,3HA2=,F8.3,
     4 12H RAD/SEC/SEC,4X,6HTHETA=,F8.3,4H DEG, 9X,6HDELTH=,F8.3,4H DEG)
C DISPLACEMENT ANALYSIS - SOLVE EQS(4.59)-(4.62)
      DO 1000 J= 1,NPT
      TH= (J-1)*DEL
      CALL RMAXIS(U0,TH,RM,2)
      CALL ROTATE(A,A0,RM,A1,A0,2)
      CALL ROTATE(B1P,A0,RM,B1,A0,2)
      CALL ROTATE(P1P,A0,RM,P1,A0,2)
      CALL ROTVEC(UA,RM,UA1,2)
      CALL PMTX(UA,PM)
      CALL QMTX(UA,QM)
      CALL QIMTX(UA,QM,QIM)
      DO 15 I= 1,3
      VU0(I)=VA0(I)=VB0(I)=AA0(I)=AB0(I)= 0.0
      T1(I)= B1P(I)-A(I)
      T2(I)= A(I)-B0(I)
   15 T3(I)= B1(I)-B0(I)
      CALL MTXVEC(T4,QIM,T1)
      E= DOT(T2,T4)
      CALL MTXVEC(T4,PM,T1)
      F= DOT(T2,T4)
      CALL MTXVEC(T4,QM,T1)
      G= DOT(T2,T4) + .5*(DOT(T1,T1)+DOT(T2,T2)-DOT(T3,T3))
```

```
C IF EFG= .FALSE. THE RRSS MECHANISM CANNOT BE ASEMBLED.
         IF(.NOT.EFG(E,F,G,ALPHA1,ALPHA2,ALPHA)) GO TO 1000
    20 CALL RMAXIS(UA,ALPHA,RM,2)
         CALL ROTATE(B,A,RM,B1P,A,2)
         CALL ROTATE(P,A,RM,P1P,A,2)
         THET= TH/CON
         ALPH= ALPHA/CON

C VELOCITY ANALYSIS - SOLVE EQS(4.63) - (4.64)
         CALL PMTX(U0,PM0)
         DO 25 I= 1,3
         T1(I)= B(I)-B0(I)
         T2(I)= B(I)-A(I)
         T3(I)= B(I)-A0(I)
         T4(I)= P(I)-A0(I)
    25 T5(I)= P(I)-A(I)
         CALL MTXVEC(T6,PM0,T3)
         CALL MTXVEC(T7,PM,T2)
C SOLVE FOR REL ANG VEL ABOUT AXIS UA
         VALPHA= - W2*DOT(T1,T6)/DOT(T1,T7)
C CALC VEL OF POINTS A, B AND P ON THE INPUT CRANK
         CALL WMTX(U0,W2,WM,2)
         CALL ROTATE(VA,VA0,WM,A,A0,2)
         CALL ROTATE(VBP,VA0,WM,B,A0,2)
         CALL ROTATE(VPP,VA0,WM,P,A0,2)
C CALC VEL OF POINTS B AND P ON THE COUPLER
         CALL WMTX(UA,VALPHA,WMA,2)
         CALL ROTATE(VB,VBP,WMA,B,A,2)
         CALL ROTATE(VP,VPP,WMA,P,A,2)
         CALL ROTVEC(VUA,WM,UA,2)
C ACCELERATION ANALYSIS - SOLVE EQS(4.65) - (4.66)
C FIRST FIND ACC OF POINTS A, B AND P ON THE INPUT CRANK
         CALL WDOT(U0,VU0,W2,A2,WD,2)
         CALL ROTATE(AA,AA0,WD,A,A0,2)
         CALL ROTATE(ABP,AA0,WD,B,A0,2)
         CALL ROTATE(APP,AA0,WD,P,A0,2)
C SOLVE FOR THE REL AND ACC AALPHA
         CALL ROTVEC(T3,WMA,T2,2)
         CALL ROTVEC(T4,WMA,T3,2)
         CALL ROTVEC(T5,WM,T3,2)
         DO 60 1= 1,3
         VUA(I)= 0.0
         T3(I)= T3(I)/VALPHA
    60 T6(I)= ABP(I) + T4(I) + 2.*T5(I)
         AALPHA= - (DOT(T1,T6) + DOT(VB,VB))/(DOT(T1,T3))
C ADD THE REL ACC COMPONENTS FOR POINTS B AND P
         CALL WDOT(UA,VUA,VALPHA,AALPHA,WD,2)
         CALL ROTATE(AB,ABP,WD,B,A,2)
         CALL ROTATE(AP,APP,WD,P,A,2)
C ADD THE CORIOLIS COMPONENT
         DO 65 I= 1,3
         AB(I)= AB(I) + 2.*T5(I)
    65 AP(I)= AP(I) + 2.*T5(I)
         WRITE(6,200) THET,W2,A2,ALPH,VALPHA,AALPHA,A,VA,AA,B,VB,AB,
    1 P,VP,AP
```

```
1000 CONTINUE
 200 FORMAT(/16X,8HPOSITION,25X,8HVELOCITY,23X,12HACCELERATION /
   1  /1X,5HTHETA,2X,F8.2,21X,E10.3,25X,E10.3
   2  /1X,5HALPHA,2X,F8.2,21X,E10.3,25X,E10.3 /
   3  /5X,1HA,      2X,3F8.2,5X,3E10.3,5X,3E10.3
   4  /5X,1HB,      2X,3F8.2,5X,3E10.3,5X,3E10.3
   5  /5X,1HP,      2X,3F8.2,5X,3E10.5X,3E10.3/)
      GO TO 5
      END
```

FIGURE 4.8   EXAMPLE 4-2. KINEMATIC ANALYSIS OF THE RRSS PROGRAM RRSSAN.
PATH GENERATION MECHANISM IN CLOSED FORM:

Program RRSSAN was first checked using the plane four-bar example given in Chapter 2. Results for a spatial problem are given below.

```
 U0=  0.     0.      1.000
UA1=  .500   0.      .867
 A0=  0.     0.      0.       B0=  2.000  0.     0.
 A1=  0.     1.000   0.       B1=  2.000  1.000  0.
 P1=  2.000  1.000   2.000
W2= 100.000 RAD/SEC     A2= 100.000 RAD/SEC/SEC     THETA= 90.000 DEG
```
DELTH= 10.000 DE

	POSITION			VELOCITY			ACCELERATION		
THETA	0.			.100E+03			.100E+03		
ALPHA	.00			−.115E+03			−.115E+03		
A	0.	1.00	0.	−.100E+03	0.	0.	−.100E+03	−.100E+05	0.
B	2.00	1.00	0.	−.100E+03	−.909E−12	0.	−.689E+02	−.100E+05	.115E+
P	2.00	1.00	2.00	−.100E+03	.115E+03	0.	.115E+05	−.988E+04	.491E+

	POSITION			VELOCITY			ACCELERATION		
THETA	10.00			.100E+03			.100E+03		
ALPHA	−11.56			−.116E+03			−.120E+04		
A	−.17	.98	0.	−.985E+02	−.174E+02	0.	.164E+04	−.987E+04	0.
B	1.83	.98	.02	−.987E+02	−.178E+02	.202E+02	.136E+04	−.106E+05	.117E+
P	1.81	1.19	2.01	−.199E+03	.961E+02	.854E+01	.133E+05	−.121E+05	.497E+

	POSITION			VELOCITY			ACCELERATION		
THETA	20.00			.100E+03			.100E+03		
ALPHA	−23.32			−.120E+03			−.367E+04		
A	−.34	.94	0.	−.940E+02	−.342E+02	0.	.333E+04	−.943E+04	0.
B	1.66	.94	.07	−.954E+02	−.381E+02	.411E+02	.205E+04	−.133E+05	.127E+
P	1.59	1.33	2.03	−.134E+03	.726E+02	.174E+02	.154E+05	−.151E+05	.541E+

The RCCC Mechanism

The analysis of the RCCC mechanism shown schematically in Figure 4.9 is somewhat more complex because of the increase in degrees of freedom at the cylindrical joints. Each cylindrical joint is defined by the location of one point such as $\mathbf{a}$, $\mathbf{c}$, or $\mathbf{e}$ plus the orientation of the associated axis $\mathbf{u}_a$, $\mathbf{u}_c$, or $\mathbf{u}_e$. There are also six unknown motion parameters s_a, s_c, s_e, λ, η, and ϕ required to define the displaced position of the mechanism. The geometry of an individual link is described in terms of the common perpendicular and the twist angle between successive joint axes [e.g., the vector $(\mathbf{a} - \mathbf{a}_0)$ and the twist angle α for the input link].

The *displacement analysis* of the RCCC mechanism can be accomplished in two phases: (1) calculation of the relative rotations λ and ϕ about axis $\mathbf{u}_a$ and $\mathbf{u}_e$, followed by (2) calculation of the linear motion components s_a, s_c, and s_e along axes $\mathbf{u}_a$, $\mathbf{u}_c$, and $\mathbf{u}_e$.

The *output angular displacement angle* ϕ is found first by noting that the twist angle β must remain constant during any displacement of the mechanism. This leads to the first constraint equation

$$(\mathbf{u}_a)^T(\mathbf{u}_c) = (\mathbf{u}_{a1})^T(\mathbf{u}_{c1}) \tag{4.67}$$

where
$$(\mathbf{u}_a) = [R_{\theta, \, \mathbf{u}_0}](\mathbf{u}_{a1})$$

$$(\mathbf{u}_c) = [R_{\phi, \, \mathbf{u}_e}](\mathbf{u}_{c1})$$

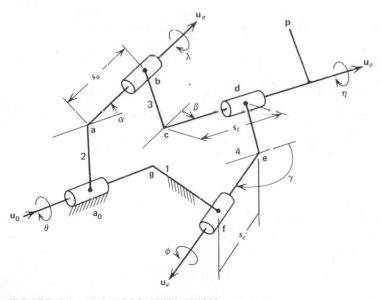

FIGURE 4.9   THE RCCC MECHANISM.

This gives two solutions for the output angle ϕ from Eq. 4.30, with

$$E = (\mathbf{u}_a)^T\{[I - Q_{\mathbf{u}_e}](\mathbf{u}_{c1})\}$$

$$F = (\mathbf{u}_a)^T\{[P_{\mathbf{u}_e}](\mathbf{u}_{c1})\}$$

$$G = (\mathbf{u}_a)^T\{[Q_{\mathbf{u}_e}](\mathbf{u}_{c_1})\} - (\mathbf{u}_{a1})^T(\mathbf{u}_{c1}) \qquad (4.68)$$

The *relative coupler displacement angle* λ is found from the condition that the twist angle γ is constant, which gives a second constraint equation

$$(\mathbf{u}_e)^T(\mathbf{u}_c) = (\mathbf{u}_{e1})^T(\mathbf{u}_{c1}) \qquad (4.69)$$

In this case the new position of the axis $\mathbf{u}_c$ is defined in terms of the relative angle λ.

$$(\mathbf{u}_c) = [R_{\lambda,\,\mathbf{u}_a}]\{[R_{\theta,\,\mathbf{u}_0}](\mathbf{u}_{c1})\} \qquad (4.70)$$

We again find two solutions for the relative displacement angle λ from Eq. 4.30 with

$$E = (\mathbf{u}_e)^T\{[I - Q_{\mathbf{u}_a}](u'_{c1})\} \qquad (4.71)$$

where $$(u'_{c1}) = [R_{\theta,\,\mathbf{u}_0}](\mathbf{u}_{c1})$$

$$F = (\mathbf{u}_e)^T\{[P_{\mathbf{u}_a}](\mathbf{u}'_{c1})\}$$

$$G = (\mathbf{u}_e)^T\{[Q_{\mathbf{u}_a}](\mathbf{u}'_{c1})\} - (\mathbf{u}_e)^T(\mathbf{u}_{c1})$$

The calculation of the *linear sliding position components* s_a, s_c, and s_e is based on alternate vector paths to locate point $\mathbf{d}$ in the mechanism. Proceeding from point $\mathbf{a}_0$, we have

$$\mathbf{d} = \mathbf{c} + s_c\,\mathbf{u}_c \qquad (4.72)$$

where $$(\mathbf{a}) = [R_{\theta,\,\mathbf{u}_0}](\mathbf{a}_1 - \mathbf{a}_0) + (\mathbf{a}_0)$$

$$(\mathbf{c}'_1) = [R_{\theta,\,\mathbf{u}_0}](\mathbf{c}_1 - \mathbf{a}_0) + (\mathbf{a}_0)$$

$$(\mathbf{c}) = [R_{\lambda,\,\mathbf{u}_a}](\mathbf{c}'_1 - \mathbf{a}) + (\mathbf{a}) + (s_a - s_{a1})(\mathbf{u}_a)$$

The second path involves link 4, and we may write

$$(\mathbf{d}) = [R_{\phi,\,\mathbf{u}_e}](\mathbf{d}_1 - \mathbf{e}_1) + (\mathbf{e})$$

$$= [R_{\phi,\,\mathbf{u}_e}](\mathbf{d}_1 - \mathbf{e}_1) + (\mathbf{f}) - s_e\,\mathbf{u}_e \qquad (4.73)$$

where $(\mathbf{f})$ is a fixed point on axis $\mathbf{u}_e$.

Equating Eqs. 4.72 and 4.73, we obtain

$$s_a\,\mathbf{u}_a + s_c\,\mathbf{u}_c + s_e\,\mathbf{u}_e = (\mathbf{f} - \mathbf{a}) + [R_{\phi,\,\mathbf{u}_e}](\mathbf{d}_1 - \mathbf{e}_1)$$

$$- [R_{\lambda,\,\mathbf{u}_a}](\mathbf{c}'_1 - \mathbf{a}) + s_{a1}(\mathbf{u}_a) \qquad (4.74)$$

Eq. 4.74 is linear in the unknown displacements s_a, s_c, and s_e and may be written in matrix form as

$$\begin{bmatrix} u_{ax} & u_{cx} & u_{ex} \\ u_{ay} & u_{cy} & u_{ey} \\ u_{az} & u_{cz} & u_{ez} \end{bmatrix} \begin{bmatrix} s_a \\ s_c \\ s_e \end{bmatrix} = \begin{bmatrix} h_x \\ h_y \\ h_z \end{bmatrix} \tag{4.75}$$

where (h) represents the known elements on the right side of Eq. 4.74. Eq. 4.75 is easily solved by Cramer's rule.

The *displacement of an arbitrary point p* on the coupler, link 3, is calculated from

$$(\mathbf{p}) = [R_{\lambda, \mathbf{u}_a}](\mathbf{p}_1' - \mathbf{a}) + (\mathbf{a}) + (s_a - s_{a1})\,\mathbf{u}_a$$
$$(\mathbf{p}') = [R_{\theta, \mathbf{u}_0}](\mathbf{p}_1 - \mathbf{a}_0) + (\mathbf{a}_0) \tag{4.76}$$

The *angular velocity analysis* follows from constraint equations obtained by differentiation of Eqs. 4.67 and 4.69. This leads to two angular velocity constraint equations:

$$(\dot{\mathbf{u}}_a)^T(\mathbf{u}_c) + (\mathbf{u}_a)^T(\dot{\mathbf{u}}_c) = 0 \tag{4.77}$$

$$(\dot{\mathbf{u}}_e)^T(\mathbf{u}_c) + (\mathbf{u}_e)^T(\dot{\mathbf{u}}_c) = 0 \tag{4.78}$$

from which

$$\dot{\phi} = -\dot{\theta}\,\frac{(\mathbf{u}_c)^T\{[P_{\mathbf{u}_0}](\mathbf{u}_a)\}}{(\mathbf{u}_a)^T\{[P_{\mathbf{u}_e}](\mathbf{u}_c)\}} \tag{4.79}$$

$$\dot{\lambda} = -\dot{\theta}\,\frac{(\mathbf{u}_e)^T\{[P_{\mathbf{u}_0}](\mathbf{u}_c)\}}{(\mathbf{u}_e)^t\{[P_{\mathbf{u}_a}](\mathbf{u}_c)\}} \tag{4.80}$$

In deriving Eqs. 4.79 and 4.80, note that in Eq. 4.79

$$(\dot{\mathbf{u}}_c) = [W_{\dot{\phi}, \mathbf{u}_e}](\mathbf{u}_c) \tag{4.81}$$

while in Eq. 4.80 we use

$$(\dot{\mathbf{u}}_c) = [W_{\dot{\theta}, \mathbf{u}_0}](\mathbf{u}_c) + [W_{\dot{\phi}, \mathbf{u}_e}](\mathbf{u}_c)$$

The *sliding velocity components* $\mathbf{s}$ are found by differentiating equation (4.72) in the form

$$\dot{\mathbf{d}} = \dot{\mathbf{c}} + \dot{s}_c\mathbf{u}_c + s_c\dot{\mathbf{u}}_c \tag{4.83}$$

where

$$\dot{\mathbf{d}} = [W_{\dot{\phi}, \mathbf{u}_e}](\mathbf{d} - \mathbf{e}) - \dot{s}_e\,\mathbf{u}_e$$
$$\dot{\mathbf{c}} = [W_{\dot{\theta}, \mathbf{u}_0}](\mathbf{c} - \mathbf{a}_0) + [W_{\dot{\lambda}, \mathbf{u}_a}](\mathbf{c} - \mathbf{a}) + \dot{s}_a\mathbf{u}_a \tag{4.84}$$

This leads to an equation in the form

$$[U](\dot{\mathbf{s}}) = (\mathbf{h}')$$ (4.85)

where

$$\mathbf{h}' = [W_{\phi,\,\mathbf{u}_e}](\mathbf{d} - \mathbf{e}) - [W_{\theta,\,\mathbf{u}_0}](\mathbf{c} - \mathbf{a}_0)$$
$$= [W_{\lambda,\,\mathbf{u}_a}](\mathbf{c} - \mathbf{a}) - s_c\,\dot{\mathbf{u}}_c$$

Eq. (4.85) can be solved for $\dot{\mathbf{s}} = (\dot{s}_a, \dot{s}_c, \dot{s}_e)^T$.

The angular acceleration analysis involves two additional constraint equations found by differentiation of Eqs. 4.77 and 4.78.

$$(\ddot{\mathbf{u}}_a)^T(\mathbf{u}_c) + 2(\dot{\mathbf{u}}_a)^T(\dot{\mathbf{u}}_c) + (\mathbf{u}_a)^T(\ddot{\mathbf{u}}_c) = 0$$ (4.86)

$$(\mathbf{u}_e)^T(\ddot{\mathbf{u}}_c) = 0$$ (4.87)

which leads to

$$\ddot{\phi} = -\frac{\dot{\phi}^2(\mathbf{u}_a)^T\{[P_{\mathbf{u}_e}][P_{\mathbf{u}_e}](\mathbf{u}_c)\} + 2(\dot{\mathbf{u}}_a)^T(\dot{\mathbf{u}}_c) + (\ddot{\mathbf{u}}_a)^T(\mathbf{u}_c)}{(\mathbf{u}_a)^T\{[P_{\mathbf{u}_e}](\mathbf{u}_c)\}}$$ (4.88)

where

$$(\ddot{\mathbf{u}}_a) = [\dot{W}_{\theta,\,\ddot{\theta},\,\mathbf{u}_0}](\mathbf{u}_a)$$
$$(\dot{\mathbf{u}}_c) = [W_{\phi,\,\mathbf{u}_e}](\mathbf{u}_c)$$

and the relative angular acceleration about $\mathbf{u}_a$ becomes

$$\ddot{\lambda} = -\frac{(\mathbf{u}_e)^T\{\ddot{\theta}[P_{\mathbf{u}_0}](\mathbf{u}_c) + \dot{\theta}[P_{\mathbf{u}_0}](\dot{\mathbf{u}}_c) + \dot{\lambda}[P_{\mathbf{u}_a}](\dot{\mathbf{u}}_c) + \dot{\lambda}[\dot{P}_{\mathbf{u}_a}](\mathbf{u}_c)\}}{(\mathbf{u}_e)^T\{[P_{\mathbf{u}_a}](\mathbf{u}_c)\}}$$ (4.89)

The sliding acceleration components $\ddot{\mathbf{s}}$ are found from a second differentiation of equation (4.72) which leads to

$$\ddot{\mathbf{d}} = \ddot{\mathbf{c}} + \ddot{s}_c\,\mathbf{u}_c + 2\dot{s}_c\,\dot{\mathbf{u}}_c + s_c\,\ddot{\mathbf{u}}_c$$ (4.90)

where

$$\ddot{\mathbf{d}} = [\dot{W}_{\phi,\,\ddot{\phi},\,\mathbf{u}_e}](\mathbf{d} - \mathbf{e}) - \ddot{s}_e\,\mathbf{u}_e$$
$$\ddot{\mathbf{c}} = [\dot{W}_{\theta,\,\ddot{\theta},\,\mathbf{u}_0}](\mathbf{c} - \mathbf{a}_0) + [\dot{W}_{\lambda,\,\ddot{\lambda},\,\mathbf{u}_a}](\mathbf{c} - \mathbf{a}) + \ddot{s}_a\,\mathbf{u}_a$$ (4.91)
$$\quad + 2[W_{\theta,\,\mathbf{u}_0}]\{[W_{\lambda,\,\mathbf{u}_a}](\mathbf{c} - \mathbf{a}) + \dot{s}_a\,\mathbf{u}_a\}$$

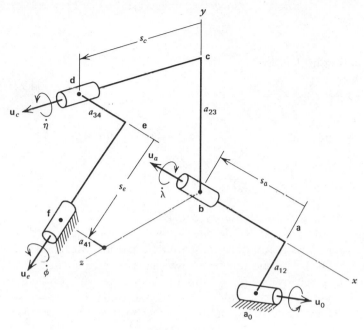

Yang data

$$a_{12} = 2. \qquad \alpha_{12} = 30°$$

$$a_{23} = 4. \qquad \alpha_{23} = 55°$$

$$a_{34} = 3. \qquad \alpha_{34} = 45°$$

$$a_{41} = 5. \qquad \alpha_{41} = 60°$$

$\mathbf{u}_0 = (.866, .499, 030)$ $\mathbf{a}_0 = (2.959, -0.117, 1.996)$

$\mathbf{u}_a = (1.00, 0.0, 0.0)$ $\mathbf{a} = (2.959, 0.0, 0.0)$

$\mathbf{u}_c = (.574, 0.0, -.819)$ $\dot{\mathbf{a}} = (-1.00, 1.729, .103)$

$\mathbf{u}_e = (.217, .668, -.712)$ $\ddot{\mathbf{a}} = (0.0, -.119, 1.996)$

$\mathbf{b} = (0.0, 0.0, 0.0)$ $\mathbf{c} - (0.0, 4.0, 0.0)$

$\mathbf{d} = (-1.042, 4.00, 1.488)$ $\mathbf{e} = (1.281, 4.979, 3.115)$

$\mathbf{f} = (.794, 3.476, 4.714)$ $s_a = 2.959$

Input motion parameters $s_c = 1.817$

$\dot{\theta} = 1.0$ rad/sec $\ddot{\theta} = 0.0$ $s_e = 2.248$

FIGURE 4.10 DATA FOR EXAMPLE 4-3. ACCELERATION ANALYSIS OF THE RCCC MECHANISM.

This leads to an equation in the form

$$[U](\ddot{\mathbf{s}}) = (\mathbf{h}'') \tag{4.92}$$

where

$$\mathbf{h}'' = [\dot{W}_{\phi, \ \dot{\phi}, \ \mathbf{u}_e}](\mathbf{d} - \mathbf{e}) - [\dot{W}_{\theta, \ \dot{\theta}, \ \mathbf{u}_0}](\mathbf{c} - \mathbf{a}_0) - [\dot{W}_{\lambda, \ \dot{\lambda}, \ \mathbf{u}_a}](\mathbf{c} - \mathbf{a})$$

$$- 2[W_{\theta, \ \mathbf{u}_0}][W_{\lambda, \ \mathbf{u}_a}](\mathbf{c} - \mathbf{a}) - 2\dot{s}_a \dot{\mathbf{u}}_a - 2\dot{s}_c \dot{\mathbf{u}}_c - s_c \ddot{\mathbf{u}}_c$$

Eq. (4.92) can be solved for $\ddot{\mathbf{s}} = \ddot{s}_a, \ddot{s}_c, \ddot{s}_e)^T$

Example 4-3. Acceleration Analysis of the RCCC
Mechanism. Figure 4.10 shows an RCCC mechanism previously analysed by Yang (A. T. Yang, "Acceleration Analysis of Spatial Four-Link Mechanisms," ASME paper 65-WA/MD-3). The coordinates shown have been calculated from the original description given by Yang in terms of the shortest perpendicular distance a_{ij} and twist angle α_{ij} for each link in the kinematic chain.

The results for velocity and acceleration analysis in the position shown were as follows.

Velocity Analysis	$\dot{\theta} = 1.0$ rad/sec
$\dot{s}_a = +1.271$	$\dot{\lambda} = 1.097$ rad/sec
$\dot{s}_c = -.807$	$\dot{\eta} = -.685$
$\dot{s}_e = +.683$	$\dot{\phi} = .747$
Acceleration Analysis	$\ddot{\theta} = 0.0$ rad/sec^2
$\ddot{s}_a = -3.711$	$\ddot{\lambda} = -.097$ rad/sec^2
$\ddot{s}_c = +.869$	$\ddot{\eta} = -.234$
$\ddot{s}_e = -3.943$	$\ddot{\phi} = -.225$

4.7 SPATIAL KINEMATIC ANALYSIS BY NUMERICAL SOLUTION OF THE CONSTRAINT EQUATIONS

A second basic method for spatial kinematic analysis does not require the algebraic solution of the constraint equations in closed form. The iterative numerical method used is less efficient in terms of computation time but is more versatile in that it will allow an arbitrary choice of input motion parameter. In addition, the time required for setting up the problem is often reduced as compared to the closed form solution.

The numerical solution of the set of nonlinear algebraic equations that result from the kinematic constraints is carried out using the Newton-Raphson method, as described in Section 6.7.

The RRSS Path Generation Mechanism

As a first example let us repeat the kinematic analysis of the RRSS mechanism. In this case we will follow the path generated by the point $\mathbf{p}$ in Figure 4.7 by incrementing one of the coordinates of $\mathbf{p}$ instead of changing the input angle θ, as was done earlier. It is clear that such a simple change in specification of the input motion would lead to major difficulties in a closed form solution. The complete set of *constraint equations* are formed as follows. The displacement of the coupler, link 3, will be described in terms of the reference point $\mathbf{p}$ plus the *absolute angular displacement* ϕ about the finite rotation axis $\mathbf{u}$ as given by Eq. 3.22. The components of $\mathbf{u}$ become unknowns in the analysis and are not to be confused with axis $\mathbf{u}_a$ in Figure 4.7.

The R-R link displacement constraint equations are written in a form that does not involve the rotation angle θ for the R-R link. Since all positions of point $\mathbf{a}$ in Figure 4.7 must lie in a plane perpendicular to axis $\mathbf{u}_0$, the first constraint equation is the *plane equation*.

$$(\mathbf{u}_0)^T(\mathbf{a} - \mathbf{a}_0) = 0 \tag{4.93}$$

Also, since point $\mathbf{a}_0$ is constrained to move relative to axis $\mathbf{u}_a$ in a plane perpendicular to $\mathbf{u}_a$, we have a second plane constraint equation

$$(\mathbf{u}_a)^T(\mathbf{a} - \mathbf{a}_0) = 0 \tag{4.94}$$

As the R-R link rotates about axis $\mathbf{u}_0$, it must maintain a constant twist angle between axes $\mathbf{u}_a$ and $\mathbf{u}_0$. This condition is assured by the third constraint equation.

$$(\mathbf{u}_a)^T(\mathbf{u}_0) = (\mathbf{u}_{a1})^T(\mathbf{u}_0) \tag{4.95}$$

Eq. 4.95 is a necessary but not sufficient condition to insure a constant twist angle. It is clear that if γ is the twist angle for the R-R link, its negative would satisfy Eq. 4.95 equally well. To eliminate this possibility, we note that the moment of the unit vector $\mathbf{u}_a$ with respect to axis $\mathbf{u}_0$ must be constant for any position of the R-R link. This leads to the fourth constraint equation.

$$((\mathbf{a} - \mathbf{a}_0) \times \mathbf{u}_a)^T(\mathbf{u}_0) = ((\mathbf{a}_1 - \mathbf{a}_0) \times \mathbf{u}_{a1})^T(\mathbf{u}_0) \tag{4.96}$$

In addition to the four basic R-R link constraint equations we note that the unknown direction cosines that comprise the components of the finite rotation axis $\mathbf{u}$ for the coupler link 3 must satisfy a fifth constraint equation.

$$(\mathbf{u})^T(\mathbf{u}) = 1.0 \tag{4.97}$$

The displacement constraint provided by the S-S link requires that link 4 maintain a constant distance between points **b** and $\mathbf{b}_0$. This gives a sixth *constant length* constraint equation

$$(\mathbf{b} - \mathbf{b}_0)^T(\mathbf{b} - \mathbf{b}_0) = (\mathbf{b}_1 - \mathbf{b}_0)^T(\mathbf{b}_1 - \mathbf{b}_0) \tag{4.98}$$

To solve for unknown displacement components, we first describe the displacement of points **a** and **b** in terms of the unknown displacement parameters ϕ and **u** for the coupler link 3. The coupler point **p** is the reference point in the displacement equation.

$$(\mathbf{a}) = [R_{\phi, \mathbf{u}}](\mathbf{a}_1 - \mathbf{p}_1) + (\mathbf{p}) \tag{4.99}$$

$$(\mathbf{b}) = [R_{\phi, \mathbf{u}}](\mathbf{b}_1 - \mathbf{p}_1) + (\mathbf{p}) \tag{4.100}$$

The displaced position of the axis $\mathbf{u}_a$ is also described by the $[R_{\phi, \mathbf{u}}]$ matrix from

$$(\mathbf{u}_a) = [R_{\phi, \mathbf{u}}](\mathbf{u}_{a1}) \tag{4.101}$$

The unknown quantities **a**, **b**, $\mathbf{u}_a$ in Eqs. 4.93 to 4.98 can now be expressed in terms of coupler motion parameters, $\mathbf{p}_1$, **p**, ϕ, and **u**. This leads to a set of six nonlinear equations with seven unknowns p_x, p_y, p_z, ϕ, u_x, u_y, and u_z. We may arbitrarily specify *any one* of the seven and solve for the remaining six unknowns.

In following a coupler point path in space it may be necessary at times to switch the coordinate of **p** being incremented. Since the solution completely describes coupler rigid body motion, we may easily follow several coupler point paths simultaneously. In the numerical solution it is convenient to use the coordinates at the previous solution as initial guesses for the iterative procedure. This not only insures rapid convergence but also tends to avoid convergence to the second of a double solution.

Once the new position of the mechanism is known we may proceed with velocity and acceleration analysis.

The *R-R link velocity constraint equations* are found by differentiation of the corresponding displacement constraint equations.

$$(\mathbf{u}_0)^T(\dot{\mathbf{a}}) = 0$$

$$(\dot{\mathbf{u}}_a)^T(\mathbf{a} - \mathbf{a}_0) + (\mathbf{u}_a)^T(\dot{\mathbf{a}}) = 0$$

$$(\dot{\mathbf{u}}_a)^T(\mathbf{u}_0) = 0$$

$$((\dot{\mathbf{a}} \times \mathbf{u}_a) + (\mathbf{a} - \mathbf{a}_0) \times \dot{\mathbf{u}}_a)^T(\mathbf{u}_0) = 0 \tag{4.102}$$

The *S-S link velocity constraint equation* becomes

$$(\dot{\mathbf{b}})^T(\mathbf{b} - \mathbf{b}_0) = 0 \tag{4.103}$$

The *R-R link acceleration constraint equations are*

$$(\mathbf{u}_0)^T(\ddot{\mathbf{a}}) = 0$$

$$(\ddot{\mathbf{u}}_a)^T(\mathbf{a} - \mathbf{a}_0) + 2(\dot{\mathbf{u}})^T(\dot{\mathbf{a}}) + (\mathbf{u}_a)^T(\ddot{\mathbf{a}}) = 0$$

$$(\ddot{\mathbf{u}}_a)^T(\mathbf{u}_0) = 0$$

$$(\ddot{\mathbf{a}}_a)^T(\mathbf{u}_0) = 0$$

$$(\ddot{\mathbf{a}} \times \mathbf{u}_a + 2\dot{\mathbf{a}} \times \dot{\mathbf{u}} + (\mathbf{a} - \mathbf{a}_0) \times \ddot{\mathbf{u}})^T(\mathbf{u}_0) = 0 \qquad (4.104)$$

The *S-S link acceleration constraint equation is*

$$(\ddot{\mathbf{b}})^T(\mathbf{b} - \mathbf{b}_0) + (\dot{\mathbf{b}})^T(\dot{\mathbf{b}}) = 0 \qquad (4.105)$$

As with the displacement analysis, the velocities $\dot{\mathbf{a}}$, $\dot{\mathbf{b}}$, and $\dot{\mathbf{u}}_a$ can be expressed in terms of coupler motion parameters $\dot{\mathbf{p}}$, $\dot{\phi}$, and $\mathbf{u}'$ from

$$(\dot{\mathbf{a}}) = [W_{\dot{\phi}, \, \mathbf{u}'}](\mathbf{a} - \mathbf{p}) + (\dot{\mathbf{p}})$$

$$(\dot{\mathbf{b}}) = [W_{\dot{\phi}, \, \mathbf{u}'}](\mathbf{b} - \mathbf{p}) + (\dot{\mathbf{p}})$$

$$(\dot{\mathbf{u}}_a) = [W_{\dot{\phi}, \, \mathbf{u}'}](\mathbf{u}_a) \qquad (4.106)$$

The instantaneous rotation axis $\mathbf{u}'$ in the angular velocity matrix $[W_{\dot{\phi}, \, \mathbf{u}'}]$ must also satisfy the direction cosine equation

$$(\mathbf{u}')^T(\mathbf{u}') = 1 \qquad (4.107)$$

Therefore the complete set of six velocity constraint equations includes Eqs. 4.102, 4.103, and 4.107. There are again seven unknowns $\dot{p}_x$, $\dot{p}_y$, $\dot{p}_z$, $\dot{\phi}$, u'_x, u'_y, and u'_z, and we may arbitrarily assume one and solve for the remaining six. The acceleration analysis would be accomplished in a similar manner with

$$(\ddot{\mathbf{a}}) = [\dot{W}_{\dot{\phi}, \, \ddot{\phi}, \, \mathbf{u}', \, \dot{\mathbf{u}}'}](\mathbf{a} - \mathbf{p}) + (\ddot{\mathbf{p}})$$

$$(\ddot{\mathbf{b}}) = [\dot{W}_{\dot{\phi}, \, \ddot{\phi}, \, \mathbf{u}', \, \dot{\mathbf{u}}'}](\mathbf{b} - \mathbf{p}) + (\ddot{\mathbf{p}})$$

$$(\ddot{\mathbf{u}}_a) = [\dot{W}_{\dot{\phi}, \, \ddot{\phi}, \, \mathbf{u}', \, \dot{\mathbf{u}}'}](\mathbf{u}_a) \qquad (4.108)$$

which leads to six equations with seven unknowns $\ddot{p}_x$, $\ddot{p}_y$, $\ddot{p}_z$, $\ddot{\phi}$, $\dot{u}'_x$, $\dot{u}'_y$, and $\dot{u}'_z$.

The *RRSC spatial slider-crank mechanism* shown in Figure 4.11 differs from its planar counterpart, the plane offset slider-crank mechanism shown in Figure 2.3, in that the output motion has two degrees of freedom ϕ and s. The four displacement constraint equations for the R-R link have already been developed in the previous example.

The S-C link displacement constraint equations include the constant length equation

$$(\mathbf{b} - \mathbf{b}'_0)^T(\mathbf{b} - \mathbf{b}'_0) = (\mathbf{b}_1 - \mathbf{b}_0)^T(\mathbf{b}_1 - \mathbf{b}_0) \qquad (4.109)$$

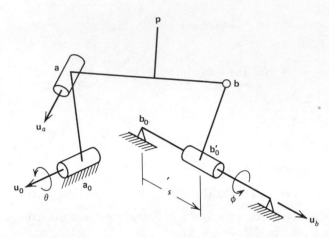

FIGURE 4.11 THE RRSC SPATIAL SLIDER-CRANK MECHANISM.

and the plane equation

$$(\mathbf{u}_b)^T(\mathbf{b} - \mathbf{b}_0') = 0 \tag{4.110}$$

where $(\mathbf{b}_0') = (\mathbf{b}_0) + s(\mathbf{u}_b)$.

Again, the displaced positions of points $\mathbf{a}$, $\mathbf{b}$, and axis $\mathbf{u}_a$ can be described in terms of coupler motion parameters with the spatial rigid body rotation matrix given by either $[R_{\alpha,\ \beta,\ \gamma}]$ from Eq. 3.9, $[R_{\phi,\ \mathbf{u}}]$ from Eq. 3.12, or in terms of Euler's angles as indicated by Eq. 3.13. If $[R_{\phi,\ \mathbf{u}}]$ is used we must include Eq. 4.97 as an additional constraint.

Example 4-4 The RRSC Path Generation Mechanism—Iterative solution of the nonlinear constraint equations. The RRSC mechanism

could also be used to generate a path in space. Assume an RRSC mechanism is specified in its first position by R-R link coordinates

$$(\mathbf{a}_0) = (2.15000,\ 7.600000,\ 2.720000)$$

$$(\mathbf{u}_0) = (0.629800,\ 0.776758,\ 0.000000)$$

$$(\mathbf{a}_1) = (0.300000,\ 9.100000,\ 10.500000)$$

$$(\mathbf{u}_{a1}) = (0.089900,\ 0.982000,\ -0.168000)$$

TABLE 4.1 EXAMPLE 4-4. THE RRSC PATH GENERATION MECHANISM

Position	Leading Coupler Point			Moving Revolute Joint Point		
	P_x	P_y	P_z	a_x	a_y	a_z
1	6.0000	7.5000	8.7000	0.3000	9.1000	10.5000
2	7.0000	6.9491	8.8060	1.2924	8.2953	10.7811
3	8.0000	6.3590	8.8201	2.2694	7.5032	10.8549
4	9.0000	5.7077	8.7430	3.2308	6.7237	10.7365
5	10.0000	4.9678	8.5565	4.1832	5.9515	10.4239

Position	Moving Sphere Joint Point			Cylinder Joint Point		
	b_x	b_y	b_z	b'_{0x}	b'_{0y}	b'_{0z}
1	11.5000	7.6000	7.2000	11.3500	5.5000	1.3000
2	12.4865	7.2652	7.2868	12.3520	5.4425	1.2950
3	13.4978	6.8295	7.3839	13.3843	5.3833	1.2898
4	14.5333	6.2512	7.4790	14.4489	5.3223	1.2845
5	15.5867	5.4821	7.5393	15.5425	5.2595	1.2790

and S-C link coordinates

$$(\mathbf{b}_0) = (11.350000, 5.500000, 1.300000)$$

$$(\mathbf{u}_b) = (0.998345, -0.057263, -0.005000)$$

$$(\mathbf{b}_1) = (11.500000, 7.600000, 7.200000)$$

The first position of the coupler point $\mathbf{p}$ is given by

$$(\mathbf{p}_1) = (6.00, 7.50, 8.70)$$

Table 4.1 lists the results of a displacement analysis for five equal increments in the x-coordinate of point $\mathbf{p}$.

A similar problem involving displacement analysis of an RRSC spatial function generation mechanism is given as Example 12-5 in Chapter 12.

Example 4-5 Displacement analysis of the RCCC mechanism by numerical solution of the constraint equations. The displacement analysis for the RCCC mechanism is based on the solution of the set of constraint equations associated with the geometry of the coupler link.

```
      FUNCTION YCOMP(X,J)
C DISPLACEMENT ANALYSIS OF THE RCCC MECHANISM
      COMMON /YCOMP/ ONCE,NOSE,V
      LOGICAL ONCE,EOF
      REAL LAM
      DIMENSION U0(3),UE(3),UA1(3),UAP(3),UA(3),UC1(3),UCP(3),UC(3),
     $ R0(3,3,2),RA(3,3,2),RC(3,3,2),RE(3,3,2),A0(3),A1(3),AJ(3),B1(3),
     $ C1(3),CP(3),CJ(3),D1(3),DP(3),DPP(3),DJ(3),E1(3),F1(3),
     $ UAPP(3),UCPP(3),BJ(3),EJ(3),T1(3),T2(3),T3(3),T4(3)
      DIMENSION X(1)
      IF(ONCE) GO TO 200
      READ(5,100) U0,UE,UA1,UC1,A0,A1,B1,C1,D1,E1,F1
  100 FORMAT(6F10.0)
      U0(3)= SIGN(SQRT(1.0-U0(1)**2-U0(2)**2),U0(3))
      UA1(3)= SIGN(SQRT(1.0-UA1(1)**2-UA1(2)**2),UA1(3))
      UC1(3)= SIGN(SQRT(1.0-UC1(1)**2-UC1(2)**2),UC1(3))
      UE(3)= SIGN(SQRT(1.0-UE(1)**2-UE(2)**2),UE(3))
      WRITE(6,110) U0,UE,UA1,UC1,A0,A1,B1,C1,D1,E1,F1
  110 FORMAT(//* U0=*3E12.5* UE=*3E12.5/* UA1=*3E12.5* UC1=*3E12.5
     $ /* A0=*3E12.5* A1=*3E12.5/* B1=*3E12.5* C1=*3E12.5/
     $ * D1=*3E12.5* E1=*3E12.5/* F1=*3E12.5//)
      CON= ATAN(1.)/45.
      ONCE= .TRUE.
  200 PHI= X(1)*CON $ SA= X(2) $ SC= X(3) $ SE= X(4)
      TH= V*CON
      CALL RMAXIS(U0,TH,R0,2)
      CALL RMAXIS(UE,PHI,RE,2)
      CALL ROTVEC(UA,R0,UA1,2)
      CALL ROTVEC(UC,RE,UC1,2)
      CALL ROTATE(AJ,A0,R0,A1,A0,2)
      DO 300 I= 1,3
      BJ(I)= AJ(I) + SA*UA(I)
  300 EJ(I)= F1(I) - SE*UE(I)
      DO 310 I= 1,3
  310 DP(I)= D1(I) + EJ(I) - E1(I)
      CALL ROTATE(DJ,EJ,RE,DP,EJ,2)
      DO 315 I= 1,3
  315 CJ(I)= DJ(I) - SC*UC(I)
      GO TO (1,2,3,4) J
    1 YCOMP= DOT(UA,UC) - DOT(UA1,UC1)
      RETURN
    2 YCOMP= PLANE(UA,BJ,CJ)
      RETURN
    3 YCOMP= PLANE(UC,BJ,CJ)
      RETURN
    4 CONTINUE
      DO 320 I= 1,3
      T1(I)= CJ(I) - BJ(I)
  320 T2(I)= C1(I) - B1(I)
      CALL CROSS(T3,T1,UC)
      CALL CROSS(T4,T2,UC1)
      YCOMP= DOT(T3,UA) - DOT(T4,UA1)
      RETURN
      END
```

FIGURE 4.12 FUNCTION YCOMP FOR EXAMPLE 4-5. DISPLACEMENT ANALYSIS OF THE RCCC MECHANISM BY NUMERICAL SOLUTION OF THE CONSTRAINT EQUATIONS.

** NEW DATA **

NTEQ NSEQ NINC NFEI NITR NSET(I), I= 1,5   V   DELV
  4   1   36   1   50   4   0   0   0   0   0.   .5000000E+01

INITIAL GUESSES FOR VARIABLES Z(I), I= 1,NTEQ
 −0.   −0.   −0.   −0.

 U0=   .86603E+00   .49912E+00   .29664E−01   UE= −.21662E+00 −.66842E+00   .71154E+00
UA1= −.10000E+01 0.   0.   UC1= −.57358E+00 0.   .81915E+00
 A0= 0.   −.11861E+00   .19965E+01   A1=   .29592E+01 0.   0.
 B1= 0.   0.   0.   C1= 0.   .40000E+01 0.
 D1= −.10419E+01   .40000E+01   .14880E+01   E1=   .12811E+01 .49787E+01 .31146E+01
 F1=   .79407E+00   .34759E+01   .47144E+01

CONVERGED IN 2 ITERATIONS
ENDING Z WITH V=   0.
.1018754E−11   .2959160E+01   .1816532E+01   .2248308E+01

CONVERGED IN 4 ITERATIONS
ENDING Z WITH V=   5.000
−.3681420E+01   .3056079E+01   .1749483E+01   .2293071E+01

CONVERGED IN 4 ITERATIONS
ENDING Z WITH V=   10.000
−.7247348E+01   .3125990E+01   .1689114E+01   .2308834E+01

CONVERGED IN 4 ITERATIONS
ENDING Z WITH V− 15.000
.1068281E+02   .3170516E+01   .1635386E+01   .2297016E+01

CONVERGED IN 4 ITERATIONS
ENDING Z WITH V=   20.000
−.1397471E+02   .3191544E+01   .1588204E+01   .2259332E+01

CONVERGED IN 4 ITERATIONS
ENDING Z WITH V=   25.000
−.1711108E+02   .3191103E+01   .1547461E+01   .2197670E+01
CONVERGED IN 4 ITERATIONS
ENDING Z WITH V=   30.000
−.2008048E+02   .3171260E+01   .1513060E+01   .2114005E+01

CONVERGED IN 4 ITERATIONS
ENDING Z WITH V=   35.000
−.2287128E+02   .3134047E+01   .1484936E+01   .2010337E+01

CONVERGED IN 4 ITERATIONS
ENDING Z WITH V=   40.000
−.2547097E+02   .3081393E+01   .1463072E+01   .1888661E+01

FIGURE 4.13   SAMPLE OUTPUT FROM PROGRAM DESIGN FOR SOLUTION OF EXAMPLE 4-5.

1. The constant twist equation.

$$(\mathbf{u}_a)^T(\mathbf{u}_c) = (\mathbf{u}_{a1})^T(\mathbf{u}_{c1})$$

2. The plane equations that require points $\mathbf{b}$ and $\mathbf{c}$ to lie on the common perpendicular between axes $\mathbf{u}_a$ and $\mathbf{u}_c$.

$$(\mathbf{u}_a)^T(\mathbf{b} - \mathbf{c}) = 0$$

$$(\mathbf{u}_c)^T(\mathbf{b} - \mathbf{c}) = 0$$

3. The constant moment equation that imposes both constant length constraint and the constant twist direction constraint.

$$[(\mathbf{c} - \mathbf{b}) \times \mathbf{u}_c]^T(\mathbf{u}_a) = [(\mathbf{c}_1 - \mathbf{b}_1) \times \mathbf{u}_{c1}]^T(\mathbf{u}_{a1})$$

This leads to a set of four analysis equations in four unknowns ϕ, s_a, s_c, And s_e. The unknowns are introduced into the equations after substitution of the relations.

$$\mathbf{u}_a = [R_\theta, \mathbf{u}_0]\mathbf{u}_{a1} \qquad \mathbf{a} = [R_\theta, \mathbf{u}_0](\mathbf{a}_1 - \mathbf{a}_0) + \mathbf{a}_0$$

$$\mathbf{b} = \mathbf{a} + s_a\mathbf{u}_a \qquad \mathbf{e} = \mathbf{f} - s_e\mathbf{u}_e$$

$$\mathbf{d}' = \mathbf{d}_1 + (\mathbf{e} - \mathbf{e}_1) \qquad \mathbf{d} = [R_\phi, \mathbf{u}_e](\mathbf{d}' - \mathbf{e}) + \mathbf{e}$$

$$\mathbf{u}_c = [R_\phi, \mathbf{u}_e]\mathbf{u}_{c1} \qquad \mathbf{c} = \mathbf{d} - s_c\mathbf{u}_c$$

These equations have been solved using program DESIGN, which solves the set of nonlinear constraint equations using the Newton-Raphson method, as described in Section 6.7. A listing of the function YCOMP is given in Figure 4.12 along with sample results in Figure 4.13 for the geometry of Figure 4.10.

Chapter 5
Mobility Analysis of Mechanisms

5.1 INTRODUCTION

The term "mobility" refers to the movability of a mechanism. The analysis of mobility requires the answers to several interrelated questions.

First, are the links capable of being assembled into a closed kinematic chain? If the links can be assembled, is it possible to create a movable mechanism by fixing one of the members to a reference link or to ground? How many independent degrees of freedom are possible? If the mechanism has one degree of freedom, what is the relationship between number of links, type of kinematic pairs at the individual joints, and the degrees of freedom of the mechanism? What range of motion is possible? Which of several possible input members is capable of continuous rotation through 360°? What is the permissible change in link lengths that will continue to permit a particular class of motion? These are some of the questions for which we will seek answers in the present chapter.

5.2 CONSTRAINT ANALYSIS

Degrees of Freedom

Grubler [1] was the first to study the relationship between the mobility of a plane four-bar linkage and the degrees of freedom of the individual members and joints.

The degrees of freedom in a plane kinematic chain can be found by adding the number of degrees of freedom for the links in the mechanism taken separately and then subtracting the degrees of freedom lost as the links are assembled. Let

F = total degrees of freedom in the mechanism

n = number of links

l = number of lower pairs (one degree of freedom)

h = number of higher pairs (two degrees of freedom)

Consider the plane four-bar linkage of Figure 5.1. Each individual link is free to move parallel to a given reference plane and requires three coordinates x, y, and θ

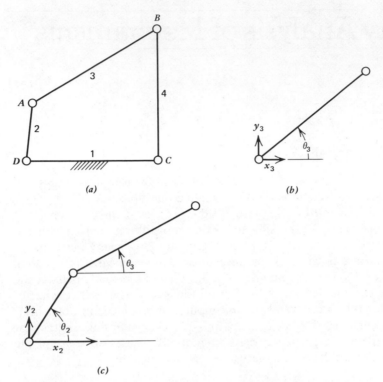

FIGURE 5.1 CONSTRAINT ANALYSIS—PLANE FOUR-BAR LINKAGE (GRÜBLER'S CRITERION). (*a*) COMPLETE LINKAGE; DEGREES OF FREEDOM $F = 1$. (*b*) EACH ISOLATED LINK HAS THREE DEGREES OF FREEDOM; TOTAL FOR LINKS, $F'' = 3n$. (*c*) WHEN TWO LINKS ARE PAIRED TOGETHER, ONE LINK LOSES TWO DEGREES OF FREEDOM; WHEN LINKS ARE CONNECTED INTO A LOOP, $F' = 3n - 2l$. (*d*) TO CREATE A USEFUL LINKAGE, ONE LINK IS FIXED TO GROUND AND LOSES THREE ADDITIONAL DEGREES OF FREEDOM. FOR THE COMPLETE LINKAGE, $F = 3n - 2l - 3 = 3(n - 1) - 2l$.

to describe its position completely (i.e., each link is free to move in a plane with three degrees of freedom).

As each link is connected into the chain, it loses two of its freedoms. After joining all four links into a closed chain, one of the links is attached to a fixed reference member and loses three degrees of freedom. This leads to the following equation for a plane mechanism with lower pairs.

$$F = 3n - 2l - 3$$

$$= 3(n - 1) - 2l \qquad (5.1)$$

Where higher pairs are involved, only one degree of freedom is lost at the connection, and Eq. 5.1 becomes

$$F = 3(n - 1) - 2l - h \qquad (5.2)$$

We note that Eq. 5.2 with $F = 1$ will predict the possibility of constrained motion with a single degree of freedom for a specific arrangement of links and pairs at the joints. Since the dimensions of the links are not involved in this equation, it cannot account for the fact that a set of real links may be impossible to assemble. Also, assuming that the links can be assembled, Eq. 5.2 gives no information about the range of useful motion possible or the rotability of the input crank.

As shown by Kutzbach [2, 3] Grubler's criterion may be generalized in the form

$$F = 6(n - 1) - \sum_{i=1}^{m} p_i c_i \qquad (5.3)$$

where

      n = number of links

      p_i = number of pairs of type i

      c_i = degrees of constraint (degrees of freedom lost) at a pair of type i

      m = number of different pair types

Eq. 5.3 for spatial mechanisms is the equivalent of eq. 5.2 as given for plane mechanisms. For single loop spatial mechanisms with one degree of freedom, the sum of the degrees of freedom of the joints must always be equal to seven.

Eq. 5.3 may also be written in the form

$$F = 6(n - 1) - \sum_{i=1}^{k} (6 - f_i) \qquad (5.4)$$

where n = number of links

                   k = total number of pairs in the mechanism

                   f_i = degrees of freedom at the ith pair

It is often convenient in constraint analysis to replace higher pairs by a kinematically equivalent combination of lower pairs. For example, in the cam mechanism of Figure 5.2, the higher pair at the point of contact between the cam and the follower can be replaced by the instantaneously equivalent coupler link AB, where A and B are located at the centers of curvature of the cam and follower at the point of contact.

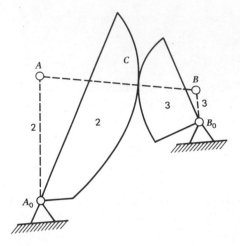

FIGURE 5.2 CONSTRAINT ANALYSIS OF
PLANE MECHANISMS WITH HIGHER PAIRS
USING AN EQUIVALENT LOWER-PAIRED
LINKAGE. (*a*) FOR THE CAM MECHANISM
$n = 3$, $l = 2$, $h = 1$, $F = 3(3 - 1) - 2(2) - 1 = 1$.
(*b*) FOR THE EQUIVALENT FOUR-BAR
LINKAGE A_oABB_o, $n = 4$, $l = 4$, $h = 0$, $F = 3(4 - 1) - 2(4) - 0 = 1$.

As an example of constraint analysis in spatial mechanisms, consider the possibility of forming a movable **RRRC** mechanism as shown in Figure 5.3. In this case

$$n = 4 \qquad p_1 = 3 \qquad p_2 = 1$$
$$c_1 = 5 \qquad c_2 = 4$$
$$F = 6(4 - 1) - 3(5) - 1(4) = -1$$

This would indicate two degrees of overconstraint. Since $F = 1$ corresponds to a movable mechanism and $F = 0$ indicates a rigid structure, $F = -1$ defines a statically indeterminate structure.

If we change to an **RRCC** arrangement with

$$n = 4 \qquad p_1 = 2 \qquad p_2 = 2$$
$$c_1 = 5 \qquad c_2 = 4$$
$$F = 6(4 - 1) - 2(5) - 2(4) = 0$$

Two other possibilities for revolute joint input are shown in Figure 5.4. For the **RRSC** mechanism,

$$n = 4 \qquad p_1 = 2 \qquad p_2 = 1 \qquad p_3 = 1$$
$$c_1 = 5 \qquad c_2 = 4 \qquad c_3 = 3$$
$$F = 6(4 - 1) - 2(5) - 1(4) - 1(3) = 1$$

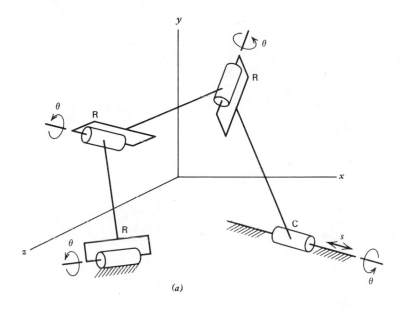

(a)

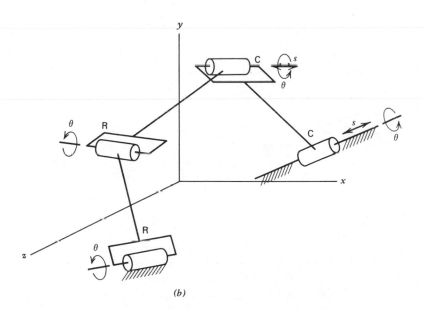

(b)

FIGURE 5.3 OVERCONSTRAINED SPATIAL LINKAGES. (a)   RRRC MECHANISM, $F = -1$. (b) RRCC MECHANISM, $F = 0$.

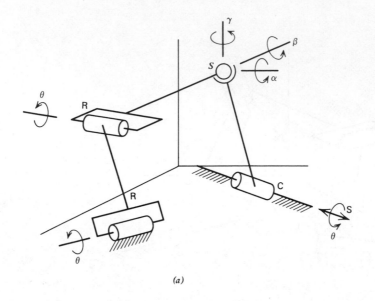

(a)

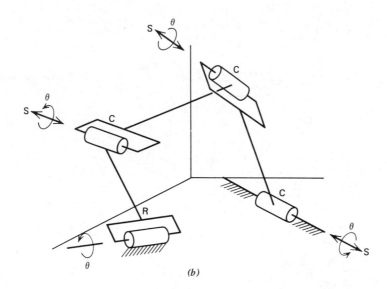

(b)

FIGURE 5.4   SPATIAL LINKAGES WITH ONE DEGREE OF FREE-
DOM. (a) RRSC MECHANISM, $F = 1$. (b) RCCC MECHANISM,
$F = 1$.

and, with the RCCC mechanism,

$$n = 4 \qquad p_1 = 1 \qquad p_2 = 3$$
$$c_1 = 5 \qquad c_2 = 4$$
$$F = 6(4 - 1) - 1(5) - 3(4) = 1$$

Therefore, both mechanisms shown in Figure 5.4 are possible one-degree-of-freedom mechanisms with rotating input and linear sliding (with rotation) output.

Maverick Mechanisms

In certain cases the answers obtained using Eqs. 5.2 and 5.3 or 5.4 can be misleading. Consider the parallel linkage of Figure 5.5a. In this case $n = 5$, $l = 6$, and $F = 3(5 - 1) - 2(6) = 0$. This would indicate a structure incapable of motion. However, we note that member 5 is kinematically redundant and may be removed without destroying the basic kinematic chain formed by links 1, 2, 3, and 4.

In Figure 5.5b the extra degree of freedom is the passive degree of freedom of the S-S link about its long axis.

The mechanism of Figure 5.5c was discovered in 1903 by G. T. Bennett [4] who proved that a four-revolute, four-link spatial mechanism could have one degree of freedom if it satisfied the following conditions.

1. Alternate (opposite) links have the same length a and the same angles of twist α between their revolute axes.
2. The twist angles for alternate links are in contrary directions. If one link is specified with twist angle α the alternate link must have a twist angle $-\alpha$.
3. The relation

$$\frac{\sin \alpha}{a} = \frac{\sin \beta}{b}$$

must be satisfied.
4. When assembled, the terminals of the central axes are brought into coincidence. The central axes thus form a skew quadrilateral with alternate sides equal.

Another special case of importance is the spherical mechanism in which all four revolute axes intersect at a common point. From Eq. 5.3,

$$F = 6(4 - 1) - 4(6 - 1) = -2$$

This is obviously in error, since the mechanism is known to have one degree of freedom.

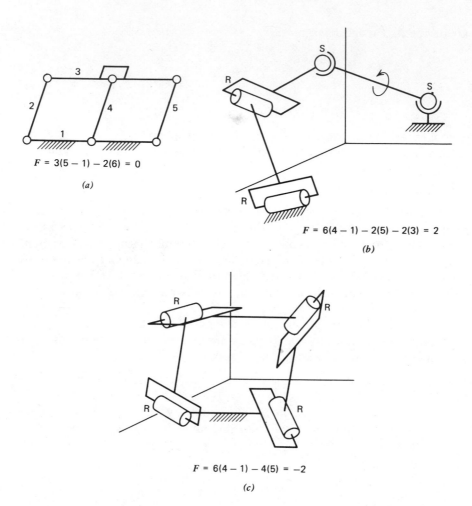

$$F = 3(5 - 1) - 2(6) = 0$$

(a)

$$F = 6(4 - 1) - 2(5) - 2(3) = 2$$

(b)

$$F = 6(4 - 1) - 4(5) = -2$$

(c)

FIGURE 5.5 MAVERICK MECHANISMS WITH CONSTRAINED MOTION THAT VIOLATE THE CONDITION $F = 1$. (a) PLANE LINKAGE WITH REDUNDANT MEMBER 5. (b) SPATIAL LINKAGE WHERE THE S-S LINK HAS AN EXTRA PASSIVE DEGREE OF FREEDOM. (c) THE BENNET MECHANISM.

Other linkages that disobey the Kutzbach criterion because of their special geometric arrangements have been reported by Goldberg [5], Bricard [6], Sarrus [7], and Franke [8]. In recent years Artobolevski [9], Dobrovolski [10], Harrisberger and Soni [11, 12], and Waldron [13] have studied the characteristics of these "overconstrained" mechanisms.

5.3   NUMBER SYNTHESIS OF PLANE LINKAGES

Assume that a plane linkage is to be synthesized with lower pairs only. What linkage arrangement will give a linkage with one degree of freedom?

We may rewrite the Grubler criterion with $F = 1$ in the form

$$l = \tfrac{3}{2}n - 2$$

Since n must be an integer, the possible combinations of links and pairs are

>2 links with 1 lower pair (trivial)
>4 links with 4 lower pairs
>6 links with 7 lower pairs
>8 links with 10 lower pairs

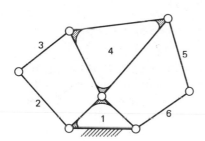

I. Watt Linkage

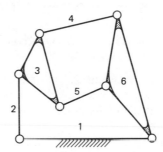

II. Stephenson Linkage

(a)

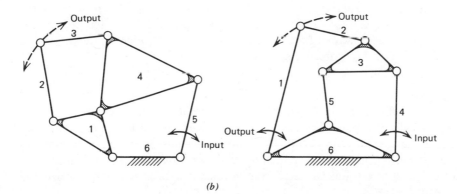

(b)

FIGURE 5.6   BASIC AND INVERTED SIX-BAR LINKAGES. (a) TWO BASIC SIX-BAR LINKAGES. (b) INVERSIONS OF BASIC SIX-BAR LINKAGES.

The four-link, four-joint plane mechanism has already been shown to have one degree of freedom.

Consider possible six-link mechanisms. It can be shown that a ternary link with three lower pairs is the most complex compound link possible in a six-link chain. This gives two possibilities for the assembly of six-link chains: type 1 (Watt's linkage), in which the two three-joint compound links are connected at a common point, and type II (Stephenson's linkage), in which they are separated by binary two-joint links, as shown in Figure 5.6.

To form a constrained mechanism, we must fix one of the links of the chain to ground. We note that it is possible to create several different mechanisms from each chain, depending on the choice of fixed number.

5.4 RANGE OF MOTION ANALYSIS

Motion Classification for the Plane Four-Bar Linkage—The Grashof Criterion

One of the earliest studies of mobility was the work of Grashof [14] in the nineteenth century. He investigated the conditions that must be satisfied in order that the input member in a four-bar linkage is capable of continuous rotation through a full 360°.

A member connected to the fixed link is known as a crank if it is possible for that member to rotate continuously. If the member is restricted to oscillation through a partial revolution, it is called a rocker. The members of a four-bar linkage may be proportioned so as to result in crank-rocker, double-crank, or double-rocker mechanisms.

To form a mechanism where one of the frame-connected links is a crank, the following relations due to Grashof must be satisfied. Let l and s represent the longest and shortest link in a four-bar chain where p and q identify the remaining two links. The Grashof criterion becomes

$$l + s < p + q \tag{5.6}$$

Under this condition, if the shortest link is the input member, two crank-rocker mechanisms are possible, one with link p as the fixed link and a second with link l as the fixed link as shown in Figure 5.7. If the shortest link is made the fixed link of the mechanism, a double-crank (the drag-link) mechanism results. When the link opposite the shortest link is fixed, the mechanism becomes a double-rocker mechanism.

In all cases when the inequality (5.6) is not satisfied, only double-rocker mechanisms can be formed.

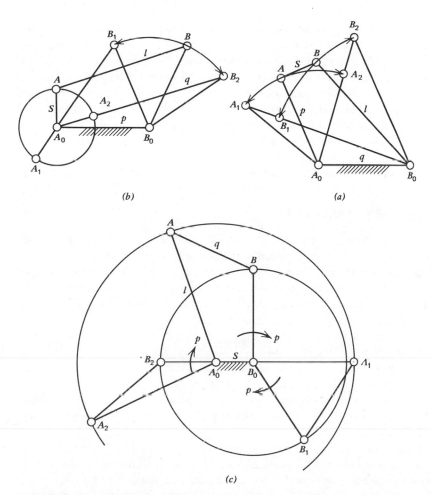

(b)   (a)

(c)

FIGURE 5.7   INVERSIONS OF THE FOUR-BAR LINKAGE THAT SATISFY
THE GRASHOF CRITERION $l + s < p + q$. (*a*) CRANK-ROCKER. (*b*) DOUBLE
ROCKER. (*c*) DRAG-LINK MECHANISM.

Displacement Analysis of Spatial Mechanisms

One obvious method for an investigation of the mobility of a specific spatial
mechanism is to carry out a kinematic analysis of the relative displacements in the
mechanism for a series of closely separated positions of the input member. A point
corresponding to a mobility limit is indicated when an imaginary solution of the
loop equations is encountered.

The displacement analysis can be accomplished in a variety of ways in addition to the methods of Chapter 4. Chace [16] has published a comprehensive account of the solution of the loop equations in vector form. Hartenberg and Denavit [17] have introduced a special coordinate transformation matrix that also can be used to define the conditions of closure of the kinematic chain. Suh [18] has discussed the description of rigid body motion using a 4×4 displacement matrix. When combined with the constraint equations, the resulting set of nonlinear equations are solved by iterative methods, as shown in Chapter 6. All of these methods involve mathematical complexity and require an enormous amount of calculation. Many investigators have sought a more direct answer to the mobility questions.

5.5 MOBILITY ANALYSIS

One of the basic questions in mobility analysis is concerned with the permissible variation in coupler link length, which will allow a movable mechanical connection between the input and output link in the mechanism. For a series of input crank positions we are interested in the minimum and maximum coupler length that will permit assembly. A coupler length between these extremes will obviously permit assembly in at least one position.

The question of maximum and minimum allowable coupler length has been investigated in recent years by Skreiner [19] and Ogawa [20] using analytical methods that find the minimum or maximum of a coupler length expressed as a function of the input and output angular position.

Hunt [21] and Jenkins, Crossley, and Hunt [22] have studied these conditions geometrically in terms of the intersections of surfaces generated by point or lines moving with linkage members.

In this chapter an alternate geometric method [23] is given for the development of mobility charts for several common plane and spatial mechanisms. The method will often give explicit solutions for the minimum and maximum coupler length that will permit the assembly of the mechanism. The classification of the motion of the input and output members as crank or rocker in the spirit of Grashof inequality is also possible in many cases.

Mobility of the Plane Four-Bar Linkage

As an example of the basic methods, we will first develop a mobility chart for the plane four-bar linkage from which we may easily derive the Grashof inequality criteria.

Consider the plane four-bar linkage $A_0 \, A \, B \, B_0$ as shown at the top of Figure 5.8. We first draw circles representing the loci of moving pivots A and B. For an arbitrary position of point A at an angle θ there will always be two unique points

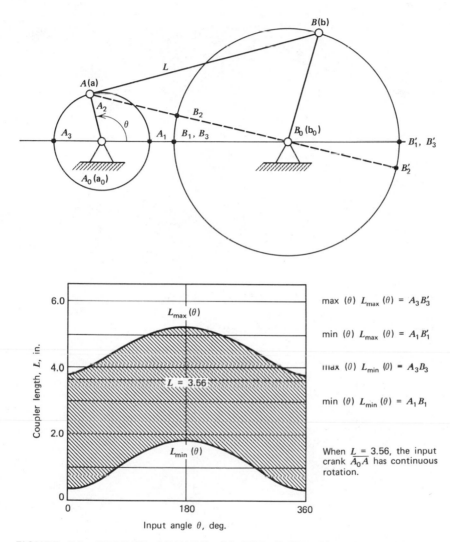

FIGURE 5.8 MOBILITY ANALYSIS OF THE PLANE FOUR-BAR LINKAGE FOR INPUT CRANK MOTION.

on the output crank circle that correspond to the length of the shortest and longest coupler length $L = AB$ that would permit assembly of the linkage. We see that these two points are at opposite ends of a diameter as defined by the line AB_0. For each value of θ these lengths L_{min} and L_{max} may be plotted to form a mobility chart, as shown in Figure 5.8.

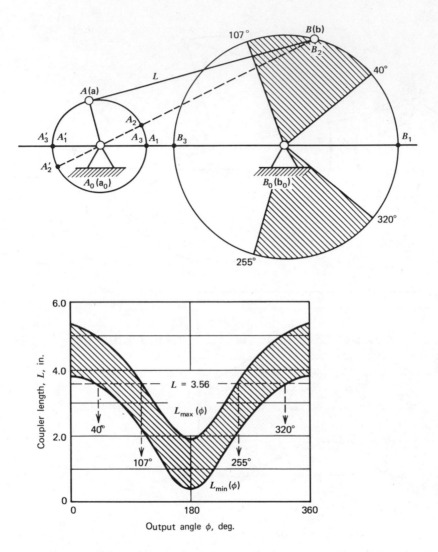

FIGURE 5.9   MOBILITY ANALYSIS OF THE PLANE FOUR-BAR LINKAGE
FOR OUTPUT CRANK MOTION.

Examination of the mobility chart reveals that it would be possible to assemble a linkage in at least one position for all values of L between $\min_\theta [L_{\min}(\theta)]$ and $\max_\theta [L_{\max}(\theta)]$. At the extreme limits, the mechanism could be assembled but would be immobile in that position. Therefore, to permit assembly, the coupler length must lie in the closed interval

$$L = \left[\min_{\theta} L_{\min}(\theta), \ \max_{\theta} L_{\max}(\theta) \right]$$
(5.7)

The shaded area of the mobility chart represents the limits of coupler length for which the mechanism would be movable for a particular input angle θ. For the example given, the actual coupler length $L = 3.56$ in. is shown as the horizontal dashed line on the chart. Since this line is well within the mobility region for a complete revolution of the input member, we conclude that the input member can be described as a *crank*. This condition can be described in terms of two inequalities:

$$L \ge \max_{\theta} (L_{\min}) \ge \overline{A_3 B_3}$$

$$L \le \min_{\theta} (L_{\max}) \le \overline{A_1 A_1}$$
(5.8)

or, in terms of link lengths $r_1 = \overline{A_0 B_0}$, $r_2 = \overline{A_0 A}$, $r_3 = \overline{AB}$, $r_d = \overline{B_0 B}$,

$$r_3 \ge r_1 + r_2 - r_4$$

$$r_3 \le r_1 - r_2 + r_4$$
(5.9)

We see that Eqs. 5.9 are equivalent to the Grashof condition if r_2 is the shortest link in the chain. A similar analysis for output link motion indicates for $L = 3.56$ in. that the mechanism can be assembled for output angles over the ranges 40 to 107° and 255 to 320° (Figure 5.9). The mechanism is therefore classified as a crank-rocker mechanism.

Mobility of the RSSR Mechanism

An RSSR mechanism will be described by a coordinate system as shown in Figure 5.10 for the first position of the mechanism. The spherical joint $B(\mathbf{b})$ rotates in a plane $\prod_B$ normal to the fixed axis $\mathbf{u}_b$ and passing through a fixed point $B_0(\mathbf{b}_0)$. The equation of the plane may be written in vector notation as

$$\mathbf{u}_b \cdot (\mathbf{b} - \mathbf{b}_0) = 0$$

where

$$\mathbf{u}_b = (\mathbf{u}_{bx}, \mathbf{u}_{by}, \mathbf{u}_{bz})$$
$$\mathbf{b} = (b_x, b_y, b_z)$$
$$\mathbf{b}_0 = (b_{0x}, b_{0y}, b_{0z})$$
(5.10)

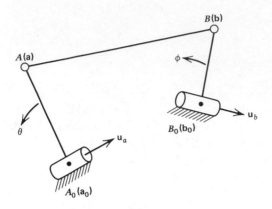

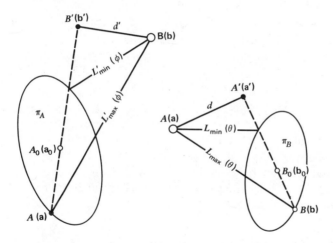

FIGURE 5.10 THE RSSR MECHANISM GEOMETRY.

The constant length of the output crank is given by

$$(\overline{BB_0})^2 = (\mathbf{b} - \mathbf{b}_0) \cdot (\mathbf{b} - \mathbf{b}_0) \tag{5.11}$$

For the position shown we note that the perpendicular to the plane $\prod_B$ is parallel to $\mathbf{u}_b$. The normal distance d is given by

$$d = \mathbf{u}_b \cdot (\mathbf{b}_0 - \mathbf{a}) \tag{5.12}$$

The distance $(\overline{B_0 A'})$ is calculated from

$$(\overline{B_0 A'})^2 = (\mathbf{a'} - \mathbf{b}_0) \cdot (\mathbf{a'} - \mathbf{b}_0)$$

where

$$\mathbf{a}' = \mathbf{a} + d\mathbf{u}_b \tag{5.13}$$

The maximum and minimum coupler length for a given position of $A(\mathbf{a})$ as specified by θ becomes

$$L_{\max}(\theta) = [(\overline{B_0 A'} + \overline{B_0 B})^2 + d^2]^{1/2}$$

$$L_{\min}(\theta) = [(\overline{B_0 A'} - \overline{B_0 B})^2 + d^2]^{1/2} \tag{5.14}$$

The calculations are repeated for a full revolution of the input crank and the results plotted as the first mobility chart of Figure 5.11.

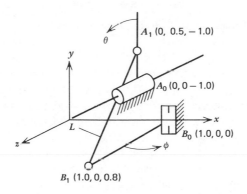

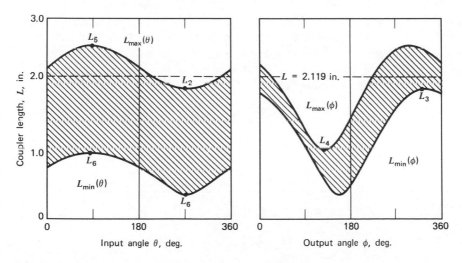

FIGURE 5.11   MOBILITY CHARTS—RSSR MECHANISM.

A similar analysis for the output crank leads to

$$L_{min}(\theta) = [(\overline{A_0 B'} - \overline{A_0 A})^2 + d'^2]^{1/2}$$
$$L_{max}(\theta) = [(\overline{A_0 B'} + \overline{A_0 A})^2 + d'^2]^{1/2} \tag{5.15}$$

where

$$\mathbf{u}_a \cdot (\mathbf{a} - \mathbf{a}_0) = 0$$
$$d' = \mathbf{u}_c \cdot (\mathbf{a}_0 - \mathbf{b})$$
$$\mathbf{b}' = \mathbf{b} + d\mathbf{u}_a$$
$$(\overline{A_0 B'})^2 = (\mathbf{b}' - \mathbf{a}_0) \cdot (\mathbf{b}' - \mathbf{a}_0)$$
$$(\overline{A_0 A})^2 = (\mathbf{a} - \mathbf{a}_0) \cdot (\mathbf{a} - \mathbf{a}_0)$$

and $B(\mathbf{b})$ is determined from a rotation displacement matrix equation in terms of $\mathbf{u}_b$, ϕ, and a given first position $B(\mathbf{b}_1)$. The results of these calculations provide information for the second chart of Figure 5.11.

The mobility charts of Figure 5.11 permit an investigation of the mobility of an RSSR mechanism with specified input and output crank length in terms of an assumed coupler length. A horizontal line extending across both charts represents an assumed coupler length. In the regions where the horizontal line remains within the shaded area between the L_{min} and L_{max} curves, the mechanism is movable. An intersection of the horizontal line with either an L_{min} or L_{max} curve indicates either a limiting position or a dead point. At points where the horizontal line passes above or below the shaded area, the linkage cannot be assembled. The input or output link is capable of 360° rotation if the horizontal line is always within the corresponding mobility region.

We may classify the possible motion for a particular mechanism in a manner similar to the Grashof criterion as follows.

Let the points of minimum and maximum value for the curves of L_{min} and L_{max} as functions of θ or ϕ be defined as

$$L_1 = \max_{\theta} [L_{min}(\theta)]$$

$$L_2 = \min_{\theta} [L_{max}(\theta)]$$

$$L_3 = \max_{\phi} [L_{min}(\phi)]$$

$$L_4 = \min_{\phi} [L_{max}(\phi)]$$

$$L_5 = \max_{\theta} [L_{max}(\theta)] = \max [L'_{max}(\theta)]$$

$$L_6 = \min_{\theta} [L_{min}(\theta)] = \min [L'_{min}(\theta)] \tag{5.16}$$

For a given coupler length L the motion of the input link can be classified as

(a) Crank—iff $L_1 \leq L \leq L_2$

(b) Rocker—if $L < L_1$ or $L > L_2$ $\qquad\qquad$ (5.17)

Similarly, the output link motion is

(c) Crank—iff $L_3 \leq L \leq L_4$

(d) Rocker—if $L < L_3$ or $L > L_4$ $\qquad\qquad$ (5.18)

The RSSR mechanism cannot be assembled if the coupler does not belong to the closed interval $[L_5, L_6]$. For a coupler length $L = L_5$ or L_6 the mechanism can be assembled in only one configuration from which it is completely immobile.

The range of motion for either crank is clearly indicated on the mobility chart. A useful spatial drag-link mechanism should avoid any of the limit positions indicated by intersections of the line L with either of the $L_{\min}$ or $L_{\max}$ curves.

Mobility of the RRSS Mechanism

The mobility analysis of the RRSS mechanism as shown in Figure 5.12 is based on the intersection of the spherical surface representing possible positions of the moving joint of the S-S link and the plane of the loci of positions of the R-S link for a given position of the R-R link.

In the synthesis of an RRSS mechanism it is often advantageous to assume the R-R link geometry such that the points $A(\mathbf{a})$ and $A_0(\mathbf{a}_0)$ define the shortest distance between the revolute axes $\mathbf{u}_a$ and $\mathbf{u}_0$. An actual mechanism may not meet these restrictions. We must therefore begin by assuming points $A'(\mathbf{a}')$ and $A_0'(\mathbf{a}_0')$, which lie on axes $\mathbf{u}_a$ and $\mathbf{u}_0$ but not necessarily on the shortest perpendicular between the axes.

Although not a prerequisite for mobility analysis, we may calculate points $A(\mathbf{a})$ and $A_0(\mathbf{a}_0)$ on the common perpendicular from the orthogonality conditions

$$\mathbf{u}_a \cdot (\mathbf{a}_0 - \mathbf{a}) = 0$$

$$\mathbf{u}_0 \cdot (\mathbf{a}_0 - \mathbf{a}) = 0 \qquad\qquad (5.19)$$

where

$$\mathbf{a} = \mathbf{a}' + d\mathbf{u}_a$$

$$\mathbf{a}_0 = \mathbf{a}_0' + e\mathbf{u}_0$$

This leads to a unique solution for the scalar distances d and e that locate points on the common normal.

$$d = \frac{mq - p}{m^2 - 1}$$

$$e = \frac{q - pm}{m^2 - 1} \qquad\qquad (5.20)$$

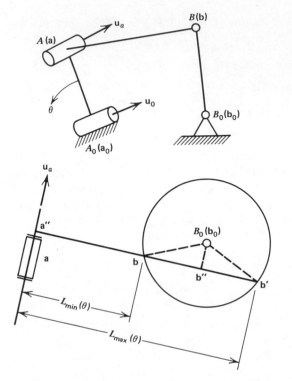

FIGURE 5.12 THE RRSS MECHANISM GEOMETRY.

where

$$p = \mathbf{u}_a \cdot (\mathbf{a}_0' - \mathbf{a}')$$

$$q = \mathbf{u}_0 \cdot (\mathbf{a}_0' - \mathbf{a}')$$

$$m = \mathbf{u}_a \cdot \mathbf{u}_0$$

With reference to Figure 5.12, the moving spheric pivot of the S-S link is restricted to the spherical surface defined by

$$(\mathbf{b} - \mathbf{b}_0) \cdot (\mathbf{b} - \mathbf{b}_0) = (\mathbf{b}_1 - \mathbf{b}_0) \cdot (\mathbf{b}_1 - \mathbf{b}_0) \tag{5.21}$$

For a position of the input crank specified by the angle θ the moving spheric pair on the R-S link must lie in a plane normal to the axis of the moving revolute axes $\mathbf{u}_a$ and passing through point $B(\mathbf{b})$. The equation of the plane is

$$\mathbf{u}_a \cdot (\mathbf{x} - \mathbf{b}) = 0 \tag{5.22}$$

where $\mathbf{x}$ (x, y, z) is any point in the plane.

We note that a point $A(\mathbf{a''})$ that also satisfies Eq. 5.22 is given by

$$\mathbf{a''} = (\mathbf{u}_a \cdot (\mathbf{b} - \mathbf{a}))\mathbf{u}_a + \mathbf{a}$$

The intersection of the plane, Eq. 5.22, and the sphere, Eq. 5.21, will be a circle. The center of the circle must lie on a line passing through the point $B_0(\mathbf{b}_0)$ and normal to the plane. The equation of the line is

$$\mathbf{x} = \mathbf{b}_0 + r\mathbf{u}_a \qquad (5.23)$$

The center of the circle, $\mathbf{b''}$ is the intersection of the line and the plane. Solving Eqs. 5.22 and 5.23 simultaneously, we obtain an expression for the normal distance r as

$$r = \mathbf{u}_a \cdot (\mathbf{b} - \mathbf{b}_0) = \mathbf{u}_a \cdot (\mathbf{a''} - \mathbf{b}_0) \qquad (5.24)$$

Therefore, the center of the circle becomes

$$\mathbf{b''} = \mathbf{b}_0 + (\mathbf{u}_a \cdot (\mathbf{a''} - \mathbf{b}_0))\mathbf{u}_a$$

The radius of the circle $R^2 = (\mathbf{b} - \mathbf{b''}) \cdot (\mathbf{b} - \mathbf{b''})$ is given by

$$R = [(\mathbf{b}_1 - \mathbf{b}_0) \cdot (\mathbf{b}_1 - \mathbf{b}_0) - (\mathbf{b''} - \mathbf{b}_0)(\mathbf{b''} - \mathbf{b}_0)]^{1/2} \qquad (5.25)$$

Thus, for a particular position of the driving link, the minimum and maximum coupler lengths are

$$L_{\min}(\theta) = |[(\mathbf{a''} - \mathbf{b''}) \cdot (\mathbf{a''} - \mathbf{b''})]^{1/2} - R|$$
$$L_{\max}(\theta) = |[(\mathbf{a''} - \mathbf{b''}) \cdot (\mathbf{a''} - \mathbf{b''})]^{1/2} + R| \qquad (5.26)$$

Again, these values, calculated for a series of input angles θ, can be combined and plotted as a mobility chart, as shown in Figure 5.13. Since the output S-S link, when disconnected from the R-S link, does not have a single degree of freedom, the mobility chart can be plotted only in terms of the input R-R link position.

Mobility of the RSRC Mechanism

The mobility analysis of the RSRC mechanism does not lead to an explicit solution for $L_{\min}$ and $L_{\max}$. In this case, referring to Figure 5.14, we again assume an input R-S link with pivots located in the first position, as shown. The point $A_0(\mathbf{a}_0)$ is assumed to lie in a plane perpendicular to axis $\mathbf{u}_0$ and passing through point $A(\mathbf{a})$. Points $B(\mathbf{b})$ and $D(\mathbf{d})$ are again assumed to lie on the shortest perpendicular between axes $\mathbf{u}_b$ and $\mathbf{u}_d$. Where these conditions are not met, points A_0, B, and D can be adjusted to positions along axes $\mathbf{u}_0$, $\mathbf{u}_b$, or $\mathbf{u}_d$, where the conditions are satisfied. Point $A(\mathbf{a})$ will not, generally, lie in a plane perpendicular to $\mathbf{u}_b$ and passing through point $B(\mathbf{b})$.

The maximum and minimum value of the allowable coupler length in this case will correspond to those positions of the R-C link where the point D is at the

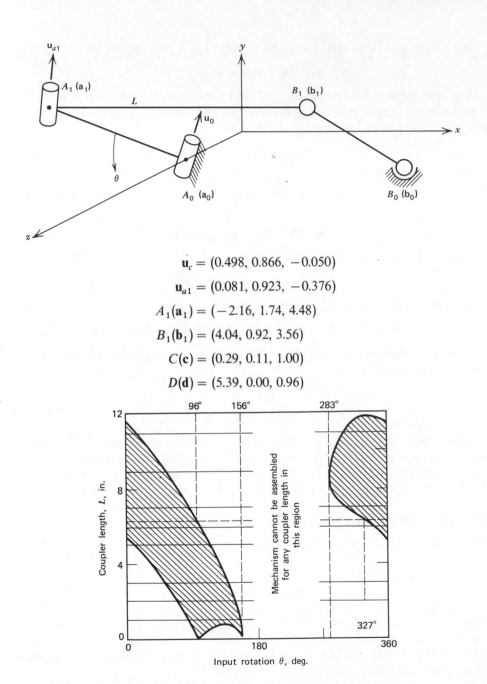

$$\mathbf{\dot{u}}_c = (0.498, 0.866, -0.050)$$

$$\mathbf{u}_{a1} = (0.081, 0.923, -0.376)$$

$$A_1(\mathbf{a}_1) = (-2.16, 1.74, 4.48)$$

$$B_1(\mathbf{b}_1) = (4.04, 0.92, 3.56)$$

$$C(\mathbf{c}) = (0.29, 0.11, 1.00)$$

$$D(\mathbf{d}) = (5.39, 0.00, 0.96)$$

FIGURE 5.13 MOBILITY CHART—RRSS MECHANISM.

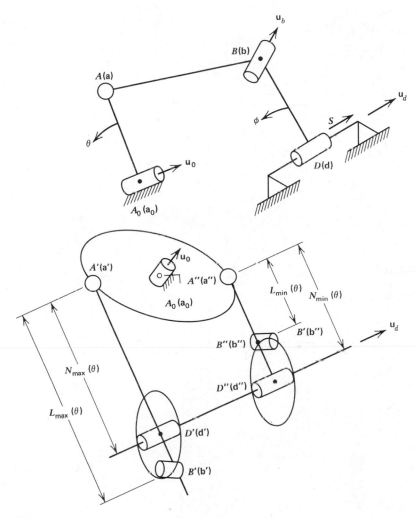

FIGURE 5.14   THE RSRC MECHANISM GEOMETRY.

shortest and longest distance from the input crank moving pivot circle. We designate the point $D'(\mathbf{d}')$ as corresponding to the maximum distance and $D''(\mathbf{d}'')$ to the
minimum. Note that in these positions the centerline of the R-C link will lie in
the plane of the R-S link as defined by $\mathbf{u}_b$, $\mathbf{b}$, and $\mathbf{a}$.

   The radius of the cylinder on which the joint $B(\mathbf{b})$ moves is

$$r_1 = \overline{DB} = [(\mathbf{b}_1 - \mathbf{d}_1) \cdot (\mathbf{b}_1 - \mathbf{d}_1)]^{1/2} \qquad (5.27)$$

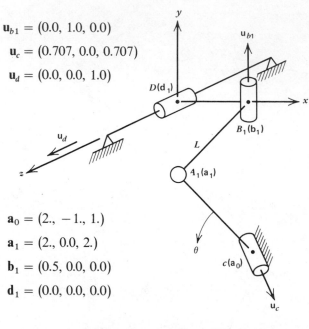

$$\mathbf{u}_{b1} = (0.0, 1.0, 0.0)$$
$$\mathbf{u}_c = (0.707, 0.0, 0.707)$$
$$\mathbf{u}_d = (0.0, 0.0, 1.0)$$

$$\mathbf{a}_0 = (2., -1., 1.)$$
$$\mathbf{a}_1 = (2., 0.0, 2.)$$
$$\mathbf{b}_1 = (0.5, 0.0, 0.0)$$
$$\mathbf{d}_1 = (0.0, 0.0, 0.0)$$

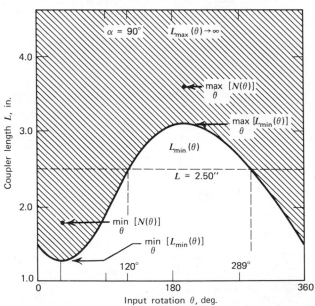

FIGURE 5.15 MOBILITY CHART RSRC MECHANISM, $\alpha = 90°$.

The circular path of the moving spherical joint on the input R-S link is described by

$$\mathbf{u}_0 \cdot (\mathbf{a} - \mathbf{a}_0) = 0 \qquad (5.28)$$

The normal distance between the circle and the line $\mathbf{u}_d$ through $D(\mathbf{d}_1)$ must satisfy the relationship

$$\mathbf{u}_d \cdot (\mathbf{a} - \mathbf{d}) = 0$$

For a particular point $A(\mathbf{a})$, corresponding to an input angle θ, the normal distance from a point on the input crank circle to the fixed line $\mathbf{u}_d$ is given by

$$N(\theta) = \{(\mathbf{a} - \mathbf{d}_1) \cdot (\mathbf{a} - \mathbf{d}_1) - [\mathbf{u}_d \cdot (\mathbf{a} - \mathbf{d}_1)]^2\}^{1/2} \qquad (5.29)$$

The minimum or maximum value of $N(\theta)$ may be found by any of several well-known methods of one-dimensional search. The Fibonacci method, available as a library subprogram HO CAL FIBO at the University of California, Berkeley, Computer Center has been used in this example.

The minimum coupler length for a given input angle θ is independent of the twist angle in the output link. The twist angle is the angle α between axes $\mathbf{u}_d$ and $\mathbf{u}_b$.

$$L_{\min}(\theta) = N(\theta) - r_1 \qquad (5.30)$$

The maximum coupler length depends on the twist angle and is given by

$$L_{\max}(\theta) = \frac{N(\theta) + r_1}{\cos \alpha} \qquad (5.31)$$

The input link is mobile for a given θ.

$$\text{iff } L_{\min}(\theta) < L < L_{\max}(\theta) \qquad (5.32)$$

Therefore, the input link is a crank.

$$\text{iff } \max_{\theta} \, [L_{\min}(\theta)] \le L \le \min_{\theta} \, [L_{\max}(\theta)] \qquad (5.33)$$

In the example shown in Figure 5.15, $\alpha = 90°$, and the upper limit curve for $L_{\max}(\theta)$ is at infinity. In general, the input link will be a rocker if a horizontal line representing a given coupler length L intersects either of the $L_{\min}(\theta)$ or the $L_{\max}(\theta)$ curves.

Chapter 6
Rigid Body Guidance

6.1 INTRODUCTION

There are many situations in the design of mechanical devices in which it is necessary either to guide a rigid body through a series of specified, finitely separated positions or to impose constraints on the velocity and/or acceleration of the moving body at a reduced number of finitely separated positions.

The specified positions are known as precision points, and the synthesized mechanism will be expected to guide the rigid body such that the position, velocity, or acceleration error is zero at the precision points. The synthesis procedures of this chapter will impose no constraints on the motion between precision points, and a small error is to be expected between precision points. The magnitude of this error can be calculated by analysis of the motions generated by the mechanism and comparison with desired motions at intermediate points. In some cases, it will be possible to reduce the maximum error over the range of the motions by respacing of precision points to place greater constraints on the design in regions where the highest error was noted in the original design. This can result in an optimum synthesis with equal values of the maximum error between successive pairs of precision points.

The fundamental problem in the kinematic synthesis of mechanical linkages is to locate a point, or set of points, fixed in the moving rigid body that will pass through a series of points in space that satisfy geometrical constraints imposed by a specific type of mechanical guiding link. For example, in the synthesis of plane four-bar linkages, the problem becomes one of locating a set of points in the moving plane which, as the plane assumes specified positions, will assume a series of positions that lie on a circular arc. The circular arc constraint is a result of assuming that the guidance is to be provided by two rigid links, each with two pivots. One pivot of each link is to be attached to the rigid body and the second to a fixed reference member. Particular points in the moving plane that satisfy this constraint are designated *circle points*. The center of the circle becomes a *center point*. Any two such links will then form a possible four-bar linkage when assembled with the fixed and moving bodies. The mobility or practicality of a particular linkage must be investigated in order to insure continuous motion, proper transmission angles, and the like.

In spatial mechanisms there are many more possibilities for the geometric form of constraining links or link-pair combinations. The basic synthesis problem, however, remains the same, and the following procedure is required.

1. Specify the position of the moving body in a number of positions. The maximum number of possible positions is determined by the type of guiding link being considered.
2. Convert the motion specifications into displacement matrix elements, usually with position 1 as the reference position.
3. Establish the geometric constraint equations at each design position as imposed by the guiding link to be synthesized.
4. Operate on the unknown coordinates in each design position using the displacement matrix equations to eliminate all parameters except those that define the synthesized linkage in the design position. The constraint equations then become design equations.
5. Solve the resulting set of nonlinear design equations using the Newton-Raphson iteration method.
6. The initial guesses required in the iteration process may be generated randomly on a high-speed digital computer in those cases where it is difficult to form estimates of the solutions systematically or intuitively. Once the computer has converged to a first solution, a family of possible solutions may be generated by incrementing one parameter and using the previous solution as initial guesses for the next solution.

6.2   PLANE RIGID BODY GUIDANCE MECHANISMS

One of the basic plane mechanisms for guidance of a rigid body through a series of specified positions is the four-bar linkage. The guided member is assumed to be attached to and move with the coupler. Synthesis requires the specification of the two fixed pivots and the first position of the two moving pivots for the pair of guiding links, often called cranks.

The Constant-Length Equations—Two Joint Cranks

Each of the guiding cranks must satisfy a condition of constant length, as shown in Figure 6.1, where points $\mathbf{a}_j = (a_{jx}, a_{jy})$ and $\mathbf{a}_0 = (a_{0x}, a_{0y})$ are representative of a typical guiding link. This leads to *crank displacement constraint equations*

$$(\mathbf{a}_j - \mathbf{a}_0)^T(\mathbf{a}_j - \mathbf{a}_0) = (\mathbf{a}_1 - \mathbf{a}_0)^T(\mathbf{a}_1 - \mathbf{a}_0) \qquad j = 2, 3, \ldots n \qquad (6.1)$$

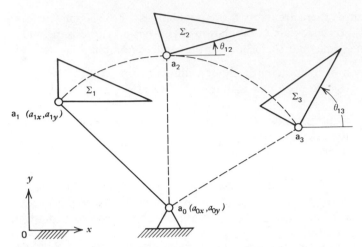

FIGURE 6.1   CRANK CONSTRAINT—THE CONSTANT LENGTH
CONDITION.

The *crank velocity constraint equation* is based on the notion that if the crank
length is constant, the rate of change of crank length is zero. Differentiation of
Eq. 6.1 leads to

$$(\dot{\mathbf{a}}_j)^T(\mathbf{a}_j - \mathbf{a}_0) = 0 \qquad j = 1, 2, 3, \ldots n \tag{6.2}$$

The *crank acceleration constraint equation* becomes

$$(\ddot{\mathbf{a}}_j)^T(\mathbf{a}_j - \mathbf{a}_0) + (\dot{\mathbf{a}}_j)^T(\dot{\mathbf{a}}_j) = 0 \qquad j = 1, 2, 3, \ldots n \tag{6.3}$$

The Constant Slope Equations—Plane Sliders

The second basic type of constraint is provided by a straight-line guide in the fixed
member along which a slider is constrained to move with one point pivoted on the
moving member, as shown in Figure 6.2. A curved path in the fixed member could
also be specified if the equation of the path is known.

The equation of the straight line path for a general point **b** may be written either
in the form of a determinant

$$\begin{vmatrix} b_{1x} & b_{1y} & 1 \\ b_{2x} & b_{2y} & 1 \\ b_{jx} & b_{jy} & 1 \end{vmatrix} = 0 \qquad j = 3, 4 \tag{6.4}$$

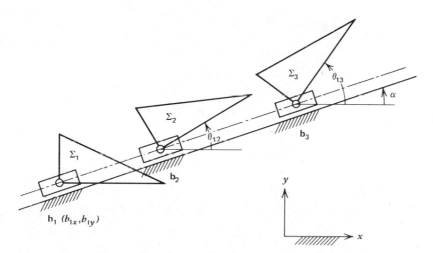

FIGURE 6.2   SLIDER CONSTRAINT—THE CONSTANT SLOPE CONDITION.

or, in the more familiar form,

$$\tan \alpha = \frac{b_{jy} - b_{1y}}{b_{jx} - b_{1x}} = \frac{b_{2y} - b_{1y}}{b_{2x} - b_{1x}} \qquad j = 3, 4 \tag{6.5}$$

The *slider velocity constraint equation* becomes

$$\tan \alpha = \frac{\dot{b}_{jy}}{\dot{b}_{jx}} \qquad j = 1, 2, \dots n \tag{6.6}$$

and the *slider acceleration constraint equation* is written as

$$\tan \alpha = \frac{\ddot{b}_{jy}}{\ddot{b}_{jx}} \qquad j = 1, 2, \dots n \tag{6.7}$$

6.3   THREE-POSITION CRANK SYNTHESIS

As a first example, let us assume that we are seeking a two-joint link (i.e., a *crank*) that is capable of guiding a plane rigid body from position 1 through a series of three positions. A point **a** on the guided body must pass through a sequence of positions, all of which are the same distance from a point $\mathbf{a}_0$ on the fixed reference member. For $n = 3$ we have two constant length constraint equations.

$$(\mathbf{a}_2 - \mathbf{a}_0)^T(\mathbf{a}_2 - \mathbf{a}_0) = (\mathbf{a}_1 - \mathbf{a}_0)^T(\mathbf{a}_1 - \mathbf{a}_0) \tag{6.8}$$

$$(\mathbf{a}_3 - \mathbf{a}_0)^T(\mathbf{a}_3 - \mathbf{a}_0) = (\mathbf{a}_1 - \mathbf{a}_0)^T(\mathbf{a}_1 - \mathbf{a}_0) \tag{6.9}$$

We may solve for the first position $\mathbf{a}_1$ by noting that

$$(\mathbf{a}_2) = [D_{12}](\mathbf{a}_1) \tag{6.10}$$

$$(\mathbf{a}_3) = [D_{13}](\mathbf{a}_1) \tag{6.11}$$

where $[D_{12}]$ and $[D_{13}]$ are known 3×3 displacement matrices that have been precalculated in terms of specified positions $\mathbf{p}_1$, $\mathbf{p}_2$, and $\mathbf{p}_3$ for one point $\mathbf{p}$ of the guided member plus the rotation angles θ_{12} and θ_{13}.

When three positions are specified, we obtain two *crank displacement design equations* by substitution of Eqs. 6.10 and 6.11 into Eqs. 6.8 and 6.9.

$$([D_{12}]\mathbf{a}_1 - \mathbf{a}_0)^T([D_{12}]\mathbf{a}_1 - \mathbf{a}_0) = (\mathbf{a}_1 - \mathbf{a}_0)^T(\mathbf{a}_1 - \mathbf{a}_0) \tag{6.12}$$

$$([D_{13}]\mathbf{a}_1 - \mathbf{a}_0)^T([D_{13}]\mathbf{a}_1 - \mathbf{a}_0) = (\mathbf{a}_1 - \mathbf{a}_0)^T(\mathbf{a}_1 - \mathbf{a}_0) \tag{6.13}$$

with four unknowns a_{1x}, a_{1y}, a_{0x}, and a_{0y}.

In this case we may specify any two of the four and obtain a set of two linear equations with two unknowns. The matrix $[D_{1j}]$ is assumed to have elements d_{ikj}. Eqs. 6.12 and 6.13 can be expanded in the form

$$(d_{11j}a_{1x} + d_{12j}a_{1y} + d_{13j} - a_{0x})(d_{11j}a_{1x} + d_{12j}a_{1y} + d_{13j} - a_{0x})$$

$$+ (d_{21j}a_{1x} + d_{22j}a_{1y} + d_{23j} - a_{0y})(d_{21j}a_{1x} + d_{22j}a_{1y} + d_{23j} - a_{0y})$$

$$= (a_{1x} - a_{0x})(a_{1x} - a_{0x}) + (a_{1y} - a_{0y})(a_{1y} - a_{0y}) \quad j = 2, 3 \tag{6.14}$$

Since, in this case, we have two equations with four unknowns, we are free to assume a fixed pivot location $\mathbf{a}_0 = (a_{0x}, a_{0y})$ and solve for the unknown first position of the moving pivot $\mathbf{a}_1 = (a_{1x}, a_{1y})$. This leads to two equations of the form

$$a_{1x}(d_{11j}d_{13j} + d_{21j}d_{23j} + (1. - d_{11j})a_{0x} - d_{21j}a_{0y})$$

$$+ a_{1y}(d_{12j}d_{13j} + d_{22j}d_{23j} + (1. - d_{22j})a_{0y} - d_{12j}a_{0x})$$

$$= d_{13j}a_{0x} + d_{23j}a_{0y} - \tfrac{1}{2}(d_{13j}^2 + d_{23j}^2)$$

which can be arranged as a set of two linear equations in the form

$$a_{1x}A_j + a_{1y}B_j = C_j \quad j = 2, 3 \tag{6.15}$$

where

$$A_j = d_{11j}d_{13j} + d_{21j}d_{23j} + (1 - d_{11j})a_{0x} - d_{21j}a_{0y}$$

$$B_j = d_{12j}d_{13j} + d_{22j}d_{23j} + (1 - d_{22j})a_{0y} - d_{12j}a_{0x}$$

$$C_j = d_{13j}a_{0x} + d_{23j}a_{0y} - \tfrac{1}{2}(d_{13j}^2 + d_{23j}^2)$$

Thus, we see that the three-position synthesis of a crank for plane rigid body guidance reduces to the solution of two linear equations in two unknowns. The solution is easily accomplished by slide rule using Cramer's rule or, to greater accuracy, by digital computer. This solution would correspond to those obtained using the pole triangle of classical graphical Burmester theory [1, 2].

The *crank velocity constraint design equations* are derived from Eq. 6.2 with

$$(\dot{\mathbf{a}}_j) = [V_j](\mathbf{a}_j) = [\dot{d}_{ikj}](\mathbf{a}_j) \tag{6.16}$$

This leads to design equations of the form

$$a_{jx} D'_j + a_{jy} E'_j = F'_j \tag{6.17}$$

where

$$D'_j = \dot{d}_{13j} - \dot{d}_{21j} a_{0y}$$
$$E'_j = \dot{d}_{23j} - \dot{d}_{12j} a_{0x}$$
$$F'_j = \dot{d}_{13j} a_{0x} + \dot{d}_{23j} a_{0y}$$

and a_{jx} and a_{jy} can be expressed in terms of a_{1x} and a_{1y} in the form $(\mathbf{a}_j) = [D_{ij}](\mathbf{a}_1)$; that is;

$$a_{jx} = d_{11j} a_{1x} + d_{12j} a_{1y} + d_{13j}$$
$$a_{jy} = d_{21j} a_{1x} + d_{22j} a_{1y} + d_{23j} \tag{6.18}$$

Substituting Eqs. 6.18 into 6.17, we obtain a design equation in terms of a_{1x} and a_{1y} as

$$a_{1x} D_j + a_{1y} E_j = F_j \tag{6.19}$$

where

$$D_j = d_{11j} D'_j + d_{21j} E'_j$$
$$E_j = d_{12j} D'_j + d_{22j} E'_j$$
$$F_j = F'_j - d_{13j} D'_j - d_{23j} E'_j$$

The *crank acceleration constraint design equations* are based on Eq. 6.3 with

$$(\ddot{\mathbf{a}}_j) = [A_j](\mathbf{a}_j) = [\ddot{d}_{ikj}](\mathbf{a}_j) \tag{6.20}$$

This again leads to a linear design equation of the form

$$a_{jx} P'_j + a_{jy} Q'_j = R'_j \tag{6.21}$$

where, as derived in appendix A

$$P'_j = \ddot{d}_{13j} - \ddot{d}_{11j}a_{0x} - \ddot{d}_{21j}a_{0y} + 2\dot{d}_{21j}\dot{d}_{23j}$$

$$Q'_j = \ddot{d}_{23j} - \ddot{d}_{12j}a_{0x} - \ddot{d}_{22j}a_{0y} + 2\dot{d}_{12j}\dot{d}_{13j}$$

$$R'_j = \ddot{d}_{13j}a_{0x} + \ddot{d}_{23j}a_{0y} - (\dot{d}^2_{13j} + \dot{d}^2_{23j})$$

and, again, we can express $\mathbf{a}_j$ in terms of $\mathbf{a}_1$ from Eqs. 6.18, which gives the linear design equation in the optional form

$$a_{1x}P_j + a_{1y}Q_j = R_j \tag{6.22}$$

where

$$P_j = d_{11j}P'_j + d_{21j}Q'_j$$

$$Q_j = d_{12j}P'_j + d_{22j}Q'_j$$

$$R_j = R'_j - d_{13j}P'_j - d_{23j}Q'_j$$

6.4   THREE-POSITION SLIDER SYNTHESIS

Expanding Eq. 6.4, we may also write it in the form

$$b_{1x}(b_{2y} - b_{3y}) - b_{1y}(b_{2x} - b_{3x}) + (b_{2x}b_{3y} - b_{3x}b_{2y}) = 0 \tag{6.23}$$

Substituting the displacement matrix equations leads to the *circle of sliders equation*

$$Ab^2_{1x} + Ab^2_{1y} + Db_{1x} + Eb_{1y} + F = 0 \tag{6.24}$$

where

$$A = Cd_{212} - Bd_{213}$$

$$B = 1 - \cos\theta_{12} = 1 - d_{112}$$

$$C = 1 - \cos\theta_{13} = 1 - d_{113}$$

$$D = Cd_{232} - Bd_{233} + d_{132}d_{213} - d_{133}d_{212}$$

$$E = Bd_{133} - Cd_{132} + d_{232}d_{213} - d_{233}d_{212}$$

$$F = d_{132}d_{233} - d_{133}d_{232}$$

d_{ikj} are elements of the displacement matrix $[D_{ij}]$

Eq. 6.24 can be rewritten in the form

$$\left(b_{1x} + \frac{D}{2A}\right)^2 + \left(b_{1y} + \frac{E}{2A}\right)^2 = \frac{D^2 + E^2 - 4AF}{4A^2} \tag{6.25}$$

which is the equation of a circle with center at $C_0 = (C_{0x}, C_{0y})$ and radius R where

$$C_{0x} = -\frac{D}{2A}$$

$$C_{0y} = -\frac{E}{2A} \tag{6.26}$$

and

$$R = \sqrt{\frac{D^2 + E^2 - 4AF}{4A^2}}$$

This circle* is the finite displacement equivalent of the inflection circle to be discussed in Chapter 10. When either θ_{12} or θ_{13} in Eq. 6.24 is zero, the value of A equals zero. Under these conditions the circle of sliders degenerates to the equation of a straight line.

$$Db_{1x} + Eb_{1y} + F = 0 \tag{6.27}$$

The *circle of sliders for specified velocity in the first position* is based on a constraint equation of the form

$$\frac{\dot{b}_{1y}}{\dot{b}_{1x}} = \frac{b_{2y} - b_{1y}}{b_{2x} - b_{1x}} \tag{6.28}$$

This again leads to a circle equation in the form of Eq. 6.24, where

$$A = \dot{d}_{211}(d_{112} - 1)$$
$$D = \dot{d}_{211}d_{132} + \dot{d}_{231}(d_{112} - 1) - \dot{d}_{131}d_{212}$$
$$E = \dot{d}_{231}d_{122} - \dot{d}_{121}d_{232} - \dot{d}_{131}(d_{222} - 1)$$
$$F = \dot{d}_{231}d_{132} - \dot{d}_{131}d_{232}$$

If the velocity is specified in the second position, the expression for A, D, E, and F must account for the inversion of the velocity terms to position 1.

A third circle of sliders is obtained with both *velocity and acceleration specified* in a particular position. In this case the constraint equation becomes

$$\frac{\dot{b}_{1y}}{\dot{b}_{1x}} = \frac{\ddot{b}_{1y}}{\ddot{b}_{1x}} \tag{6.29}$$

---

* The existence of the circle of sliders, derived in terms of displacement matrix elements, was suggested by A. Volino in private correspondence with the authors.

and the constants in Eq. 6.24 become

$$A = \ddot{d}_{111} \dot{d}_{211}$$

$$D = \dot{d}_{211} \ddot{d}_{131} + \dot{d}_{231} \ddot{d}_{111} - \dot{d}_{131} \ddot{d}_{211}$$

$$E = \dot{d}_{231} \ddot{d}_{121} - \dot{d}_{121} \ddot{d}_{231} - \dot{d}_{131} \ddot{d}_{221}$$

$$F = \dot{d}_{231} \ddot{d}_{131} - \dot{d}_{131} \ddot{d}_{231}$$

In all cases, once the equation of the circle is known, we may specify either b_{1x} or b_{1y} to locate the first position of a possible slider. The second position is then calculated from the matrix equation

$$(\mathbf{b}_2) = [D_{12}](\mathbf{b}_1) \tag{6.30}$$

and the direction of the slider motion is found from

$$\tan \alpha = \frac{b_{2y} - b_{1y}}{b_{2x} - b_{1x}} \tag{6.31}$$

The direction of the slider motion can be specified if desired. We then write a set of two linear equations where the equations can be any two of the following.

$$(b_{jy} - b_{1y}) = \tan \alpha (b_{jx} - b_{1x}) \tag{6.32}$$

$$\dot{b}_{jy} = \tan \alpha \dot{b}_{jx} \tag{6.33}$$

$$\ddot{b}_{jy} = \tan \alpha \ddot{b}_{jx} \tag{6.34}$$

When combined with the appropriate displacement matrix equation, this leads to a set of two equations with unknowns b_{1x} and b_{1y}.

6.5   INPUT CRANK MOTION PARAMETERS

When velocity or acceleration of the moving plane is specified, the synthesis equations lead to an infinite variety of possible guiding cranks (or sliders). Each of these input members must have a unique motion to be compatible with the desired motion of the moving plane. This compatibility is assured by equating expressions for the motion of the moving crank pivot. The *input crank velocity* is found from

$$\dot{\mathbf{b}} = \begin{bmatrix} 0 & -\dot{\alpha} \\ \dot{\alpha} & 0 \end{bmatrix} (\mathbf{b} - \mathbf{b}_0) = \begin{bmatrix} 0 & -\dot{\theta} \\ \dot{\theta} & 0 \end{bmatrix} (\mathbf{b} - \mathbf{p}) + (\dot{\mathbf{p}}) \tag{6.35}$$

where $\mathbf{p}$ is the reference point on the moving plane. Equating expressions for $\dot{b}_x$,

$$-\dot{\alpha}(b_y - b_{0y}) = -\dot{\theta}(b_y - p_y) + \dot{p}_x$$

from which

$$\dot{\alpha} = -\frac{\dot{p}_x - \dot{\theta}(b_y - p_y)}{(b_y - b_{0y})} \tag{6.36}$$

The *input crank angular acceleration* is found in a similar manner from

$$\ddot{\mathbf{b}} = \begin{bmatrix} -\dot{\alpha}^2 & -\ddot{\alpha} \\ \ddot{\alpha} & -\dot{\alpha}^2 \end{bmatrix}(\mathbf{b} - \mathbf{b}_0) = \begin{bmatrix} -\dot{\theta}^2 & -\ddot{\theta} \\ \ddot{\theta} & -\dot{\theta}^2 \end{bmatrix}(\mathbf{b} - \mathbf{p}) + (\ddot{\mathbf{p}}) \tag{6.37}$$

Equating expressions for $\ddot{b}_x$,

$$-\dot{\alpha}^2(b_x - b_{0x}) - \ddot{\alpha}(b_y - b_{0y}) = -\dot{\theta}^2(b_x - p_x) - \ddot{\theta}(b_y - p_y) + \ddot{p}_x$$

from which

$$\ddot{\alpha} = -\frac{\ddot{p}_x - \dot{\theta}^2(b_x - p_x) - \ddot{\theta}(b_y - p_y) + \dot{\alpha}^2(b_x - b_{0x})}{(b_y - b_{0y})} \tag{6.38}$$

6.6 EXAMPLE PROBLEMS: THREE-POSITION SYNTHESIS

The following numerical examples are given as test problems that may be useful in checking computer programs. The six-digit results are from FORTRAN programs using a CDC 6400 computer.

Example 6-1 Crank Synthesis—Three finitely separated positions of a moving plane. The finitely separated positions of a plane are specified by

$$\mathbf{p}_1 = (1.0, 1.0)$$

$$\mathbf{p}_2 = (2.0, 0.5) \qquad \theta_{12} = 0.0°$$

$$\mathbf{p}_3 = (3.0, 1.5) \qquad \theta_{13} = 45.0$$

For a fixed pivot assumed at $\mathbf{b}_0 = (0.0, 0.0)$, Eqs. 6.15 give the first position of the corresponding moving pivot at $\mathbf{b}_1 = (0.994078, 3.238155)$. A second fixed pivot assumed at $\mathbf{c}_0 = (5.0, 0.0)$ results in a moving pivot at $\mathbf{c}_1 = (3.547722, -1.654555)$, as shown in Figure 6.3.

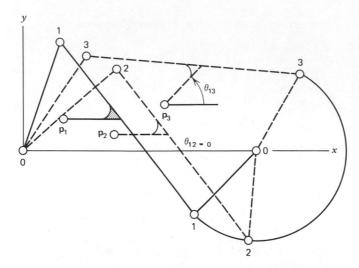

FIGURE 6.3 EXAMPLE 6.1. CRANK SYNTHESIS—THREE
FINITELY SEPARATED POSITIONS OF A PLANE.

Example 6-2 Slider Synthesis—Finite Displacements. Three positions
are specified:

$$\mathbf{p}_1 = (1.0, \ 1.0)$$

$$\mathbf{p}_2 = (2.0, \ 0.0) \qquad \theta_{12} = 30.0°$$

$$\mathbf{p}_3 = (3.0, \ 2.0) \qquad \theta_{13} = 60.0$$

Using Eqs. 6.25 and 6.26, the circle of sliders is found with center

$$\mathbf{c}_0 = (c_{0x}, \ c_{0y}) = (3.866025, \ 6.964101)$$

and radius $R = 4.625181$.

Example 6-3 Slider Synthesis with $\theta_{12} = 0.0$. The three positions of
Example 6-1 are specified with $\theta_{12} = 0.0$. The circle of sliders degenerates to the
straight line

$$b_{1y} = 0.86730b_{1x} + 2.45308$$

which intersects the y-axis at $(0.0, 2.45308)$. Selecting the y-axis intercept as the
first position for a slider and adding the second crank of Example 6-1, a possible
slider-crank is formed, as shown in Figure 6.4.

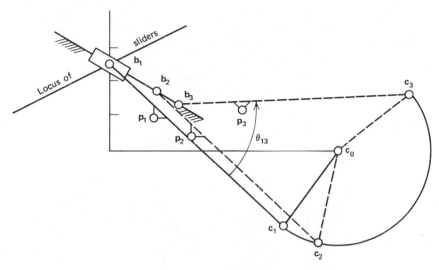

FIGURE 6.4   EXAMPLE 6-3 SLIDER SYNTHESIS WITH $\theta_{12} = 0.0$.

Example 6-4 Slider that moves in a specified direction. Using the data of Example 6-2, synthesize a double-slider mechanism. The first slider moves with $\alpha_1 = 30°$ and the second with $\alpha_2 = 102.3°$. Using Eq. 6.32, we find the first position of the slider pivot **b** at

$$\mathbf{b}_1 = (1.133975,\ 10.696152) \qquad \text{for} \qquad \alpha_1 = 30°$$

and

$$\mathbf{b}_1' = (3.931090,\ 2.339378) \qquad \text{for} \qquad \alpha_2 = 102.3°$$

which are shown in Figure 6.5.

Example 6-5 Crank synthesis—Velocity specified in first position plus displacement to a second position. The motion of the moving plane is specified by

$$\mathbf{P}_1 = (p_{1x},\ p_{1y}) = (1.0,\ 1.0)$$

$$\dot{\mathbf{P}}_1 = (\dot{p}_{1x},\ \dot{p}_{1y}) = (100.,\ 50.) \qquad \dot{\theta}_1 = 104.7 \text{ rad/sec}$$

$$\mathbf{P}_2 = (p_{2x},\ p_{2y}) = (3.0,\ 3.0) \qquad \theta_{12} = 45.0°$$

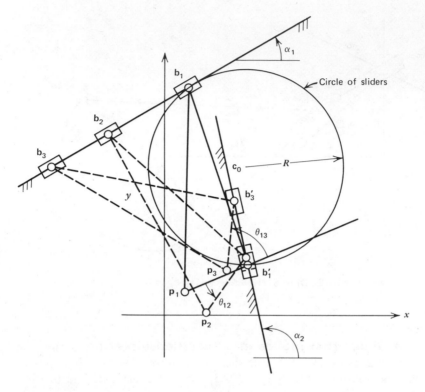

FIGURE 6.5 EXAMPLE 6.4. SLIDERS THAT MOVE IN SPECIFIED DIRECTIONS.

From Eqs. 6.15 and 6.17 we find the following combination of moving pivots $\mathbf{a}_1$ for a series of specified fixed pivots $\mathbf{a}_0$.

Fixed Pivot Coordinates, $\mathbf{a}_0$	Moving Pivot Coordinates, $\mathbf{a}_1$
$\mathbf{a}_0 = (5.0, 0.0)$	$\mathbf{a}_1 = (1.029, 1.733)$
$\mathbf{a}_0' = (5.0, -5.0)$	$\mathbf{a}_1' = (-0.250, 3.156)$
$\mathbf{a}_0'' = (1.0, -5.0)$	$\mathbf{a}_1'' = (0.187, 6.835)$

Three possible four-bar linkages have been identified.

Example 6-6 Crank synthesis—Velocity and acceleration specified in one position. The motion of the moving plane is specified by

$$\mathbf{p}_1 = (26.05, -18.44)$$

$$\dot{\mathbf{p}}_1 = (-6.74, 1.24) \qquad \dot{\theta}_1 = 0.05 \text{ rad/sec}$$

$$\ddot{\mathbf{p}}_1 = (0.0, 0.0) \qquad\quad \ddot{\theta}_1 = 0.30 \text{ rad/sec/sec}$$

In this case we see that point $\mathbf{p}_1$ is the acceleration pole. From Eqs. 6.17 and 6.21 we find that for a fixed pivot,

$$\mathbf{a}_0 = (0.0, 0.0)$$

The corresponding moving pivot is located at

$$\mathbf{a}_1 = (-0.068, 8.327)$$

Assuming a second fixed pivot at

$$\mathbf{a}_0' = (16.0, 0.0)$$

we calculate a moving pivot at

$$\mathbf{a}_1' = (18.024, 21.027)$$

6.7   FOUR-POSITION SYNTHESIS—CRANK CONSTRAINT

When four precision points are specified, the crank design equations are nonlinear and must be solved by iterative techniques.

The Newton-Raphson Method

As a first example, let us consider the problem of finding a root of a nonlinear function of one variable as shown in Figure 6.6. The function $f(x)$ can be approximated by a Taylor's series expansion, truncated after the first derivative term in the linear form

$$f(x + \delta x) = f(x) + \frac{d}{dx} f(x) \cdot \delta x \tag{6.39}$$

where δx is the change in x that will lead to a new value of x where $f(x + \delta x)$ approaches zero.

Since the left side approaches zero, we may calculate the correction δx from

$$\delta x = \frac{-f(x)}{\dfrac{df(x)}{dx}} \tag{6.40}$$

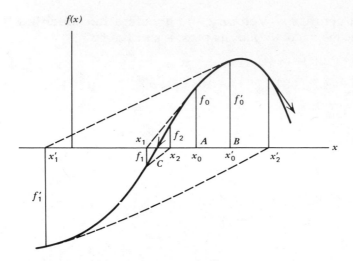

FIGURE 6.6 THE NEWTON-RAPHSON METHOD.

In Figure 6.6 we may assume x_0 at point A as a first approximation to a root. The next approximation is then found from

$$x_1 = x_0 + \delta x = x_0 - \frac{f(x_0)}{\dfrac{df(x_0)}{dx}}$$

Similarly,

$$x_2 = x_1 - \frac{f(x_1)}{\dfrac{df(x_1)}{dx}}$$

and so forth.

If the iterative process begins at point A in Figure 6.6, the solution will converge rapidly to the vicinity of a root at point C. On the other hand, if we begin at point B, the iterative solution is seen to be unstable and may either diverge or tend to converge to a second root to the right of point B, as shown in Figure 6.6. Therefore, we see that the basic Newton-Raphson method is sensitive to initial guesses, and convergence is not guaranteed.

In kinematic synthesis we must deal often with a system of n nonlinear equations in n unknowns. In this case we may generalize the Newton-Raphson method as follows.

Assume a set of n equations in the form

$$f_1(x_1, x_2, x_3, \ldots x_n) = 0$$
$$f_2(x_1, x_2, x_3, \ldots x_n) = 0$$
$$\vdots \qquad \vdots$$
$$f_n(x_1, x_2, x_3, \ldots x_n) = 0$$

Each of these equations can be expanded in Taylor's series form as

$$f_i(\mathbf{x} + \boldsymbol{\delta}) = f_i(\mathbf{x}) + \frac{\partial f_i}{\partial x_1}\delta_1 + \frac{\partial f_i}{\partial x_2}\delta_2 + \frac{\partial f_i}{\partial x_3} + \cdots \frac{\partial f_i}{\partial x_n}\delta_n$$

$$i = 1, 2, \ldots n \qquad (6.41)$$

where, at a solution,

$$f_i(\mathbf{x} + \boldsymbol{\delta}) \rightarrow 0$$

Eqs. 6.41 can be arranged as a set of linear equations in the matrix form $[\partial f / \partial x]\boldsymbol{\delta} = -\mathbf{f}$; that is,

$$\begin{bmatrix} \dfrac{\partial f_1}{\partial x_1} & \dfrac{\partial f_1}{\partial x_2} & \cdots & \dfrac{\partial f_1}{\partial x_n} \\ \vdots & \vdots & & \vdots \\ \dfrac{\partial f_n}{\partial x_1} & \dfrac{\partial f_n}{\partial x_2} & \cdots & \dfrac{\partial f_n}{\partial x_n} \end{bmatrix} \begin{bmatrix} \delta_1 \\ \vdots \\ \delta n \end{bmatrix} = \begin{bmatrix} -f_1 \\ \vdots \\ -f_n \end{bmatrix} \qquad (6.42)$$

The solution of Eq. 6.42 is accomplished numerically using the Gauss elimination method. Once the correction vector $\boldsymbol{\delta}$ is known, the next approximation to the unknowns vector $\mathbf{x}$ is found from the recursion relation

$$\mathbf{x}_{K+1} = \mathbf{x}_K + \boldsymbol{\delta}_K \qquad (6.43)$$

The process of forming partial derivatives, solving Eq. 6.42, and calculating new values for $\mathbf{x}$ is repeated until, for all values of x_j, if

$$|x_j| \geq 10^{-7} \qquad \left|\frac{\delta_j}{x_j}\right| \leq 10^{-7}$$

or, if,

$$|x_j| \leq 10^{-7} \qquad |\delta_j| \leq 10^{-7} \qquad j = 1, 2, \ldots n \qquad (6.44)$$

The problem of instability can be avoided in many cases as follows. As each new point $\mathbf{x}_{K+1}$ is found, we compare the norm of the function vector at the new point to the previous value. If the norm of the new function vector is greater than the old value in absolute value, we assume that the full step δ_K would not be productive and we take a partial step $\delta_K/5$. In the single variable problem of Figure 6.6 we see that an initial guess x_0' where the function equals f_0' would predict a root at point x_1'. The functional value f_1' is clearly greater in absolute magnitude than f_0'; therefore, we do not take the full step calculated from Eq. 6.40 but, instead, use $\delta x/5$. This leads us near to point A, from which rapid convergence is assured.

In solving sets of nonlinear equations we compare the norm at successive points where the norm is defined as

$$\text{norm}_K = \sum_{i=1}^{n} |f_i(\mathbf{x})|_K \tag{6.45}$$

This process of comparison of norm and dividing the correction vector δ by an arbitrary constant where indicated is termed *damping* and has proved to be very successful in assuring convergence even with very poor initial guesses. Subroutine SIMEQS in program DESIGN, as discussed in Chapter 12, includes this feature. All sets of nonlinear design equations derived from the kinematic constraint equations as given in the following sections have been solved using program DESIGN. Refer to Chapter 12 for detailed instructions for the use of this and other useful FORTRAN programs.

The partial derivatives $\partial f_i/\partial x_j$ are formed numerically from

$$\frac{\partial f_i}{\partial x_j} = \frac{f_i(x_1, x_2, x_j + \varepsilon, \cdots x_n) - f_i(x_1, x_2, x_j, \cdots x_n)}{\varepsilon} \tag{6.46}$$

where

$$\varepsilon = 10^{-5} |x_j| \quad \text{for} \quad |x_j| > 1.0$$
$$\varepsilon = 10^{-5} \quad \text{if} \quad |x_j| \leq 1.0$$

Crank Constraint Equations—Four-Position Plane Rigid Body Guidance

The constant length condition for a guiding crank leads to a set of three equations.

$$(\mathbf{a}_j - \mathbf{a}_0)^T(\mathbf{a}_j - \mathbf{a}_0) - (\mathbf{a}_1 - \mathbf{a}_0)^T(\mathbf{a}_1 - \mathbf{a}_0) = 0 \qquad j = 2, 3, 4 \tag{6.47}$$

where

$$(\mathbf{a}_j) = [D_{1j}](\mathbf{a}_1)$$

Since the displacement matrices $[D_{1j}]$ are known in terms of the specified positions of the moving plane, this leads to a set of three nonlinear algebraic equations

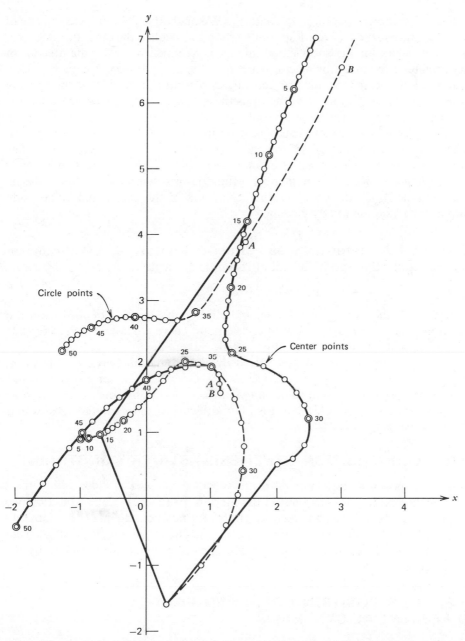

FIGURE 6.7 EXAMPLE 6.7. CENTER-POINT AND CIRCLE-POINT CURVES. FOUR-POSITION RIGID BODY GUIDANCE.

with four unknowns a_{0x}, a_{0y}, a_{1x}, and a_{1y}. We may arbitrarily assume any one of the four and use the Newton-Raphson method to calculate the other three. Since a series of values for the specified parameter is assumed, we calculate a series of coordinated fixed and moving crank pivots in the first position. Any pair of these cranks used together would theoretically constrain the motion of the moving plane to move sequentially through positions 1, 2, 3, and 4. However, often the linkage does not exhibit mobility through all positions in sequence, and each combination of two guiding links must be investigated for mobility separately. The locus of all possible fixed pivots is known as the *center-point curve*. The corresponding curve of coordinated moving pivots shown in the first position is the *circle-point curve*. This numerical technique of solving the nonlinear design equations leads to the simultaneous generation of the center and circle point curves for plane rigid body guidance.

Example 6-7 Center-point and Circle-Point Curves. A fourth position is added to the three-position data of Example 6-1 with

$$\mathbf{p}_4 = (2.0,\ 2.0) \qquad \theta_{14} = 90°$$

Figure 6.7 shows the center-point and circle-point curves obtained from the solution of Eqs. 6.47. Sections of the circle-point curve are not defined in regions where the center points tend to move off the paper.

A five-position problem is best solved by a combination of two four-position problems. The intersection of a 1234 set of curves with a 1235 set may or may not exist. A unique five-position solution may also be found by solving a set of four design equations (Eqs. 6.47), but it is often difficult to specify initial guesses that insure convergence.

6.8   FOUR-POSITION SYNTHESIS—SLIDER CONSTRAINT

With four positions of a moving plane specified we seek a slider with a fixed straight line guide for constraint of the motion. In this case there would be two circle equations that give the locus of sliders for positions 123 and 124. Either of these two circles may degenerate to a straight line if the angular displacement $\theta_{1j} = 0$. A possible unique slider for four position guidance is defined by the intersection of the two circles.

6.9   FOUR-POSITION COMBINED FINITE-DIFFERENTIAL SYNTHESIS

Combinations of constraint equations may be formed at will so long as the number of constraint equations does not exceed the number of unknown linkage parameters.

Example 6-8 Combined Displacement-Velocity-Acceleration Rigid Body Guidance. The motion of a reference point and the angular motion of a plane rigid body are specified by

$$\mathbf{p}_1 = (p_{1x}, p_{1y}) = (0.0, 0.0) \qquad \mathbf{p}_2 = (p_{2x}, p_{2y}) = (1.0, 1.0)$$

$$\dot{\mathbf{p}}_1 = (\dot{p}_{1x}, \dot{p}_{1y}) = (0.0, 0.0) \qquad \theta_{12} = +30°$$

$$\ddot{\mathbf{p}}_1 = (\ddot{p}_{1x}, \ddot{p}_{1y}) = (0.0, 1.0) \qquad \dot{\theta}_1 = +1.0, \ddot{\theta}_1 = 0.0$$

The first three of the four positions are infinitesimally separated followed by a final finite displacement.

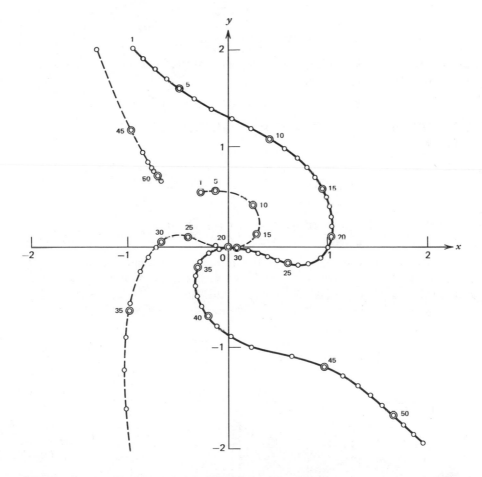

FIGURE 6.8   EXAMPLE 6.8. CENTER-POINT AND CIRCLE-POINT CURVES. FOUR-POSITION COMBINED FINITE-DIFFERENTIAL SYNTHESIS.

The set of three constraint equations become

$$(\mathbf{a}_2 - \mathbf{a}_0)^T(\mathbf{a}_2 - \mathbf{a}_0) = (\mathbf{a}_1 - \mathbf{a}_0)^T(\mathbf{a}_1 - \mathbf{a}_0)$$

$$(\dot{\mathbf{a}}_1)^T(\mathbf{a}_1 - \mathbf{a}_0) = 0$$

$$(\ddot{\mathbf{a}}_1)^T(\mathbf{a}_1 - \mathbf{a}_0) + (\dot{\mathbf{a}}_1)^T(\dot{\mathbf{a}}_1) = 0$$

These equations are converted to three design equations with four unknowns a_{0x}, a_{0y}, a_{1x}, a_{1y}, when we substitute

$$(\mathbf{a}_2) = [D_{12}](\mathbf{a}_1)$$

$$(\dot{\mathbf{a}}_1) = [V_1](\mathbf{a}_1)$$

$$(\ddot{\mathbf{a}}_1) = [A_1](\mathbf{a}_1)$$

Solving the design equations using program DESIGN, we obtain the center-point and circle-point curves of Figure 6.8.

6.10 SPHERICAL RIGID BODY GUIDANCE MECHANISMS

Spherical rigid body motion requires that all points in the moving body maintain a constant distance from a fixed reference point during a finite displacement.

Spherical Displacement Matrices by Direct Numerical Inversion

It is sometimes convenient to form the numerical elements of the spherical displacement matrix by specification of the displacement of any three arbitrary points in the rigid body (plus the origin). In this case the displacement matrix is found from the change in position of three points $\mathbf{p}$, $\mathbf{q}$, and $\mathbf{r}$ in the form

$$\begin{bmatrix} p_{2x} & q_{2x} & r_{2x} & 0 \\ p_{2y} & q_{2y} & r_{2y} & 0 \\ p_{2z} & q_{2z} & r_{2z} & 0 \\ 1 & 1 & 1 & 1 \end{bmatrix} = [D_{12}] \begin{bmatrix} p_{1x} & q_{1x} & r_{1x} & 0 \\ p_{1y} & q_{1y} & r_{1y} & 0 \\ p_{1z} & q_{1z} & r_{1z} & 0 \\ 1 & 1 & 1 & 1 \end{bmatrix} \qquad (6.48)$$

from which

$$[D_{12}] = \begin{bmatrix} p_{2x} & q_{2x} & r_{2x} & 0 \\ p_{2y} & q_{2y} & r_{2y} & 0 \\ p_{2z} & q_{2z} & r_{2z} & 0 \\ 1 & 1 & 1 & 1 \end{bmatrix} \begin{bmatrix} p_{1x} & q_{1x} & r_{1x} & 0 \\ p_{1y} & q_{1y} & r_{1y} & 0 \\ p_{1z} & q_{1z} & r_{1z} & 0 \\ 1 & 1 & 1 & 1 \end{bmatrix}^{-1} \qquad (6.49)$$

Design Equations for Spherical Rigid Body Guidance

In seeking a possible guiding crank we note that all axes, both fixed and moving, must intersect at a common point. We assume the origin of the coordinate system at this point. Any point on a guiding crank must pass through a series of positions that form a circular arc about the fixed axis, and the spherical constraint equations are developed as follows. Assume four positions are specified for a point $\mathbf{p}$ on a moving body whose motion is assumed to be spherical. The equation of the sphere is found from the determinant

$$
\begin{vmatrix}
(x^2 + y^2 + z^2) & x & y & z & 1 \\
(p_{1x}^2 + p_{1y}^2 + p_{1z}^2) & p_{1x} & p_{1y} & p_{1z} & 1 \\
(p_{2x}^2 + p_{2y}^2 + p_{2z}^2) & p_{2x} & p_{2y} & p_{2z} & 1 \\
(p_{3x}^2 + p_{3y}^2 + p_{3z}^2) & p_{3x} & p_{3y} & p_{3z} & 1 \\
(p_{4x}^2 + p_{4y}^2 + p_{4z}^2) & p_{4x} & p_{4y} & p_{4z} & 1
\end{vmatrix} = 0 \qquad (6.50)
$$

which can be expanded in the form

$$
x^2 + y^2 + z^2 + Ax + By + Cz + D = 0 \qquad (6.51)
$$

Reducing Eq. 6.51 to the normal form

$$
(x - c_x)^2 + (y - c_y)^2 + (z - c_z)^2 = R^2 \qquad (6.52)
$$

we can calculate the coordinates of the center of the spherical motion $\mathbf{c} = (c_x, c_y, c_z)$ and the spherical radius R for the motion of point $\mathbf{p}$.

Eq. 6.52 is next normalized by dividing by R^2 and shifting the spherical center to the origin by transforming all coordinates by

$$
\left(-\frac{c_x}{R}, \ -\frac{c_y}{R}, \ -\frac{c_z}{R} \right)
$$

The coordinates of all specified points are normalized in a similar manner and the synthesis can then be carried out on the surface of a unit sphere with center at the origin.

The *constraint equations for spherical cranks* must satisfy the constant length condition

$$
(\mathbf{a}_2 - \mathbf{a}_0)^T (\mathbf{a}_2 - \mathbf{a}_0) = (\mathbf{a}_1 - \mathbf{a}_0)^T (\mathbf{a}_1 - \mathbf{a}_0) \qquad j = 2, 3, \ldots n \qquad (6.53)
$$

plus two equations of the unit sphere

$$
(\mathbf{a}_1)^T (\mathbf{a}_1) = 1.0 \qquad (6.54)
$$

$$
(\mathbf{a}_0)^T (\mathbf{a}_0) = 1.0
$$

For n specified positions of a rigid body we see that there will always be $n + 1$ constraint equations with six unknowns $a_{0x}, a_{0y}, a_{0z}, a_{1x}, a_{1y}, a_{1z}$. The maximum number of rigid body positions that can be specified is therefore equal to five with no arbitrary choice of parameter. A four-position specification would lead to spherical center-point and circle-point curves with an arbitrary choice of any one of the six coordinates.

Example 6.9 Synthesis of a Spherical Rigid Body Guidance Mechanism. The relative motion at the human hip joint approximates spherical motion. This example was part of a study of the feasibility of constructing an

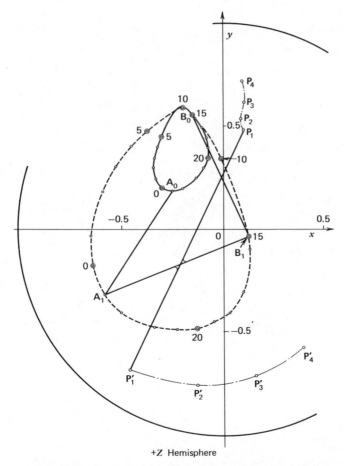

+Z Hemisphere

FIGURE 6.9 EXAMPLE 6.9 SPHERICAL CENTER-POINT AND CIRCLE POINT CURVES.

external mechanical linkage that might serve as an orthopedic brace for guidance of the femur relative to the pelvis.

With the pelvis as the reference, the relative position of two points on the femur $P(\mathbf{p})$ and $P'(\mathbf{p}')$ were measured then normalized to give

$\mathbf{P}_1 = (0.105040,\ 0.482820,\ 0.869397)$ $\mathbf{P}'_1 = (-0.464640,\ -0.676760,\ 0.571057)$

$\mathbf{P}_2 = (0.090725,\ 0.541283,\ 0.835931)$ $\mathbf{P}'_2 = (-0.133748,\ -0.751642,\ 0.645868)$

$\mathbf{P}_3 = (0.104155,\ 0.620000,\ 0.777658)$ $\mathbf{P}'_3 = (0.161113,\ -0.702067,\ 0.693646)$

$\mathbf{P}_4 = (0.096772,\ 0.725698,\ 0.681173)$ $\mathbf{P}'_4 = (0.400762,\ -0.564306,\ 0.721769)$

The spherical center-point and circle-point curves are calculated from Eqs. 6.53 and 6.54, which leads to a system of $n + 1$ design equations with six unknowns a_{0x}, a_{0y}, a_{0z}, a_{1x}, a_{1y}, and a_{1z}.

With four positions specified, there are five equations, hence a single parameter family of solutions. The curves shown in Figure 6.9, projected on the $x - y$ plane, are the center-point and circle-point curves. It is interesting to note that spherical center-point and center-point curves typically form closed loops on the surface of the unit sphere. Each point defines an axis of rotation that intersects the sphere at two diametrically opposite points.

One solution has been selected and displayed in Figure 6.9 with

$$\mathbf{a}_0 = (-0.2500,\ 0.1900,\ 0.9494) \quad \mathbf{a}_1 = (-0.5839,\ -0.3149,\ 0.7482)$$
$$\mathbf{b}_0 = (-0.1500,\ 0.5595,\ 0.8151) \quad \mathbf{b}_1 = (0.1214,\ -0.0275,\ 0.9922)$$

Three-position spherical rigid body guidance

A somewhat special technique is useful in three-position problems. A plane π containing three specified points $\mathbf{a}_1$, $\mathbf{a}_2$, and $\mathbf{a}_3$ on the surface of the sphere is defined by the determinant form of the plane equation as

$$\begin{vmatrix} x & y & z & 1 \\ a_{1x} & a_{1y} & a_{1z} & 1 \\ a_{2x} & a_{2y} & a_{2z} & 1 \\ a_{3x} & a_{3y} & a_{3z} & 1 \end{vmatrix} = 0 \tag{6.55}$$

A plane π', parallel to π and tangent to the sphere, will locate the fixed pivot $\mathbf{a}_0 = (a_{0x}, a_{0y}, a_{0z})$ at the point of tangency.

This results in the pair of equations

$$
\frac{\begin{vmatrix} a_{1x} & a_{1y} & 1 \\ a_{2x} & a_{2y} & 1 \\ a_{3x} & a_{3y} & 1 \end{vmatrix}}{a_{0z}} = -\frac{\begin{vmatrix} a_{1x} & a_{1z} & 1 \\ a_{2x} & a_{2z} & 1 \\ a_{3x} & a_{3z} & 1 \end{vmatrix}}{a_{0y}} = \frac{\begin{vmatrix} a_{1y} & a_{1z} & 1 \\ a_{2y} & a_{2z} & 1 \\ a_{3y} & 2_{3z} & 1 \end{vmatrix}}{a_{0x}}
\tag{6.56}
$$

The design equations become

$$
\frac{A}{a_{0z}} = -\frac{B}{a_{0y}} = \frac{C}{a_{0x}}
\tag{6.57}
$$

$$
a_{0x}^2 + a_{0y}^2 + a_{0z}^2 = 1
\tag{6.58}
$$

which leads to

$$
a_{0x} = \frac{A}{\pm D} \qquad a_{0y} = \frac{B}{\pm D} \qquad a_{0z} = \frac{C}{\pm D}
\tag{6.59}
$$

where

$$
D = \sqrt{A^2 + B^2 + C^2}
$$

The $\pm$ sign accounts for the possibility of two locations for $\mathbf{a}_0$ corresponding to the two intersections of the fixed axis with the surface of the unit sphere.

6.11 GENERAL SPATIAL RIGID BODY GUIDANCE

An almost infinite variety of spatial mechanisms can be formed to guide a rigid body through a series of specified positions in space. The number of possible positions that can be prescribed depends on the constraints provided by the guiding link used. In this section we will consider the constraint equations for a set of potentially useful guiding links. The complete rigid body guidance mechanism will often include links with different degrees of constraint. The maximum number of allowable specified positions (i.e., precision points) will always be dictated by the particular link that has the larger number of constraints. This often will allow greater freedom of choice of parameters for a second link that may have fewer constraints.

The Sphere-Sphere (S-S) Link

The sphere-sphere link shown in Figure 6.10 is the most simple combination of two kinematic pairs useful in spatial rigid body guidance. Note that the S-S link must always be used in combination with at least one additional guiding link of

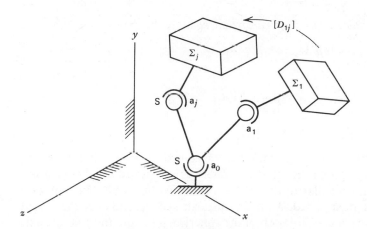

FIGURE 6.10   THE SPHERE-SPHERE (S-S) Link.

higher degree of constraint. The S-S link must satisfy the constant length condition only. Assuming a fixed pivot $\mathbf{a}_0$ and corresponding moving pivot $\mathbf{a}$, this leads
to the *S-S link displacement constraint equation.*

$$(\mathbf{a}_j - \mathbf{a}_0)^T(\mathbf{a}_j - \mathbf{a}_0) = (\mathbf{a}_1 - \mathbf{a}_0)^T(\mathbf{a}_1 - \mathbf{a}_0) \qquad j = 2, 3, \ldots n \qquad (6.60)$$

where

$$\mathbf{a}_0 = (a_{0x}, a_{0y}, a_{0z})$$

$$\mathbf{a}_1 = (a_{1x}, a_{1y}, a_{1z})$$

$$(\mathbf{a}_j) = [D_{1j}](\mathbf{a}_1)$$

Since there are six unknowns a_{0x}, a_{0y}, a_{0z}, a_{1x}, a_{1y}, and a_{1z}, we see that a
maximum of seven finitely separated positions of a rigid body can be specified.

The *S-S link velocity constraint equation* is found by differentiating Eq. 6.60 to
give

$$(\dot{\mathbf{a}}_j)^T(\mathbf{a}_j - \mathbf{a}_0) = 0 \qquad (6.61)$$

where

$$(\dot{\mathbf{a}}_j) = [V_j](\mathbf{a}_j)$$

$$(\mathbf{a}_j) = [D_{1j}](\mathbf{a}_1)$$

and $[D_{1j}]$ and $[V_j]$ are known in terms of the specified rigid body motions.

Similarly, the *S-S link acceleration constraint equation* becomes

$$(\ddot{\mathbf{a}}_j)^T(\mathbf{a}_j - \mathbf{a}_0) + (\dot{\mathbf{a}}_j)^T(\dot{\mathbf{a}}_j) = 0 \tag{6.62}$$

with

$$(\ddot{\mathbf{a}}_j) = [A_j](\mathbf{a}_j)$$

The Revolute-Sphere (R-S) Link

The R-S link shown in Figure 6.11 may be considered as an S-S link with one additional constraint imposed such that the spherical joint is restricted to rotation in a plane that is perpendicular to the axis $\mathbf{u}_0$ of the revolute joint. One point on the revolute axis must be specified. We will arbitrarily specify a point $\mathbf{a}_0$ that lies at the intersection of the plane of rotation of the spherical joint and the revolute axis $\mathbf{u}_0$. All other possible reference points on $\mathbf{u}_0$ would be kinematically equivalent. The *R-S link displacement constraint equations* become

$$(\mathbf{a}_j - \mathbf{a}_0)^T(\mathbf{a}_j - \mathbf{a}_0) = (\mathbf{a}_1 - \mathbf{a}_0)^T(\mathbf{a}_1 - \mathbf{a}_0) \qquad j = 2, 3, \ldots n$$

$$(\mathbf{u}_0)^T(\mathbf{a}_j - \mathbf{a}_0) = 0 \qquad\qquad j = 1, 2, 3, \ldots n$$

$$(\mathbf{u}_0)^T(\mathbf{u}_0) = 1 \tag{6.63}$$

After substitution of $(\mathbf{a}_j) = [D_{1j}](\mathbf{a}_1)$, we obtain a set of nonlinear design equations with nine unknowns u_{0x}, u_{0y}, u_{0z}, a_{0x}, a_{0y}, a_{0z}, a_{1x}, a_{1y}, and a_{1z}. We see

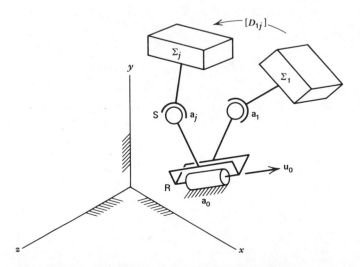

FIGURE 6.11   THE REVOLUTE-SPHERE (R-S) LINK.

that a maximum of four specified positions of the rigid body will give eight design equations. Five specified positions would result in ten equations and, therefore, no solution. With eight design equations for four specified positions we have free choice of one of the nine unknowns.

The *R-S link velocity constraint equations* become

$$(\dot{\mathbf{a}}_j)^T(\mathbf{a}_j - \mathbf{a}_0) = 0$$

$$(\mathbf{u}_0)^T(\dot{\mathbf{a}}_j) = 0 \tag{6.64}$$

and the *R-S link acceleration constraint equations* are

$$(\ddot{\mathbf{a}}_j)^T(\mathbf{a}_j - \mathbf{a}_0) + (\dot{\mathbf{a}}_j)^T(\dot{\mathbf{a}}_j) = 0$$

$$(\mathbf{u}_0)^T(\ddot{\mathbf{a}}_j) = 0 \tag{6.65}$$

The last of Eqs. 6.63 may be required with Eqs. 6.64 or 6.65 if the displacement constraint equations are not active.

The Revolute-Revolute (R-R) Link

The R-R link shown in Figure 6.12 must satisfy all the constraint equations for the R-S link plus two additional requirements. The angle of twist between the fixed axis $\mathbf{u}_0$ and the moving axis $\mathbf{u}$ must remain constant during a displacement, since both axes are fixed in the R-R link. In addition we require a second plane equation that takes note of the fact that point $\mathbf{a}_0$ must have a relative rotation about the moving axis $\mathbf{u}_j$. This constrains $\mathbf{a}_0$ to lie in a plane perpendicular to $\mathbf{u}_j$ at all times.

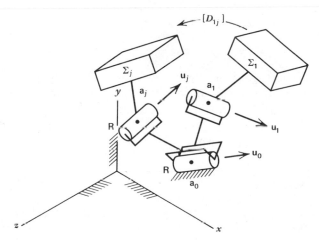

FIGURE 6.12   THE REVOLUTE-REVOLUTE (R-R) LINK.

Hence, the *R-R link displacement constraint* equations can be written as follows. *The plane equations:*

$$(\mathbf{u}_0)^T(\mathbf{a}_j - \mathbf{a}_0) = 0 \qquad j = 1, 2, 3$$

$$(\mathbf{u}_j)^T(\mathbf{a}_j - \mathbf{a}_0) = 0 \qquad j = 1, 2, 3 \tag{6.66}$$

The direction cosine equations:

$$(\mathbf{u}_0)^T(\mathbf{u}_0) = 1$$

$$(\mathbf{u}_1)^T(\mathbf{u}_1) = 1 \tag{6.67}$$

The constant twist equations:

$$(\mathbf{u}_j)^T(\mathbf{u}_0) = (\mathbf{u}_1)^T(\mathbf{u}_0) \qquad j = 2, 3 \tag{6.68}$$

The constant moment equations:

$$(\mathbf{u}_0)^T((\mathbf{a}_j - \mathbf{a}_0) \times \mathbf{u}_j) = (\mathbf{u}_0)^T((\mathbf{a}_1 - \mathbf{a}_0) \times \mathbf{u}_1) \qquad j = 2, 3 \tag{6.69}$$

Eq. 6.69 is required because Eq. 6.68 is a necessary but not sufficient condition to insure a constant twist angle α in all positions. Either a positive or negative value for the twist angle would satisfy Eq. 6.68. The moment equation (Eq. 6.69), when combined with the constant twist equation (Eq. 6.68), insures that the moment of vector $\mathbf{u}_j$ about axis $\mathbf{u}_0$ is a constant. Since $\mathbf{u}_j$ is a unit vector that makes a constant angle α with $\mathbf{u}_0$, Eq. 6.69 also substitutes for the constant length equation used with the S-S or R-S links.

Eqs. 6.66 to 6.69 constitute a set of twelve constraint equations for three finitely separated positions of a rigid body to be guided by an R-R link. When combined with

$$(\mathbf{a}_j) = [D_{1j}](\mathbf{a}_1)$$

and

$$(\mathbf{u}_j) = [R_{1j}](\mathbf{u}_1)$$

they form a set of twelve design equations with twelve unknown scalar components of $\mathbf{u}_0$, $\mathbf{u}_1$, $\mathbf{a}_0$ and $\mathbf{a}_1$. Therefore, the maximum number of positions which can be specified for an R-R link to be used for rigid body guidance is three, with no arbitrary choice of parameter.

The constant twist angle and constant moment equations can be replaced with *two constant length equations*

$$(\mathbf{a}_j - \mathbf{a}_0)^T(\mathbf{a}_j - \mathbf{a}_0) = (\mathbf{a}_1 - \mathbf{a}_0)^T(\mathbf{a}_1 - \mathbf{a}_0) \qquad j = 2, 3 \tag{6.70}$$

and

$$[(\mathbf{a}_j + \mathbf{u}_j) - (\mathbf{a}_0 + \mathbf{u}_0)]^T[(\mathbf{a}_j + \mathbf{u}_j) - (\mathbf{a}_0 + \mathbf{u}_0)]$$

$$= [(\mathbf{a}_1 + \mathbf{u}_1) - (\mathbf{a}_0 + \mathbf{u}_0)]^T[(\mathbf{a}_1 + \mathbf{u}_1) - (\mathbf{a}_0 + \mathbf{u}_0)] \qquad j = 2, 3 \tag{6.71}$$

Equation 6.71 insures, for an initial twist angle α_1, that α_j is not allowed to assume a value equal to $\alpha_1 + \pi$. This second constant length equation may be thought of as adding an imaginary line L', rigidly attached to the R-R link, which must maintain a fixed relationship to the line L between points $\mathbf{a}_0$ and $\mathbf{a}$.

Although the constant twist equations (Eqs. 6.68 and 6.69) are perhaps more basic, the constant length equations (Eqs. 6.70 and 6.71) have also been successfully applied in R-R link synthesis.

The *R-R link velocity constraint equations* may be written by differentiation of the displacement constraint equations in either of the forms mentioned previously. In the first case they become

$$(\mathbf{u}_0)^T(\dot{\mathbf{a}}_j) = 0 \tag{6.72}$$

$$(\dot{\mathbf{u}}_j)^T(\mathbf{a}_j - \mathbf{a}_0) + (\mathbf{u}_j)^T(\dot{\mathbf{a}}_j) = 0 \tag{6.73}$$

$$(\dot{\mathbf{u}}_j)^T(\mathbf{u}_0) = 0 \tag{6.74}$$

$$[\dot{\mathbf{a}}_j \times \mathbf{u}_j + (\mathbf{a}_j - \mathbf{a}_0) \times \dot{\mathbf{u}}_j]^T(\mathbf{u}_0) = 0 \tag{6.75}$$

Equations 6.74 and 6.75 may be replaced by their equivalent found by differentiating eqs. 6.70 and 6.71.

$$(\dot{\mathbf{a}}_j)^T(\mathbf{a}_j - \mathbf{a}_0) = 0 \tag{6.76}$$

$$(\dot{\mathbf{a}}_j + \dot{\mathbf{u}}_j)^T[(\mathbf{a}_j + \mathbf{u}_j) - (\mathbf{a}_0 + \mathbf{u}_0)] = 0 \tag{6.77}$$

The *R-R link acceleration constraint equations* become

$$(\mathbf{u}_0)^T(\ddot{\mathbf{a}}_j) = 0 \tag{6.78}$$

$$(\ddot{\mathbf{u}}_j)^T(\mathbf{a}_j - \mathbf{a}_0) + 2(\dot{\mathbf{u}}_j)^T(\dot{\mathbf{a}}_j) + (\mathbf{u}_j)^T(\ddot{\mathbf{a}}_j) = 0 \tag{6.79}$$

$$(\ddot{\mathbf{u}}_j)^T(\mathbf{u}_0) = 0 \tag{6.80}$$

$$[\ddot{\mathbf{a}}_j \times \mathbf{u}_j + 2\dot{\mathbf{a}}_j \times \dot{\mathbf{u}}_j + (\mathbf{a}_j - \mathbf{a}_0) \times \ddot{\mathbf{u}}_j]^T(\mathbf{u}_0) = 0 \tag{6.81}$$

Again, Eqs. 6.80 and 6.81 may be replaced by an alternate pair found by differentiation of Eqs. 6.76 and 6.77.

$$(\ddot{\mathbf{a}}_j)^T(\mathbf{a}_j - \mathbf{a}_0) + (\dot{\mathbf{a}}_j)^T(\dot{\mathbf{a}}_j) = 0 \tag{6.82}$$

$$(\ddot{\mathbf{a}}_j + \ddot{\mathbf{u}}_j)^T[(\mathbf{a}_j + \mathbf{u}_j) - (\mathbf{a}_0 + \mathbf{u}_0)] + (\dot{\mathbf{a}}_j + \dot{\mathbf{u}}_j)^T(\dot{\mathbf{a}}_j + \dot{\mathbf{u}}_j) = 0 \tag{6.83}$$

The Revolute-Cylindrical (R-C) Link

The R-C link shown in Figure 6.13 must satisfy all of the constraints imposed by an R-R link plus an additional displacement constraint equation, which accounts for the translational degree of freedom of the cylindrical joint. In this case the two

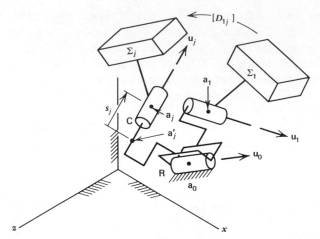

FIGURE 6.13   THE REVOLUTE-CYLINDRICAL (R-C) LINK.

constant length equations (Eqs. 6.70 and 6.71) are written with the coordinates of the intermediate point $\mathbf{a}'_j$ replacing $\mathbf{a}_j$ where

$$\mathbf{a}'_j = \mathbf{a}_j - s_j\,\mathbf{u}_j \tag{6.84}$$

R-C link synthesis for three specified rigid body positions leads to a set of twelve design equations. The two added translations s_2 and s_3 give a total of fourteen unknowns. Therefore, the maximum number of precision points that can be specified is three with arbitrary choice of any of two scalar parameters.

Velocity and acceleration constraint equations are also written in terms of point $\mathbf{a}'_j$ with the added equations

$$\dot{\mathbf{a}}'_j = \dot{\mathbf{a}}_j - [W]s_j\,\mathbf{u}_j - \dot{s}_j\,\mathbf{u}_j \tag{6.85}$$

$$\ddot{\mathbf{a}}'_j = \dot{\mathbf{a}}_j - [\dot{W}]s_j\,\mathbf{u}_j - \ddot{s}_j\,\mathbf{u}_j - 2[W]\dot{s}_j\,\mathbf{u}_j \tag{6.86}$$

where $[W]$ and $[\dot{W}]$ are the specified angular velocity and angular acceleration matrices as defined by Eqs. 3.60 and 3.67.

The Cylindrical-Cylindrical (C-C) Link

The C-C link shown in Figure 6.14 again must satisfy all of the displacement constraints imposed by the R-R link. In this case a translation s'_j, measured from the first position, is allowed along the fixed cylindrical axis $\mathbf{u}_0$. A similar translation s_j is assumed along the moving joint axis $\mathbf{u}_j$. The constraint equations are written in terms of points $\mathbf{a}'_0$ and $\mathbf{a}'_j$, as shown in Figure 6.14 where

$$\mathbf{a}'_0 = \mathbf{a}_0 + s'_j\,\mathbf{u}_0 \tag{6.87}$$

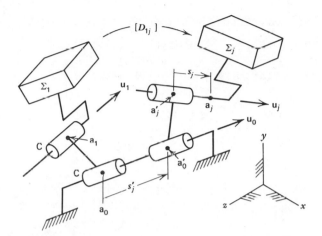

FIGURE 6.14   THE CYLINDRICAL-CYLINDRICAL (C-C) LINK.

and

$$\mathbf{a}'_j = \mathbf{a}_j - s_j \mathbf{u}_j \qquad (6.88)$$

A maximum of five specified rigid body positions is possible with a C-C link. This results in a total of twenty equations with twenty unknowns from the three scalar components of $\mathbf{u}_0$, $\mathbf{u}_1$, $\mathbf{a}_0$, $\mathbf{a}_1$ plus the eight translation components s'_2, s'_3, s'_4, s'_5, s_2, s_3, s_4, and s_5. There is no arbitrary choice of parameter if five positions are specified.

The *velocity and acceleration constraint equations for the C-C link* are somewhat more complex than for the R-R link, since they must account for the velocity and acceleration of point $\mathbf{a}'_0$ plus both the relative motion and Coriolis component at point $\mathbf{a}'_j$.

Example 6.10 Three-Position Rigid Body Guidance Using R-R, S-S, C-S, or R-C Links.

Data from Roth (reference 7, p. 72) have been converted into equivalent screw motion data in the form

$$\mathbf{p}_{12} = (0.941, \, 0.521, \, -0.419)$$

$$\mathbf{u}_{12} = (0.3600, \, 0.0965, \, 0.9280)$$

$$s_{12} = 1.3844 \qquad \phi_{12} = 133.20°$$

$$\mathbf{p}_{13} = (-0.610, \, 0.837, \, 0.123)$$

$$\mathbf{u}_{13} = (0.1457, \, -0.0390, \, 0.9886)$$

$$s_{13} = 1.8991 \qquad \phi_{13} = 70.62°$$

This data results in screw displacement matrices for rigid body motion given by

$$[D_{12}] = \begin{bmatrix} -0.4662298 & -0.6179617 & 0.6331190 & 2.466281 \\ 0.7350041 & -0.6688602 & -0.1115741 & 0.265085 \\ 0.4924280 & 0.4132833 & 0.7661579 & 0.507641 \\ 0. & 0. & 0. & 1.0 \end{bmatrix}$$

$$[D_{13}] = \begin{bmatrix} 0.3460161 & -0.9363812 & 0.0594521 & 0.654105 \\ 0.9287878 & 0.3328481 & -0.1632059 & 1.071321 \\ 0.1330325 & 0.1116829 & 0.9848526 & 1.867013 \\ 0. & 0. & 0. & 1.0 \end{bmatrix}$$

Program DESIGN was used to solve for possible guiding links. A general purpose function YCOMP for this purpose is given as Example 12-6 in Section 12.3.

R-R link synthesis Eqs. 6.66 to 6.69 form a set of 12 design equations in 12 variables with no free variables. With initial guesses

$$\mathbf{a}_0 = (-0.5, 2.0, -1.5) \qquad \mathbf{a}_1 = (1.5, 0.0, -0.75)$$

$$\mathbf{u}_0 = (-0.4, -0.6, 0.7) \qquad \mathbf{u}_1 = (0.5, 0.0, 1.0)$$

program DESIGN converged in five iterations to a solution

$$\mathbf{a}_0 = (-4.4484, 3.8498, -3.5881) \qquad \mathbf{a}_1 = (-1.0609, -0.3334, -5.2130)$$

$$\mathbf{u}_0 = (-0.400863, -0.595143, 0.696501) \quad \mathbf{u}_1 = (0.478231, 0.046617, 0.876996)$$

S-S link synthesis For three position synthesis of an S-S guiding link there are two design equations with six variables. a_{0x}, a_{0y}, a_{0z}, and a_{12} will be specified and designated by an asterisk in the results. With initial guesses,

$$\mathbf{a}_0 = (1.0^*, 0.0^*, 0.0^*) \qquad \mathbf{a}_1 = (1.0, 1.0, 0.0^*)$$

program DESIGN converged in three iterations to the solution

$$\mathbf{a}_0 = (1.0^*, 0.0^*, 0.0^*) \qquad \mathbf{a}_1 = (-0.90695, 0.05569, 0.0^*)$$

C-S link synthesis In this case there are a total of six design equations with 11 variables. The unknown sliding displacements d_{12} and d_{13} are added to the nine geometrical parameters in $\mathbf{a}_0$, $\mathbf{u}_0$, $\mathbf{a}_1$. a_{0x}, a_{0y}, a_{0z}, and u_{0y} will be specified.

With initial guesses,

$$\mathbf{a}_0 = (0.0^*, 0.0^*, 0.0^*) \qquad \mathbf{u}_0 = (-0.935^*, 0.041^*, 1.00)$$

$$\mathbf{a}_1 = (1.0, 1.0, 0.0) \qquad d_{12} = 1.0 \qquad d_{13} = 2.0$$

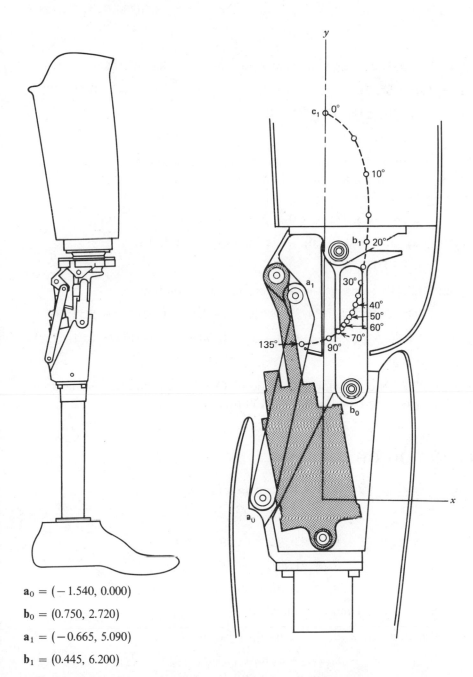

$\mathbf{a}_0 = (-1.540, 0.000)$

$\mathbf{b}_0 = (0.750, 2.720)$

$\mathbf{a}_1 = (-0.665, 5.090)$

$\mathbf{b}_1 = (0.445, 6.200)$

FIGURE 6.15 FIXED CENTRODE FOR A PROSTHETIC KNEE MECHANISM. LINKAGE ARRANGEMENT AND PATH OF KNEE CENTER FOR THE UC-BL FOUR-BAR POLYCENTRIC KNEE.

program DESIGN converged in eight iterations to a solution

$\mathbf{a}_0 = (0.0^*, 0.0^*, 0.0^*)$ $\qquad\qquad\qquad$ $\mathbf{u}_0 = (-0.935^*, 0.041^*, 0.352258)$

$\mathbf{a}_1 = (0.417901, 2.937692, 0.766478)$

$d_{12} = -0.055053$ $\qquad\qquad\qquad\qquad\qquad$ $d_{13} = 2.936367$

R-C link synthesis There are a total of 12 design equations with 14 variables $\mathbf{a}_0$, $\mathbf{u}_0$, $\mathbf{a}_1$, $\mathbf{u}_1$, d_{12}, d_{13}. $\mathbf{u}_{0x}$ and $\mathbf{u}_{0y}$ will be specified.

With initial guesses

$\mathbf{a}_0 = (-0.3, 40.0, -6.0)$ $\qquad$ $\mathbf{u}_0 = (-0.935^*, 0.041^*, 0.40)$

$\mathbf{a}_1 = (4.0, 1.0, -2.0)$ $\qquad\qquad$ $\mathbf{u}_1 = (0.5, -0.3, 0.8)$

$d_{12} = 1.0$ $\qquad\qquad\qquad\qquad$ $d_{13} = 2.0$

program DESIGN converged in five iterations to a solution.

$\mathbf{a}_0 = (0.5938, 43.1806, -6.2975)$ $\qquad$ $\mathbf{u}_0 = (-0.935^*, 0.041^*, 0.352258)$

$\mathbf{a}_1 = (-5.0870, 6.7169, -17.1216)$ $\qquad$ $\mathbf{u}_1 = (0.483697, -0.317473, 0.8156727)$

$d_{12} = -33.1262$ $\qquad\qquad\qquad\qquad$ $d_{13} = -20.5951$

The FORTRAN coding for the function YCOMP, which generated the above examples, is listed in Example 12-6 in Chapter 12.

APPENDIX A
DERIVATION OF EQ. 6.21

From $\qquad\qquad\qquad\qquad$ $(\ddot{\mathbf{a}}_j)^T(\mathbf{a}_j - \mathbf{a}_0) + (\dot{\mathbf{a}}_j)^T(\dot{\mathbf{a}}_j) = 0$ $\qquad\qquad\qquad$ (a)

$$\ddot{a}_{jx} = \ddot{d}_{11j}a_{jx} + \ddot{d}_{12j}a_{jy} + \ddot{d}_{13j}$$

$$\ddot{a}_{jy} = \ddot{d}_{21j}a_{jx} + \ddot{d}_{22j}a_{jy} + \ddot{d}_{23j}$$

$$\dot{a}_{jx} = \dot{d}_{11j}a_{jx} + \dot{d}_{12j}a_{jy} + \dot{d}_{13j}$$

$$\dot{a}_{jy} = \dot{d}_{21j}a_{jx} + \dot{d}_{22j}a_{jy} + \dot{d}_{23j}$$

Substitute into Eq. (a) and expand, which gives

$$a_{jx}^2\ddot{d}_{11j} + a_{jx}a_{jy}\ddot{d}_{12j} + a_{jx}\ddot{d}_{13j} - a_{jx}a_{0x}\ddot{d}_{11j} - a_{jy}a_{0x}\ddot{d}_{12j}$$

$$- a_{0x}\ddot{d}_{13j} + a_{jx}a_{jy}\ddot{d}_{21j} + a_{jy}\ddot{d}_{22j} + a_{jy}\ddot{d}_{23j} - a_{jx}a_{0y}\ddot{d}_{21j}$$

$$- a_{jy}a_{0y}\ddot{d}_{22j} - a_{0y}\ddot{d}_{23j} = -(\dot{d}_{12j}\dot{d}_{jy} + 2\dot{d}_{12j}a_{jy}\dot{d}_{13j} + \dot{d}_{13j}^2)$$

$$- (\dot{d}_{21j}^2a_{jx}^2 + 2\dot{d}_{21j}a_{jx}\dot{d}_{23j} + \dot{d}_{23j}^2)$$

Collect terms.

$$a_{jx}(\ddot{d}_{13j} - a_{0x}\ddot{d}_{11j} - a_{0y}\ddot{d}_{21j} + 2\dot{d}_{21j}\dot{d}_{23j})$$
$$+ a_{jy}(\ddot{d}_{23j} - a_{0x}\ddot{d}_{12j} - a_{0y}\ddot{d}_{22j} + 2\dot{d}_{12j}\dot{d}_{13j})$$
$$= a_{0x}\ddot{d}_{13j} + a_{0y}\ddot{d}_{23j} - (\dot{d}_{13j}^2 + \dot{d}_{23j}^2)$$

Chapter 7

Function Generation

7.1 INTRODUCTION

The term function generator refers to a class of mechanisms in which an output motion that is a specified function of the input motion is created. The input and output members may be either a crank or slider, each with motion specified with respect to a fixed reference member. When properly designed, it is often possible to mechanize smooth continuous mathematical functions with very small error over a limited range of the independent variable.

The general method of synthesis will be similar to that used for the design of rigid body guidance mechanisms. It will be shown that the function generator synthesis problem can always be converted to an equivalent rigid body guidance problem by employing the principle of kinematic inversion. In this case the motion of the guided rigid body is described by the relative motion of the input member with respect to the output member. This inversion principle will be demonstrated by application to a wide variety of plane and spatial function generator mechanisms. However, before beginning a discussion of the actual methods to be employed, we must outline some necessary preliminary calculations.

7.2 PROBLEM FORMULATION—SPECIFICATION OF PRECISION POINTS

Assume a function is to be mechanized by a double-crank mechanism such that the input rotation θ is proportional to the independent variable x in a mathematical function $y = f(x)$. The output rotation ϕ is proportional to the dependent variable y. The problem can be stated in the form

Mechanize $y = f(x)$

such that the input motion θ is proportional to x

and the output motion ϕ is proportional to y

for a range in the variables

$$x_0 \leq x \leq x_f$$

$$f(x_0) \leq y \leq f(x_f)$$

and a range in input and output motion

$$\theta_0 \le \theta \le \theta_f$$

$$\phi_0 \le \phi \le \phi_f$$

Except in certain very special cases it will not be possible to generate the mechanized function with zero error over the range of interest. We are restricted to a limited number of points of zero error (i.e., *precision* points), depending on the number of kinematic constraint equations that must be satisfied by the particular mechanism to be employed. The difference between the mechanized output motion ϕ and that calculated from the mathematical function is known as the *structural error*. The maximum structural error between precision points is

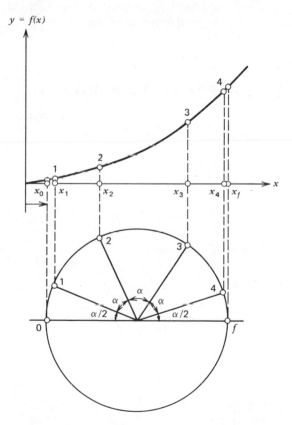

FIGURE 7.1 CHEBYSHEV SPACING AS A FIRST APPROXIMATION TO OPTIMAL PRECISION POINT SPACING.

often less than 1.0% over the full range of the mechanism. The actual error curve is highly dependent on the choice of precision point spacing for the independent variable. The optimal precision point spacing is that which results in equal values of maximum absolute error between each successive pair of precision points as well as at the beginning and end of the range of motion. To this end, it is generally not desirable to specify precision points at the minimum and maximum values for the independent variable but, instead, to specify values within the range of θ or x, as shown in Figure 7.1. This procedure tends to reduce the maximum error to be encountered over the full range of motion.

Chebyshev spacing, as indicated in Figure 7.1, may be used as a first approximation to optimal precision point spacing. Freudenstein [1] has shown a method for systematic respacing of precision points to arrive at the optimal design with equal error between precision points. A different approach to optimal synthesis, based on minimization of the integral of the structural error over the full range of motion, is discussed in Chapter 9.

7.3 CHEBYSHEV SPACING OF PRECISION POINTS

The graphical basis for Chebyshev spacing is shown in Figure 7.1. In numerical calculations it is convenient to use the equivalent analytical expression

$$x_j = x_0 + \frac{\Delta x}{2}\left[1 - \cos\left(j\alpha - \frac{\alpha}{2}\right)\right] \qquad j = 1, 2, \ldots, n \qquad (7.1)$$

where

$$\Delta x = x_f - x_0$$

$$n = \text{number of precision points}$$

$$\alpha = \frac{180}{n} \text{ degrees}$$

For *three precision points* $n = 3$, $\alpha = 180/3 = 60$,

$$x_1 = x_0 + \frac{\Delta x}{2}[1 - \cos(60 - 30)] = x_0 + 0.067\,\Delta x$$

$$x_2 = x_0 + \frac{\Delta x}{2}[1 - \cos(120 - 30)] = x_0 + 0.500\,\Delta x$$

$$x_3 = x_0 + \frac{\Delta x}{2}[1 - \cos(180 - 30)] = x_0 + 0.933\,\Delta x \qquad (7.2)$$

Similarly, for *four precision points*,

$$x_1 = x_0 + 0.038 \, \Delta x$$

$$x_2 = x_0 + 0.309 \, \Delta x$$

$$x_3 = x_0 + 0.692 \, \Delta x$$

$$x_4 = x_0 + 0.962 \, \Delta x \tag{7.3}$$

7.4 SCALE FACTORS FOR INPUT AND OUTPUT MOTION

To insure that the mechanized variables θ and ϕ are proportional to the functional variables x and y, we define scale factors

$$k_\theta = \frac{\Delta \theta}{\Delta x} = \frac{\theta_f - \theta_0}{x_f - x_0} \tag{7.4}$$

$$k_\phi = \frac{\Delta \phi}{\Delta y} = \frac{\phi_f - \phi_0}{f(x_f) - f(x_0)} \tag{7.5}$$

The specific angles corresponding to precision points are then calculated from

$$\theta_i = \theta_0 + k_\theta(x_j - x_0) \tag{7.6}$$

$$\phi_j = \phi_0 + k_\phi[f(x_j) - f(x_0)] \tag{7.7}$$

The actual angles used in synthesis are often the displacement angles referred to the first position. In this case we must be careful to calculate

$$\theta_{1j} = \theta_j - \theta_1 \tag{7.8}$$

and

$$\phi_{1j} = \phi_j - \phi_1 \tag{7.9}$$

7.5 THREE-POSITION FUNCTION GENERATOR MECHANISMS

The Four-Bar Linkage Function Generator

Figure 7.2 shows the geometry of the plane four-bar function generator where θ_{1j} and ϕ_{1j} are the specified crank rotations calculated from Eqs. 7.4 to 7.9. We can convert these specified crank rotations into an equivalent relative rigid body motion of the input crank $\mathbf{a}_0 \mathbf{a}$ with respect to the output crank $\mathbf{b}_0 \mathbf{b}$. This

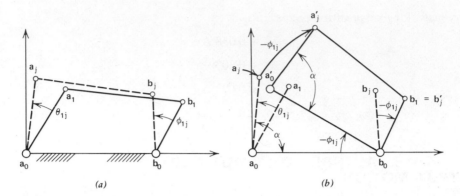

FIGURE 7.2 KINEMATIC INVERSION OF THE PLANE FOUR-BAR FUNCTION GENERATOR.

procedure of *kinematic inversion* about the first position of the output crank is described analytically as follows. We first allow the function generator to assume its jth position, as shown in Figure 7.2a. The rotation of point a_1 about a_0 is described by a 3×3 matrix with $a_0 = (0., 0.)$ as the reference point in the form

$$[D_{\theta_{1j}}] = \begin{bmatrix} \cos \theta_{1j} & -\sin \theta_{1j} & 0 \\ \sin \theta_{1j} & \cos \theta_{1j} & 0 \\ 0 & 0 & 1 \end{bmatrix} \qquad (7.10)$$

We next *invert* the entire mechanism in its jth position, holding all joint angles constant, such that the jth position of the output link is rotated through an angle $-\phi_{1j}$ to bring line $\overline{b_0 b_j}$ into coincidence with $\overline{b_0 b_1}$. During the inversion we note that point a_0 is inverted to a_0' and point a_j moves to a_j'. The second displacement components are described by a displacement matrix with a negative rotation angle $-\phi_{1j}$ in the form

$$[D_{-\phi_{1j}}] = \begin{bmatrix} \cos \phi_{1j} & \sin \phi_{1j} & (1 - \cos \phi_{1j}) \\ -\sin \phi_{1j} & \cos \phi_{1j} & \sin \phi_{1j} \\ 0 & 0 & 1 \end{bmatrix} \qquad (7.11)$$

where the reference point in this case is $b_0 = (1., 0.)$.

The total relative motion is described by the *relative displacement matrix*

$$[D_R] = [D_{-\phi_{1j}}][D_{\theta_{1j}}]$$
$$= \begin{bmatrix} \cos(\theta_{1j} - \phi_{1j}) & -\sin(\theta_{1j} - \phi_{1j}) & (1 - \cos \phi_{1j}) \\ \sin(\theta_{1j} - \phi_{1j}) & \cos(\theta_{1j} - \phi_{1j}) & \sin \phi_{1j} \\ 0 & 0 & 1 \end{bmatrix} \qquad (7.12)$$

In the inversion the coupler $\overline{\mathbf{ab}}$ becomes the equivalent of a guiding crank in a rigid body guidance problem where the input crank is now the guided body with motion specified by matrix (7.12). The fixed link $\mathbf{a}_0\,\overline{\mathbf{b}_0}$ is also a guiding crank in the inverted motion. This is often helpful, since $\mathbf{a}_0\,\overline{\mathbf{b}_0}$ may be used as an initial guess for iterative solutions.

For the three-position, four-bar function generator we may proceed in a manner analogous to Example 6.1, which employed linear Eqs. 6.15 for three-position crank synthesis. In the present case the displacement matrix elements d_{ikj} are calculated from matrix (7.12). Coordinates $\mathbf{a}_0 = (a_{0x}, a_{0y})$ in Eq. 6.15 are replaced by specified coordinates for the first position of the output crank pivot $\mathbf{b}_1 = (b_{1x}, b_{1y})$, which locates the center for the relative motion of point $\mathbf{a}$ with respect to the output link $\overline{\mathbf{b}_0\,\mathbf{b}}$.

Example 7-1 The Three-Position Four-Bar Function Generator. Rotation angles for the input and output cranks are specified as

$$\theta_{12} = -30.0° \qquad \phi_{12} = 15.0°$$

$$\theta_{13} = -60.0 \qquad \phi_{13} = 45.0$$

The first position of the output crank moving pivot is assumed at $\mathbf{b}_1 = (0.3750, 0.1875)$. Eqs. 6.15 give $\mathbf{a}_1 = (0.346046, 0.529250)$. The resulting linkage is shown in its three positions in Figure 7.3.

The Slider-Crank Function Generator

The kinematic inversion technique can also be used in the synthesis of a slider-crank function generator. In this case we assume the input crank angle θ_{1j} to be proportional to the independent variable x in the functional relationship $y = f(x)$. The slider displacement s_{1j} is proportional to y and is assumed in a direction that makes an angle α with the horizontal axis.

We assume the fixed pivot of the input crank at $\mathbf{a}_0 = (0., 0.)$ and the first position of the slider at $\mathbf{b}_1 = (1., 0.)$. In this case the relative motion of the input crank is inverted about the first position of the output slider.

The relative displacement matrix becomes

$$[D_R] = [D_{-s_{1j}}][D_{\theta_{1j}}] = \begin{bmatrix} \cos\theta_{1j} & -\sin\theta_{1j} & -s_{1j}\cos\alpha \\ \sin\theta_{1j} & \cos\theta_{1j} & -s_{1j}\sin\alpha \\ 0 & 0 & 1 \end{bmatrix} \qquad (7.13)$$

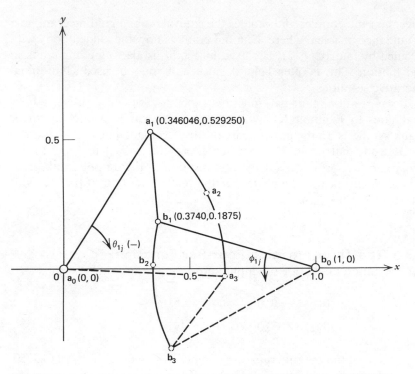

FIGURE 7.3 EXAMPLE 7.1. THREE-POSITION, FOUR-BAR FUNCTION
GENERATOR.

Recall that we have assumed a location for point $\mathbf{b}_1$ that becomes the center for the relative rotation of point $\mathbf{a}_1$. The coordinates of point $\mathbf{a}_1$ are again calculated from Eq. 6.15, with displacement matrix elements given by matrix (7.13) in this case.

Example 7-2 The Three-Position Slider-Crank Function
Generator. Input crank rotations and output slider displacements are specified as

$$\theta_{12} = 30.0° \qquad s_{12} = 0.5 \text{ in.}$$

$$\theta_{13} = 45.0 \qquad s_{13} = 0.875$$

The slider angle is given as $\alpha = 60.0°$. Matrix (7.13) and Eq. 6.15 give the first position of the moving crank pivot $\mathbf{a}_1 = (-4.390884, -6.752166)$.

The synthesized mechanism is shown in Figure 7.4.

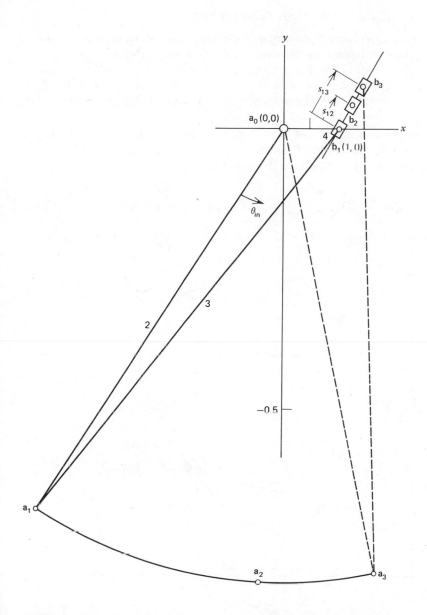

**FIGURE   7.4   EXAMPLE   7.2.   THREE-POSITION   SLIDER-CRANK
FUNCTION GENERATOR.**

The Double Slider Function Generator

In this case both input and output motions are specified as slider displacements at specified angles with respect to the fixed y-axis. We may arbitrarily specify the angle for the input slider such that it moves along the y-axis with its first position $\mathbf{a}_1$ at the origin. The angle β for the output slider is also arbitrary, with the first position of the output slider located at point $\mathbf{b}_1$ with unknown coordinates (b_{1x}, b_{1y}).

The motion of the output slider with respect to the input slider is described by the relative displacement matrix (7.14), where r_{1j} is the input slider displacement and s_{1j} describes the output motion (Figure 7.5).

$$[D_R] = [D_{-r_{1j}}][D_{s_{1j}}] = \begin{bmatrix} 1 & 0 & (s_{1j} \sin \alpha) \\ 0 & 1 & (s_{1j} \cos \alpha - r_{1j}) \\ 0 & 0 & 1 \end{bmatrix} \tag{7.14}$$

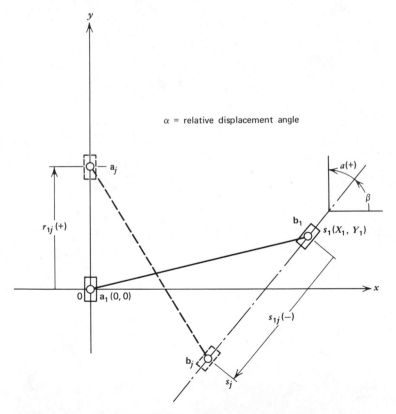

α = relative displacement angle

FIGURE 7.5   DOUBLE-SLIDER FUNCTION GENERATOR.

The coordinates of point b_1 are found from the constant length condition for the coupler, link $\overline{ab}$. In the inversion, in this case referred to the first position of the input slider, point a_1 becomes the fixed pivot in Eq. 6.15, and we solve for the coordinates of the moving pivot b_1.

Example   7-3 The   Three-Position   Double-Slider   Function Generator. Translation displacements for two sliders are specified as

$$r_{12} = 1.0 \qquad s_{12} = 2.0$$

$$r_{13} = 2.0 \qquad s_{13} = 3.0$$

and the relative slider angle $\alpha = 30.0°$.
  The relative displacement matrices becomes

$$[D_{12}] = \begin{bmatrix} 1 & 0 & 1.000 \\ 0 & 1 & 0.732 \\ 0 & 0 & 1 \end{bmatrix} \qquad [D_{13}] = \begin{bmatrix} 1 & 0 & 1.500 \\ 0 & 1 & 0.598 \\ 0 & 0 & 1 \end{bmatrix}$$

In this case, substituting a_1 for a_0 and b_1 for a_1 in Eq. 6.15, we obtain

$$b_1 = (-0.9904, 0.3038) \qquad .$$

7.6 FUNCTION GENERATORS WITH COMBINED FINITE AND DIFFERENTIAL MOTION SPECIFICATIONS

The Plane Four-Bar Function Generator

The relative differential motion of the input crank with respect to the output crank is described by the *relative velocity matrix* in 3×3 form:

$$[V_R] = \begin{bmatrix} 0 & -(\dot{\theta}_j - \dot{\phi}_j) & 0 \\ (\dot{\theta}_j - \dot{\phi}_j) & 0 & \dot{\phi}_j \\ 0 & 0 & 0 \end{bmatrix} \tag{7.15}$$

and/or the *relative acceleration matrix*

$$[A_R] = \begin{bmatrix} -(\dot{\theta}_j - \dot{\phi}_j)^2 & -(\ddot{\theta}_j - \ddot{\phi}_j) & \dot{\phi}_j^2 \\ (\ddot{\theta}_j - \ddot{\phi}_j) & -(\dot{\theta}_j - \dot{\phi}_j)^2 & \ddot{\phi}_j \\ 0 & 0 & 0 \end{bmatrix} \tag{7.16}$$

Example 7-4 Velocity-Acceleration Synthesis of a Plane Four-Bar Function Generator. Assume coordinated input-output crank motions are given as

$$\dot{\theta} = 8.0 \text{ rad/sec} \qquad \dot{\phi} = -3.0$$
$$\ddot{\theta} = 0.0 \text{ rad/sec}^2 \qquad \ddot{\phi} = 0.0$$

The moving pivot for the output crank is arbitrarily specified at $\mathbf{b}_1 = (0.639, -0.262)$, and we require the corresponding position for the first position of the input crank moving pivot $\mathbf{a}_1$.

The *velocity constraint equation* is based on Eq. 6.2, with $\mathbf{b}_1$ replacing $\mathbf{a}_0$ as the center of relative rotation. This leads to the first design equation

$$a_{1x}A + a_{1y}B = C \tag{7.17}$$

where

$$A = \dot{d}_{13} - \dot{d}_{21}b_{1y} = -(\dot{\theta} - \dot{\phi})b_{1y}$$
$$B = \dot{d}_{23} - \dot{d}_{12}b_{1x} = \dot{\phi} + (\dot{\theta} - \dot{\phi})b_{1x}$$
$$C = \dot{d}_{13}b_{1x} + \dot{d}_{23}b_{1y} = \dot{\phi}b_{1y}$$

The *acceleration constraint* (Eq. 6.3) gives a second design equation, with $\mathbf{b}_1$ replacing $\mathbf{a}_0$ and matrix elements from Eqs. 7.15 and 7.16.

$$a_{1x}D + a_{1y}E = F$$

where

$$D = \dot{\phi}^2 + (\dot{\theta} - \dot{\phi})^2 b_{1x} - (\ddot{\theta} - \ddot{\phi})b_{1y} + 2\dot{\phi}(\dot{\theta} - \dot{\phi})$$
$$E = \ddot{\phi} + (\ddot{\theta} - \ddot{\phi})b_{1x} + (\dot{\theta} - \dot{\phi})^2 b_{1y}$$
$$F = \dot{\phi}^2 a_{0x} + \ddot{\phi}a_{0y} - \dot{\phi}^2 \tag{7.18}$$

Solving Eqs. 7.17 and 7.18 simultaneously, we obtain, for the specified $\dot{\theta}$, $\ddot{\theta}$, $\dot{\phi}$, and $\ddot{\phi}$,

$$a_{1x} = \frac{CE - BF}{AE - BD} \tag{7.19}$$

$$a_{1y} = \frac{AF - CD}{AE - BD} \tag{7.20}$$

from which $\mathbf{a}_1 = (0.068, 0.146)$.

The Slider-Crank Function Generator

The *relative velocity matrix* for input crank motion relative to the output slider becomes

$$[V_R] = \begin{bmatrix} 0 & -\dot{\theta}_j & -\dot{s}_j \cos \alpha \\ \dot{\theta}_j & 0 & -\dot{s}_j \sin \alpha \\ 0 & 0 & 0 \end{bmatrix} \tag{7.21}$$

The *relative acceleration matrix* is

$$[A_R] = \begin{bmatrix} -\dot{\theta}_j^2 & -\ddot{\theta}_j & -\ddot{s}_j \cos \alpha \\ \ddot{\theta}_j & -\dot{\theta}_j^2 & -\ddot{s}_j \sin \alpha \\ 0 & 0 & 0 \end{bmatrix} \tag{7.22}$$

Again, from Eq. 6.17, with $\mathbf{b}_1$ replacing $\mathbf{a}_0$, the *velocity design equation* becomes

$$a_{1x} A + a_{1y} B = C$$

where

$$A = -\dot{s}_j \cos \alpha - \dot{\theta}_j b_{1y}$$
$$B = -\dot{s}_j \sin \alpha + \dot{\theta}_j b_{1x}$$
$$C = -\dot{s}_j \cos \alpha b_{1x} - \dot{s}_j \sin \alpha b_{1y}$$
$$= -\dot{s}_j(b_{1x} \cos \alpha + b_{1y} \sin \alpha) \tag{7.23}$$

The *acceleration design equation* from Eq. 6.21 is

$$a_{1x} D + a_{1y} E = F \tag{7.24}$$

where

$$D = -\ddot{s}_j \cos \alpha + \dot{\theta}_j^2 b_{1x} - \ddot{\theta}_j b_{1y} - 2\dot{\theta}_j \dot{s}_j \sin \alpha$$
$$E = -\ddot{s}_j \sin \alpha + \ddot{\theta}_j b_{1x} + \dot{\theta}_j^2 b_{1y} + 2\dot{\theta}_j \dot{s}_j \cos \alpha$$
$$F = -\ddot{s}_j \cos \alpha b_{1x} - \ddot{s}_j \sin \alpha b_{1y} - \dot{s}_j^2$$

Example   7-5 Velocity-Acceleration   Synthesis   of   a   Plane Slider-Crank Function Generator. For an input motion

$$\dot{\theta} = 1.0 \text{ rad/sec}$$
$$\ddot{\theta} = 3.5 \text{ rad/sec}^2$$

we require an output motion

$$\dot{s} = 2.0 \text{ in./sec}$$

$$\ddot{s} = 3.0 \text{ in./sec}^2$$

The first position of the slider pivot is specified at $\mathbf{b}_1 = (1.0, 0.0)$ with an angle $\alpha = 180°$. Solving Eqs. 7.23 and 7.24, we obtain the first position of the moving crank pivot at

$$\mathbf{a}_1 = (0.0, \ 2.000)$$

The Double-Slider Function Generator

Differentiating the $[D_R]$ matrix (7.14), we obtain the *relative velocity matrix* for the velocity of the output slider with respect to the input slider in the form

$$[V_R] = \begin{bmatrix} 0 & 0 & \dot{s}_j \sin \alpha \\ 0 & 0 & (\dot{s}_j \cos \alpha - \dot{r}_j) \\ 0 & 0 & 0 \end{bmatrix} \tag{7.25}$$

The *relative acceleration matrix* becomes

$$[A_R] = \begin{bmatrix} 0 & 0 & \ddot{s}_j \sin \alpha \\ 0 & 0 & (\ddot{s}_j \cos \alpha - \ddot{r}_j) \\ 0 & 0 & 0 \end{bmatrix} \tag{7.26}$$

The design equations are again based on the constant length of the coupler $\overline{\mathbf{ab}}$. The *velocity design equation* becomes, with $\mathbf{a}_1 = (0., 0.)$,

$$Ab_{1x} + Bb_{1y} = C \tag{7.27}$$

where

$$A = \dot{s}_j \sin \alpha$$

$$B = \dot{s}_j \cos \alpha - \dot{r}_j$$

$$C = 0$$

The *acceleration design equation* is

$$Db_{1x} + Eb_{1y} = F$$

$$D = \ddot{s}_j \sin \alpha$$

$$E = (\ddot{s}_j \cos \alpha - \ddot{r}_j)$$

$$F = -(\dot{s}_j \sin \alpha)^2 - (\dot{s}_j \cos \alpha - \dot{r}_j)^2 \tag{7.28}$$

Example   7-6 Velocity-Acceleration   Synthesis   of   a   Plane Double-Slider Function Generator. Synthesize a double-slider mechanism to meet the following specifications:

$$\dot{r} = 1.0 \text{ in./sec} \qquad \ddot{r} = 2.0 \text{ in./sec}^2$$

$$\dot{s} = 3.0 \qquad\qquad \ddot{s} = 4.0$$

The first slider moves along the y-axis, as shown in Figure 7.5. The second slider direction is assumed parallel to the x-axis (i.e., $\alpha = 90°$). The point $\mathbf{a}_1$ was assumed at the origin in the derivation of the relative motion matrices.
Substituting into Eqs. 7.27 and 7.28 with

$$A = 3.0 \qquad B = -1.0 \qquad C = 0.$$

$$D = 4.0 \qquad E = -2.0 \qquad F = -9.0 - 1.0 = -10.0$$

the coordinates of the second slider pivot become

$$b_{1x} = \frac{CE - BF}{AE - BD} = \frac{-(-1.)(-10.)}{(3.)(-2.) - (-1.)(4.)} = \frac{-10.}{-2.} = 5.0$$

$$b_{1y} = \frac{AF - CD}{AE - BD} = \frac{(3.)(-10.)}{-2.} = \frac{-30.}{-2.} = 15.0$$

7.7   FOUR-POSITION PLANE FUNCTION GENERATORS

When four coordinated values of input and output motion are to be coordinated with a specified functional relationship, the design procedure is again analogous to that used for rigid body guidance mechanisms. The input and output motions are transformed into relative displacement, velocity, and acceleration matrices, using the relationships derived in Sections 7.5 and 7.6. In general it will not be possible to obtain a set of linear design equations. The nonlinear design equations must be solved using numerical iterative methods.

The Four-Bar Linkage Function Generator

The constraint equations are again based on the constant length of the coupler as the linkage assumes a series of four positions relative to the first position of the output crank. Using the coordinate system of Figure 7.2, we obtain a set of three design equations.

$$(\mathbf{a}_j' - \mathbf{b}_1)^T(\mathbf{a}_j' - \mathbf{b}_1) = (\mathbf{a}_1 - \mathbf{b}_1)^T(\mathbf{a}_1 - \mathbf{b}_1) \qquad j = 2, 3, 4 \qquad (7.29)$$

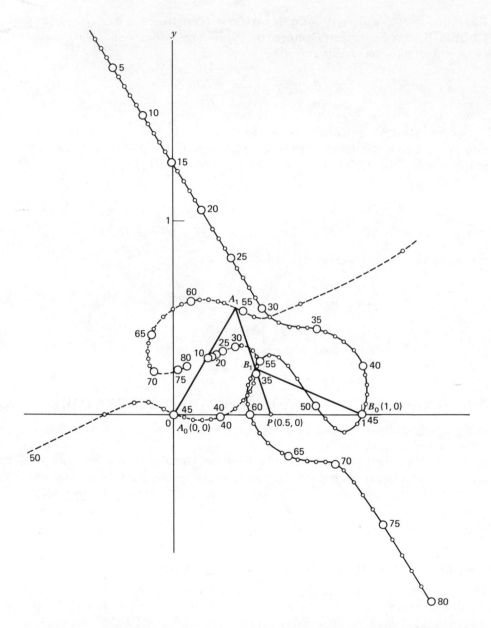

FIGURE 7.6 EXAMPLE 7.7. THE SELECTED LINKAGE WITH A_1(0.316397, 0.553513) AND B_1(0.422429, 0.233854).

There are four unknown coordinates a_{1x}, a_{1y}, b_{1x}, and b_{1y}. The coordinates of point $\mathbf{a}'_j$ can be calculated in terms of $\mathbf{a}_1$ from

$$(\mathbf{a}'_j) = [D_R](\mathbf{a}_1)$$

Example   7-7 The   Crossed   Linkage   Four-Bar   Function Generator. Figure 7.6 shows the relative center point and circle point curves for a linkage designed to generate the function $y = e^x$, $0 \leq x \leq 1.2$ with precision points at $x = 0.$, 0.4, 0.8, 1.2. A particular solution has been selected such that the velocity ratio in the first position equals -1.0.

The velocity ratio condition could have been imposed as an additional design constraint equation to be solved simultaneously with the set of three constant length equations. This would lead to a unique solution. The velocity ratio condition can be imposed in the form of the equation of a straight line

$$\begin{vmatrix} a_{jx} & a_{jy} & 1 \\ p_{jx} & p_{jy} & 1 \\ b_{jx} & b_{jy} & 1 \end{vmatrix} = 0 \tag{7.30}$$

where $\mathbf{p}_j$ defines the location of the relative velocity pole and $\mathbf{a}_j$ the input crank pivot in the jth position.

7.8   THE SPHERICAL FOUR-BAR FUNCTION GENERATOR

Figure 7.7 illustrates the coordinate system used in the synthesis of a spherical linkage four-bar function generator. The fixed input axis is assumed to lie on the x-axis of a set of orthogonal axes that intersect at the center of the unit sphere. The output axis lies in the x-y plane at an arbitrary angle α as shown.

The synthesis will again involve transforming the function generator problem to an equivalent rigid body guidance problem by describing the relative motion of the input link with respect to the first position of the output link. The relative displacement matrix is defined by

$$[D_R] = [D_{-\phi_{1j}}][D_{\theta_{1j}}] \tag{7.31}$$

The elements of the relative displacement matrix become

$$d_{11j} = \cos^2 \alpha + \sin^2 \alpha \cos \phi_{1j}$$

$$d_{12j} = \sin \alpha \cos \alpha \cos \theta_{1j}(1 - \cos \phi_{1j}) - \sin \alpha \sin \theta_{1j} \sin \phi_{1j}$$

$$d_{13j} = -\sin \alpha \cos \alpha \sin \theta_{1j}(1 - \cos \phi_{1j}) - \sin \alpha \cos \theta_{1j} \sin \phi_{1j}$$

$$d_{14j} = 0.0$$

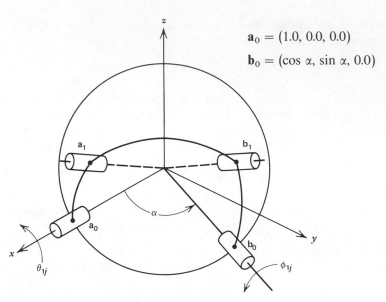

$$\mathbf{a}_0 = (1.0,\ 0.0,\ 0.0)$$
$$\mathbf{b}_0 = (\cos \alpha,\ \sin \alpha,\ 0.0)$$

FIGURE 7.7 THE SPHERICAL FOUR-BAR FUNCTION GENERATOR.

$$d_{21j} = \sin \alpha \cos \alpha (1 - \cos \phi_{1j})$$

$$d_{22j} = \sin^2 \alpha \cos \theta_{1j} + \cos^2 \alpha \cos \phi_{1j} \cos \theta_{1j} + \cos \alpha \sin \theta_{1j} \sin \phi_{1j}$$

$$d_{23j} = -\sin^2 \alpha \sin \theta_{1j} - \cos^2 \alpha \sin \theta_{1j} \cos \phi_{1j} + \cos \alpha \cos \theta_{1j} \sin \phi_{1j}$$

$$d_{24j} = 0.0$$

$$d_{31j} = \sin \alpha \sin \phi_{1j}$$

$$d_{32j} = \sin \theta_{1j} \cos \phi_{1j} - \cos \alpha \cos \theta_{1j} \sin \phi_{1j}$$

$$d_{33j} = \cos \theta_{1j} \cos \phi_{1j} + \cos \alpha \sin \theta_{1j} \sin \phi_{1j}$$

$$d_{34j} = d_{41j} = d_{42j} = d_{43j} = 0.0$$

$$d_{44j} = 1.$$

With the relative displacement matrix (7.31) available in numerical form, the constant length equation (Eq. 7.29) plus two additional relationships that restrict points $\mathbf{a}_1$ and $\mathbf{b}_1$ to the unit sphere

$$(\mathbf{a}_1)^T(\mathbf{a}_1) = 1 \tag{7.32}$$

$$(\mathbf{b}_1)^T(\mathbf{b}_1) = 1 \tag{7.33}$$

are available as constraint equations.

For five precision points this would lead to a set of six design equations in the six scalar coordinates of points $\mathbf{a}_1$ and $\mathbf{b}_1$. If we assume the angle α as a free variable a maximum of six precision points is possible with the $[D_R]$ matrix as a function of α.

7.9   THE RSSR SPATIAL FOUR-BAR FUNCTION GENERATOR

The RSSR mechanism is perhaps the simplest of spatial mechanisms and yet potentially one of the most useful. The kinematic inversion method will again be applied in the synthesis procedures. The same principles are easily extended to other spatial function generator mechanisms. For example, the input could be either the rotation about a revolute joint or the sliding of a prismatic pair. The output motion could be either revolute, prismatic, or the two degrees of freedom of a cylindrical pair.

For the RSSR mechanism it is assumed that the input and output axes do not intersect and that the common perpendicular between the axes is of unit length, as shown in Figure 7.8. All other dimensions are ratios of this basic dimension, and the synthesized linkage may be scaled to any desired size by application of a constant scale factor to all linear dimensions.

The input axis $\mathbf{u}_a$ is arbitrarily located along the z-axis so that its direction cosines become $\mathbf{u}_a = (0., 0., 1.0)$. The direction cosines of the output axis are $\mathbf{u}_b = (0., -\sin \alpha, \cos \alpha)$. The mechanism will be completely specified in its first

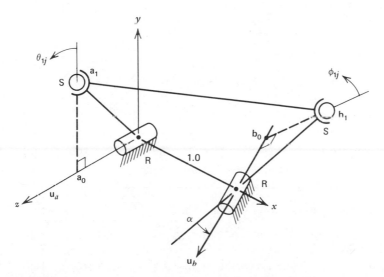

FIGURE 7.8   THE RSSR SPATIAL FUNCTION GENERATOR.

position by the six scalar coordinates of points $\mathbf{a}_1$ and $\mathbf{b}_1$ plus the angle α. Once these quantities have been determined points $\mathbf{a}_0$ and $\mathbf{b}_0$ on the normals to axes $\mathbf{u}_a$ and $\mathbf{u}_b$ may be determined by analytic geometry.

We again invert the various positions of the mechanism about the first position of the output crank $\overline{\mathbf{b}_1 \mathbf{b}_0}$. The relative displacement matrix is calculated from

$$[D_R] = [D_{-\phi_{1j}}][D_{\theta_{1j}}] \tag{7.34}$$

We note that the displacement matrix in each case is expressed in the form of Eq. 3.22, where the rotation submatrix is given by Eq. 3.12.

This leads to the following expressions for the elements of the relative displacement matrix $[D_R]$ for the RSSR mechanism.

$$d_{11j} = \cos \theta_{1j} \cos \phi_{1j} + \sin \theta_{1j} \cos \alpha \sin \phi_{1j}$$

$$d_{12j} = -\sin \theta_{1j} \cos \phi_{1j} + \cos \theta_{1j} \cos \alpha \sin \phi_{1j}$$

$$d_{13j} = \sin \alpha \sin \phi_{1j}$$

$$d_{14j} = \text{vers } \phi_{1j} = 1. - \cos \phi_{1j}$$

$$d_{21j} = -\cos \theta_{1j} \cos \alpha \sin \phi_{1j} + \sin \theta_{1j}(\sin^2 \alpha \text{ vers } \phi_{1j} + \cos \phi_{1j})$$

$$d_{22j} = \sin \theta_{1j} \cos \alpha \sin \phi_{1j} + \cos \theta_{1j}(\sin^2 \alpha \text{ vers } \phi_{1j} + \cos \phi_{1j})$$

$$d_{23j} = -\sin \alpha \cos \alpha \text{ vers } \phi_{1j}$$

$$d_{24j} = \cos \alpha \sin \phi_{1j}$$

$$d_{31j} = -\cos \theta_{1j} \sin \alpha \sin \phi_{1j} - \sin \theta_{1j} \sin \alpha \cos \alpha \text{ vers } \phi_{1j}$$

$$d_{32j} = \sin \theta_{1j} \sin \alpha \sin \phi_{1j} - \cos \theta_{1j} \sin \alpha \cos \alpha \text{ vers } \phi_{1j}$$

$$d_{33j} = \cos^2 \alpha \text{ vers } \phi_{1j} + \cos \phi_{1j}$$

$$d_{34j} = \sin \alpha \sin \phi_{1j}$$

$$d_{41j} = d_{42j} = d_{43j} = 0.$$

$$d_{44j} = 1.0$$

In the case of the RSSR function generator we do not have the extra conditions imposed by Eqs. 7.31 and 7.32; hence the constant length condition is the only basic constraint.

$$(\mathbf{a}_j' - \mathbf{b}_1)^T(\mathbf{a}_j' - \mathbf{b}_1) = (\mathbf{a}_1 - \mathbf{b}_1)^T(\mathbf{a}_1 - \mathbf{b}_1) \qquad j = 2, 3 \ldots \tag{7.35}$$

When the angle α is specified, this leads to a unique solution for the six coordinates of points $\mathbf{a}_1$ and $\mathbf{b}_1$ for a maximum of seven precision points. If α is left as a free variable a maximum of eight precision points is possible, but the relative displacement matrix $[D_R]$ must be recalculated for each variation in α.

Example 7-8 Synthesis of an RSSR Function Generator with Specified Plane of Rotation for Both Input and Output Cranks.

Function: $y = e^x$

Range of motion:

1. Independent variable $0 \le x \le 1.2$.
2. Input crank angle $-90° \le \theta \le 0$ (clockwise).
3. Output crank angle $0 \le \phi \le 90°$.

Precision points: $x = 0., 0.4, 0.8, 1.2$

Plane of rotation:

1. input crank on plane $z = 1.5$.
2. output crank on plane $y = 2.0$.

Angle between input and output axes, $\alpha = 90°$

The displacement angles are computed as

$$\theta_{12} = -30.0° \qquad \phi_{12} = 19.078428°$$
$$\theta_{13} = -60.0 \qquad \phi_{13} = 47.540045$$
$$\theta_{14} = -90.0 \qquad \phi_{14} = 90.0$$

The constraint equations become

$$(\mathbf{a}_j' - \mathbf{b}_1)^T(\mathbf{a}_j' - \mathbf{b}_1) = (\mathbf{a}_1 - \mathbf{b}_1)^T(\mathbf{a}_1 - \mathbf{b}_1) \qquad j = 2, 3, 4$$
$$a_{1z} = 1.5$$
$$b_{1y} = 2.0$$

The coordinates of points $\mathbf{a}_j'$ are written in terms of $\mathbf{a}_1$ and the specified relative displacement matrix in the form

$$\mathbf{a}_j' = [D_R]\mathbf{a}_1$$

This leads to a set of five design equations in the six unknowns $a_{1x}, a_{1y}, a_{1z}, b_{1x}, b_{1y},$ and b_{1z}. We arbitrarily select b_{1x} as the specified parameter and allow b_{1x} to vary from 0.1 to 3.6 in increments of 0.1. This leads to points $\mathbf{b}_1$ on the *output crank sphere point curve*, which are plotted on the plane $y = 2.0$. The solution of the nonlinear design equations leads to the simultaneous location of corresponding points $\mathbf{a}_1$ on the *input crank sphere point curve*. The results of these calculations are shown in Figure 7.9. A particular solution, corresponding to

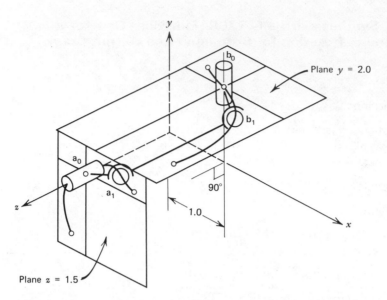

FIGURE 7.9 EXAMPLE 7.8. SOLUTION FOR THE RSSR FUNCTION GENERATOR.

TABLE 7.1 RESULTS FOR THE FOUR PRECISION POINT RSSR FUNCTION GENERATOR OF EXAMPLE 7-8

Point	Point Coordinates on the Sphere Point Curves					
	$\mathbf{a}_1 = (a_{1x}, a_{1y}, a_{1z})$			$\mathbf{b}_1 = (b_{1x}, b_{1y}, b_{1z})$		
1	-0.410	-1.334	1.500	0.200	2.000	-0.284
2	-0.345	-0.699	1.500	0.400	2.000	-0.137
3	-0.243	-0.359	1.500	0.600	2.000	-0.066
4	-0.125	-0.141	1.500	0.800	2.000	-0.028
5	0.0	0.0	1.500	1.000	2.000	0.0
6	0.291	0.131	1.500	1.500	2.000	0.159
7	0.510	0.114	1.500	2.000	2.000	0.562
8	0.667	0.073	1.500	2.500	2.000	1.179
9	0.782	0.041	1.500	3.000	2.000	1.943
10	0.868	0.019	1.500	3.500	2.000	2.810

$\mathbf{a}_1 = (0.510, 0.114, 1.500)$ and $\mathbf{b}_1 = (2.000, 2,000, 0.562)$, is shown in Figure 7.9. A complete set of coordinates for selected points on the sphere point curves is given in Table 7.1.

The method employed in Example 7-8 may be generalized to constrain the sphere point curves to lie on any specified surface. For example, instead of planes perpendicular to one of the coordinate axes, we might specify the equation of arbitrary planes

$$Ab_{1x} + Bb_{1y} + Cb_{1z} + D = 0$$

$$Ea_{1x} + Fa_{1y} + Ga_{1z} + H = 0 \tag{7.36}$$

or the more general equations of arbitrary surfaces in the form

$$f(a_{1x}, a_{1y}, a_{1z}) = 0$$

$$g(b_{1x}, b_{1y}, b_{1z}) = 0 \tag{7.37}$$

Example 7-9 Design of an RSSR Function Generator with Six Precision Points.

Function: $y = \cos x$

Range of motion

1. Independent variable $0 \le x \le 180°$.
2. Input crank angle $0 \le \theta \le 150°$.
3. Output crank angle $-100° \le \phi \le 0$.

Precision points using six point Chebyshev spacing $x = 3.069°$,

26.361°, 66.708°, 113.292°, 153.639°, 176.931°

Angle between input and output axes, $\alpha = 90°$
The calculated crank displacement angles are

Point	Input θ	Output ϕ
1	0.00	0.00
2	19.41	− 5.13
3	53.03	− 30.17
4	91.85	− 69.70
5	125.47	− 94.74
6	144.88	− 99.86

In this case the constant length Eq. 7.35 for $j = 2, 3, 4, 5, 6$ leads to a set of five equations with six scalar components of $\mathbf{a}_1$ and $\mathbf{b}_1$ as unknowns. Since we have more variables than equations, we are free to choose any one of the six coordinates arbitrarily and solve the nonlinear design equations for the remaining five. This leads to the simultaneous generation of two spatial curves for the sphere point curves. Table 7.2 lists the coordinates of points on two sections of the curves. The first section of Table 7.2 gives results for a specification of b_{1z} over the range from $+0.200$ to -1.500. The second section gives results found by specifying b_{1x} for values between -4.000 and $+3.000$.

TABLE 7.2 RESULTS FOR THE SIX PRECISION POINT RSSR FUNCTION GENERATOR OF EXAMPLE 7-9

	Point Coordinates on the Sphere Point Curves					
Point	$\mathbf{a}_1 = (a_{1x}, a_{1y}, a_{1z})$			$\mathbf{b}_1 = (b_{1x}, b_{1y}, b_{1z})$		
	First Section					
1	0.360	−0.366	−0.832	2.910	−3.068	0.200
2	0.377	−0.320	−0.839	2.316	−2.075	0.0
3	0.461	−0.225	−0.907	1.823	−0.973	−0.200
4	0.731	−0.111	−1.259	1.838	−0.378	−0.400
5	1.015	−0.014	−1.710	2.034	−0.141	−0.600
6	1.294	0.077	−2.173	2.257	0.004	−0.800
7	1.570	0.166	−2.641	2.487	0.115	−1.000
8	1.843	0.253	−3.108	2.721	0.210	−1.200
9	2.115	0.339	−3.575	2.957	0.294	−1.400
10	2.250	0.382	−3.809	3.075	0.334	−1.500
	Second Section					
11	−5.638	−2.113	10.028	−4.000	−1.216	4.448
12	−4.512	−1.757	8.066	−3.000	−0.954	3.606
13	−3.384	−1.400	6.103	−2.000	−0.682	2.766
14	−2.251	−1.042	4.137	−1.000	−0.388	1.927
15	−1.099	−0.680	2.167	0.0	−0.002	1.098
16	−0.485	−0.498	1.187	0.500	0.438	0.738
17	−0.077	−0.513	0.830	0.900	3.356	2.106
18	−0.027	−0.564	0.809	1.500	8.368	5.650
19	−0.016	−0.576	0.800	2.000	12.263	8.486
20	−0.007	−0.587	0.792	3.000	19.912	14.090

7.10   THE RSSR FUNCTION GENERATOR WITH SPECIFIED VELOCITY AND ACCELERATION

The elements of the relative velocity and acceleration matrices are found by differentiating the corresponding element of the relative displacement matrix (7.34) and then setting all finite displacement angles to zero.

The elements of the *relative velocity matrix* become

$$[V_R] = \begin{bmatrix} 0 & -(\dot{\theta}_j - \dot{\phi}_j \cos \alpha) & \dot{\phi}_j \sin \alpha & 0 \\ (\dot{\theta}_j - \dot{\phi}_j \cos \alpha) & 0 & 0 & \dot{\phi}_j \cos \alpha \\ -\dot{\phi}_j \sin \alpha & 0 & 0 & \dot{\phi}_j \sin \alpha \\ 0 & 0 & 0 & 0 \end{bmatrix}$$

$$(7.38)$$

The elements of the relative acceleration matrix are given by

$$[A_R] = [a_{ij}] \qquad (7.39)$$

where

$$a_{11} = -(\dot{\theta}_j^2 + \dot{\phi}_j^2 - 2\dot{\theta}_j\dot{\phi}_j \cos \alpha)$$

$$a_{12} = -(\ddot{\theta}_j - \ddot{\phi}_j \cos \alpha)$$

$$a_{13} = (\ddot{\phi}_j \sin \alpha)$$

$$a_{14} = (\dot{\phi}_j^2)$$

$$a_{21} = (\ddot{\theta}_j - \ddot{\phi}_j \cos \alpha)$$

$$a_{22} = -(\dot{\theta}_j^2 + \dot{\phi}_j^2 \cos^2 \alpha - 2\dot{\theta}_j\dot{\phi}_j \cos \alpha)$$

$$a_{23} = (-\dot{\phi}_j^2 \sin \alpha \cos \alpha)$$

$$a_{24} = (\ddot{\phi}_j \cos \alpha)$$

$$a_{31} = (-\ddot{\phi}_j \sin \alpha)$$

$$a_{32} = -(\dot{\phi}_j^2 \sin \alpha \cos \alpha - 2\dot{\theta}_j\dot{\phi}_j \sin \alpha)$$

$$a_{33} = -(\dot{\phi}_j^2 \sin^2 \alpha)$$

$$a_{34} = (\ddot{\phi}_j \sin \alpha)$$

$$a_{41} = a_{42} = a_{43} = a_{44} = 0.$$

Example 7-10 Synthesis of the RSSR Function Generator with Specified Velocity and Acceleration

$$\dot{\theta} = -1.0 \qquad \dot{\phi} = 0.415 \qquad \alpha = -90°$$

$$\ddot{\theta} = 0.0 \qquad \ddot{\phi} = 0.160$$

Input crank length $= 1.0$

The relative velocity matrix is computed as

$$[V_R] = \begin{bmatrix} 0 & 1.000 & 0.415 & 0 \\ -1.000 & 0 & 0 & 0 \\ -0.415 & 0 & 0 & 0.415 \\ 0 & 0 & 0 & 0 \end{bmatrix} \tag{7.40}$$

and the relative acceleration matrix becomes

$$[A_R] = \begin{bmatrix} -(0.415^2 + 1) & 0 & -0.16 & 0.415^2 \\ 0 & -1.0 & 0 & 0 \\ 0.16 & -2(0.415) & -0.415^2 & -0.16 \\ 0 & 0 & 0 & 0 \end{bmatrix} \tag{7.41}$$

The constraint equations are

$$(\dot{\mathbf{a}}_1)^T(\mathbf{a}_1 - \mathbf{b}_1) = 0 \tag{7.42}$$

$$(\ddot{\mathbf{a}}_1)^T(\mathbf{a}_1 - \mathbf{b}_1) + (\dot{\mathbf{a}}_1)^T(\dot{\mathbf{a}}_1) = 0 \tag{7.43}$$

where

$$\dot{\mathbf{a}}_1 = [V_R]\mathbf{a}_1 \qquad \ddot{\mathbf{a}}_1 = [A_R]\mathbf{a}_1$$

The unit length condition on the input crank adds an additional constraint equation. In this case we note that the input axis is specified on the z-axis and, since point $\mathbf{a}$ will rotate in a plane perpendicular to the z-axis, the constant unit length is specified by

$$a_{1x}^2 + a_{1y}^2 = 1 \tag{7.44}$$

We now have three design equations with six unknowns $a_{1x}, a_{1y}, a_{1z}, b_{1x}, b_{1y}$, and b_{1z}. We may impose three additional constraints as follows: (1) assume the input crank makes an angle of $60°$ with the x-axis. This requires that

$$\frac{a_{1y}}{a_{1x}} = \tan 60° \tag{7.45}$$

(2) specify that the input crank is to rotate in a plane perpendicular to the z-axis with

$$a_{1z} = 2.500 \tag{7.46}$$

and (3) specify that the output crank rotates in a plane defined by

$$b_{1y} = 1.500 \tag{7.47}$$

Solving the set of six design equations (Eqs. 7.42 to 7.47), we find the coordinates of the spherical joints

$$\mathbf{a}_1 = (-0.500, \ -0.866, \ 2.500)$$
$$\mathbf{b}_1 = (-4.240, \ 1.500, \ 1.630)$$

The general method illustrated for the RSSR function generator is applicable to a wide variety of spatial function generators once that expressions for the relative motion matrices have been derived for the particular mechanism under consideration. The design equations will, in general, be nonlinear algebraic equations but, when confidence in the Newton-Raphson method has been achieved, the solution is relatively straightforward.

Chapter 8
Path Generation

8.1 INTRODUCTION

A path generation mechanism is designed to guide one point on a moving rigid body such that it passes through a specified sequence of points on a path in space. It is often important that these specified path points are coordinated with specified displacement angles of an input crank. In the formulation of a typical problem the rigid body rotation angles of the guided body that contains the moving point are included as unknowns to be specified. This results in an advantage as compared to the synthesis of rigid body guidance mechanisms since, in this case, the designer has the option of either increasing the number of specified path points or, alternatively, there is greater freedom of choice of design parameters.

8.2 THE PLANE FOUR-BAR PATH GENERATION LINKAGE

In this case the constraint equations are based on the constant length condition for each of the two guiding links $\overline{\mathbf{a}\mathbf{a}_0}$ and $\overline{\mathbf{b}\mathbf{b}_0}$ as shown in Figure 8.1. The constraint equations become

$$(\mathbf{a}_j - \mathbf{a}_0)^T(\mathbf{a}_j - \mathbf{a}_0) = (\mathbf{a}_1 - \mathbf{a}_0)^T(\mathbf{a}_1 - \mathbf{a}_0) \qquad j = 2, 3, 4, \ldots$$

$$(\mathbf{b}_j - \mathbf{b}_0)^T(\mathbf{b}_j - \mathbf{b}_0) = (\mathbf{b}_1 - \mathbf{b}_0)^T(\mathbf{b}_1 - \mathbf{b}_0) \qquad j = 2, 3, 4, \ldots \quad (8.1)$$

The coordinates of points $\mathbf{a}_j$ and $\mathbf{b}_j$ are calculated from

$$(\mathbf{a}_j) = [D_{1j}](\mathbf{a}_1)$$

$$(\mathbf{b}_j) = [D_{1j}](\mathbf{b}_1)$$

where $[D_{1j}]$ is defined by Eq. 3.17 with $(\mathbf{p}_j), j = 1, 2, 3, \ldots$ as the specified points on a plane path and θ_{1j} as rotation angles to be determined along with the unknown coordinates of points $\mathbf{a}_0$, $\mathbf{a}_1$, $\mathbf{b}_0$ and $\mathbf{b}_1$.

There is considerable freedom of choice in the specification of design parameters, depending on the number of specified path precision points. Table 8.1 lists some possible combinations of specified and unspecified design parameters for various numbers of specified path points. As noted in Table 8.1, the theoretical

190

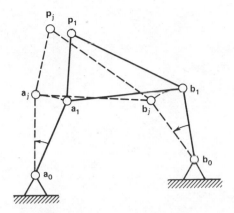

FIGURE   8.1   THE   PLANE   FOUR-BAR GENERATOR.

maximum number of specified path points is equal to nine. A practical limit is probably six or seven with specification of at least three of the sixteen variables.

We note that the unknowns include the eight coordinates $a_{0x}, a_{0y}, a_{1x}, a_{1y}, b_{0x}, b_{0y}, b_{1x}$, and b_{1y}, plus the added rotation angles $\theta_{1j}, j = 2, 3, \ldots n$. In solving the set of nonlinear design equations we may arbitrarily assume any of the unknowns to make the number of unknowns equal to the number of equations.

Additional constraint equations may be combined with Eqs. 8.1. For example, assume that we are given two links of length L_2 and L_4 to be used as cranks in the linkage. We require the location of two fixed pivots and the first position of the

TABLE 8.1   PARAMETRIC RELATIONSHIPS   THE PLANE FOUR-BAR PATH GENERATOR LINKAGE

Path Points	Design Equations	Unknown Variables	Specified Variables	Selected Examples	
				Specify	Calculate
2	2	9	7	$a_0, a_1, b_1, \theta_{12}$	b_0
3	4	10	6	a_0, a_1, b_0	$b_1, \theta_{12}, \theta_{13}$
4	6	11	5	a_0, a_1, b_{0x}	$b_{0y}, b_1, \theta_{12}, \theta_{13}, \theta_{14}$
5	8	12	4	a_0, b_0	$a_1, b_1, \theta_{12}, \theta_{13}, \theta_{14}, \theta_{15}$
6	10	13	3	a_0, b_{0x}	All others
7	12	14	2	a_0	All others
8	14	15	1	a_{0x}	All others
9	16	16	0	None—possible unique solution	

Beyond six path points, solutions are highly dependent on initial guesses, numerical accuracy, and choice of path point coordinates.

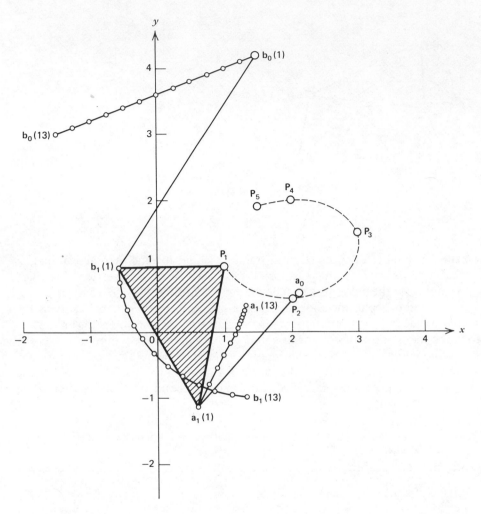

FIGURE 8.2 EXAMPLE 8.1. FOUR-BAR LINKAGE FOR PATH GENERATION WITH FIVE PRECISION POINTS AND ARBITRARY CHOICE OF BOTH FIXED PIVOTS. ONE FIXED PIVOT VARIED TO DISPLAY LOCI OF TWO MOVING PIVOTS.

two moving pivots to be attached to the coupler of the linkage. Two additional constraint equations must be added to the basic set (8.1)

$$L_2^2 = (\mathbf{a}_1 - \mathbf{a}_0)^T(\mathbf{a}_1 - \mathbf{a}_0) \tag{8.2}$$
$$L_4^2 = (\mathbf{b}_1 - \mathbf{b}_0)^T(\mathbf{b}_1 - \mathbf{b}_0)$$

For five path precision points this would lead to a set of ten design equations with eight unknowns.

Example 8-1 Design of a plane four-bar path generator with five path precision points. The five path precision points are specified as

$$\mathbf{p}_1 = (1.00,\ 1.00)$$

$$\mathbf{p}_2 = (2.00,\ 0.50)$$

$$\mathbf{p}_3 = (3.00,\ 1.50)$$

$$\mathbf{p}_4 = (2.00,\ 2.00)$$

$$\mathbf{p}_5 = (1.50,\ 1.90)$$

From Table 8.1 we see that with five specified path precision points we may arbitrarily specify four parameters. We arbitrarily decide to specify both fixed pivots $\mathbf{a}_0 = (a_{0x},\ a_{0y})$ and $\mathbf{b}_0 = (b_{0x},\ b_{0y})$. Let

$$\mathbf{a}_0 = (2.10,\ 0.60)$$

$$\mathbf{b}_0 = (1.50,\ 4.20)$$

The solution of the design equations (Eq. 8.1) gives

$$\mathbf{a}_1 = (0.607,\ -1.127)$$

$$\mathbf{b}_1 = (-0.586,\ 0.997)$$

The results are shown in Figure 8.2, which displays a family of solutions for an arbitrary series of assumed positions for $\mathbf{b}_0$ along a straight line as shown. Note that the rotation angles for the coupler would also be included in the unknowns but are not listed here.

Example 8-2 Design of a Plane Four-Bar Path Generator with Specified Length for Both Guiding Cranks. The path precision points are the same as given for Example 8.1. In this case we assume

$$L_2 = 1.0$$

$$L_4 = 2.0$$

Since two new constraint equations have been added, we must reduce the number of specified parameters by two.

In this case we decide to specify

$$\mathbf{a}_0 = (2.10,\ 0.50)$$

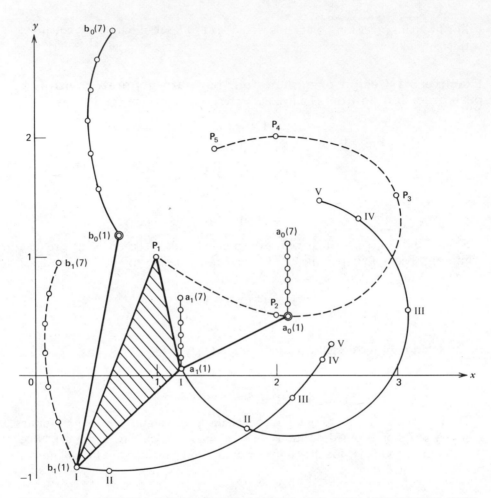

FIGURE 8.3 EXAMPLE 8.2. FOUR-BAR LINKAGE FOR PATH GENERATION WITH FIVE PRECISION POINTS. ARBITRARY SPECIFICATION OF ONE FIXED PIVOT AND LENGTHS OF BOTH GUIDING CRANKS. ONE FIXED PIVOT VARIED TO DISPLAY LOCI OF TWO MOVING PIVOTS AND SECOND FIXED PIVOT.

and solve the set of ten design equations for the unknowns $b_{0x}, b_{0y}, a_{1x}, a_{1y}, b_{1x}, b_{1y}, \theta_{12}, \theta_{13}, \theta_{14}$, and θ_{15}. This leads to

$$\mathbf{b}_0 = (0.693, \ 1.184)$$

$$\mathbf{a}_1 = (1.207, \ 0.050)$$

$$\mathbf{b}_1 = (0.334, \ -0.783)$$

Since the location of $\mathbf{a}_0$ is arbitrary, it may be varied to display a locus of possible solutions, as shown in Fig. 8.3.

8.3 THE SPHERICAL FOUR-BAR PATH GENERATOR

The spherical linkage, where all axes must intersect at a common point, is also useful as a possible means for guiding one point on a moving rigid body through a series of spatial path precision points. If all path points lie on a spherical surface, Table 8.2 indicates that a maximum of nine precision points is theoretically possible.

TABLE 8.2 PARAMETRIC RELATIONSHIPS—THE SPHERICAL FOUR-BAR PATH GENERATOR LINKAGE

Path Points	Design Equations	Unknown Variables	Specified Variables	Selected Example of Specified Variables
2	6	13	7	$a_{0x}, a_{0y}, a_{1x}, a_{1y}, b_{0x}, b_{0y}, \beta_{12}$
3	8	14	6	$a_{0x}, a_{0y}, b_{0x}, b_{0y}, \beta_{12}, \beta_{13}$
4	10	15	5	$a_{0x}, a_{0y}, \beta_{12}, \beta_{13}, \beta_{14}$
5	12	16	4	$\beta_{12}, \beta_{13}, \beta_{14}, \beta_{15}$
6	14	17	3	$\beta_{12}, \beta_{13}, \beta_{14}$
7	16	18	2	β_{12}, β_{13}
8	18	19	1	β_{12}
9	20	20	0	None—possible unique solution

Beyond six path points, solutions are highly dependent on initial guesses, numerical accuracy, and choice of path point coordinates.

The displacement of the rigid body (i.e., the coupler of the spherical four-bar linkage) will be described in terms of the displacement of the coupler point following the specified path plus the rotation angles α_{1j} and β_{1j}, as shown in Figure 8.4. The angle α_{1j} corresponds to a great circle rotation about an axis $\mathbf{u}_j$ normal to the plane defined by the center of the circle $\mathbf{c}_0$ and the two positions of point $\mathbf{p}$ (i.e., $\mathbf{p}_1$ and $\mathbf{p}_j$). The angle β_{1j} is assumed about the axis $\mathbf{c}_0 \mathbf{p}_j$.

For two specified positions of the path point $\mathbf{p}_1$ and $\mathbf{p}_j$ we see that the spherical angle α_{1j} is defined by

$$\cos \alpha_{1j} = (\mathbf{p}_1)^T(\mathbf{p}_j) = p_{1x}p_{jx} + p_{1y}p_{jy} + p_{1z}p_{jz} \tag{8.3}$$

The direction of axis $\mathbf{u}_j$ is defined by

$$\mathbf{u}_j = \frac{\mathbf{r}_1 \times \mathbf{r}_j}{[(r_1)^T(r_1)]^{1/2}} = \frac{\mathbf{r}_1 \times \mathbf{r}_j}{\|\mathbf{r}_1\|} \tag{8.4}$$

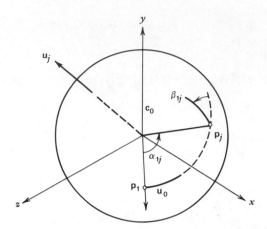

FIGURE 8.4 SPHERICAL ANGLES FOR DESCRIPTION OF THE COUPLER DISPLACE-MENT IN A SPHERICAL FOUR-BAR LINKAGE.

where

$$\mathbf{r}_1 = (\mathbf{p}_1 - \mathbf{c}_0)$$

$$\mathbf{r}_j = (\mathbf{p}_j - \mathbf{c}_0)$$

The angular displacement β_{1j} is accounted for first. A rotation about axis $\mathbf{u}_0 = (\mathbf{p}_1 - \mathbf{c}_0)$, where

$$\mathbf{u}_0 = \frac{(\mathbf{p}_1 - \mathbf{c}_0)}{[(\mathbf{p}_1 - \mathbf{c}_0)^T(\mathbf{p}_1 - \mathbf{c}_0)]^{1/2}} = \frac{\mathbf{r}_1}{[(\mathbf{r}_1)^T(\mathbf{r}_1)]^{1/2}} = \frac{\mathbf{r}_1}{\|\mathbf{r}_1\|} \tag{8.5}$$

The rigid body displacement matrix is then formed as

$$[D_{1j}] = [D_{u_j,\,\alpha_{1j}}][D_{u_0,\,\beta_{1j}}] \tag{8.6}$$

where $[D_{u_j,\,\alpha_{1j}}]$ and $[D_{u_0,\,\beta_{1j}}]$ are given by matrix (3.22).

Figure 8.5 shows a typical spherical four-bar path generation linkage. In Table 8.2 we see that for up to a maximum of five precision path points, the guiding cranks may be determined independently. With more than five path precision points the crank pivot coordinates for the two guiding cranks are mutually depen-dent on the specification of angles β_{1j}. A maximum of nine path precision points is theoretically possible, although it is unlikely that convergence to such a solution would be possible for arbitrary path precision points.

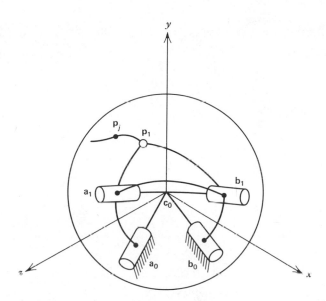

FIGURE 8.5   THE SPHERICAL FOUR-BAR PATH GEN-ERATOR.

Example 8-3 Design of a Spherical Path Generation Mechanism with Four Path Precision Points.

As indicated in Table 8.2, we may arbitrarily assume five of the unknown linkage parameters $a_{0x}, a_{0y}, a_{0z}, a_{1x}, a_{1z}$, and b_{1x} plus the three angular displacement angles β_{12}, β_{13}, and β_{14}. The four path precision points are specified as

$$\mathbf{p}_1 = (1.00,\ 1.00,\ 2.00)$$

$$\mathbf{p}_2 = (2.50,\ 1.75,\ 2.50)$$

$$\mathbf{p}_3 = (4.00,\ 2.00,\ 2.00)$$

$$\mathbf{p}_4 = (4.50,\ 1.00,\ 1.00)$$

A spherical surface that contains these four points is then defined by expanding Eq. 6.50 in the form of Eq. 6.51:

$$x^2 + y^2 + z^2 - 4.1875x - 5.4375y + 1.59375z + 0.4375 = 0 \qquad (8.7)$$

which can be rearranged in the normal form (Eq. 6.52) as

$$(x - 2.094)^2 + (y - 2.719)^2 + (z + 0.797)^2 = (3.460)^2 \qquad (8.8)$$

Hence, the coordinates of points on a unit sphere are found from the transformation relations

$$x' = \frac{x}{3.460} - 0.605$$

$$y' = \frac{y}{3.460} - 0.786$$

$$z' = \frac{z}{3.460} + 0.230 \tag{8.9}$$

where

$$(x')^2 + (y')^2 + (z')^2 = 1.0$$

The coordinates of the path precision points transformed to the surface of a unit sphere become

$$\mathbf{p}'_1 = (-0.316, -0.497, 0.808)$$
$$\mathbf{p}'_2 = (0.117, -0.280, 0.953)$$
$$\mathbf{p}'_3 = (0.551, -0.208, 0.808)$$
$$\mathbf{p}'_4 = (0.695, -0.497, 0.519)$$

As indicated in Table 8.2, we may specify five coordinates arbitrarily. We specify the first four as

$$a_{0x} = 0.100 \qquad a_{1x} = -0.600$$
$$a_{0z} = 0.200 \qquad a_{1z} = 0.400$$

Then we may calculate

$$a_{0y} = \sqrt{1. - a_{0x}^2 - a_{0z}^2} = -0.975$$
$$a_{1y} = \sqrt{1. - a_{1x}^2 - a_{1z}^2} = -0.693$$

which completes the specification of one crank $\mathbf{a}_0\,\mathbf{a}_1$ on the unit sphere in the first position of the linkage.

From the condition that the spherical distance $\overline{\mathbf{a}\mathbf{p}'}$ is constant we obtain the coordinates of points

$$\mathbf{a}_1 = (-0.600, -0.693, 0.400)$$
$$\mathbf{a}_2 = (-0.277, -0.586, 0.761)$$
$$\mathbf{a}_3 = (0.124, -0.528, 0.840)$$
$$\mathbf{a}_4 = (0.252, -0.520, 0.816)$$

With points $\mathbf{a}_j$ and $\mathbf{p}'_j$ known we can calculate the spherical rigid body displacement matrices $[D_{\mathbf{u}_j, \alpha_{1j}}]$. We finally specify the fifth arbitrary coordinate as
$b_{1x} = 0.0$ and introduce the three unknown angles β_{12}, β_{13}, and β_{14} into matrices
$[D_{\mathbf{u}_0, \beta_{1j}}]$. The coupler displacement is now described by a rigid body displacement
matrix as defined by Eq. 8.6. The constant length constraint equations for link $\overline{\mathbf{b}_0 \mathbf{b}}$
lead to a set of design equations

$$(\mathbf{b}_j - \mathbf{b}_0)^T(\mathbf{b}_j - \mathbf{b}_0) = (\mathbf{b}_1 - \mathbf{b}_0)^T(\mathbf{b}_1 - \mathbf{b}_0) \qquad j = 2, 3, 4$$

$$(\mathbf{b}_0)^T(\mathbf{b}_0) = 1$$

$$(\mathbf{b}_1)^T(\mathbf{b}_1) = 1$$

which may be solved for the remaining five unknown coordinates b_{0x}, b_{0y}, b_{0z},
b_{1y}, and b_{1z}.

An alternate method for solving the second crank coordinates is based on the
fact that four positions of a spherical circle point must lie in a common plane.
Therefore, we may write

$$\begin{vmatrix} b_{1x} & b_{1y} & b_{1z} & 1 \\ b_{2x} & b_{2y} & b_{2z} & 1 \\ b_{3x} & b_{3y} & b_{3z} & 1 \\ b_{4x} & b_{4y} & b_{4z} & 1 \end{vmatrix} = 1.0$$

We again select $b_{1x} = 0.0$ and calculate the coordinates of points $\mathbf{b}_j$ in terms of
the elements of $[D_{1j}]$ and $\mathbf{b}_1$. Expansion of the determinant leads to the third-order
polynomial

$$0.002628b_{1y}^3 + 0.003725b_{1y}^2 b_{1z} - 0.026118b_{1y}b_{1z}^2 + 0.025681b_{1z}^3 = 0$$

and, since $b_{1x}^2 + b_{1y}^2 + b_{1z}^2 = 1$, we obtain a cubic in b_{1z}^2 with a single real solution

$$b_{1z}^2 = 0.052720 \qquad b_{1z} = \pm 0.2$$

$$b_{1y} = \pm 0.973$$

Selecting the negative value for b_{1y} such that the moving pivot lies in the $-y$
hemisphere, we note that only the positive value for b_{1z} is consistent with the
polynomial equation. Once the first position $\mathbf{b}_1$ is known, the positions $\mathbf{b}_2, \mathbf{b}_3$,
and $\mathbf{b}_4$ are calculated from the displacement matrices.

A digital computer solution resulted in the spherical center-point and circle-
point curves shown in Figure 8.6. Note that a single closed curve is continuous on
both hemispheres. A plot on either hemisphere would display all solutions because of the dual nature of points on a common diameter of the sphere.

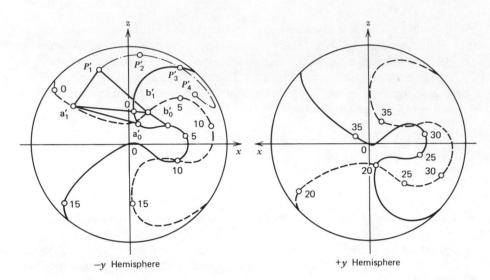

FIGURE 8.6 SPHERICAL PATH GENERATOR FOR FOUR POINTS IN SPACE. SPHERICAL CENTER-POINT (SOLID LINE) AND CIRCLE-POINT CURVES FROM COMPUTER SOLUTION. ONE SELECTED LINKAGE IS SHOWN WITH NUMERICAL COORDINATES:

a'_0 = (0.100000, −0.974679, 0.200000)
a'_1 = (−0.600000, −0.692820, 0.400000)
b'_0 = (0.400000, −0.8986859, 0.1798990)
b'_1 = (0.1967558, −0.9219490, 0.3336124)

8.4 DISPLACEMENT MATRICES FOR SPATIAL PATH GENERATION MECHANISMS

In the synthesis of spatial path generation mechanisms we must be careful to select an analytical form for the spatial displacement matrix that insures that the unspecified motion parameters are independent. For example, matrix (3.12) has four parameters u_x, u_y, u_z, and ϕ. The direction cosines must also satisfy the relationship

$$u_x^2 + u_y^2 + u_z^2 = 1 \tag{8.10}$$

It is clear that given u_x and u_y, either $+$ or $-u_z$ would satisfy Eq. 8.10; hence the direction cosines are not independent.

On the other hand, either of matrices (3.9) or (3.13), which involve only rotation angles taken in a definite specified sequence, are expressed in terms of three independent rotation angles (i.e., α, β, γ or ψ, θ, ϕ). In each case three unknown displacement angles are added for a new specified path precision point.

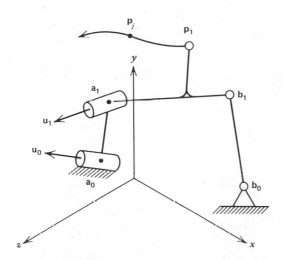

FIGURE 8.7 THE RRSS PATH GENERATION
MECHANISM.

8.5 THE RRSS PATH GENERATION MECHANISM

As shown in Figure 8.7, eighteen dimensional parameters are required to define an RRSS mechanism in its first position, namely the components of $\mathbf{u}_0$, $\mathbf{a}_0$, $\mathbf{u}_1$, $\mathbf{a}_1$, $\mathbf{b}_0$, and $\mathbf{b}_1$. For two precision points three additional unknowns in the form of three Euler displacement angles ψ_{1j}, θ_{1j}, and ϕ_{1j} must be added, giving a total of twenty-one unknowns. Table 8.3 summarizes these relationships for up to the theoretical maximum of eight path precision points.

**TABLE 8.3 PARAMETRIC RELATIONS—THE
RRSS PATH GENERATION MECHANISM**

Path Points	Design Equations	Total Variables	Specified Variables
2	9	21	12
3	14	24	10
4	19	27	8
5	24	30	6
6	29	33	4
7	34	36	2
8	39	39	0

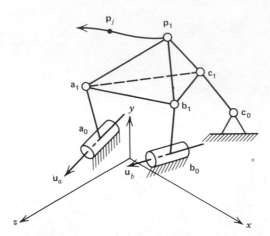

FIGURE 8.8   THE RSSR-SS PATH GENERATION
MECHANISM.

The design equations are based on the simultaneous solution of the constraint equations for both the R-R and S-S guiding links, as given in Section 6.11. For two specified path points this leads to a set of nine design equations, eight for the R-R link and one for the S-S link. With three specified path points we have fourteen design equations with twenty-four variables; we may arbitrarily assume ten of the variables. The selection must be done with care. For example, if we decide to specify all six of the displacement angles, we have converted the path generation problem to one of rigid body guidance, and a unique solution for the R-R link is possible. The remaining four free variables may be associated with the S-S link (e.g., b_{0x}, b_{0y}, b_{0z}, and b_{1x}, as in Figure 8.7).

If the six displacement angles were specified along with the fixed point a_0 the R-R link would be overconstrained and there would be no solution.

If four general path points in space are specified, the spherical four-bar, which involves only revolute joints, may be considered as an alternative to the RRSS mechanism.

For five path points an RRSS mechanism may be synthesized with the rigid body displacement angles as unknown parameters. As indicated in Table 8.1, we are free to specify six of the unknowns. It will be found convenient in this case to specify the six scalar coordinates of the fixed points a_0 and b_0.

8.6   THE RSSR-SS PATH GENERATION MECHANISM

The RSSR-SS mechanism as shown in Figure 8.8 is a particularly simple spatial mechanism that may be used as a path generator in those cases where small displacements are possible at the spherical joints.

As shown in Section 6.11, for two specified path points there would be a total of nine constraint equations. Each of the R-S links must satisfy four equations, as indicated by Eqs. 6.63. The S-S link adds the ninth equation in the form of a single constant length equation. The number of geometric parameters required to specify the mechanism in its first position equals twenty-four, the six scalar components of the two axes $\mathbf{u}_a$ and $\mathbf{u}_b$ plus the eighteen coordinates of points $\mathbf{a}_0, \mathbf{a}_1, \mathbf{b}_0, \mathbf{b}_1, \mathbf{c}_0$, and $\mathbf{c}_1$. To these unknowns we must add the three unknown displacement angles for the coupler to give a total of twenty-seven unknowns for a two-position problem. Since we have nine design equations, we are free to specify eighteen parameters.

For three path precision points three additional displacement angles are added giving a total of thirty unknowns with fourteen design equations and an arbitrary choice of sixteen parameters. A summary of these parametric relations, up to the theoretical maximum of eleven path precision points is given in Table 8.4.

TABLE 8.4   PARAMETRIC RELATIONSHIPS—THE RRSS-SS PATH GENERATION LINKAGE

Path Points	Design Equations	Total Variables	Specified Variables
2	9	27	18
3	14	30	16
4	19	33	14
5	24	36	12
6	29	39	10
7	34	42	8
8	39	45	6
9	44	48	4
10	49	51	2
11	54	54	0

For four specified path precision points it is possible to specify all nine of the displacement angles. This again converts the path generation problem to one of rigid body guidance, and we may solve for each of the R-S links and the S-S link pivots as independent problems. As discussed in Section 6.11, R-S link synthesis allows a maximum of four specified rigid body positions with free choice of one geometric parameter. Specification of a_{0x} and b_{0x} would lead to a unique pair of R-S links. The remaining three free parameters might be specified as the coordinates of the fixed pivot $\mathbf{c}_0 = (c_{0x}, c_{0y}, c_{0z})$ for the S-S link.

For five specified path points we allow the displacement angles to become free parameters and specify the twelve coordinates of points $\mathbf{a}_0, \mathbf{a}_1, \mathbf{b}_0$, and $\mathbf{b}_1$.

8.7 DISPLACEMENT ANALYSIS—PATH GENERATION MECHANISMS

The design equations used in the synthesis of any of the path generation mechanisms will also be found useful in the analysis of the motions generated by the mechanism. This procedure has already been outlined in detail in Section 4.7 for the RRSS path generation mechanism.

Example 8-4 Displacement Analysis of a Plane Four-Bar Path Generation Mechanism. The geometry of a plane four-bar path generation mechanism is specified by the point coordinates

$$\mathbf{a}_0 = (2.100, \, 0.600)$$

$$\mathbf{a}_1 = (0.607, \, -1.127)$$

$$\mathbf{b}_0 = (1.500, \, 4.200)$$

$$\mathbf{b}_1 = (-0.586, \, 0.997)$$

$$\mathbf{p}_1 = (1.000, \, 1.000)$$

TABLE 8.5 PATH GENERATION
ANALYSIS FOR A PLANE FOUR-BAR
LINKAGE—EXAMPLE 8.4

Path Point		Coupler Angle
p_{jx}	p_{jy}	θ_{1j} (radians)
1.000	1.000	0.000
1.100	0.927	−0.007
1.200	0.859	−0.012
1.300	0.796	−0.017
1.400	0.739	−0.021
1.500	0.687	−0.024
1.600	0.640	−0.026
1.700	0.597	−0.027
1.800	0.560	−0.027
1.900	0.527	−0.026
2.000	0.500	−0.024

   The analysis involves the solution of two simultaneous equations (Eq. 8.1) for $j = 2$ with three unknowns p_{jx}, p_{jy}, and θ_{1j}. In this case, we arbitrarily decide to follow the path by specification of p_{jx} and solve for p_{jy} and θ_{1j}. Once the new position $\mathbf{p}_j$ is known, along with the displacement angle θ_{1j}, we may form the coupler displacement matrix $[D_{1j}]$ and calculate the new positions $\mathbf{a}_j$ and $\mathbf{b}_j$. Table 8.5 lists the results for p_{jx}, p_{jy}, and θ_{1j} with equal increments $\Delta p_x = [p_{jx} - p_{(j-1)x}]$ specified.

Chapter 9
Optimal Synthesis of Mechanisms

9.1 INTRODUCTION

In previous chapters we described methods for the synthesis of plane and spatial mechanisms that must satisfy a finite number of constraint equations at a prescribed set of precision points. The difference between the specified and generated motion is known as *structural error* and, in precision point synthesis, the objective was to reduce the structural error to zero at all precision points. We have seen that the maximum error between precision points is a function of precision point spacing. Freudenstein [1] has discussed a method of systematic respacing of precision points such that the maximum structural error between each successive pair of precision points is equalized. This leads to an optimum for a precision point synthesis. Chebyshev spacing is used as an initial approximation to the optimal precision point spacing.

9.2 MINIMIZATION OF THE INTEGRATED ERROR

A different approach to the problem is to establish a function of the structural error integrated over the full range of motion of the mechanism. A specific set of linkage parameters that would lead to a minimum of the error function could then be considered an optimal kinematic design. The results obtained in this way are often superior in terms of the maximum structural error computed over the full range of motion, even though there may be fewer points of zero error than for a comparable precision point synthesis.

Design Versus Precision Points

A design point will be defined as any position of the mechanism where the motion is specified. There is no requirement that the structural error be zero at all design points, only that the sum of the squares of the structural errors, used as an approximation of the integrated error function, should be a minimum. The generated motion then tends to approach the specified motion closely over the full range of motion. Many design points may be specified, although experience has shown that numbers larger than three or four times the maximum number of precision points do not lead to a reduction in the maximum error.

9.3 OPTIMAL KINEMATIC SYNTHESIS AS A PROBLEM IN NONLINEAR PROGRAMMING

The optimal synthesis of mechanisms can be formulated as a problem in nonlinear programming. This branch of mathematical programming has developed rapidly in recent years. The recent book by Himmelblau [2] is an excellent account of methods available and includes many applications and computer codes.

The nonlinear programming problem can be stated in very simple terms

$$\text{Minimize } f(\mathbf{x})$$

where $(\mathbf{x}) = (x_1, x_2, \ldots x_n)^T$ is the vector of the design parameters subject to the inequality (regional) constraints

$$g_i(\mathbf{x}) \leq 0 \qquad i = 1, 2, \ldots p$$

and the equality (functional) constraints

$$h_j(\mathbf{x}) = 0 \qquad j = 1, 2, \ldots q$$

Thus, if the vector $(\mathbf{x}^*)$ corresponds to the minimum (optimal) value of the function,

$$f(\mathbf{x}^*) = \min_{(\mathbf{x}) \in R} f(\mathbf{x})$$

where the domain R is determined by the constraints.

9.4 THE OBJECTIVE FUNCTION

The objective function $f(\mathbf{x})$ is the mathematical function of the variables $(\mathbf{x}) = (x_1, x_2, \ldots x_n)^T$ for which there exists one or more points $\mathbf{x}^*$ at which the function is a minimum. Each of these *local minima* will correspond to an optimal set of the design variables. The *best* design must then be selected from the local minima. The lowest value of these local minima is known as the *global minimum*. In the synthesis of mechanisms there is no assurance that the global minimum corresponds to the *optimal design*, and each possible design corresponding to a local minimum of the objective function should be examined on its own merits.

The objective function can be formulated in a variety of ways. One obvious method would be to extend the set of constraint equations associated with the precision point synthesis of particular guiding links such that the number of design equations exceeds the number of unknown linkage parameters. The minimization procedure then involves reducing the sum of the residuals of the design equations to a minimum. In the case where the number of equations equals the number of variables we would expect to reduce the residuals to zero, and the result should correspond to those obtained for precision point synthesis. It is characteristic of this type of formulation that the required kinematic analysis is carried out numerically as part of the error minimization procedure.

A second basically different approach is to form the objective function directly as the sum of the squares of the structural error at the design points with the structural error computed separately by a kinematic analysis program in closed form. For example, in the case of an RRSS path generation mechanism, Eqs. 4.59 to 4.62 would generate path point coordinates $\mathbf{q}_j$, $j = 1, 2, \ldots m$, which can be compared to specified path points $\mathbf{p}_j, j = 1, 2, \ldots m$. The objective function is then formed as a function of the path error in the form

$$f(\mathbf{x}) = \sum_{j=1}^{m} (\mathbf{p}_j - \mathbf{q}_j)^T (\mathbf{p}_j - \mathbf{q}_j) \tag{9.1}$$

A total of eighteen coordinates corresponding to $\mathbf{a}_0$, $\mathbf{u}_0$, $\mathbf{a}_1$, $\mathbf{u}_1$, $\mathbf{b}_1$, and $\mathbf{b}_0$ are required to specify the RRSS mechanism in its first position. The optimal solution would correspond to the minimum value of the function given by Eq. 9.1.

Specific examples of the formulation of objective functions will be given in later sections.

9.5 INEQUALITY CONSTRAINTS

Inequality constraints are often described as *regional constraints* in that they form upper and lower limits on the variables. The mathematical description of this type of constraint can be in either of the two inequality forms

$$g_i(\mathbf{x}) \geq 0$$

or

$$g_i(\mathbf{x}) \leq 0$$

In either case two regions of the search space are defined. A region where *all* inequalities are satisfied is described as a *feasible* region. A region where *any* inequality is violated is *nonfeasible*.

We will adopt the convention that a current set of variables $\mathbf{x}$ is feasible if, for all values of i,

$$g_i(\mathbf{x}) \leq 0 \qquad i = 1, 2, \ldots p \tag{9.2}$$

In linkage synthesis the inequality constraints may limit the solution in several ways. For example, in the case of the RRSS path generator, we may wish to restrict the location of the fixed point $\mathbf{a}_0$ to a localized area in the region of a specified point $\mathbf{d} = (d_{0x}, d_{0y}, d_{0z})$ such that $\mathbf{a}_0$ lies within a sphere of radius equal to r units with center at $\mathbf{d}$. This condition can be imposed as an inequality constraint in the form

$$(\mathbf{a}_0 - \mathbf{d})^T(\mathbf{a}_0 - \mathbf{d}) - r^2 \leq 0 \tag{9.3}$$

The input crank length could be restricted to values between a specified minimum value l_{min} and a maximum value l_{max} by the double inequality

$$l_{min}^2 \leq (\mathbf{a} - \mathbf{a}_0)^T(\mathbf{a} - \mathbf{a}_0) \leq l_{max}^2 \tag{9.4}$$

This would lead to two inequality constraint functions to be satisfied in the form

$$l_{min}^2 - (\mathbf{a} - \mathbf{a}_0)^T(\mathbf{a} - \mathbf{a}_0) \leq 0 \tag{9.5}$$

$$(\mathbf{a} - \mathbf{a}_0)^T(\mathbf{a} - \mathbf{a}_0) - l_{max}^2 \leq 0 \tag{9.6}$$

We often require that certain variables must be positive. The condition $0 \leq x \leq 10.$ would give two inequality constraints:

$$-x \leq 0 \tag{9.7}$$

$$x - 10 \leq 0 \tag{9.8}$$

9.6   EQUALITY CONSTRAINTS

Equality constraints are sometimes referred to as *functional constraints*, since they impose a definite functional relationship between variables that must be satisfied. For example, in mechanism synthesis the direction cosine equation leads to the equality constraint function

$$h_1(\mathbf{x}) = (\mathbf{u}_0)^T(\mathbf{u}_0) - 1 = 0 \tag{9.9}$$

which must always be satisfied if the direction cosines form a valid set of numbers.

9.7   GEOMETRICAL REPRESENTATION OF THE NONLINEAR PROGRAMMING PROBLEM

The minimization of a function of n variables can be visualized as finding a minimum on a hypersurface plotted in an $n + 1$ dimensional coordinate system with the objective function $f(\mathbf{x})$ as the $(n + 1)$th coordinate. Unfortunately, we cannot visualize n-dimensional space, and we must restrict the geometrical representation of the nonlinear programming problem to a two-variable problem with the objective function plotted as the third coordinate in three-dimensional

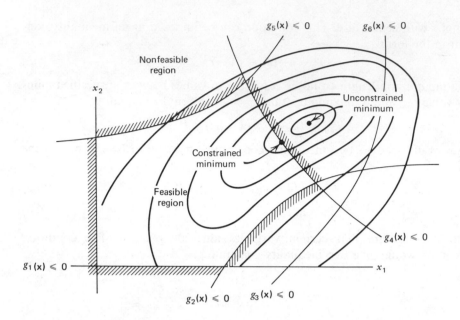

FIGURE 9.1 GEOMETRICAL REPRESENTATION OF A TWO-DIMENSIONAL OPTIMIZATION PROBLEM.

space, as shown in Figure 9.1. The most convenient graphical representation of the function $f(\mathbf{x})$ is in the form of contours of constant value for $f(x)$ plotted on the basis plane with coordinates x_1 and x_2.

The inequality constraints define the boundary between a feasible and a non-feasible region. Each inequality forms one side of a possible closed polygon, as shown in Figure 9.1. The number of inequality constraint functions is not limited to any specific number, and one or more constraints may be redundant, as indicated by $g_6(\mathbf{x}) \leq 0$ in Figure 9.1.

An equality constraint defines a particular curve on the basis plane. The constrained minimum must satisfy the definite functional relationship described by the curve. Hence, in the presence of an equality constraint, the constrained minimum must lie somewhere on the equality constraint line. A second equality constraint in a two-dimensional problem would force the minimum to be located at the intersection of the equality constraints, and the problem would no longer involve an optimization. Therefore, we see that each equality constraint reduces the number of degrees of freedom by one, and the maximum number of equality constraints in an n-dimensional problem is always one less than the number of variables.

9.8   UNCONSTRAINED MINIMIZATION

Mathematical Properties of a Minimum

A function $f(\mathbf{x})$ has a minimum at $\mathbf{x}^*$ if

$$f(\mathbf{x}^*) \leq f(\mathbf{x}) \qquad \text{for all } \mathbf{x} \tag{9.10}$$

If $f(\mathbf{x})$ has continuous first derivatives the minimum will be at a point where

$$\nabla f(\mathbf{x}^*) = 0 \tag{9.11}$$

where

$$\nabla f(\mathbf{x}) = \begin{bmatrix} \dfrac{\partial f}{\partial x_1} \\[2em] \dfrac{\partial f}{\partial x_n} \end{bmatrix}$$

The point $\mathbf{x}^*$ is a minimum (in contrast to a maximum) if the Hessian of the function $f(\mathbf{x})$ is positive at $\mathbf{x}^*$; that is,

$$\nabla^2 f(\mathbf{x}^*) > 0 \tag{9.12}$$

where

$$[H] = \nabla^2 f(\mathbf{x}^*) = \begin{bmatrix} \dfrac{\partial^2 f}{\partial x_1, \partial x_1} & \cdots & \dfrac{\partial^2 f}{\partial x_1, \partial x_n} \\[2em] \dfrac{\partial^2 f}{\partial x_n, \partial x_1} & \cdots & \dfrac{\partial^2 f}{\partial x_n, \partial x_n} \end{bmatrix}$$

Local versus Global Minima

There is no known method for locating the global minimum directly. The best that can be done at present is to compare local minima found from a random selection of starting points. This is not a serious disadvantage in many real problems with regional constraints that limit the solutions to a particular region of interest.

Approximation of Functions

Many of the mathematical methods employed in the search for the minimum of a function insure convergence if the function is of quadratic form. Any function with continuous first and second derivatives can be approximated in the form of a

Taylor's series. In the vicinity of a minimum $\mathbf{x}^*$ the approximation takes the quadratic form

$$f(\mathbf{x}) = f(\mathbf{x}^*) + [\nabla f(\mathbf{x}^*)]^T(\mathbf{x} - \mathbf{x}^*)$$
$$+ \tfrac{1}{2}(\mathbf{x} - \mathbf{x}^*)^T[\nabla^2 f(\mathbf{x}^*)](\mathbf{x} - \mathbf{x}^*) \tag{9.13}$$

where the Taylor's series has been truncated after the second derivative term.

The Search for a Minimum

In the search for a minimum we first must decide on a search direction that is likely to lead toward the minimum. After the direction is established, the second question involves the determination of the step size that would lead to the minimum of the function along the current search direction.

The classical method is the *method of steepest descent*. In this method the search direction is along the negative gradient of the function; that is, the unit search direction vector $\mathbf{s}$ is given by

$$\mathbf{s} = - \frac{[\nabla f(x)]}{\{[\nabla f(x)]^T[\nabla f(x)]\}^{1/2}}$$

$$= - \frac{[\nabla f(\mathbf{x})]}{\|\nabla f(\mathbf{x})\|} \tag{9.14}$$

The step length is the scalar length λ in the direction $\mathbf{s}$ that would lead to a point where

$$\frac{\partial f(\mathbf{x}_0 + \lambda \mathbf{s})}{\partial \lambda} = [\nabla f(\mathbf{x}_0)]^T(\mathbf{s}) + (\mathbf{s})^T[\nabla^2 f(\mathbf{x}_0)](\lambda \mathbf{s}) = 0 \tag{9.15}$$

Thus, the step length is given by

$$\lambda^* = - \frac{[\nabla f(\mathbf{x}_0)]^T(\mathbf{s})}{(\mathbf{s})^T[\nabla^2 f(\mathbf{x}_0)](\mathbf{s})} \tag{9.16}$$

where $\mathbf{x}_0$ defines the variables at the beginning of the step.

It is obvious that the next search direction to be employed will be orthogonal to the current search direction; that is, the directions $\mathbf{s}^K$ and $\mathbf{s}^{K+1}$ must satisfy the orthogonality condition

$$(\mathbf{s}^K)^T(\mathbf{s}^{K+1}) = 0 \tag{9.17}$$

Although the method of steepest descent seems attractive at first glance, it is not a practical method for most real problems. The method tends to terminate at all stationary points where $\nabla f(\mathbf{x}) = 0$, including saddle points and maxima and,

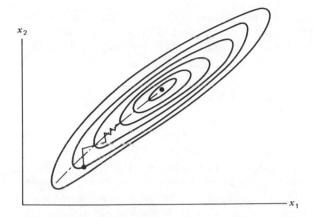

FIGURE 9.2   OSCILLATION OF THE METHOD OF STEEPEST DESCENT IN THE PRESENCE OF A NARROW VALLEY IN THE OBJECTIVE FUNCTION SURFACE.

unless the problem is scaled very carefully, will tend to oscillate in the presence of a long narrow steep-sided valley in the objective function surface, as indicated in Figure 9.2. This can lead to an exceedingly slow rate of convergence. Himmelblau [2] gives an excellent discussion of the method.

Conjugate Directions

The slow convergence problems of the method of steepest descent can be improved by using successive search directions that are conjugate with respect to the Hessian matrix; that is,

$$(s^{K+1})^{T}[\nabla^2 f(x^{K})](s^{K}) = 0 \qquad (9.18)$$

An important property of conjugate search directions can be stated as follows.

If a set of conjugate search directions (s_i), $i = 1, 2, \ldots n$ are employed sequentially, convergence to the minimum of a quadratic function is assured in n searches, one in each of n conjugate directions. The proof of this property is given in Appendix A at the end of this chapter.

Powell [3] has suggested a method for generating a set of conjugate directions based on the following theorem.

If, starting at an arbitrary base point x^0, the point x^a is found at the minimum of the function along the direction s^0, and if, starting from a second point $x^1 \neq x^0$, a minimum is found at point x^b along a line parallel to s^0, then if $f(x^b) < f(x^a)$, the direction $(x^b - x^a)$ is conjugate to s^0. The proof is given in Appendix B.

In multivariable problems if we begin with a set of conjugate directions, and if, starting from $\mathbf{x}^0$ we find $\mathbf{x}^a$ after p searches $(p < r)$ and, similarly, starting from $\mathbf{x}^1$ we find $\mathbf{x}^b$ after p searches, then the direction $(\mathbf{x}^b - \mathbf{x}^a)$ will be conjugate to all p directions.

Nonderivative Search for a Minimum

As we have seen in Eq. 9.16, the classical gradient method of steepest descent requires the calculation of both the first and second partial derivatives of the function before the step λ can be calculated. This requires either the formation of the partial derivatives in analytical form or their numerical approximation by finite differences. In either case this can lead to an excessive amount of computation in multivariable problems. An interesting alternative is the use of a quadratic approximation of the function along the current search direction with successive approximation to the minimum of the function.

Assume that the function can be approximated in the current search direction by a quadratic function of the step length λ in the form

$$F(\lambda) = c_0 + c_1 \lambda + c_2 \lambda^2 \tag{9.19}$$

with minimum where

$$\frac{dF(\lambda)}{d\lambda} = c_1 + 2c_2 \lambda = 0 \tag{9.20}$$

or

$$\lambda^* = -\frac{c_1}{2c_2} \tag{9.21}$$

The constants c_1 and c_2 can be evaluated and used to predict the minimum along the line of the current search direction using Powell's nonderivative technique. Assume that the value of the function is known for three values of λ (i.e., f_a at $\lambda = a$, f_b at $\lambda = b$, and f_c at $\lambda = c$).

Therefore, from Eq. 9.19,

$$f_a = c_0 + c_1 a + c_2 a^2$$
$$f_b = c_0 + c_1 b + c_2 b^2 \tag{9.22}$$
$$f_c = c_0 + c_1 c + c_2 c^2$$

Solving for c_1 and c_2,

$$c_1 = \frac{\begin{vmatrix} 1 & f_a & a^2 \\ 1 & f_b & b^2 \\ 1 & f_c & c^2 \end{vmatrix}}{\det} \qquad c_2 = \frac{\begin{vmatrix} 1 & a & f_a \\ 1 & b & f_b \\ 1 & c & f_c \end{vmatrix}}{\det}$$

From Eq. 9.21,

$$\lambda^* = d = \frac{1}{2}\frac{f_a(b^2 - c^2) + f_b(c^2 - a^2) + f_c(a^2 - b^2)}{f_a(b - c) + f_b(c - a) + f_c(a - b)} \tag{9.23}$$

Point d will define a minimum (not a maximum) if

$$\frac{f_a(b - c) + f_b(c - a) + f_c(a - b)}{(a - b)(b - c)(c - a)} < 0 \tag{9.24}$$

The value of f_d is calculated and compared with f_a, f_b, and f_c. The point corresponding to the largest value is discarded and a new approximation d is calculated. The search is terminated when successive points d agree within a specified accuracy ε.

9.9 POWELL'S NONDERIVATIVE DIRECT SEARCH METHOD

Powell's direct search method combines the basic simplicity of the quadratic extrapolation method for predicting the minimum along a line of search with the rapid convergence characteristics of a conjugate directions search. The method is of general usefulness and is particularly attractive, since it does not require the derivation or use of partial derivatives.

The actual Powell algorithm is incorporated in **PROGRAM PCON** as listed in Chapter 12. The search procedure is carried out as follows.

1. From a base point x_0^K a set of linearly independent directions s_i for $i = 1, 2, \ldots n$ are searched in sequence. (The coordinate directions are used in the first iteration.) Each search is carried out using Powell's nonderivative method and leads to successive minima at x_i^K where i is the search index and k is the iteration counter.

2. As a final step, search the direction $(x_n^K - x_0^K)$. This direction is tested to see if it would be advantageous to replace one of the current directions s_i^K with the new direction $(x_n^K - x_0^K)$. A step equal to $2(x_n^K - x_0^K)$ is taken from x_0^K, yielding the point $(2x_n^K - x_0^K)$. A new direction is introduced only if the determinant of the matrix of search directions increases (each search direction s_i has n components in the coordinate directions).

3. Assume that $\Delta^K =$ the largest reduction in $f(x)$ noted during the mth search. Let $f_1 = f(x_0^K)$, $f_2 = f(x_n^K)$, and $f_3 = f(2x_n^K - x_0)$. If $f_3 \geq f_2$ and/or $(f_1 - 2f_2 + f_3)(f_1 - f_2 - \Delta^K)^2 \geq 0.5\,\Delta^K(f_1 - f_3)^2$, then the new direction is not advantageous and the current search directions are repeated in the $(k + 1)$th iteration.

4. If the tests in step 3 are not satisfied, we replace the mth search direction by adding the new direction $(\mathbf{x}_n^K - \mathbf{x}_0^K)$ as the last in the set of new search directions $[S] = (\mathbf{s}_1, \mathbf{s}_2, \ldots, \mathbf{s}_{m-1}, \mathbf{s}_{m+1}, \ldots (\mathbf{x}_n^K - \mathbf{x}_0^K))$. In this case the last search direction in the next iteration is always parallel to the search direction introduced in step 2. Thus, the final step in the $(k+1)$th iteration is a search along a line between minima along two parallel search directions. As shown in Appendix A, as the search leads into a region where the function can be closely approximated by a quadratic function, this final search will tend to be directed toward a minimum of the function.

5. The search is terminated after any iteration in which the change in each variable in $\mathbf{x}$ is less than a specified accuracy ε_i for $i = 1, 2, \ldots n$ or when $\|\mathbf{x}_n^K - \mathbf{x}_0^K\| \leq 0.1\varepsilon$.

9.10 THE METHOD OF LEAST SQUARES

Assume an objective function of the form

$$F(\mathbf{x}) = \sum_{k=1}^{m} [f^k(\mathbf{x})]^2 \tag{9.25}$$

where the superscript refers to the kth function of $\mathbf{x} = (x_1, x_2, \ldots, x_n)$.
Partial derivatives are designated as

$$g_i^k(\mathbf{x}) = \frac{\partial f^k(\mathbf{x})}{\partial x_i} \tag{9.26}$$

$$G_{ij}^k(\mathbf{x}) = \frac{\partial^2 f^k(\mathbf{x})}{\partial x_i \, \partial x_j} \tag{9.27}$$

The search is initiated at a base point $\mathbf{x}_0$. The minimum is assumed to exist at a nearby point $\mathbf{x}^*$. At the minimum

$$g_i^k(\mathbf{x}^*) = 0 \qquad i = 1, 2, \ldots, m \tag{9.28}$$

where

$$\mathbf{x}^* = \mathbf{x}_0 + \boldsymbol{\delta} \tag{9.29}$$

Substituting Eq. 9.29 into Eq. 9.25 and forming the partial derivatives,

$$\sum_{k=1}^{m} f^k(\mathbf{x}_0 + \boldsymbol{\delta}) \frac{\partial}{\partial x_i} f^k(\mathbf{x}_0 + \boldsymbol{\delta}) = 0 \tag{9.30}$$

Expanding Eq. 9.30 as a truncated Taylor series,

$$\sum_{k=1}^{m} \left\{ f^k(\mathbf{x}_0) \frac{\partial f^k(\mathbf{x}_0)}{\partial x_i} + \sum_{j=1}^{n} \left[f^k(\mathbf{x}_0) \frac{\partial^2 f^k(\mathbf{x}_0)}{\partial x_i \, \partial x_j} + \frac{\partial f^k(\mathbf{x}_0)}{\partial x_i} \frac{\partial f^k(\mathbf{x}_0)}{\partial x_j} \right] \delta \right\} = 0$$

$$i - 1, 2, \ldots n \qquad (9.31)$$

The least squares method is dependent on the further approximation that the second term in Eq. 9.31 can be assumed negligible. The correction vector is then calculated from

$$\sum_{j=1}^{n} \left[\sum_{k=1}^{m} g_i^k(\mathbf{x}_0)g_j^k(\mathbf{x}_0) \right] \delta = - \sum_{k-1} f^k(\mathbf{x}_0)g_i^k(\mathbf{x}_0)$$

$$i = 1, 2, \ldots n \qquad (9.32)$$

The solution of Eq. 9.32 can be expressed in the concise matrix notation

$$\delta = -[[g_i^k]^T[g_i^k]]^{-1}[g_i^k]^T f^k(\mathbf{x}_0) \qquad (9.33)$$

or Eq. 9.32 may be solved directly by Gauss' elimination. Once the correction vector δ is known, a search is made in the direction of δ to find the scalar λ that minimizes $F(\mathbf{x}_0 + \lambda \, \delta)$. The point $\mathbf{x}_0 + \lambda \, \delta$ becomes the base point for the next iteration.

9.11 POWELL'S NONDERIVATIVE LEAST SQUARES METHOD

Powell [4] has also applied his nonderivative search technique to the least squares minimization problem. The derivative of the kth function along the ith direction $\mathbf{s}_i$ is given by

$$\gamma_i^k \approx \sum_{j=1}^{m} g_j^k(\mathbf{x})s_{ij} \qquad i = 1, 2, \ldots n \qquad k = 1, 2, \ldots m \qquad (9.34)$$

where s_{ij} is the jth component of $\mathbf{s}_i$.

The components of γ_i^k are scaled such that

$$\sum_{k=1}^{m} (\gamma_i^k)^2 = 1 \qquad i - 1, 2, \ldots n \qquad (9.35)$$

A correction vector $\mathbf{q}$ is calculated from an equation analogous to Eq. 9.32.

$$\sum_{j=1}^{n} \left(\sum_{k=1}^{m} \gamma_i^k \gamma_j^k \right) q_j = - \sum_{k=1}^{m} \gamma_i^k f^k(\mathbf{x}_0) \qquad i = 1, 2, \ldots n \qquad (9.36)$$

A direction δ is then calculated from

$$\delta = \sum_{i=1}^{n} (\mathbf{q}_i)^T (\mathbf{s}_i) \qquad (9.37)$$

and the iteration is extended to minimize

$$F(\mathbf{x}_0 + \lambda \, \delta)$$

Numerical Approximation of the Partial Derivatives

The function values $f^k(\mathbf{x}_0 + \lambda_1 \, \delta)$ and $f^k(\mathbf{x}_0 + \lambda_2 \, \delta)$ that yield the lowest and next to lowest values for $f^k(\mathbf{x})$ are noted. Then we define

$$\mathbf{u}^k(\delta) \approx \frac{\partial f^k(\mathbf{x}_0 + \lambda \, \delta)}{\partial \lambda} \approx \frac{f^k(\mathbf{x}_0 + \lambda_1 \, \delta) - f^k(\mathbf{x}_0 + \lambda_2 \, \delta)}{(\lambda_1 - \lambda_2)} \qquad (9.38)$$

The first approximation is improved by

$$\mathbf{v}^k(\delta) = \mathbf{u}^k(\delta) - \mu f^k(\mathbf{x}_0 + \lambda_1 \, \delta) \qquad (9.39)$$

where

$$\mu = \frac{\displaystyle\sum_{k=1}^{m} \mathbf{u}^k(\delta) f^k(\mathbf{x}_0 + \lambda_1 \, \delta)}{\displaystyle\sum_{k=1}^{m} [f^k(\mathbf{x}_0 + \lambda_1 \, \delta)]^2}$$

$\mathbf{v}^k(\delta)$ and δ are scaled so that

$$\sum_{k=1}^{m} [\mathbf{v}^k(\delta)]^2 = 1 \qquad (9.40)$$

In the next iteration one of the directions $\mathbf{s}_i$ is replaced by δ. The direction $\mathbf{s}_i$ is replaced such that

$$(\mathbf{q}_t)^T (\mathbf{p}_t) = \max(\mathbf{q}_i)^T (\mathbf{p}_i) \qquad (9.41)$$

where

$$\mathbf{p}_t = - \sum_{k=1}^{m} \gamma_i^k f^k(\mathbf{x}_0)$$

$\mathbf{x}_0 + \lambda_m \, \delta$ then replaces $\mathbf{x}_0$, and the next iteration proceeds with a new set of directions

$$\mathbf{s} = (\mathbf{s}_1, \mathbf{s}_2, \ldots \mathbf{s}_t = \delta \cdots \mathbf{s}_n)$$

The most time-consuming step is the solution of Eqs. 9.36, which can be carried out by Gauss elimination. Powell suggests using Rosen's method of partial matrix inversion.

9.12 CONSTRAINED MINIMIZATION—PENALTY FUNCTIONS

The constrained minimization problem outlined in Section 9.1 can be converted into an artificial unconstrained problem using the *created response surface technique* first proposed by Carroll [5].

The Interior Penalty Function

The inequality constraints can be accounted for by forming a modified objective function $U(\mathbf{x})$ in the form

$$U(\mathbf{x}) = f(\mathbf{x}) - r \sum_{i=1}^{p} \frac{1}{g_i(\mathbf{x})} \tag{9.42}$$

The effect of the added term is best seen by a graphical plot of the variation in $f(\mathbf{x})$ or $U(\mathbf{x})$ as a function of a particular search direction $\mathbf{s}_i$, as shown in Figure 9.3. The lower line represents the unconstrained function $f(x)$ with a minimum in a nonfeasible region. The constrained minimum would clearly lie on the boundary between the feasible and nonfeasible region determined by the inequality constraint $g_i(\mathbf{x}) \le 0$.

The minimization procedure begins with the factor $r \le 1.0$ at a feasible point in the search space, and a minimum of the function $U(\mathbf{x})$ is found for that value of r. The process is then repeated for a sequence of decreasing values of r. In Figure 9.3 we see that as r decreases, the function $U(\mathbf{x})$ approaches $f(\mathbf{x})$ in the feasible region but increases abruptly as $g_i(\mathbf{x}) \to 0$ near the constraint boundary. As r approaches zero, the search eventually encounters a very sharp change in curvature and is

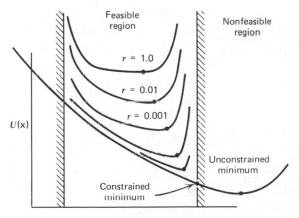

FIGURE 9.3 AN INTERIOR PENALTY FUNCTION.

unable to approach the boundary more closely. This procedure has been developed and applied extensively by Fiacco and McCormick [6] in their Sequential Unconstrained Minimization Technique, the SUMT algorithm. Various proposals have been made for a rational basis of change in the value of r for successive minimizations. In PROGRAM PCON, as listed in Chapter 12,

$$r = (FAC)^{1 - IR} \tag{9.43}$$

where $IR = 1$ at the start and is incremented by 1 after each successful suboptimum is found. The factor FAC can be set arbitrarily; 10. is suggested for normal use.

The Exterior Penalty Function

As an alternative, a penalty function can be formed as

$$U(\mathbf{x}) = f(\mathbf{x}) + \frac{1}{r} \sum_{i=1}^{p} [g_i(\mathbf{x})]^2 \tag{9.44}$$

where $g_i(\mathbf{x})$ is set equal to zero if the inequality is satisfied. In this case for $r \geq 1.0$ the constrained minimum would lie in a nonfeasible region, as shown in Figure 9.4. As r is decreased, the multiplier $1/r$ increases and the modified function tends to tilt upward and create a series of suboptimal points that approach the constraint boundary from outside the feasible region.

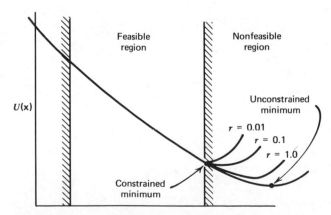

FIGURE 9.4 AN EXTERIOR PENALTY FUNCTION.

Equality Constraints

Equality constraints are accounted for by a combined penalty function. Assuming an interior inequality penalty, the combined modified objective function becomes

$$U(\mathbf{x}) = f(\mathbf{x}) - r \sum_{i=1}^{p} \frac{1}{g_i(\mathbf{x})} + \frac{1}{r} \sum_{j=1}^{q} [h_j(\mathbf{x})]^2 \tag{9.45}$$

The equality penalty function tends to create a series of intersecting, narrow, steep-sided valleys in the modified objective function surface. The optimization code must be capable of following such contours efficiently. Both programs PCON and LSTCON have this capability.

Scaling

In preparing a problem for optimization situations should be avoided where large differences exist in the sensitivity of $U(\mathbf{x})$ to a change in a particular x_i as compared to another x_j. One method of avoiding such difficulties would be to establish normalized variables in terms of the expected range in each variable. Where it is inconvenient to scale variables in this manner, programs PCON and LSTCON allow specification of individual convergence tolerances for each variable as a possible substitute procedure.

It is also important that a particular inequality or equality constraint not be allowed to dominate the combined objective function. This problem is accounted for automatically in programs PCON and LSTCON, which set weighting factors for each constraint such that they are all normalized at the start of each suboptimization.

9.13   CASE STUDIES IN OPTIMAL DESIGN OF MECHANISMS

Example 9-1 Optimal Synthesis of a Plane Four-Bar Function Generator. A four-bar linkage is to be synthesized such that the input and output crank rotations are proportional to the variables x and y in an arbitrary function $y = f(x)$, $x_0 < x < x_f$.

The objective function is formed using the method of kinematic inversion about the first position of the output crank. The function to be minimized is the square of the residuals of the constant length equation for the coupler with coordinates as defined in Figure 7.2. The problem is stated as:

Minimize $f(\mathbf{x})$

where

$$f(\mathbf{x}) = \sum_{j=2}^{m} [(\mathbf{a}'_j - \mathbf{b}_1)^T (\mathbf{a}'_j - \mathbf{b}_1) - (\mathbf{a}_1 - \mathbf{b}_1)^T (\mathbf{a}_1 - \mathbf{b}_1)]^2$$

and

$$(\mathbf{a}'_j) = [D_R](\mathbf{a}_1)$$

with $[D_R]$ computed from Eq. 7.12.

As a specific example, consider a five design point solution for the function

$$y = e^x \qquad 0 \le x \le 1.2$$
$$\Delta\theta = -90° \qquad \Delta\phi = 90°$$

Equally spaced values of the independent variable at $x = 0.0, 0.3, 0.6, 0.9$, and 1.2 led to solution number 1 in Table 9.1. The solution, using Powell's direct search method, gave what appears to be a five precision point solution with

TABLE 9.1 OPTIMUM SYNTHESIS OF A PLANE FOUR-BAR FUNCTION GENERATOR FOR

$$y = e^x \qquad 0 < x < 1.2 \qquad \Delta\theta = -90° \qquad \Delta\phi = +90°$$

Solution 1	Five equally spaced design points at
	$\theta = 0, -22.5, -45.0, -67.5, -90.0°$
Initial guesses	$a_{1x} = -0.30 \; a_{1y} = 0.40 \; b_{1x} = 0.90 \; b_{1y} = -0.50$
Initial value of objective function	$U_0 = 0.9428$
Converged in 50 moves	$U_{min} = 0.7005E - 15$
Final values	$a_{1x} = -0.13818784$
	$a_{1y} = 0.27323550$
	$b_{1x} = 0.76440524$
	$b_{1y} = -0.24412105$
Zero error at	$\theta = 0, -22.5, -45.0, -67.5, -90.0°$
maximum error in output angle ϕ, is 0.019° at $\theta = -8.5°$	
Solution 2	Twenty-one equally spaced design points from
	$\theta = 0$ to $\theta = -90°$
Initial guesses	$a_{1x} = -0.30 \; a_{1y} = 0.40 \; b_{1x} = 0.90 \; b_{1y} = -0.50$
Initial value of objective function	$U_0 = 2.8062$
Converged in 46 moves	$U_{min} = 0.1777E - 06$
Final values	$a_{1x} = -0.13858297$
	$a_{1y} = 0.27423277$
	$b_{1x} = 0.76293115$
	$b_{1y} = -0.24376526$
Zero error at	$\theta = 0, -20.3, -45.1, -68.3, -85.4°$
maximum error in output angle ϕ, is $-0.017°$ at $\theta = -8.0°$	

maximum error in the output angle ϕ over the full range equal to $0.019°$. Attempts to find a precision point solution using the methods of Chapter 7 converged to the degenerate solution corresponding to the link $\mathbf{a}_0 \mathbf{b}_0$ of unit length.

The equivalent of respacing of precision points can sometimes be accomplished by using a larger number of design points. Solution number 2 in Table 9.1 shows a result obtained for twenty-one equally spaced design points. The results indicate five points of zero error and a maximum error over the full range equal to $-0.017°$.

Different solutions can be achieved, depending on the initial guesses used to start the search. A family of solutions for a similar problem with

$$y = e^x 0 \le x \le 1.0$$

$$\Delta\theta = -90 \qquad \Delta\phi = 90°$$

were generated using Powell's least squares search method. Twenty-six random combinations of a_{1x}, a_{1y}, b_{1x}, and b_{1y} in the region of the fixed pivots were used as initial guesses. Convergence occurred in twenty-four cases with fifteen solutions corresponding to the degenerate case of unit fixed link. Unfortunately, this represents the global minimum for the problem and cannot be avoided. Of the remaining nine local minima, the solution with

$$\mathbf{a}_1 = (-0.129, 0.229)$$

$$\mathbf{b}_1 = (0.802, -0.202)$$

with a maximum output angle error $= -0.013°$ at $\theta = -7.5°$ was found to represent the most practical linkage.

Example 9-2 Optimal Synthesis of a Plane Four-Bar Linkage for Combined Path Generation and Rigid Body Guidance.
In this example we wish to synthesize a four-bar linkage that will simultaneously guide a rigid body through a series of prescribed positions while at the same time tracing a specified path. Such problems have many practical implications but have been difficult to solve in the past because of mathematical complexity.

We will specify *nine* design points on the point path and *five* rigid body positions to be coordinated with specific input crank angles as listed in Table 9.2. Fixed pivots for the input and output cranks are arbitrarily specified as shown in Figure 9.5 at

$$\mathbf{a}_0 = (0.0, 0.0)$$

$$\mathbf{b}_0 = (4.0, 0.0)$$

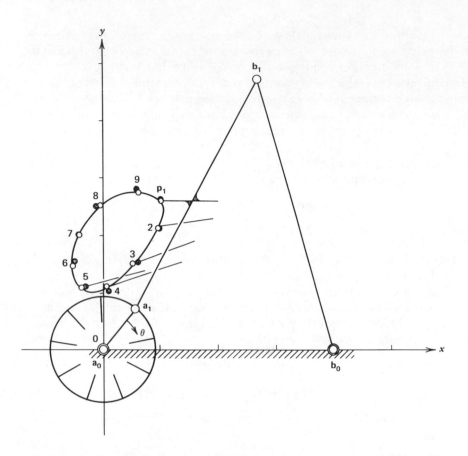

FIGURE 9.5   EXAMPLE 9.2. COORDINATE SYSTEM FOR THE PLANE FOUR-BAR COMBINED PATH GENERATION-GUIDANCE MECHANISM.

There are a total of ten unspecified parameters to be determined such that the objective function is minimized. To describe the first position of the linkage, the six coordinates of points a_1, b_1, and p_1 are unknown. In addition we must include the four rigid body rotation angles α_{16}, α_{17}, α_{18}, and α_{19} that are not included in the specification of the five point rigid body guidance.

The objective function has two parts: the first is concerned with the path errors, and the second involves the rigid body guidance. The rigid body guidance constraint is imposed as a constant length condition on the guiding crank $b_0 b$.

The objective function takes the form

$$f(\mathbf{x}) = f_1(\mathbf{x}) + f_2(\mathbf{x})$$

TABLE 9.2   SPECIFIED AND GENERATED MOTIONS FOR OPTIMUM SYNTHESIS OF A PLANE FOUR-BAR MECHANISM FOR COMBINED PATH GENERATION AND RIGID BODY GUIDANCE

Point Number	Input Crank Angle (Degrees)	Coupler Point Coordinates				Displacement Angle α (degrees)	
		Px		Py			
		Specified	Generated	Specified	Generated	Specified	Generated
1	0.0	1.00	1.01856	2.60	2.59481	0.00	0.00000
2	−40.0	1.00	0.95491	2.10	2.12288	12.00	11.73141
3	−80.0	0.60	0.52889	1.50	1.49085	21.00	21.80045
4	−120.0	0.10	0.04059	1.00	1.07209	22.00	22.61295
5	−160.0	−0.30	−0.35743	1.10	1.07245	15.00	16.01340
6	−200.0	−0.50	−0.53801	1.50	1.44224	—	6.88456
7	−240.0	−0.40	−0.39811	2.00	1.99723	—	−1.21383
8	−280.0	−0.10	−0.04966	2.50	2.49575	—	−6.06110
9	−320.0	0.60	0.62645	2.80	2.72715	—	−6.11201

where, with $\mathbf{p}_j$ as a generated path point and $\mathbf{q}_j$ as a specified path point,

$$f_1(\mathbf{x}) = \sum_{j=1}^{9} (\mathbf{q}_j - \mathbf{p}_j)^T (\mathbf{q}_j - \mathbf{p}_j)$$

and

$$f_2(\mathbf{x}) = \sum_{j=2}^{9} [(\mathbf{b}_j - \mathbf{b}_0)^T (\mathbf{b}_j - \mathbf{b}_0) - (\mathbf{b}_1 - \mathbf{b}_0)^T (\mathbf{b}_1 - \mathbf{b}_0)]$$

where

$$(\mathbf{a}_j) = [R_{\theta_{1j}}](\mathbf{a}_1 - \mathbf{a}_0) + (\mathbf{a}_0)$$

$$(\mathbf{b}_j) = [R_{\alpha_{1j}}](\mathbf{b}_1 - \mathbf{a}_1) + (\mathbf{a}_j)$$

$$(\mathbf{p}_j) = [R_{\alpha_{1j}}](\mathbf{p}_1 - \mathbf{a}_1) + (\mathbf{a}_j)$$

All angles $\theta_{1j}, j = 2, 9$ are specified along with $\alpha_{12}, \ldots \alpha_{15}$, as listed in Table 9.2.

This problem has also been solved using Powell's least squares method. Random initial guesses in the region of the specified fixed pivots gave several local minima. An excellent solution was selected with

$$\mathbf{a}_1 = (0.5586, 0.7141)$$

$$\mathbf{b}_1 = (2.6385, 4.7551)$$

$$\mathbf{p}_1 = (1.0186, 2.5948)$$

The minimum value of the combined objective function $f(\mathbf{x}) = 0.05$. Table 9.2 includes a comparison of specified and generated motions in this example. This same problem is used as an example in Chapter 12.

Example 9-3 The Auto Window Glass Guidance

Mechanism. Gustafson [7] has described the optimal design of a cross-arm automobile window guidance linkage. To enter the door cavity properly, the window must be guided such that one point on the window follows a prescribed path while the window translates with minimum angular motion. The cross-arm linkage shown in Figure 9.6 offers the possibility of a compact linkage in such cases.

The objective function has three parts, and the problem can be formulated such that the optimization can proceed in three stages. In the first stage the specified

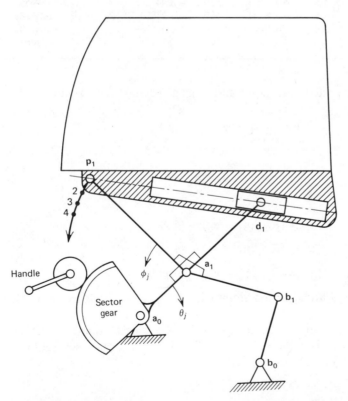

FIGURE 9.6 EXAMPLE 9.3. COORDINATE SYSTEM FOR THE OPTIMUM SYNTHESIS OF AN AUTOMOBILE WINDOW-GLASS GUIDANCE MECHANISM.

path points $\mathbf{p}_j$ are coordinated with specified input crank angles θ_j. The objective function is written as the sum of the errors in the constant length requirement for line $\overline{\mathbf{ap}}$.

$$f_1(\mathbf{x}) = \sum_{j=2}^{m} [(\mathbf{p}_j - \mathbf{a}_j)^T(\mathbf{p}_j - \mathbf{a}_j) - (\mathbf{p}_1 - \mathbf{a}_1)^T(\mathbf{p}_1 - \mathbf{a}_1)]^2$$

where

$$(\mathbf{a}_j) = [R_{\theta_{1j}}](\mathbf{a}_1 - \mathbf{a}_0) + (\mathbf{a}_0)$$

and

$$\mathbf{p}_j, \ j = 1, 2, \ldots m$$

are specified.

There are four unknowns at this stage, a_{0x}, a_{0y}, a_{1x}, and a_{1y}. Note that θ_{1j} is negative in this example.

With a_{0x}, a_{0y}, a_{1x}, and a_{1y} specified, we may calculate the coupler rotation angles

$$\phi_{1j} = \tan^{-1}\frac{(p_{jy} - a_{jy})}{(p_{jx} - a_{jx})} - \tan^{-1}\frac{(p_{1y} - a_{1y})}{(p_{1x} - a_{1x})}$$

The second part of the objective function is concerned with the actual generation of the path points as points on the rigid coupler with additional constraint provided by the crank $\mathbf{b}_0\,\mathbf{b}$. This leads to

$$f_2(\mathbf{x}) - \sum_{j=2}^{m} [(\mathbf{b}_j - \mathbf{b}_0)^t(\mathbf{b}_j - \mathbf{b}_0) - (\mathbf{b}_1 - \mathbf{b}_0)^T(\mathbf{b}_1 - \mathbf{b}_0)]^2$$

where

$$(\mathbf{b}_j) = [R_{\phi_{1j}}](\mathbf{b}_1 - \mathbf{p}_1) + (\mathbf{p}_j)$$

A second minimization can be carried out using $f_2(\mathbf{x})$ alone or the combined function $f_1(\mathbf{x}) + f_2(\mathbf{x})$. If the combined function is minimized as an eight-variable problem with unknowns a_{0x}, a_{0y}, a_{1x}, a_{1y}, b_{0x}, b_{0y}, b_{1x}, and b_{1y}, we see that the results of the first stage provide an excellent set of initial guesses for the first four variables.

The third part of the objective function is concerned with the parallel motion requirement for the window glass. In this example the centerline of the slot that guides point $\mathbf{d}_j$ is assumed to pass through point $\mathbf{p}_j$ in all positions. If desired, additional parameters could be introduced by specification of a desired offset or the coordinates of a nearby point $\mathbf{p}_1'$. The third part of the objective function becomes

$$f_3(\mathbf{x}) = \sum_{j=2}^{m} \left[\left(\frac{p_{jy} - d_{jy}}{p_{jx} - d_{jx}}\right) - \left(\frac{p_{1y} - d_{1y}}{p_{1x} - d_{1x}}\right) \right]^2$$

In the actual solution it is often more practical for the designer to specify certain linkage parameters. In this example we specify

$$a_{0x} = 30.8$$

$$b_{0x} = 37.3$$

$$d_{1x} = 35.0$$

The twelve path points given in Table 9.3 are to be coordinated with twelve sector gear rotations with $\Delta\theta = -82.896°$ in equal increments of $-7.536°$.

TABLE 9.3 SPECIFIED PATH POINT COOR-DINATES (P_j) TO BE COORDINATED WITH INPUT ANGLES θ_j FOR THE AUTO WINDOW MECHANISM OF EXAMPLES 9.3

Point Number	Px_j	Py_j	θ_j
1	27.970	21.270	0.000
2	27.542	19.970	−7.536
3	27.218	18.685	−15.072
4	26.928	17.368	−22.608
5	26.650	16.000	−30.144
6	26.386	14.569	−37.680
7	26.149	13.079	−45.216
8	25.959	11.550	−52.752
9	25.834	10.027	−60.288
10	25.773	8.584	−67.824
11	25.739	7.307	−75.360
12	25.670	6.270	−82.896

Window to remain parallel to its initial position in all subsequent positions.

Solving for the remaining seven parameters, we obtain an optimal design specified in its first position by

$$\mathbf{a}_0 = (30.8, 13.613) \qquad \mathbf{a}_1 = (34.310, 15.563)$$

$$\mathbf{b}_0 = (37.3, 11.604) \qquad \mathbf{b}_1 = (36.253, 12.602)$$

$$\mathbf{d}_1 = (35.0, 29.673)$$

with $f(\mathbf{x}^*) = 5.08$.

When all ten variables are left as free parameters, the optimal solution was found with initial guesses specified as

$$\mathbf{a}'_0 = (30.0, 5.0) \qquad \mathbf{a}'_1 = (35.0, 15.0)$$

$$\mathbf{b}'_0 = (37.0, 7.0) \qquad \mathbf{b}'_1 = (35.0, 15.0)$$

$$\mathbf{d}'_1 = (35.0, 20.0)$$

The objective function was reduced to

$$f(\mathbf{x}^*) = 0.066 \text{ in } 39 \text{ iterations}$$

The optimal design corresponded to

$$\mathbf{a}_0 = (30.824, 13.559) \qquad \mathbf{a}_1 - (34.328, 15.530)$$

$$\mathbf{b}_0 = (37.330, 7.560) \qquad \mathbf{b}_1 = (37.492, 12.175)$$

$$\mathbf{d}_1 - (39.342, 21.120)$$

Example 9-4 The RRSS Path Generation Mechanism (Least Squares Solution). In this case we wish to design a path generation mechanism in which a point on the RS link follows a spatial curve with the path point coordinates corresponding to specified input crank angular displacements. We will specify nine path points and input crank displacements as listed in Table 9.4. A total of twenty-one coordinates of axes $\mathbf{u}_0$, $\mathbf{u}_1$ and points $\mathbf{a}_0$, $\mathbf{a}_1$, $\mathbf{b}_0$, $\mathbf{b}_1$, and $\mathbf{q}_1$ must be specified to describe the mechanism in its first position.

We will arbitrarily assume

$$\mathbf{u}_0 = (0.0, 0.0, 1.00)$$

$$\mathbf{a}_0 = (0.0, 0.0, 0.0)$$

$$\mathbf{b}_0 = (0.80, 0.0, 0.0)$$

and determine the remaining twelve coordinates such as to minimize the objective function.

The objective function has three parts; that is,

$$f(\mathbf{x}) = f_1(\mathbf{x}) + f_2(\mathbf{x}) + f_3(\mathbf{x})$$

where

> $f_1(\mathbf{x})$ restricts $\mathbf{a}_j$ to a plane normal to $\mathbf{u}_0$
>
> $f_2(\mathbf{x})$ maintains constant length $\overline{\mathbf{b}_0\,\mathbf{b}}$
>
> $f_3(\mathbf{x})$ minimizes the path error

TABLE 9.4 EXAMPLE 9.4—SPECIFIED PATH POINTS AND RESULTS FOR THE
OPTIMUM SYNTHESIS OF THE RRSS PATH GENERATING MECHANISM
$\mathbf{a}_0 = (0.0, 0.0, 0.0)$ $\mathbf{u}_0 = (0.0, 0.0, 0.1)$ $\mathbf{b}_0 = (0.8, 0.0, 0.0)$

Mechanism Coordinates	Point Number	Path Coordinates		
		Px	Py	Pz
Specified motion	1	1.000	0.50	0.50
	2	0.74	0.34	0.74
Input angle	3	0.16	0.27	0.72
	4	−0.25	0.30	0.50
Incremented	5	−0.44	0.48	0.33
−40° for each of	6	−0.38	0.75	0.25
nine design points	7	−0.08	0.97	0.23
	8	0.38	1.00	0.26
	9	0.80	0.79	0.34
Solution				
$\mathbf{a}_1 = (0.303, 0.232, 0.0)$	1	0.999	0.499	0.501
	2	0.740	0.339	0.742
$\mathbf{u}_1 = (0.232, -0.302, 0.925)$	3	0.162	0.270	0.716
	4	−0.252	0.302	0.502
$\mathbf{b}_1 = (0.468, -0.484, -0.277)$	5	−0.438	0.481	0.330
	6	−0.382	0.749	0.247
$\mathbf{q}_1 = (0.999, 0.499, 0.501)$	7	−0.079	0.970	0.232
	8	0.377	0.997	0.261
$U_{\min}(x) = 0.0001$	9	0.802	0.793	0.337

The coupler motion is described in terms of the specified input crank motion θ_{1j} and the unknown relative rotation α_{1j} about the moving revolute joint $\mathbf{u}_j$. The generated path point $\mathbf{q}_j$ is given by

$$(\mathbf{q}'_1) = [R_{\alpha_{1j}}, \mathbf{u}_1](\mathbf{q}_1 - \mathbf{a}_1) + (\mathbf{a}_1)$$

$$(\mathbf{q}_j) = [R_{\theta_{1j}}, \mathbf{u}_0](\mathbf{q}'_1 - \mathbf{a}_0) + (\mathbf{a}_0)$$

The combined objective function is formed as the sum of

$$f_1(\mathbf{x}) = \sum_{j=1}^{m} (\mathbf{u}_j)^T (\mathbf{a}_j - \mathbf{a}_0)$$

$$f_2(\mathbf{x}) = \sum_{j=2}^{m} [(\mathbf{b}_j - \mathbf{b}_0)^T (\mathbf{b}_j - \mathbf{b}_0) - (\mathbf{b}_1 - \mathbf{b}_0)^T (\mathbf{b}_1 - \mathbf{b}_0)]$$

$$f_3(\mathbf{x}) = \sum_{j=1}^{m} (\mathbf{q}_j - \mathbf{p}_j)^T (\mathbf{q}_j - \mathbf{p}_j)$$

where $\mathbf{q}_1 = (q_{1x}, q_{1y}, q_{1z})$ can add three additional unknowns. Each relative rotation angle α_{1j} also adds an unknown.

The total number of unknowns, after specification of $\mathbf{u}_0$, $\mathbf{a}_0$, and $\mathbf{b}_0$, equals eighteen, distributed as follows.

$$\mathbf{u}_1 = (u_{1x}, u_{1y}, u_{1z}) \qquad (2) \qquad u_{1z}^2 = 1 - u_{1x}^2 - u_{1y}^2$$

$$\mathbf{a}_1 = (a_{1x}, a_{1y}, a_{1z}) \qquad (2) \qquad a_{1z} = 0.0$$

$$\mathbf{b}_1 = (b_{1x}, b_{1y}, b_{1z}) \qquad (3)$$

$$\mathbf{q}_1 = (q_{1x}, q_{1y}, q_{1z}) \qquad (3)$$

$$\alpha_{1j}; j = 2, 9 \qquad (8)$$

$$\text{Total unknowns} \qquad (18)$$

This problem was also solved using Powell's least squares method. For nine design points the combined objective function contains twenty-six residuals. For a series of random initial guesses a number of local minima were found. One of these solutions is given in Table 9.4. The match between specified and generated path points is indicated in Figure 9.7, which displays the x-y projection of both specified and generated point paths.

Example 9-5 The RRSS Path Generation Mechanism (Direct Search Solution). The data of Example 9-4 is used again in this example. The method used in Example 9-4 is very dependent on the initial guesses for the relative rotation angles α_{1}, $j = 2, 9$. This dependency can be reduced by formulating the objective function in another form. In this case the number of unknowns is reduced to a total of eight after specification of

$$\mathbf{a}_0 = (0., 0., 0.) \qquad \mathbf{u}_0 = (0., 0., 1.0)$$

$$\mathbf{b}_0 = (0.8, 0., 0.) \qquad \mathbf{p}_1 = (1., 0.5, 0.5)$$

The angles $\alpha_{1j} j = 2, 9$ are not required in this formulation. Point $\mathbf{p}_1$ could be included in the vector of unknown parameters, but this would introduce three additional unknowns into the problem.

The objective function contains three terms; that is,

$$f(\mathbf{x}) = f_1(\mathbf{x}) + f_2(\mathbf{x}) + f_3(\mathbf{x})$$

where $f_1(\mathbf{x})$ is concerned with the constant length constraint on the distance between the coupler path point $\mathbf{p}_j$ and the moving input crank pivot $\mathbf{a}_j$. $f_2(\mathbf{x})$ is the constant angle condition between axis $\mathbf{u}_j$ and the vector $(\mathbf{p}_j - \mathbf{a}_j)$. $f_3(\mathbf{x})$ insures constant length for the S-S link.

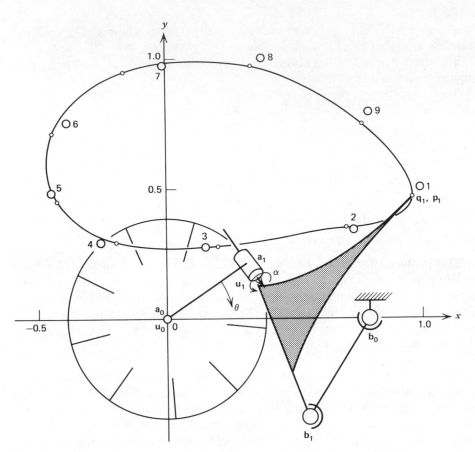

FIGURE 9.7 EXAMPLE 9.4. THE RRSS PATH GENERATING MECHANISM.

Both $f_1(\mathbf{x})$ and $f_2(\mathbf{x})$ are a consequence of the required rigid body motion for the coupler. It is also important to note that a check for continuous mobility is not made in this formulation.

The equality constraint

$$h(\mathbf{x}) = (\mathbf{u}_0)^T(\mathbf{a}_1 - \mathbf{a}_0)$$

is added to require that points $\mathbf{a}_1$ and $\mathbf{a}_0$ lie on a line perpendicular to axis $\mathbf{u}_0$.

The problem is solved using Powell's direct search conjugate directions method, as coded in program PCON. The objective function is formed in function FFUN(X) with eight unknowns in the $\mathbf{x}$ vector $(u_{1x}, u_{1y}, a_{1x}, a_{1y}, a_{1z}, b_{1x}, b_{1y},$ and $b_{1z})$. The steps are as follows.

1. Using the current values of u_{1x} and u_{1y}
   If $(u_{1x}^2 + u_{1y}^2)$ is greater than 1.
   Let $u_{1x} = u_{1x}/(u_{1x}^2 + u_{1y}^2)^{1/2}$ $u_{1y} = u_{1y}/(u_{1x}^2 + u_{1y}^2)^{1/2}$
   then $u_{1z} = (1. - u_{1x}^2 - u_{1y}^2)^{1/2}$
2. Calculate $\mathbf{u}_j = [R_{\theta, \mathbf{u}_0}]\mathbf{u}_1$.
3. Calculate $\mathbf{a}_j = [R_{\theta, \mathbf{u}_0}](\mathbf{a}_1 - \mathbf{a}_0) + \mathbf{a}_0$.
4. Calculate a unit direction vector $\mathbf{v}_j = (\mathbf{p}_j - \mathbf{a}_j)/\|(\mathbf{p}_j - \mathbf{a}_j)\|$.
5. Calculate the rigid body displacement matrix for the R-S link using the three points $\mathbf{a}_j$, $\mathbf{a}_j + \mathbf{u}_j$, and $\mathbf{a}_j + \mathbf{v}_j$. (Use subroutine DM3PT in SPAPAC.)
6. Calculate a new position for point $\mathbf{b}_j$ from $\mathbf{b}_j = [D_{1j}]\mathbf{b}_1$.
7. Finally,

$$f(\mathbf{x}) = f_1(\mathbf{x}) + f_2(\mathbf{x}) + f_3(\mathbf{x})$$

with

$$f_1(\mathbf{x}) = \sum [(\mathbf{p}_j - \mathbf{a}_j)^T(\mathbf{p}_j - \mathbf{a}_j) - (\mathbf{p}_1 - \mathbf{a}_1)^T(\mathbf{p}_1 - \mathbf{a}_1)]^2$$
$$f_2(\mathbf{x}) = \sum [\mathbf{u}_j^T\mathbf{v}_j - \mathbf{u}_1^T\mathbf{v}_1]^2 \qquad\qquad j = 2, 9$$
$$f_3(\mathbf{x}) = \sum [(\mathbf{b}_j - \mathbf{b}_0)^T(\mathbf{b}_j - \mathbf{b}_0) - (\mathbf{b}_1 - \mathbf{b}_0)^T(\mathbf{b}_1 - \mathbf{b}_0)]^2$$

The results of the error minimization were as follows:
   With initial guesses

$$\mathbf{a}_1 = (0.331, 0.196, 0.109) \qquad \mathbf{u}_1 = (0.232, -0.301, --)$$
$$\mathbf{b}_1 = (0.3, -0.6, 0.0)$$

the objective function was reduced from an initial value $f_0 = 0.365$ to a value $f = 0.000216$ in eighteen iterations corresponding to the solution

$$\mathbf{a}_1 = (0.3038, 0.2322, -0.00095)$$
$$\mathbf{b}_1 = (0.4795, -0.4788, -0.2781)$$
$$\mathbf{u}_1 = (0.2331, -0.3012, 0.9246)$$
$$\mathbf{q}_1 = (1.0, 0.5, 0.5)$$
$$h_1(\mathbf{x}) = 2.035 \times 10^{-6}$$

Displacement analysis of this mechanism gave the following results. The vector $\mathbf{e}$ is the error comparing generated and specified points $\mathbf{p}_j$.

Point Number	p_x	e_x	p_y	e_y	p_z	e_z
1	1.0	0.	0.5	0.	0.5	0.
2	0.740	0.	0.341	0.001	0.742	0.002
3	0.163	0.003	0.269	−0.001	0.716	−0.004
4	−0.252	−0.002	0.302	0.002	0.503	0.003
5	−0.438	+0.002	0.481	0.001	0.331	0.001
6	−0.382	−0.002	0.750	0.	0.247	−0.003
7	−0.078	0.002	0.971	0.001	0.231	0.001
8	0.379	−0.001	0.997	−0.003	0.260	0.
9	0.804	0.004	0.793	0.003	0.335	−0.005

Example 9-6 Optimal Synthesis of the Plane Four-Bar Function Generation Mechanism with Inequality and Equality Constraints.

The twenty-one design point problem solved as Example 9-1 will be extended to include regional (inequality) and functional (equality) constraints. The problem is restated in the following form.

Minimize

$$f(\mathbf{x}) = \sum_{j=2}^{21} [(\mathbf{a}'_j - \mathbf{b}_1)^T(\mathbf{a}'_j - \mathbf{b}_1) - (\mathbf{a}_1 - \mathbf{b}_1)^T(\mathbf{a}_1 - \mathbf{b}_1)]^2$$

where

$$\mathbf{a}'_j = [D_{rel}]\mathbf{a}_1$$

subject to inequality constraints that limit the acceptable locations for $\mathbf{a}_1$ and $\mathbf{b}_1$ to a circle of radius $r = 0.1$ about points $\mathbf{a}_c$ and $\mathbf{b}_c$, respectively. This leads to two inequality constraint functions

$$g_1(\mathbf{x}) = (a_{1x} - a_{cx})^2 + (a_{1y} - a_{cy})^2 - r^2 \leq 0.0$$

$$g_2(\mathbf{x}) = (b_{1x} - b_{cx})^2 + (b_{1y} - b_{cy})^2 - r^2 \leq 0.0$$

where

$$\mathbf{a}_c = (-.12, 0.25)$$

$$\mathbf{b}_c = (0.75, -0.25)$$

$$r = 0.25$$

An equality constraint has been added which limits solutions to those where $a_{1y} = -b_{1y}$; that is,

$$h_1(\mathbf{x}) = a_{1y} + b_{1y} = 0.0$$

This problem was solved twice using program PCON. The first solution imposed only the inequality constraints. The second solution involved both inequality and equality constraint functions. A nonfeasible initial guess vector $\mathbf{x}_0$ was used in both cases, with

$$x_1 = a_{1x} = -0.3$$
$$x_2 = a_{1y} = 0.4$$
$$x_3 = b_{1x} = 0.9$$
$$x_4 = b_{1y} = -0.5$$

Program PCON was executed with parameter ISRCH = 0 and located a feasible starting point at

$$\mathbf{x}_0' = (-1.133, 0.317, 0.814, -0.320)$$

where

$$f_0' = 2.324$$

With the inequality constraints only (MG = 2, MH = 0) PCON converged with IR = 7 and EPS1 = 1.E-5 at a solution

$$x_1^* = (-0.139, 0.274, 0.763, -0.244)$$

where

$$f_1^* = 1.78\text{E-06} \qquad \text{and} \qquad U_1^* = 8.24\text{E-06}$$

and

$$g_1(\mathbf{x}) = -0.616\text{E-01}$$
$$g_2(\mathbf{x}) = -0.623\text{E-01}$$

With both inequality and equality constraints active (MG = 2, MH = 1), PCON converged with IR — 6 at a solution

$$\mathbf{x}_2^* = (-0.122, 0.262, 0.792, -0.262)$$

where

$$f_2^* = 0.872\text{E-05} \qquad \text{and} \qquad U_2^* = 0.152\text{E-04}$$

and

$$g_1(\mathbf{x}) = -0.624\text{E-01}$$
$$g_2(\mathbf{x}) = -0.606\text{E-01}$$
$$h_1(\mathbf{x}) = 0.604\text{E-13}$$

It is interesting to note that the second solution at $\mathbf{x}_2^*$ has a higher value for the objective function $f(\mathbf{x})$ as a consequence of the influence of $h_1(\mathbf{x})$ on the minimum.

APPENDIX A
PROOF OF CONVERGENCE FOR CONJUGATE DIRECTIONS SEARCH

If conjugate search directions are employed sequentially, in any order, the minimum of a quadratic function of n variables will be found in n searches.

Proof

Assume

$$f(\mathbf{x}) = a + \mathbf{b}^T\mathbf{x} + \tfrac{1}{2}\mathbf{x}^T[H]\mathbf{x} \tag{a}$$

so that

$$\nabla f(\mathbf{x}) = \mathbf{b} + [H]\mathbf{x} \tag{b}$$

where $[H] \equiv \nabla^2 f(\mathbf{x})$. At the minimum $\mathbf{x}^*$,

$$\nabla f(\mathbf{x}^*) = 0 \tag{c}$$

that is,

$$\mathbf{x}^* = -[H]^{-1}\mathbf{b} \tag{d}$$

At the end of the nth search,

$$\mathbf{x}^n = \mathbf{x}^0 + \sum_{k=0}^{n-1} \lambda^k \mathbf{s}^k \tag{e}$$

For each search we minimize $f(\mathbf{x}^k + \lambda^k \mathbf{s}^k)$; that is,

$$\frac{df(\mathbf{x}^k + \lambda^k \mathbf{s}^k)}{d\lambda^k} = \nabla^T f(\mathbf{x}^k)\mathbf{s}^k + (\mathbf{s}^k)^T \nabla^2 f(\mathbf{x}^k)\lambda^k \mathbf{s}^k = 0$$

which leads to

$$\lambda^k = -\frac{\nabla^T f(\mathbf{x}^k)\mathbf{s}^k}{(\mathbf{s}^k)^T \nabla^2 f(\mathbf{x}^k)\mathbf{s}^k}$$

Hence,

$$\mathbf{x}^n = \mathbf{x}^0 - \sum_{k=0}^{n=1} \left[\frac{(\mathbf{s}^k)^T \nabla f(\mathbf{x}^k)}{(\mathbf{s}^k)^T[H]\mathbf{s}^k}\right]\mathbf{s}^k \tag{f}$$

Substituting Eq. b into Eq. f,

$$(\mathbf{s}^k)^T \nabla f(\mathbf{x}) = (\mathbf{s}^k)^T([H]\mathbf{x} + \mathbf{b}) = (\mathbf{s}^k)^T\left\{[H]\left(\mathbf{x}^0 + \sum_{i=1}^{k-1} \lambda^i \mathbf{s}^i\right) + \mathbf{b}\right\} \tag{g}$$

We remember that $\mathbf{s}^k$ and $\mathbf{s}^i$ are conjugate directions, hence $(\mathbf{s}^k)^T[H]\mathbf{s}^i = 0$ for all i, and Eq. g reduces to

$$(\mathbf{s}^k)^T \, \nabla f(\mathbf{x}) = (\mathbf{s}^k)^T([H]\mathbf{x}^0 + \mathbf{b})$$

Substituting into Eq. f,

$$\mathbf{x}^n = \mathbf{x}^0 - \sum_{k=0}^{n-1} \frac{(\mathbf{s}^k)^T([H]\mathbf{x}^0 + \mathbf{b})\mathbf{s}^k}{(\mathbf{s}^k)^T[H]\mathbf{s}^k}$$

$$= \mathbf{x}^0 - \sum_{k=0}^{n-1} \frac{\{(\mathbf{s}^k)^T[H]\mathbf{s}^k\}\mathbf{x}^0}{(\mathbf{s}^k)^T[H]\mathbf{s}^k} - \sum_{k=0}^{n-1} \frac{(\mathbf{s}^k)^T\mathbf{b}\mathbf{s}^k}{(\mathbf{s}^k)^T[H]\mathbf{s}^k}$$

The second term on the right $= -\mathbf{x}^0$. The third term, when multiplied by $[H]\times$ $[H]^{-1} = [I]$, leads to

$$x^n = -[H]^{-1}\mathbf{b}$$

which shows that $\mathbf{x}^n = \mathbf{x}^*$.

APPENDIX B
GENERATION OF CONJUGATE DIRECTIONS IN THE POWELL SEARCH METHOD

For the first search, where $\mathbf{s}$ is the search direction leading to a minimum at $\mathbf{x}^a$,

$$\mathbf{s}^T \, \nabla f(\mathbf{x}^a) = \mathbf{s}^T([H]\mathbf{x}^a + \mathbf{b}) = 0$$

and for the second search in a parallel direction from a different initial point,

$$\mathbf{s}^T \, \nabla f(\mathbf{x}^b) = \mathbf{s}^T([H]\mathbf{x}^b + \mathbf{b}) = 0$$

Subtracting gives

$$\mathbf{s}^T[[H](\mathbf{x}^b - \mathbf{x}^a)] = 0$$

Hence, we see that a vector $(\mathbf{x}^b - \mathbf{x}^a)$ connecting the two minima $\mathbf{x}^a$ and $\mathbf{x}^b$ is $[H]$ conjugate with respect to the original search direction $\mathbf{s}$.

Chapter 10
Differential Geometry of Motion

10.1 INTRODUCTION

In this chapter we will investigate certain fundamental geometrical properties of the motion of rigid bodies. We begin with an introduction to screw calculus, followed by a discussion of the geometry of screw-axis surfaces, mechanism synthesis in terms of geometrical parameters, and higher-order path curvature analysis for spatial mechanisms.

10.2 INSTANTANEOUS SCREW PARAMETERS

Consider a point on the instantaneous screw axis at the intersection of the screw axis and the x-y plane ($z = 0.0$). The x-y plane intersection will be found convenient when expressions for spatial motion are verified or visualized in terms of a special two-dimensional case. In situations where the screw axis is parallel to the x-y plane we may assume an intersection with either the y-z or x-z plane without loss of generality.

We assume a point $\mathbf{p}_0 = (p_{0x}, p_{0y}, p_{0z})$ on the instantaneous screw axis that moves with the rigid body. The point $\mathbf{p}_0$ has a linear velocity $\dot{s}\mathbf{u}$, where $\mathbf{u}$ is the direction cosine vector for the screw axis. The velocity of point $\mathbf{p}_0$ is also given by Eq. 3.72 where $\mathbf{p}_0$ replaces the point of interest $\mathbf{q}$ and point $\mathbf{p}$ is an arbitrary point in the rigid body with known velocity $\dot{\mathbf{p}}$. This leads to

$$
\dot{\mathbf{p}}_0 = \begin{bmatrix} \dot{s}u_x \\ \dot{s}u_y \\ \dot{s}u_z \\ 0 \end{bmatrix} = \begin{bmatrix} 0 & -u_z\dot{\phi} & u_y\dot{\phi} & (\dot{p}_x + u_z\dot{\phi}p_y - u_y\dot{\phi}p_z) \\ u_z\dot{\phi} & 0 & -u_x\dot{\phi} & (\dot{p}_y - u_z\dot{\phi}p_x + u_x\dot{\phi}p_z) \\ -u_y\dot{\phi} & u_x\dot{\phi} & 0 & (\dot{p}_z + u_y\dot{\phi}p_x - u_x\dot{\phi}p_y) \\ 0 & 0 & 0 & 0 \end{bmatrix} \begin{bmatrix} p_{0x} \\ p_{0y} \\ p_{0z} \\ 1 \end{bmatrix}
$$

$$
= \begin{bmatrix} -u_z\dot{\phi}p_{0y} + u_y\dot{\phi}p_{0z} + \dot{p}_x + u_z\dot{\phi}p_y - u_y\dot{\phi}p_z \\ u_z\dot{\phi}p_{0x} - u_x\dot{\phi}p_{0z} + \dot{p}_y - u_z\dot{\phi}p_x + u_x\dot{\phi}p_z \\ -u_y\dot{\phi}p_{0x} + u_x\dot{\phi}p_{0y} + \dot{p}_z + u_y\dot{\phi}p_x - u_x\dot{\phi}p_y \\ 0 \end{bmatrix} \tag{10.1}
$$

We assume $p_{0z} = 0.0$ and the intersection of the instantaneous screw axis with the x-y plane is found by equating expressions for $\dot{p}_{0x}$ and $\dot{p}_{0y}$ from the two sides of

238

Eq. 10.1. This leads to

$$p_{0x} = \frac{1}{u_z \dot\phi} (\dot s u_y - \dot p_y + u_z \dot\phi p_x - u_x \dot\phi p_z)$$

$$p_{0y} = \frac{-1}{u_z \dot\phi} (\dot s u_x - \dot p_x - u_z \dot\phi p_y + u_y \dot\phi p_z) \tag{10.2}$$

In the plane case where $\dot s = u_x = u_y = 0$, $u_z = 1$ we see that Eq. 10.2 reduces to Eq. 10.38, which determines the coordinates of the instantaneous velocity pole for plane motion.

The equivalent of the *pole velocity* in plane motion would be the velocity of the point of intersection of the instantaneous screw axis as measured in the x-y plane. Differentiating Eq. 10.2, we obtain

$$\dot p_{0x} = \frac{1}{(u_z \dot\phi)^2} \{ u_z \dot\phi [(\ddot s u_y + \dot s \dot u_y - \ddot p_y + \dot u_z (\dot\phi p_x) + u_z (\ddot\phi p_x + \dot\phi \dot p_x)$$

$$- \dot u_x (\dot\phi p_z) - u_x (\ddot\phi p_z + \dot\phi \dot p_z)] + (\dot s u_y - \dot p_y + u_z \dot\phi p_x - u_x \dot\phi p_z)(\dot u_z \dot\phi + u_z \ddot\phi) \}$$

$$\dot p_{0y} = \frac{-1}{(u_z \dot\phi)^2} \{ u_z \dot\phi [(\ddot s u_x + \dot s \dot u_x - \ddot p_x - \dot u_z (\dot\phi p_y) - {}_z (\ddot\phi p_y + \dot\phi \dot p_y)$$

$$+ \dot u_y (\dot\phi p_z) + u_y (\ddot\phi p_z + \dot\phi \dot p_z)] + (\dot s u_x - \dot p_x - u_z \dot\phi p_y + u_y \dot\phi p_z)(\dot u_z \dot\phi + u_z \ddot\phi) \}$$

$$\tag{10.3}$$

Again, with $\dot s = \ddot s = 0$, $u_x = u_y = 0$, $u_z = 1$, $\dot u_x = \dot u_y = \dot u_z = 0$, Eqs. 10.3 become the expressions for pole velocity components in plane motion (Eqs. 10.41).

The *velocity matrix* in terms of the instantaneous screw parameters $\mathbf{u}$, $\dot\phi$, $\dot s$, and p_0 is given by

$$[V] = \begin{bmatrix} 0 & -u_z \dot\phi & u_y \dot\phi & (\dot s u_x + u_z \dot\phi p_{0y} - u_y \dot\phi p_{0z}) \\ u_z \dot\phi & 0 & -u_x \dot\phi & (\dot s u_y - u_z \dot\phi p_{0x} + u_x \dot\phi p_{0z}) \\ -u_y \dot\phi & u_x \dot\phi & 0 & (\dot s u_z + u_y \dot\phi p_{0x} - u_x \dot\phi p_{0y}) \\ 0 & 0 & 0 & 0 \end{bmatrix} \tag{10.4}$$

where $\mathbf{p}_0 = (p_{0x}, p_{0y}, p_{0z})$ is an arbitrary point on the instantaneous screw axis.

In deriving Eq. 10.4 we note that the same result would be obtained by differentiation of Eq. 3.26, which leads to

$$\begin{bmatrix} \dot{\mathbf{q}} \\ 0 \end{bmatrix} = \begin{bmatrix} [W] & (\dot{\mathbf{p}}_1 + \dot s \mathbf{u} + s \dot{\mathbf{u}} - [W]\mathbf{p}_1) \\ 0 & 1 \end{bmatrix} \begin{bmatrix} \mathbf{q} \\ 1 \end{bmatrix} \tag{10.5}$$

In Eq. 3.26 point $\mathbf{p}_1$ becomes the reference point (i.e., $\mathbf{p}_1 = \mathbf{p}_0$ and $\dot{\mathbf{p}}_1 = 0$). The term $\dot{s}\mathbf{u}$ equals zero, since $\dot{s}$ by definition is along the instantaneous screw direction $\mathbf{u}$. With these assumptions Eq. 10.5 reduces to

$$\begin{bmatrix} \dot{\mathbf{q}} \\ 0 \end{bmatrix} = \begin{bmatrix} [W] & (\dot{s}\mathbf{u} - [W]\mathbf{p}_0) \\ 0 & 1 \end{bmatrix}\begin{bmatrix} \mathbf{q} \\ 1 \end{bmatrix}$$

$$= [V]\begin{bmatrix} \mathbf{q} \\ 1 \end{bmatrix} \tag{10.6}$$

Differentiating Eq. 10.6, we obtain

$$\begin{bmatrix} \ddot{\mathbf{q}} \\ 0 \end{bmatrix} = \begin{bmatrix} [\dot{W}] & (\ddot{s}\mathbf{u} + \dot{s}\dot{\mathbf{u}} - [\dot{W}]\mathbf{p}_0 - [W]\dot{\mathbf{p}}_0) \\ 0 & 0 \end{bmatrix}\begin{bmatrix} \mathbf{q} \\ 1 \end{bmatrix} \tag{10.7}$$

which, when expanded, gives the *acceleration matrix* in terms of instantaneous screw motion parameters.

$$[A] = \begin{bmatrix} \ddot{d}_{11} & \ddot{d}_{12} & \ddot{d}_{13} & (\ddot{s}u_x + \dot{s}\dot{u}_x + u_z\dot{\phi}\dot{p}_{0y} - u_y\dot{\phi}\dot{p}_{0z} - \ddot{d}_{11}p_{0x} - \ddot{d}_{12}p_{0y} - \ddot{d}_{13}p_{0z}) \\ \ddot{d}_{21} & \ddot{d}_{22} & \ddot{d}_{23} & (\ddot{s}u_y + \dot{s}\dot{u}_y - u_z\dot{\phi}\dot{p}_{0x} + u_x\dot{\phi}\dot{p}_{0z} - \ddot{d}_{21}p_{0x} - \ddot{d}_{22}p_{0y} - \ddot{d}_{23}p_{0z}) \\ \ddot{d}_{31} & \ddot{d}_{32} & \ddot{d}_{33} & (\ddot{s}u_z + \dot{s}\dot{u}_z + u_y\dot{\phi}\dot{p}_{0x} - u_x\dot{\phi}\dot{p}_{0y} - \ddot{d}_{31}p_{0x} - \ddot{d}_{32}p_{0y} - \ddot{d}_{33}p_{0z}) \\ 0 & 0 & 0 & 0 \end{bmatrix}$$

$$\tag{10.8}$$

where $\ddot{d}_{11}$, $\ddot{d}_{12}$, and so on are as defined in Eq. 3.67.

Comparing terms in the fourth column with the acceleration matrix in the general form of Eq. 3.73, we see that

$$\ddot{p}_{0x} = \ddot{s}u_x + \dot{s}\dot{u}_x + u_z\dot{\phi}\dot{p}_{0y} - u_y\dot{\phi}\dot{p}_{0z}$$

$$\ddot{p}_{0y} = \ddot{s}u_y + \dot{s}\dot{u}_y - u_z\dot{\phi}\dot{p}_{0x} + u_x\dot{\phi}\dot{p}_{0z}$$

$$\ddot{p}_{0z} = \ddot{s}u_z + \dot{s}\dot{u}_z + u_y\dot{\phi}\dot{p}_{0x} - u_x\dot{\phi}\dot{p}_{0y} \tag{10.9}$$

where

$$\dot{\mathbf{p}}_0 = (\dot{p}_{0x}, \dot{p}_{0y}, \dot{p}_{0z}) = (\dot{s}u_x, \dot{s}u_y, \dot{s}u_z)$$

10.3 GEOMETRY OF SCREW AXIS SURFACES

When given the velocity matrix elements that describe the velocity of a rigid body in a series of finitely separated positions, it is possible to locate a series of positions for the instantaneous screw axis. These lines in space form a ruled surface that may be designated as the *fixed screw axis surface* or *fixed axode*. Attached to, and moving with the moving body, is a second system of lines that sequentially rolls into contact and lies along corresponding lines of the fixed axode system. The system of moving lines will define the *moving screw axis surface* or *moving axode*. The motion of the moving rigid body is completely and uniquely specified by the rolling of these two surfaces, combined with a specified sliding motion along the screw axis.

Instant Pitch—A Unique Geometrical Property of Rigid Body Motion

Recall that the location of the velocity pole for a plane motion mechanism can be defined in purely geometrical terms. For example, the velocity pole or instant center for the coupler in a four-bar plane mechanism is located at the intersection of the line of centers of the two guiding cranks. This leads to the location of a special point of the moving plane that has zero velocity in the given position of the linkage.

In spatial mechanisms we may also locate the instantaneous screw axis based on purely geometrical considerations. However, once the screw axis is known, we note that the angular velocity $\dot{\phi}$ and sliding motion component $\dot{s}u$ are not independent.

We define the instant pitch $\dot{S}_{IP}$ as the ratio of sliding to angular motion components (i.e., the value of $\dot{S}$ when $\dot{\phi} = 1.0$). With these assumptions we may rewrite Eq. 3.72 in terms of the instant pitch $\dot{S}_{IP}$ in the form

$$[V] = \begin{bmatrix} 0 & -u_z & u_y & (\dot{S}_{IP}u_x + u_z p_{0y} - u_y p_{0z}) \\ u_z & 0 & -u_x & (\dot{S}_{IP}u_y - u_z p_{0x} + u_x p_{0z}) \\ -u_y & u_x & 0 & (\dot{S}_{IP}u_z + u_y p_{0x} - u_x p_{0y}) \\ 0 & 0 & 0 & 0 \end{bmatrix}$$

(10.10)

Equation 10.10 defines the *velocity matrix of instant pitch*. For a given screw axis defined by a point $\mathbf{p}_0$ and axis $\mathbf{u}$, the instant pitch S_{IP} completes the definition of the matrix elements.

Geometric Properties of Screw Axis Surfaces

A special curve of reference on the screw axis surface, the *directrix*, is useful in the description of the geometrical properties of a ruled surface as formed by the locus of the instantaneous screw axes. The directrix can be any curve that intersects all generators of the surface. We will arbitrarily assume the directrix as the curve formed by the intersection of the fixed screw axis surface with the x-y plane, as shown in Figure 10.1. In the special case of plane motion parallel to the x-y plane the directrix becomes the fixed polode.

The displacement of the screw axis itself can also be represented as a screw motion. Hence, the original rigid body motion will be described by a *first-order*

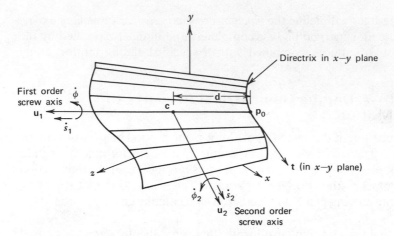

FIGURE 10.1   GEOMETRY OF THE FIXED SCREW AXIS SURFACE.

screw, the displacement of first-order screw by a *second-order* screw, and so forth. The order of the screw motion will be described by the subscript of the screw motion parameter—$\mathbf{u}_1, \dot{S}_1$, and $\dot{\phi}_1$, for the first-order screw, and $\mathbf{u}_2, \dot{S}_2$, and $\dot{\phi}_2$ for the second-order screw.

Point **c**, as shown in Figure 10.1, is the intersection of axis $\mathbf{u}_1$ and $\mathbf{u}_2$ and is designated as the *central point* of the generator of the surface. It is obvious that these two axes are perpendicular.

The *parameter of distribution* corresponds to the instant pitch of the second-order instantaneous screw, $\dot{S}_{2IP}$.

Instantaneous Screw Calculus

The following discussion is based on the work of Weatherburn [1]. In this case all derivatives are with respect to time. Similar expressions could also be developed with derivatives taken with respect to arc length, input angle, and the like.

The distance d from the point $\mathbf{p}_0$ on the directrix to the central point **c** may be expressed in terms of differential displacement matrix elements as follows.

$$
\begin{aligned}
d &= -\frac{(\mathbf{t})^T(\dot{\mathbf{u}}_1)}{(\dot{\mathbf{u}}_1)^T(\dot{\mathbf{u}}_1)} \\[2mm]
&= -\frac{(\dot{p}_{0x}, \dot{p}_{0y}, \dot{p}_{0z})^T(\dot{u}_{1x}, \dot{u}_{1y}, \dot{u}_{1z})}{(\dot{u}_{1x}, \dot{u}_{1y}, \dot{u}_{1z})^T(\dot{u}_{1x}, \dot{u}_{1y}, \dot{u}_{1z})}
\end{aligned} \qquad (10.11)
$$

Since $p_{0z} = \text{constant} = 0$, $\dot{p}_{0z} = 0$; therefore

$$d = \frac{\dot{p}_{0x}\dot{u}_{1x} + \dot{p}_{0y}\dot{u}_{1y}}{\dot{u}_{1x}^2 + \dot{u}_{1y}^2 + \dot{u}_{1z}^2} \tag{10.12}$$

The coordinates of the central point $\mathbf{c}$ become

$$c_x = p_{0x} + u_{1x}\, d$$
$$c_y = p_{0y} + u_{1y}\, d$$
$$c_z = 0.0 + u_{1z}\, d \tag{10.13}$$

where d is positive in the $\mathbf{u}_1$ direction.

The direction cosines for the second-order screw axis $\mathbf{u}_2$ are found from

$$\mathbf{u}_2 = \frac{\mathbf{u}_1 \times \dot{\mathbf{u}}_1}{\sqrt{\dot{u}_{1x}^2 + \dot{u}_{1y}^2 + \dot{u}_{1z}^2}} \tag{10.14}$$

which gives

$$u_{2x} = \frac{u_{1y}\dot{u}_{1z} - \dot{u}_{1y}u_{1z}}{\sqrt{\dot{u}_{1x}^2 + \dot{u}_{1y}^2 + \dot{u}_{1z}^2}} = \frac{u_{1y}\dot{u}_{1z} - \dot{u}_{1y}u_{1z}}{|\dot{\mathbf{u}}_1|}$$

$$u_{2y} = \frac{-u_{1x}\dot{u}_{1z} + \dot{u}_{1x}u_{1z}}{|\dot{\mathbf{u}}_1|}$$

$$u_{2z} = \frac{u_{1x}\dot{u}_{1y} - \dot{u}_{1x}u_{1y}}{|\dot{\mathbf{u}}_1|}$$

The angular velocity $\dot{\phi}_2$ is given by

$$\dot{\phi}_2 = |\dot{\mathbf{u}}_1| \tag{10.15}$$

Projection of the vector $\mathbf{t}$ in the $\mathbf{u}_2$ direction gives the sliding velocity component $\dot{S}_2$ as

$$\dot{S}_2 = (\mathbf{t})^T\left(\frac{\mathbf{u}_1 \times \dot{\mathbf{u}}_1}{|\dot{\mathbf{u}}_1|}\right) \tag{10.16}$$

which may be expanded as

$$\dot{S}_2 = \frac{\dot{p}_{0x}(u_{1y}\dot{u}_{1z} - \dot{u}_{1y}u_{1z}) + \dot{p}_{0y}(u_{1z}\dot{u}_{1x} - \dot{u}_{1z}u_{1x})}{|\dot{\mathbf{u}}_1|} \tag{10.17}$$

The instant pitch of the second-order screw axis is defined by

$$\dot{S}_{2\text{IP}} = \frac{\dot{S}_2}{\dot{\phi}_2} = \frac{\dot{p}_{0x}(u_{1y}\dot{u}_{1z} - \dot{u}_{1y}u_{1z}) + \dot{p}_{0y}(u_{1z}\dot{u}_{1x} - \dot{u}_{1z}u_{1x})}{\dot{u}_{1x}^2 + \dot{u}_{1y}^2 + \dot{u}_{1z}^2} \tag{10.18}$$

Example 10-1 Instantaneous Screw Motion Parameters for the Coupler of a Spatial Double-Slider Mechanism. We will analyse the spatial double-slider mechanism shown in Figure 10.2. The input velocity $\dot{\mathbf{a}}$ and acceleration $\ddot{\mathbf{a}}$ are specified as follows.

$$\mathbf{a} = (0.0,\ 3.0,\ 0.0)$$

$$\dot{\mathbf{a}} = (0.0,\ 1.0,\ 0.0)$$

$$\ddot{\mathbf{a}} = (0.0,\ 1.0,\ 0.0)$$

The output motion is calculated for a specified position for the output slider

$$\mathbf{b} = (8.66,\ 0.0,\ 4.00)$$

The coordinates of point $\mathbf{c}$ are specified as

$$\mathbf{c} = (-16.00/8.66,\ 3.00,\ 4.00)$$

From these motion specifications we are able to write seven constraint equations in terms of the seven unknown screw parameters $p_{0x}, p_{0y}, u_{1x}, u_{1y}, u_{1z}, \dot{\phi}_1$, and $\dot{s}_1$ in the form

$$\dot{a}_x = 0$$

$$\dot{a}_z = 0$$

$$\dot{b}_y = 0$$

$$\dot{b}_z = 0$$

$$\dot{a}_y = \dot{c}_y = 1$$

$$u_{1x}^2 + u_{1y}^2 + u_{1z}^2 = 1 \qquad\qquad (10.19)$$

Rewriting Eqs. 10.19 in terms of unknown velocity matrix elements,

$$\dot{d}_{12}a_y + \dot{d}_{14} = 0$$

$$\dot{d}_{32}a_y + \dot{d}_{34} = 0$$

$$\dot{d}_{21}b_x + \dot{d}_{23}b_z + \dot{d}_{24} = 0$$

$$\dot{d}_{31}b_x + \dot{d}_{34} = 0$$

$$u_x c_z - u_z c_x = 0 \qquad [\text{from } \dot{\phi}\mathbf{u} \times (\mathbf{c} - \mathbf{a})]$$

$$\dot{d}_{24} - 1.0 = 0$$

$$u_{1x}^2 + u_{1y}^2 + u_{1z}^2 - 1.0 = 0 \qquad\qquad (10.20)$$

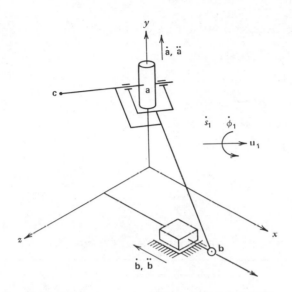

FIGURE 10.2   THE SPATIAL DOUBLE-SLIDER
MECHANISM.

where the velocity matrix elements include the screw motion parameters as indicated by Eqs. 3.60 and 3.72.

$$\dot{d}_{12} = -u_{1z}\dot{\phi} \qquad \dot{d}_{21} = -\dot{d}_{12}$$

$$\dot{d}_{14} = u_{1z}\dot{\phi}p_{0y} \qquad \dot{d}_{24} = u_{1z}p_{0x}$$

$$\dot{d}_{32} = u_{1x}\dot{\phi} \qquad \dot{d}_{23} = -\dot{d}_{32}$$

$$\dot{d}_{34} = u_{1y}p_{0x} \qquad \dot{d}_{31} = u_{1y}\dot{\phi} \qquad (10.21)$$

Therefore, we may solve Eqs. 10.20 for the unknown seven parameters of the
first-order screw.

The second-order screw parameters $\dot{p}_{0x}$, $\dot{p}_{0y}$, $\dot{u}_{1x}$, $\dot{u}_{1y}$, $\dot{u}_{1z}$, $\ddot{\phi}_1$, and $\ddot{s}_1$ are
calculated from constraint equations based on

$$\ddot{a}_x = 0$$

$$\ddot{a}_z = 0$$

$$\ddot{b}_y = 0$$

$$\ddot{b}_z = 0$$

$$\ddot{a}_y = \ddot{c}_y = 1$$

$$u_{1x}\dot{u}_{1x} + u_{1y}\dot{u}_{1y} + u_{1z}\dot{u}_{1z} = 0 \qquad (10.22)$$

which may be rewritten in terms of acceleration matrix elements as

$$\ddot{d}_{12}a_y + \ddot{d}_{14} = 0$$

$$\ddot{d}_{32}a_y + \ddot{d}_{34} = 0$$

$$\ddot{d}_{21}b_x + \ddot{d}_{23}b_z + \ddot{d}_{24} = 0$$

$$\ddot{d}_{31}b_x + \ddot{d}_{33}b_z + \ddot{d}_{34} = 0$$

$$\ddot{d}_{21}c_x + \ddot{d}_{23}c_z = 0$$

$$\ddot{d}_{22}a_y + \ddot{d}_{24} = 1$$

$$u_{1x}\dot{u}_{1x} + u_{1y}\dot{u}_{1y} + u_{1z}\dot{u}_{1z} = 0 \qquad (10.23)$$

TABLE 10.1   EXAMPLE 10.1, FIRST-ORDER SCREW AXIS PARAMETERS

	Input		First-order Screw Axis							
	Velocity	Acceleration	Screw Position			Screw Direction			Velocity		Instant Pitch
Number	$\dot{a}_y$	$\ddot{a}_y$	P_{0x}	P_{0y}	P_{0z}	u_{x_1}	u_{y_1}	u_{z_1}	$\dot{\psi}_1$	$\dot{S}_1$	$\dot{S}_{1\mathrm{IP}}$
1	1.0	1.0	10.290435	2.3731889	0.0000000	−0.41496877	0.14375361	0.89840738	−0.10593127	0.14375361	−1.3570460
2	2.0	3.0							−0.21186254	0.28750723	
3	3.0	5.0							−0.31779382	0.43126084	
4	4.0	7.0							−0.42372509	0.57501446	
5	5.0	9.0							−0.52965636	0.71876807	

TABLE 10.2   EXAMPLE 10.1, SECOND-ORDER SCREW AXIS PARAMETERS

	Input		Second-Order Screw Axis							
	Velocity	Acceleration	Screw Position			Screw Direction			Velocity		Instant Pitch
Number	$\dot{a}_y$	$\ddot{a}_y$	c_x	c_y	c_z	u_{x_2}	u_{y_2}	u_{z_2}	$\dot{\phi}_2$	$\dot{s}_2$	$\dot{s}_{2\mathrm{IP}}$
1	1.0	1.0	15.810808	0.46081956	−11.951607	−0.89367829	−0.24966517	−0.37283566	0.059733263	0.20532835	3.4374206
2	2.0	3.0							0.11946653	0.41065669	
3	3.0	5.0							0.17919979	0.61598504	
4	4.0	7.0							0.23893305	0.82131339	
5	5.0	9.0							0.29866632	1.0266417	

TABLE 10.3 EXAMPLE 10.1, ACCELERATION PARAMETERS

Number	Input		Acceleration Parameters							
	Velocity	Acceleration	Velocity of Screw Position			Velocity of Screw Direction			Acceleration	
	$\dot{a}_y$	$\ddot{a}_y$	$\dot{P}_{0x}$	$\dot{P}_{0y}$	$\dot{P}_{0z}$	$\dot{u}_{x_1}$	$\dot{u}_{y_1}$	$\dot{u}_{z_1}$	$\ddot{\phi}_1$	$\ddot{s}_1$
1	1.0	1 0	−0.43968504	0.75144095	0.0000000	−0.010196740	0.057200711	−0.013862461	−0.11031311	0.20095433
2	2.0	3 0	−0.87937008	1.5028819		−0.020393480	0.11440142	−0.027724923	−0.33532118	0.66006369
3	3.0	5 0	−1.3190551	2.2543228		−0.030590220	0.17160213	−0.041587384	−0.56909293	1.2335745
4	4.0	7 0	−1.7587402	3.0057638		−0.040786960	0.22880285	−0.055449846	−0.81162837	1.9214867
5	5.0	9 0	−2.1984252	3.7572047		−0.050983700	0.28600356	−0.069312307	−1.0629275	2.723803

After substitution of the appropriate expressions for acceleration matrix elements $\ddot{d}_{ij}$ in terms of screw motion parameters, we may solve Eqs. 10.23 for the seven unknown second-order screw parameters as listed before in Eq. 10.22.

Tables 10.1, 10.2, and 10.3 summarize the results for different values of input velocity and acceleration. It is interesting to note that the geometric properties of the first-order and second-order screw axes are independent of the magnitude of the input motion components.

10.4 DIFFERENTIAL DISPLACEMENT MATRICES IN TERMS OF GEOMETRICAL PARAMETERS

Kinematics is often described as the *geometry of motion*. Hence, it is natural to describe the motion of mechanisms in terms of purely geometric parameters.

The Phi Matrix

We will use the angular displacement ϕ about the spatial rotation axis **u** as the independent geometric parameter. This will provide a description of the rigid body displacement except in the special case of pure translation, where the angle ϕ is undefined. Thus, we see that the description of rigid body motion in terms of the angle ϕ alone is somewhat limited and cannot be quaranteed to describe the motion.

We may derive the *phi matrices* directly from the previous differential matrices in terms of time by making the following substitutions and assumptions. Assume that

$$\dot{\phi} = \frac{d\phi}{dt} = 1 \qquad \ddot{\phi} = \dddot{\phi} = 0$$

Then, adopting the notation

$$\frac{d\mathbf{a}}{d\phi} = \mathbf{a}' \qquad \frac{d^2\mathbf{a}}{d\phi^2} = \mathbf{a}'' \text{ etc.}$$

we have,

$$\dot{\mathbf{a}} = \frac{d\mathbf{a}}{d\phi}\frac{d\phi}{dt}$$

$$\ddot{\mathbf{a}} = \frac{d^2\mathbf{a}}{d\phi}\frac{d\phi}{dt}\frac{1}{dt} + \frac{d\mathbf{a}}{d\phi}\frac{d^2\phi}{dt^2} = \frac{d^2\mathbf{a}}{d\phi^2}\left(\frac{d\phi}{dt}\right)^2 + \frac{d\mathbf{a}}{d\phi}\frac{d^2\phi}{dt^2}$$

$$\dddot{\mathbf{a}} = \frac{d^3\mathbf{a}}{d\phi^2}\frac{1}{dt}\left(\frac{d\phi}{dt}\right)^2 + 2\frac{d^2\mathbf{a}}{d\phi^2}\frac{d\phi}{dt}\frac{d^2\phi}{dt^2} + \frac{d^2\mathbf{a}}{d\phi}\frac{d^2\phi}{dt}\frac{1}{dt^2} + \frac{d\mathbf{a}}{d\phi}\frac{d^3\phi}{dt^3}$$

$$= \frac{d^3\mathbf{a}}{d\phi^3}\left(\frac{d\phi}{dt}\right)^3 + 3\frac{d^2\mathbf{a}}{d\phi^2}\frac{d\phi}{dt}\frac{d^2\phi}{dt^2} + \frac{d\mathbf{a}}{d\phi}\frac{d^3\phi}{dt^3} \tag{10.24}$$

Hence, for $\dot{\phi} = 1$, $\ddot{\phi} = \dddot{\phi} = 0$, we see that $\dot{\mathbf{a}} = \mathbf{a}'$, $\ddot{\mathbf{a}} = \mathbf{a}''$, and $\dddot{\mathbf{a}} = \mathbf{a}'''$.

Substituting into equations of the form of Eq. 3.75, we obtain the following first-order, second-order, and third-order phi matrices in the form given below, where $\dot{\phi}$ is assumed equal to unity.

$$[V'] = \begin{bmatrix} [R'] & (\mathbf{p}' - [R']\mathbf{p}) \\ 0 & 0 \end{bmatrix} \tag{10.25}$$

$$[A'] = \begin{bmatrix} [R''] & (\mathbf{p}'' - [R'']\mathbf{p}) \\ 0 & 0 \end{bmatrix} \tag{10.26}$$

$$[J'] = \begin{bmatrix} [R'''] & (\mathbf{p}''' - [R''']\mathbf{p}) \\ 0 & 0 \end{bmatrix} \tag{10.27}$$

The *first-order phi rotation matrix* is given by

$$[R'] = [W] = [P_u]$$
$$\dot{\phi} = 1$$

that is,

$$[R'] = \begin{bmatrix} 0 & -u_z & u_y \\ u_z & 0 & -u_x \\ -u_y & u_x & 0 \end{bmatrix} \tag{10.28}$$

The *second-order phi rotation matrix* becomes

$$[R''] = [\dot{W}] = [P'_u] + [P_u][P_u]$$
$$\dot{\phi} = 1, \ddot{\phi} = 0$$

which leads to

$$[R''] = \begin{bmatrix} (u_x^2 - 1) & (u_x u_y - u'_z) & (u_x u_z + u'_y) \\ (u_x u_y + u'_z) & (u_y^2 - 1) & (u_y u_z - u'_x) \\ (u_x u_z - u'_y) & (u_y u_z + u'_x) & (u_z^2 - 1) \end{bmatrix} \tag{10.29}$$

The *third-order phi rotation* matrix is found from $[R'''] = [\ddot{W}]$ with $\dot{\phi} = 1$, $\ddot{\phi} = 0$, and $\dddot{\phi} = 0$. The elements of the matrix may be found from Eq. 3.68 in the form

$$[R'''] = [P''_u] + 2[P'_u][P_u] + [P_u][P'_u] + [P_u][P_u][P_u] \tag{10.30}$$

or from Eq. 3.70 with $\ddot{\phi} = \dddot{\phi} = 0$.

This leads to the following expressions for the $[R''']$ matrix elements.

$$d'''_{11} = 3u_x u'_x$$

$$d'''_{12} = 2u_x u'_y + u'_x u_y - u''_z + u_z$$

$$d'''_{13} = 2u_x u'_z + u'_x u_z + u''_y - u_y$$

$$d'''_{21} = 2u'_x u_y + u_x u'_y + u''_z - u_z$$

$$d'''_{22} = 3u_y u'_y$$

$$d'''_{23} = 2u_y u'_z + u'_y u_z - u''_x + u_x$$

$$d'''_{31} = 2u'_x u_z + u_x u'_z - u''_y + u_y$$

$$d'''_{32} = 2u'_y u_z + u_y u'_z + u''_x - u_x$$

$$d'''_{33} = 3u_z u'_z \tag{10.31}$$

The Plane Phi Matrices

For plane motion $u_x = u_y = 0$, $u_z = 1$, $u'_x = u''_x = 0$, and so on. In this case, the third row and column of each spatial phi matrix equals zero, and the plane phi matrices become

$$[D'] = \begin{bmatrix} 0 & -1 & p'_x + p_y \\ 1 & 0 & p'_y - p_x \\ 0 & 0 & 0 \end{bmatrix} \tag{10.32}$$

$$[D''] = \begin{bmatrix} -1 & 0 & p''_x + p_y \\ 0 & -1 & p''_y + p_x \\ 0 & 0 & 0 \end{bmatrix} \tag{10.33}$$

$$[D'''] = \begin{bmatrix} 0 & 1 & p'''_x - p_y \\ -1 & 0 & p'''_y + p_x \\ 0 & 0 & 0 \end{bmatrix} \tag{10.34}$$

where $\mathbf{p}$, $\mathbf{p}' = [(dp_x/d\phi), (dp_y/d\phi)]$, and so forth, refer to the specified motion of an arbitrary point of the moving plane.

Canonical Coordinate Systems

If we select a particular coordinate system such that the moving reference point $\mathbf{p}$ is located at the origin [i.e., $\mathbf{p} = (0.0, 0.0)$], then

$$[D'] = \begin{bmatrix} 0 & -1 & p'_x \\ 1 & 0 & p'_y \\ 0 & 0 & 0 \end{bmatrix} \qquad (10.35)$$

$$[D''] = \begin{bmatrix} -1 & 0 & p''_x \\ 0 & -1 & p''_y \\ 0 & 0 & 0 \end{bmatrix} \qquad (10.36)$$

$$[D'''] = \begin{bmatrix} 0 & 1 & p'''_x \\ -1 & 0 & p'''_y \\ 0 & 0 & 0 \end{bmatrix} \qquad (10.37)$$

These matrices are equivalent to the canonical forms used by others [2, 3, 4] in studies of higher-order plane curvature theory.

10.5   PLANE PATH CURVATURE

The Velocity Pole, $\mathbf{p}_0$

A point of the moving plane that instantaneously has zero velocity may be located as follows. We write an expression for $\dot{\mathbf{p}}_0$ in the form

$$(\dot{\mathbf{p}}_0) = \begin{bmatrix} 0 & -\dot{\theta} & (\dot{p}_x + p_y\dot{\theta}) \\ \dot{\theta} & 0 & (\dot{p}_y - p_x\dot{\theta}) \\ 0 & 0 & 0 \end{bmatrix} \begin{bmatrix} p_{0x} \\ p_{0y} \\ 1 \end{bmatrix} = \begin{bmatrix} 0 \\ 0 \\ 0 \end{bmatrix} \qquad (10.38)$$

where $\mathbf{p}$ is an arbitrary point of the plane with specified motion.
   This leads to

$$-\dot{\theta}p_{0y} + (\dot{p}_x + p_y\dot{\theta}) = 0$$

$$\dot{\theta}p_{0x} + (\dot{p}_y - p_x\dot{\theta}) = 0 \qquad (10.39)$$

from which the coordinates of the velocity pole $\mathbf{p}_0 = (p_{0x}, p_{0y})$ can be calculated as

$$p_{0x} = p_x - \frac{\dot{p}_y}{\dot{\theta}}$$

$$p_{0y} = p_y + \frac{\dot{p}_x}{\dot{\theta}} \qquad (10.40)$$

Pole Velocity, $\dot{\bar{\mathbf{p}}}_0$

The term *pole velocity* refers to the rate of change of location of the velocity pole in the fixed coordinate system. It does *not* refer to a point that is attached to the moving plane. The overbar will be used to designate the pole velocity components as compared to the motion components of a point in the moving plane with coordinates $\mathbf{p}_0$. The pole velocity components are found by differentiating Eqs. 10.40.

$$\dot{\bar{p}}_{0x} = \dot{p}_x + \frac{\dot{p}_y\dot{\theta} - \ddot{p}_y\theta}{\theta^2}$$

$$\dot{\bar{p}}_{0y} = \dot{p}_y + \frac{\ddot{p}_x\theta - \dot{p}_x\dot{\theta}}{\theta^2} \tag{10.41}$$

As a special case, assume that the reference point $\mathbf{p}$ is located at the velocity pole $\mathbf{p}_0$ and the velocity $\dot{\mathbf{p}}_0 = (0.0, 0.0)$ and acceleration $\ddot{\mathbf{p}}_0 = (\ddot{p}_{0x}, \ddot{p}_{0y})$ of the point $\mathbf{p}_0$ are specified. The pole velocity components become

$$\dot{\bar{p}}_{0x} = -\frac{\ddot{p}_{0y}}{\dot{\theta}} \qquad \dot{\bar{p}}_{0y} = \frac{\ddot{p}_{0x}}{\dot{\theta}} \tag{10.42}$$

Equations 10.42 are equivalent to the vector equation

$$-\dot{\theta} \times \dot{\bar{p}}_0 = \ddot{\mathbf{p}}_0 \tag{10.43}$$

If we introduce a canonical coordinate system such that the x-axis is parallel to the pole velocity vector (i.e., $p_{0x} = p_{0y} = 0$, $\dot{p}_{0x} = \dot{p}_{0y} = 0$, $\ddot{p}_{0x} = 0$, $\ddot{p}_{0y} = -1$, and $\dot{\theta} = 1$, as shown in Figure 10.3), then we see from Eqs. 10.42 that the pole velocity is related to the angular velocity of the moving plane by the relationships

$$\dot{\bar{p}}_{0x} = -\frac{\ddot{p}_{0y}}{\dot{\theta}} = -\left(\frac{-1}{1}\right) = +1 \qquad \dot{\bar{p}}_{0y} = 0 \tag{10.44}$$

The Acceleration Pole

A point of the moving plane with zero acceleration (i.e., the acceleration pole $\mathbf{q}_0$) is found from

$$[A](\mathbf{q}_0) = \begin{bmatrix} -\dot{\theta}^2 & -\ddot{\theta} & (\ddot{p}_x + p_x\dot{\theta}^2 + p_y\ddot{\theta}) \\ \ddot{\theta} & -\dot{\theta}^2 & (\ddot{p}_y - p_x\ddot{\theta} + p_y\dot{\theta}^2) \\ 0 & 0 & 0 \end{bmatrix}\begin{bmatrix} q_{0x} \\ q_{0y} \\ 1 \end{bmatrix} = \begin{bmatrix} 0 \\ 0 \\ 0 \end{bmatrix} \tag{10.45}$$

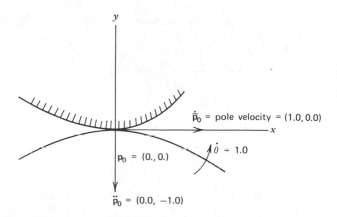

FIGURE 10.3   POLE VELOCITY IN CANONICAL COORDINATES.

which leads to

$$q_{0x} = \frac{\dot\theta^2(\ddot{p}_x + p_x\dot\theta^2) - \ddot\theta(\ddot{p}_y - p_x\dot\theta)}{\dot\theta^4 + \ddot\theta^2}$$

$$q_{0y} = \frac{\dot\theta^2(\ddot{p}_y + p_y\dot\theta^2) + \ddot\theta(\ddot{p}_x + p_y\dot\theta)}{\dot\theta^4 + \ddot\theta^2} \tag{10.46}$$

The Inflection Circle—Zero Normal Acceleration

The normal acceleration of a point **p** moving on a plane curved path can be expressed in terms of the velocity $\mathbf{v} = \dot{\mathbf{p}}$ and the acceleration $\mathbf{a} = \ddot{\mathbf{p}}$ of the point in the form

$$va_n = va \sin \alpha = \left| \mathbf{v} \times \mathbf{a} \right|$$

$$= \dot{p}_x\ddot{p}_y - \dot{p}_y\ddot{p}_x \tag{10.47}$$

For a nonzero velocity $\dot{\mathbf{p}}$ the normal acceleration will be zero if

$$\dot{p}_x\ddot{p}_y - \dot{p}_y\ddot{p}_x = 0 \tag{10.48}$$

Points that satisfy Eq. 10.48 must be at a point in their path with zero curvature (i.e., at an *inflection point*). We will express the motion of the plane in terms of displacement matrices with the velocity pole as the reference point with its position $\mathbf{p}_0$, velocity $\dot{\mathbf{p}}_0$, and acceleration $\ddot{\mathbf{p}}_0$ specified. The motion of point $\mathbf{p}$ is given by

$$
\begin{bmatrix} \dot{p}_x \\ \dot{p}_y \\ 0 \end{bmatrix} = \begin{bmatrix} 0 & -\dot{\theta} & (0 + p_{0y}\dot{\theta}) \\ \dot{\theta} & 0 & (0 - p_{0x}\dot{\theta}) \\ 0 & 0 & 0 \end{bmatrix} \begin{bmatrix} p_x \\ p_y \\ 1 \end{bmatrix}
\tag{10.49}
$$

$$
\begin{bmatrix} \ddot{p}_x \\ \ddot{p}_y \\ 0 \end{bmatrix} = \begin{bmatrix} -\dot{\theta}^2 & -\ddot{\theta} & (\ddot{p}_{0x} + p_{0x}\dot{\theta}^2 + p_{0y}\ddot{\theta}) \\ \ddot{\theta} & -\dot{\theta}^2 & (\ddot{p}_{0y} - p_{0x}\ddot{\theta} + p_{0y}\dot{\theta}^2) \\ 0 & 0 & 0 \end{bmatrix} \begin{bmatrix} p_x \\ p_y \\ 1 \end{bmatrix}
\tag{10.50}
$$

Substitution of the resulting expressions for $\dot{p}_x$, $\dot{p}_y$, $\ddot{p}_x$, and $\ddot{p}_y$ into Eq. 10.48 leads to

$$
\dot{\theta}^2[(p_x - p_{0x})^2 + (p_y - p_{0y})^2] - \ddot{p}_{0x}(p_x - p_{0x}) + \ddot{p}_{0y}(p_y - p_{0y}) = 0 \tag{10.51}
$$

which reduces to the equation of a circle

$$
p_x^2 + p_y^2 + Ap_x + Bp_y + C = 0 \tag{10.52}
$$

where

$$
A = -\left(2p_{0x} + \frac{\ddot{p}_{0x}}{\dot{\theta}^2}\right)
$$

$$
B = -\left(2p_{0y} + \frac{\ddot{p}_{0y}}{\dot{\theta}^2}\right)
$$

$$
C = p_{0x}^2 + p_{0y}^2 + \frac{p_{0x}\ddot{p}_{0x} + p_{0y}\ddot{p}_{0y}}{\dot{\theta}^2}
$$

Note that Eq. 10.52 is independent of the angular acceleration $\ddot{\theta}$. Eq. 10.52 defines the *inflection circle*, which is the locus of all points in the moving plane that are instantaneously passing through a point on their paths of motion with zero curvature. We note that the center of the circle is located at

$$
\mathbf{c} = (c_x, c_y) = \left(-\frac{A}{2}, -\frac{B}{2}\right) \tag{10.53}
$$

and the radius of the circle is given by

$$
R = \tfrac{1}{2}\sqrt{A^2 + B^2 - 4C} \tag{10.54}
$$

If the origin of the coordinate system is located at the velocity pole $\mathbf{p}_0$, Eq. 10.52 takes the simplified form

$$
p_x^2 + p_y^2 - \frac{\ddot{p}_{0x}}{\dot{\theta}^2}p_x - \frac{\ddot{p}_{0y}}{\dot{\theta}^2}p_y = 0 \tag{10.55}
$$

The Zero Tangential Acceleration Circle

The tangential acceleration of a general point $\mathbf{p}$ is given in terms of the variables defined for Eq. 10.47 as

$$a_t = a \cos \alpha$$

Hence

$$va_t = va \cos \alpha = |\mathbf{v} \cdot \mathbf{a}|$$
$$= \dot{p}_x \ddot{p}_x + \dot{p}_y \ddot{p}_y \tag{10.56}$$

Therefore, all points of the moving plane that instantaneously have zero tangential acceleration along their paths of motion must satisfy the relationship

$$\dot{p}_x \ddot{p}_x + \dot{p}_y \ddot{p}_y = 0 \tag{10.57}$$

Substituting Eqs. 10.49 and 10.50, we obtain a second circle equation

$$p_x^2 + p_y^2 + Dp_x + Ep_y + F = 0 \tag{10.58}$$

where

$$D = -\left(2p_{0x} - \frac{\ddot{p}_{0y}}{\ddot{\theta}}\right)$$

$$E = -\left(2p_{0y} + \frac{\ddot{p}_{0x}}{\ddot{\theta}}\right)$$

$$F = p_{0x}^2 + p_{0y}^2 + \frac{\ddot{p}_{0x} p_{0y}}{\ddot{\theta}} - \frac{\ddot{p}_{0y} p_{0x}}{\ddot{\theta}}$$

The *zero tangential acceleration circle* is seen to have its center at

$$\mathbf{d} = (d_x, d_y) = \left(-\frac{D}{2}, -\frac{E}{2}\right) \tag{10.59}$$

and its radius S is given by

$$S = \tfrac{1}{2}\sqrt{D^2 + E^2 - 4F} \tag{10.60}$$

As with the inflection circle, a simplification results if the origin coincides with the velocity pole. In this case the circle equation becomes

$$p_x^2 + p_y^2 - \frac{\ddot{p}_{0x} p_y}{\ddot{\theta}} + \frac{\ddot{p}_{0y} p_x}{\ddot{\theta}} = 0 \tag{10.61}$$

The intersection of the inflection and zero tangential acceleration circles defines a point that has both zero normal and tangential acceleration components. Hence, we have an alternate means of locating the acceleration pole $\mathbf{q}_0$ previously defined by Eqs. 10.46.

There is an interesting relationship between the coordinates of the center of the inflection circle $\mathbf{c}$ and the center of the zero tangential acceleration circle $\mathbf{d}$. We define

$$k = \frac{\dot\theta^2}{\ddot\theta} \tag{10.62}$$

Then, if $\mathbf{p}_0 = (0.0, 0.0)$,

$$d_x = -kc_y$$
$$d_y = kc_x$$

and

$$S = kR \tag{10.63}$$

A further simplification of the circle equations is possible if a canonical coordinate system is adopted such that $p_{0x} = p_{0y} = \dot p_{0x} = \dot p_{0y} = \ddot p_{0x} = 0$ and $\ddot p_{0y} = -1$. Under these conditions the equation of the *inflection circle* becomes

$$p_x^2 + p_y^2 + p_y = 0 \tag{10.64}$$

and the corresponding canonical form of the *zero tangential acceleration circle* becomes

$$p_x^2 + p_y^2 - kp_x = 0 \tag{10.65}$$

The Cubic of Stationary Curvature

Curvature for a point moving on a plane path is defined by

$$\frac{1}{\rho} = \frac{|\mathbf{v} \times \mathbf{a}|}{|\mathbf{v}^3|} = \frac{|\dot p_x \ddot p_y - \dot p_y \ddot p_x|}{(\dot p_x^2 + \dot p_y^2)^{3/2}} \tag{10.66}$$

For *stationary* curvature,

$$\frac{d}{dt}\left(\frac{1}{\rho}\right) = \frac{d}{ds}\left(\frac{1}{\rho}\right) = 0$$

that is,

$$\frac{|(|\mathbf{v}^3|)(\dot{\mathbf{v}} \times \mathbf{a} + \mathbf{v} \times \dot{\mathbf{a}}) - (\mathbf{v} \times \mathbf{a}) \cdot 3|\mathbf{v}^2| \cdot |\dot{\mathbf{v}}||}{|\mathbf{v}^6|} = 0$$

Which, since $\dot{\mathbf{v}} \times \mathbf{a} = 0$, gives

$$|\mathbf{v}|(\mathbf{v} \times \dot{\mathbf{a}}) - (\mathbf{v} \times \mathbf{a}) \cdot 3|\dot{\mathbf{v}}| = 0 \qquad (10.67)$$

Making the substitutions

$$\mathbf{v} \times \mathbf{a} = \dot{p}_x \ddot{p}_y - \dot{p}_y \ddot{p}_x$$

$$\mathbf{v} \times \dot{\mathbf{a}} = \dot{p}\dddot{p}_y - \dot{p}_y \dddot{p}_x$$

$$|\mathbf{v}| = (\dot{p}_x^2 + \dot{p}_y^2)^{1/2}$$

$$|\dot{\mathbf{v}}| = \frac{|\dot{p}_x \ddot{p}_x + \dot{p}_y \ddot{p}_y|}{(\dot{p}_x^2 + \dot{p}_y^2)^{1/2}}$$

we obtain

$$(\dot{p}_x^2 + \dot{p}_y^2)(\dot{p}_x \dddot{p}_y - \dot{p}_y \dddot{p}_x) - 3(\dot{p}_x \ddot{p}_y - \dot{p}_y \ddot{p}_x)|\dot{p}_x \ddot{p}_x + \dot{p}_y \ddot{p}_y| = 0 \qquad (10.68)$$

Points that satisfy Eq. 10.68 have stationary curvature for an infinitesimal displacement ds along their path. Such points are seen to have fourth-order contact with a portion of a circular arc.

Adopting a coordinate system with origin at the velocity pole $\mathbf{p}_0$ and x-direction parallel to the pole velocity vector $\dot{\bar{\mathbf{p}}}_0$, Eq. 10.68 reduces to the cubic form given by Roth and Yang (6).

$$(p_x^2 + p_y^2)[(p_{0x}''' + 3p_{0y}'') p_x + p_{0y}''' p_y] - 3(p_{0y}'')^2 p_x p_y = 0 \qquad (10.69)$$

The *instantaneous invariants* p_{0y}'', p_{0x}''', p_{0y}''' can be calculated from the results of a second-acceleration analysis. Rewrite Eqs. 10.24 in the form

$$p_{0y}'' = \frac{1}{\dot{\theta}^2} \ddot{p}_{0y} \qquad p_{0x}''' = \frac{1}{\dot{\theta}^3} \dddot{p}_{0x} \qquad p_{0y}''' = \frac{1}{\dot{\theta}^3} \dddot{p}_{0y} - 3 \frac{\ddot{\theta}}{\dot{\theta}^2} p_{0y}''$$

For the canonical coordinate system of Eqs. 10.64 and 10.65 the mechanism is scaled to give $\ddot{p}_{0y} = -1$, and the cubic equation can be written in a form similar to that given by Veldkamp (5).

$$(p_x^2 + p_y^2)[(p_{0x}''' - 3)p_x + p_{0y}''' p_y] - 3 p_x p_y = 0 \qquad (10.70)$$

Geometric Analysis of Plane Mechanisms

The inflection circle and the cubic of stationary curvature have been shown to be geometrical instead of kinematic properties of plane motion. The zero tangential acceleration circle (the Bresse circle) requires the additional specification of the angular acceleration of the plane. The higher-order motion components of the point in the moving plane, which is instantaneously coincident with the velocity pole, become the *instantaneous invariants* of the plane motion. The following examples demonstrate the method of calculation of the invariants on a purely geometric basis. In this section we will adopt the notation of Veldkamp (5). A point $\mathbf{p}$ of the moving plane will have coordinates $\mathbf{p} = (x, y)$. The higher-order motion components of the point at the velocity pole $\mathbf{p}_0 = (a, b) = (0, 0)$ will be given by

$$\mathbf{p}_0' = (a_1, b_1) \qquad \mathbf{p}_0'' = (a_2, b_2) \qquad \mathbf{p}_0''' = (a_3, b_3)$$

The phi rotation matrices can be extracted from Equations 10.32 to 10.34 in the form

$$[R'] = \begin{bmatrix} 0 & -1 \\ 1 & 0 \end{bmatrix} \quad [R''] = \begin{bmatrix} -1 & 0 \\ 0 & -1 \end{bmatrix} \quad [R'''] = \begin{bmatrix} 0 & 1 \\ -1 & 0 \end{bmatrix}$$

Example 10.2 The Rolling Cylinder. The geometric constraint provided by the rolling cylinder with coordinate system as shown in Figure 10.4, leads to the higher-order motion components for the center of the cylinder, point $\mathbf{a}$, in the form

$$a_x' = \frac{da_x}{d\phi} = \frac{rd\phi}{d\phi} = r \qquad a_y' = \frac{da_y}{d\phi} = 0$$

$$a_x'' = \frac{d}{d\phi}(a_x') = 0 \qquad a_y'' = 0$$

$$a_x''' = 0 \qquad a_y''' = 0$$

The instantaneous invariants are calculated from

$$\mathbf{p}_0' = [R'](\mathbf{p}_0 - \mathbf{a}) + \mathbf{a}' = \begin{bmatrix} 0 & -1 \\ 1 & 0 \end{bmatrix}\begin{pmatrix} 0 \\ r \end{pmatrix} + \begin{pmatrix} r \\ 0 \end{pmatrix} = \begin{pmatrix} 0 \\ 0 \end{pmatrix}$$

$$\mathbf{p}_0'' = [R''](\mathbf{p}_0 - \mathbf{a}) + \mathbf{a}'' = \begin{bmatrix} -1 & 0 \\ 0 & -1 \end{bmatrix}\begin{pmatrix} 0 \\ r \end{pmatrix} + \begin{pmatrix} 0 \\ 0 \end{pmatrix} = \begin{pmatrix} 0 \\ -r \end{pmatrix}$$

$$\mathbf{p}_0''' = [R'''](\mathbf{p}_0 - \mathbf{a}) + \mathbf{a}''' = \begin{bmatrix} 0 & 1 \\ -1 & 0 \end{bmatrix}\begin{pmatrix} 0 \\ r \end{pmatrix} + \begin{pmatrix} 0 \\ 0 \end{pmatrix} = \begin{pmatrix} r \\ 0 \end{pmatrix}$$

From Eq. 10.55, the equation of the inflection circle can be written using Veldkamp's notation in the form

$$x^2 + y^2 - b_2 y = 0$$

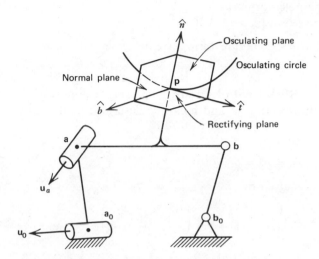

FIGURE 10.4   THE ROLLING CYLINDER. EXAMPLE 10.2.

For the rolling cylinder we see that $b_2 = -r$, which leads to a circle equation in the form

$$(x - 0)^2 + \left(y + \frac{r}{2}\right)^2 = \left(\frac{r}{2}\right)^2$$

The inflection circle has its center at point $(0, -r/2)$ and a radius equal to one half the radius of the cylinder.

From Eq. 10.69, the equation of the cubic of stationary curvature can be expressed in the form

$$(x^2 + y^2)[(a_3 + 3b_2)x + b_3 y] - 3b_2^2 xy = 0$$

which, with $b_2 = -r$, $a_3 = r$, and $b_3 = 0$, becomes

$$(x^2 + y^2)[(r - 3r)x + 0] - 3r^2 xy = 0$$

When rewritten in the form

$$x\left[(x - 0)^2 + \left(y + \frac{3}{4}r\right)^2 - \left(\frac{3}{4}r\right)^2\right] = 0$$

we note that the cubic equation degenerates to:

1. The straight line $x = 0$ (the pole normal).
2. A circle of radius $= \frac{3}{4}r$ centered at $[0, -(3/4r)]$.

It is interesting to note that a degenerate form is to be expected whenever $b_3 = 0$ or $a_3 = -3b_2$.

Example 10.3 The Double-Slider Mechanism. Referring to Figure 10.5, the constraints provided by the sliders at $\mathbf{a}$ and $\mathbf{b}$ can be described analytically in the form

$$a'_x = a''_x = a'''_x = 0$$
$$b'_y = b''_y = b'''_y = 0$$

The motion of $\mathbf{b}$ in terms of $\mathbf{a}$ is given by

$$\mathbf{b}' = [R'](\mathbf{b} - \mathbf{a}) + \mathbf{a}' = \begin{bmatrix} 0 & -1 \\ 1 & 0 \end{bmatrix} \begin{pmatrix} e \\ f \end{pmatrix} + \begin{pmatrix} 0 \\ a'_y \end{pmatrix} = \begin{pmatrix} -f \\ e + a'_y \end{pmatrix}$$

$$\mathbf{b}'' = [R''](\mathbf{b} - \mathbf{a}) + \mathbf{a}'' = \begin{bmatrix} -1 & 0 \\ 1 & -1 \end{bmatrix} \begin{pmatrix} e \\ f \end{pmatrix} + \begin{pmatrix} 0 \\ a''_y \end{pmatrix} = \begin{pmatrix} -e \\ a''_y - f \end{pmatrix}$$

$$\mathbf{b}''' = [R'''](\mathbf{b} - \mathbf{a}) + \mathbf{a}''' = \begin{bmatrix} 0 & 1 \\ -1 & 0 \end{bmatrix} \begin{pmatrix} e \\ f \end{pmatrix} + \begin{pmatrix} 0 \\ a'''_y \end{pmatrix} = \begin{pmatrix} f \\ a'''_y - e \end{pmatrix}$$

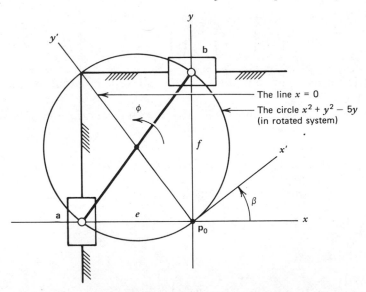

The line $x = 0$

The circle $x^2 + y^2 - 5y$ (in rotated system)

FIGURE 10.5 THE DOUBLE SLIDER. EXAMPLE 10.3.

which leads to

$$b'_x = -f \qquad b''_x = -e \qquad b'''_x = f$$
$$a'_y = -e \qquad a''_y = f \qquad a'''_y = e$$

The motion of point $\mathbf{p}_0$ becomes

$$\mathbf{p}'_0 = [R'](\mathbf{p}_0 - \mathbf{a}) + \mathbf{a}' = \begin{bmatrix} 0 & -1 \\ 1 & 0 \end{bmatrix} \begin{pmatrix} e \\ 0 \end{pmatrix} + \begin{pmatrix} 0 \\ -e \end{pmatrix} = \begin{pmatrix} 0 \\ 0 \end{pmatrix}$$

$$\mathbf{p}''_0 = [R''](\mathbf{p}_0 - \mathbf{a}) + \mathbf{a}'' = \begin{bmatrix} -1 & 0 \\ 0 & -1 \end{bmatrix} \begin{pmatrix} e \\ 0 \end{pmatrix} + \begin{pmatrix} 0 \\ f \end{pmatrix} = \begin{pmatrix} -e \\ f \end{pmatrix}$$

$$\mathbf{p}'''_0 = [R'''](\mathbf{p}_0 - \mathbf{a}) + \mathbf{a}''' = \begin{bmatrix} 0 & 1 \\ -1 & 0 \end{bmatrix} \begin{pmatrix} e \\ 0 \end{pmatrix} + \begin{pmatrix} 0 \\ e \end{pmatrix} = \begin{pmatrix} 0 \\ 0 \end{pmatrix}$$

Rotate the coordinate system through an angle β such that P''_{0x} in the rotated system equals zero.

$$\begin{pmatrix} 0 \\ p''_{0y} \end{pmatrix} = \begin{bmatrix} \cos \beta & \sin \beta \\ -\sin \beta & \cos \beta \end{bmatrix} \begin{pmatrix} -e \\ f \end{pmatrix}$$

$$0 \quad = -e \cos \beta + f \sin \beta, \ \tan \beta = \frac{e}{f}$$

$$p''_{0y} = e \sin \beta + f \cos \beta$$

Assume

$$e = 3 \qquad f = 4 \qquad \tan \beta = \frac{3}{4} \qquad \beta = 36.87° \qquad p''_{0y} = 3 \sin \beta + 4 \cos \beta = 5$$

$$\begin{pmatrix} p'''_{0x} \\ p'''_{0y} \end{pmatrix} - \begin{bmatrix} \cos \beta & \sin \beta \\ -\sin \beta & \cos \beta \end{bmatrix} \begin{pmatrix} 0 \\ 0 \end{pmatrix} = \begin{pmatrix} 0 \\ 0 \end{pmatrix}$$

In the rotated coordinate system the equation of the inflection circle becomes

$$x^2 + y^2 - 5y = 0$$

and the cubic of stationary curvature takes the form

$$(x^2 + y^2)[(0 + 15)x + 0] - 75xy = 0$$

which degenerates to:

1. The equation of the straight line $x = 0$.
2. The circle $x^2 + y^2 - 5y = 0$.

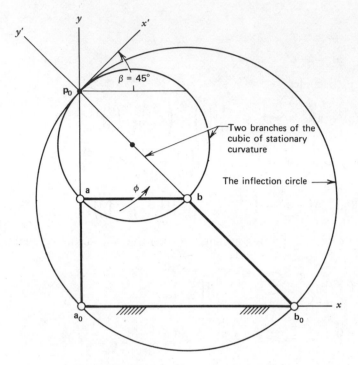

FIGURE 10.6 THE FOUR-BAR LINKAGE. EXAMPLE 10.4.

Example 10.4 The Four-Bar Linkage. The four-bar linkage shown in Figure 10.6 will be analyzed on a "geometric" basis.
First-order analysis (origin at $a_0 = (0, 0)$)
Constraint equations

$$(\mathbf{a}')^T(\mathbf{a} - \mathbf{a}_0) = 0 \qquad\qquad (\mathbf{b}')^T(\mathbf{b} - \mathbf{b}_0) = 0$$

$$\mathbf{a}' = [R'](\mathbf{a} - \mathbf{p}_0) + \mathbf{p}_0' \qquad\qquad \mathbf{b}' = [R'](\mathbf{b} - \mathbf{p}_0) + \mathbf{p}_0'$$

$$= \begin{bmatrix} 0 & -1 \\ 1 & 0 \end{bmatrix}\begin{pmatrix} 0 - p_{0x} \\ 1 - p_{0y} \end{pmatrix} + \begin{pmatrix} 0 \\ 0 \end{pmatrix} \qquad\qquad = \begin{bmatrix} 0 & -1 \\ 1 & 0 \end{bmatrix}\begin{pmatrix} 1 - p_{0x} \\ 1 - p_{0y} \end{pmatrix} + \begin{pmatrix} 0 \\ 0 \end{pmatrix}$$

$$= \begin{pmatrix} p_{0y} - 1 \\ -p_{0x} \end{pmatrix} \qquad\qquad = \begin{pmatrix} p_{0y} - 1 \\ 1 - p_{0x} \end{pmatrix}$$

Substitute into the first-order constraint equations:

$$\begin{pmatrix} p_{0y} - 1 \\ -p_{0x} \end{pmatrix}^T \begin{pmatrix} 0 \\ 1 \end{pmatrix} = 0 \qquad\qquad \begin{pmatrix} p_{0y} - 1 \\ 1 - p_{0x} \end{pmatrix}^T \begin{pmatrix} -1 \\ 1 \end{pmatrix} = 0$$

from which the velocity pole coordinates become $\mathbf{p}_0 = (0, 2)$. Transfer the origin to point $\mathbf{p}_0$. The new coordinate system becomes

$$\mathbf{p}_0 = (0, 0) \qquad \mathbf{a} = (0, -1) \qquad \mathbf{b} = (1, -1)$$
$$\mathbf{a}_0 = (0, -2) \qquad \mathbf{b}_0 = (2, -2)$$

In the new coordinate system

$$\mathbf{a}' = \begin{bmatrix} 0 & -1 \\ 1 & 0 \end{bmatrix} (\mathbf{a} - \mathbf{p}_0) + \mathbf{p}_0' = \begin{pmatrix} 1 \\ 0 \end{pmatrix}$$

$$\mathbf{b}' = \begin{bmatrix} 0 & -1 \\ 1 & 0 \end{bmatrix} (\mathbf{b} - \mathbf{p}_0) + \mathbf{p}_0' = \begin{pmatrix} 1 \\ 1 \end{pmatrix}$$

Second-order analysis
Constraint equations

$$(\mathbf{a}'')^T (\mathbf{a} - \mathbf{a}_0) + (\mathbf{a}')^T (\mathbf{a}') = 0$$
$$(\mathbf{b}'')^T (\mathbf{b} - \mathbf{b}_0) + (\mathbf{b}')^T (\mathbf{b}') = 0$$
$$\mathbf{b}'' = [R''](\mathbf{b} - \mathbf{a}) + \mathbf{a}''$$

$$= \begin{bmatrix} -1 & 0 \\ 0 & -1 \end{bmatrix} \begin{pmatrix} 1 \\ 0 \end{pmatrix} + \begin{pmatrix} a_x'' \\ a_y'' \end{pmatrix} = \begin{pmatrix} a_x'' & -1 \\ a_y'' \end{pmatrix}$$

Substitute into the constant length constraint equations:

$$\begin{pmatrix} a_x'' \\ a_y'' \end{pmatrix}^T \begin{pmatrix} 0 \\ 1 \end{pmatrix} + \begin{pmatrix} 1 \\ 0 \end{pmatrix}^T \begin{pmatrix} 1 \\ 0 \end{pmatrix} = 0 \qquad \begin{pmatrix} a_x'' & -1 \\ a_y'' \end{pmatrix}^T \begin{pmatrix} -1 \\ 1 \end{pmatrix} + \begin{pmatrix} 1 \\ 1 \end{pmatrix}^T \begin{pmatrix} 1 \\ 1 \end{pmatrix} = 0$$

which leads to $\mathbf{a}'' = \begin{pmatrix} 2 \\ -1 \end{pmatrix}$ $\mathbf{b}'' = \begin{pmatrix} 1 \\ -1 \end{pmatrix}$.

Third-order analysis
Constraint equations

$$(\mathbf{a}''')^T (\mathbf{a} - \mathbf{a}_0) + 3(\mathbf{a}')^T (\mathbf{a}'') = 0$$
$$(\mathbf{b}''')^T (\mathbf{b} - \mathbf{b}_0) + 3(\mathbf{b}')^T (\mathbf{b}'') = 0$$
$$\mathbf{b}''' = [R'''](\mathbf{b} - \mathbf{a}) + \mathbf{a}'''$$

$$= \begin{bmatrix} 0 & 1 \\ -1 & 0 \end{bmatrix} \begin{pmatrix} 1 \\ 0 \end{pmatrix} + \mathbf{a}''' = \begin{pmatrix} a_x''' \\ a_y''' & -1 \end{pmatrix}$$

Substitute into the constraint equations:

$$\begin{pmatrix} a'''_x \\ a'''_y \end{pmatrix}^T \begin{pmatrix} 0 \\ 1 \end{pmatrix} + 3\begin{pmatrix} 1 \\ 0 \end{pmatrix}^T \begin{pmatrix} 2 \\ -1 \end{pmatrix} = 0$$

$$\begin{pmatrix} a'''_x \\ a'''_y & -1 \end{pmatrix}^T \begin{pmatrix} -1 \\ 1 \end{pmatrix} + 3\begin{pmatrix} 1 \\ 1 \end{pmatrix}^T \begin{pmatrix} 1 \\ -1 \end{pmatrix} = 0$$

which leads to $\mathbf{a}''' = \begin{pmatrix} -7 \\ -6 \end{pmatrix}$ $\mathbf{b}''' = \begin{pmatrix} -7 \\ -7 \end{pmatrix}$.

Motion of point $\mathbf{p}_0$ *at the velocity pole*

$$\mathbf{p}''_0 = [R''](\mathbf{p}_0 - \mathbf{a}) + \mathbf{a}'' = \begin{bmatrix} -1 & 0 \\ 0 & -1 \end{bmatrix}\begin{pmatrix} 0 \\ 1 \end{pmatrix} + \begin{pmatrix} 2 \\ -1 \end{pmatrix} = \begin{pmatrix} 2 \\ -2 \end{pmatrix}$$

$$\mathbf{p}'''_0 = [R'''](\mathbf{p}_0 - \mathbf{a}) + \mathbf{a}''' = \begin{bmatrix} 0 & 1 \\ -1 & 0 \end{bmatrix}\begin{pmatrix} 0 \\ 1 \end{pmatrix} + \begin{pmatrix} -7 \\ -6 \end{pmatrix} = \begin{pmatrix} -6 \\ -6 \end{pmatrix}$$

Rotate the coordinate system such that $p''_{0x} = 0$ in the new system

$$\begin{pmatrix} 0 \\ p''_{0y} \end{pmatrix} = \begin{bmatrix} \cos\beta & \sin\beta \\ -\sin\beta & \cos\beta \end{bmatrix}\begin{pmatrix} 2 \\ -2 \end{pmatrix}$$

$$2\cos\beta - 2\sin\beta = 0, \qquad \tan\beta = 1, \qquad \beta = 45°$$
$$p''_{0y} = -2\sin\beta - 2\cos\beta, \qquad p''_{0y} = -2\sqrt{2}$$

The new third-order components become

$$\begin{pmatrix} p'''_{0x} \\ p'''_{0y} \end{pmatrix} = \begin{bmatrix} \cos\beta & \sin\beta \\ -\sin\beta & \cos\beta \end{bmatrix}\begin{pmatrix} -6 \\ -6 \end{pmatrix} = \begin{pmatrix} -6\sqrt{2} \\ 0 \end{pmatrix}$$

The instantaneous invariants are

$$b_2 = -2\sqrt{2} \qquad a_3 = -6\sqrt{2} \qquad b_3 = 0$$

The equation of the inflection circle is

$$x^2 + y^2 + 2\sqrt{2}\, y = 0$$

The cubic of stationary curvature takes the degenerate form

$$x = 0$$
$$(x^2 + y^2) + \sqrt{2}\, y = 0$$

Example 10.5 Higher-order kinematic synthesis from the instantaneous invariants of the coupler motion. Data are from example 3 in references 6. Specify the motion of one point in a moving plane in the form

$$\mathbf{p} = (0, 0) \qquad \dot{\mathbf{p}} = (3, 0) \qquad \ddot{\mathbf{p}} = (-3, 0) \qquad \dddot{\mathbf{p}} = (5, 0)$$
$$\dot{\phi} = 5 \qquad \ddot{\phi} = 10 \qquad \dddot{\phi} = -5$$

1. Locate the velocity pole $\mathbf{p}_0$ as shown in Fig. 10.7 (Eqs. 10.40)

$$p_{0x} = 0 - 0 = 0 \qquad p_{0y} - 0 + \frac{3}{5} = 0.6$$

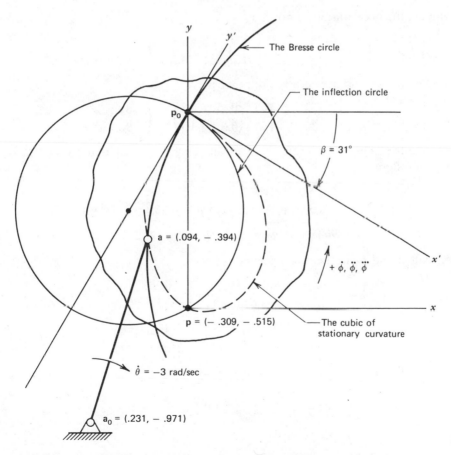

FIGURE 10.7 CRANK SYNTHESIS USING INSTANTANEOUS INVARIANTS.

2. Transform the origin to the velocity pole.

$$\mathbf{p}_0 = (0, 0) \qquad \mathbf{p} = (0, -0.6)$$

3. Form the differential rotation matrices.

$$[W] = \begin{bmatrix} 0 & -\dot{\phi} \\ \dot{\phi} & 0 \end{bmatrix} = \begin{bmatrix} 0 & -5. \\ 5. & 0 \end{bmatrix}$$

$$[\dot{W}] = \begin{bmatrix} -\dot{\phi}^2 & -\ddot{\phi} \\ \ddot{\phi} & -\dot{\phi}^2 \end{bmatrix} = \begin{bmatrix} -25. & -10. \\ 10. & -25. \end{bmatrix}$$

$$[\ddot{W}] = \begin{bmatrix} -3\dot{\phi}\ddot{\phi} & -(\dddot{\phi} - \dot{\phi}^3) \\ (\dddot{\phi} - \dot{\phi}^3) & -3\dot{\phi}\ddot{\phi} \end{bmatrix} = \begin{bmatrix} -150. & 130. \\ -130. & -150. \end{bmatrix}$$

4. Compute the acceleration of point $\mathbf{p}_0$.

$$\ddot{\mathbf{p}}_0 = \begin{bmatrix} -25. & -10. \\ 10. & -25. \end{bmatrix}\begin{pmatrix} 0. \\ 0.6 \end{pmatrix} + \begin{pmatrix} -3. \\ 0. \end{pmatrix} = \begin{pmatrix} -9. \\ -15. \end{pmatrix}$$

5. Compute the second acceleration of point $\mathbf{p}_0$.

$$\dddot{\mathbf{p}}_0 = \begin{bmatrix} -150. & 130. \\ -130. & -150. \end{bmatrix}\begin{pmatrix} 0 \\ 0.6 \end{pmatrix} + \begin{pmatrix} 5. \\ 0 \end{pmatrix} = \begin{pmatrix} 83. \\ -90. \end{pmatrix}$$

6. Calculate the angle β to rotate the coordinate system such that $\ddot{p}_{0x} = 0$ in the new system.

$$\begin{pmatrix} 0 \\ \ddot{p}_{0y} \end{pmatrix} = \begin{bmatrix} \cos \beta & \sin \beta \\ -\sin \beta & \cos \beta \end{bmatrix}\begin{pmatrix} -9. \\ -15. \end{pmatrix}$$

which gives $\beta = -30.9638°$ $\qquad \ddot{p}_{0y} = -17.4929$

7. Calculate $\mathbf{p}$ and $\ddot{\mathbf{p}}_0$ in the new system.

$$\begin{pmatrix} p_x \\ p_y \end{pmatrix} = \begin{bmatrix} \cos \beta & \sin \beta \\ -\sin \beta & \cos \beta \end{bmatrix}\begin{pmatrix} 0. \\ 0.6 \end{pmatrix} = \begin{pmatrix} -0.3087 \\ -0.5145 \end{pmatrix}$$

$$\begin{pmatrix} \dddot{p}_{0x} \\ \dddot{p}_{0y} \end{pmatrix} = \begin{bmatrix} \cos \beta & \sin \beta \\ -\sin \beta & \cos \beta \end{bmatrix}\begin{pmatrix} 83. \\ -90. \end{pmatrix} = \begin{pmatrix} 117.4766 \\ -34.4711 \end{pmatrix}$$

8. Calculate the instantaneous invariants (Eq. 10.69).

$$b_2 = p''_{0y} = \frac{\ddot{p}_{0y}}{\dot{\phi}^2} = \frac{-17.4929}{25} = -0.6997$$

$$a_3 = p'''_{0x} = \frac{\dddot{p}_{0x}}{\dot{\phi}^3} = \frac{117.4766}{125.} = 0.9398$$

$$b_3 = p_{0y}''' = \frac{\dddot{p}_{0y}}{\dot{\phi}^3} - 3\frac{\ddot{\phi}}{\dot{\phi}^2}p_{0y}'' = \frac{(-34.4711)}{125.} - 3.\left(\frac{10.}{25.}\right)(-0.6977)$$

$$= 0.5638$$

9. The equation of the inflection circle (Eq. 10.55) becomes [with $\mathbf{p} = (x, y)$].

$$(x^2 + y^2) - b_2 y = 0$$
$$(x^2 + y^2) + 0.6997y = 0$$

10. The cubic of stationary curvature is

$$(x^2 + y^2)[(a_3 + 3b_2)x + b_3 y] - 3(b_2)^2 xy = 0$$

which gives

$$(x^2 + y^2)[-1.1593x + 0.5638y] - 1.4687xy = 0$$

or

$$(x^2 + y^2)[0.7893x - 0.3839y] + xy = 0$$

11. Synthesize an input crank to provide the required rigid-body guidance. A moving crank pivot $\mathbf{a} = (x, y)$ must satisfy the constraint equation

$$(\dot{\mathbf{a}})^T(\ddot{\mathbf{a}}) = 0$$

where

$$\dot{\mathbf{a}} = [W](\mathbf{a} - \mathbf{p}_0) + \dot{\mathbf{p}}_0 = \begin{pmatrix} -5y \\ 5x \end{pmatrix}$$

$$\ddot{\mathbf{a}} = [W](\mathbf{a} - \mathbf{p}_0) + \ddot{\mathbf{p}}_0 = \begin{pmatrix} -25.x - 10.y \\ 10.x - 25.y - 17.4929 \end{pmatrix}$$

which leads to the equation of the zero tangential acceleration circle in the form

$$(x^2 + y^2) - 1.74929x - 0$$

The intersection of this circle (the Bresse circle) and the cubic of stationary curvature gives

$$\mathbf{a} = \begin{pmatrix} x \\ y \end{pmatrix} = \begin{pmatrix} 0.0937 \\ -0.3938 \end{pmatrix}$$

The corresponding fixed pivot $\mathbf{a}_0$ must satisfy the constraint equations

$$(\mathbf{a}')^T(\mathbf{a} - \mathbf{a}_0) = 0$$
$$(\mathbf{a}'')^T(\mathbf{a} - \mathbf{a}_0) + (\mathbf{a}')^T(\mathbf{a}') = 0$$

which gives
$$\mathbf{a}_0 = \begin{pmatrix} 0.231 \\ -0.9713 \end{pmatrix}$$

The required angular velocity of the input crank is found by noting that

$$\dot{\mathbf{a}} = [W](\mathbf{a} - \mathbf{p}_0) = \begin{bmatrix} 0 & -5. \\ 5. & 0 \end{bmatrix} \begin{pmatrix} 0.0937 \\ -0.3938 \end{pmatrix} = \begin{pmatrix} 1.969 \\ 0.4685 \end{pmatrix}$$

and, when $\mathbf{a}$ is considered on the input crank,

$$\dot{\mathbf{a}} = \begin{bmatrix} 0 & -\dot{\theta} \\ \dot{\theta} & 0 \end{bmatrix} (\mathbf{a} - \mathbf{a}_0)$$

Therefore, equating the x-components,

$$1.969 = -\dot{\theta}(a_y - a_{0y}) = -\dot{\theta}(0.6248)$$

$$\dot{\theta} = \frac{1.969}{0.6248} = -3.1514 \text{ rad/sec}$$

10.6 HIGHER-ORDER PATH CURVATURE IN SPATIAL COUPLER CURVES

Parametric Equations

We assume that a curve in space is generated by one point on the coupler of a spatial mechanism. The point coordinates are described by three parametric equations in terms of a single independent variable. The variable could be either the coupler rotation angle ϕ, the input crank rotation θ, time t, or the arc length along the curve S. Differential geometry of space curves is usually discussed in terms of the arc length S. In the case of spatial mechanisms either time t or the coupler rotation ϕ will be found more convenient. The arc length S can be introduced as necessary from either of the relationships

$$S = \int_{t_0}^{t} ds = \int_{t_0}^{t} \sqrt{\dot{x}^2 + \dot{y}^2 + \dot{z}^2} \; dt \tag{10.71}$$

$$S = \int_{\phi_0}^{\phi} \sqrt{x'^2 + y'^2 + z'^2} \; d\phi \tag{10.72}$$

where
$$x = f_1(t) \qquad y = f_2(t) \qquad z = f_3(t)$$

or
$$x = g_1(\phi) \qquad y = g_2(\phi) \qquad z = g_3(\phi)$$

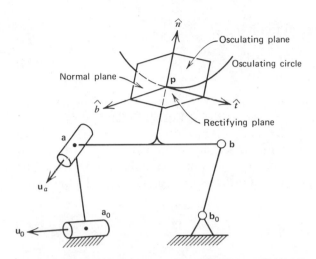

FIGURE 10.8 GEOMETRY OF SPATIAL COUPLER
CURVE FOR THE RRSS MECHANISM.

Geometric Properties of Spatial Curves

With reference to Figure 10.8, we define the following well known geometric
terms.

1. *The tangent,* $\hat{t}$, to a regular arc C at the point $\mathbf{p} = (p_x, p_y, p_z)$ is the limit of a
   straight line (the secant) through point $\mathbf{p}$ and another point $\mathbf{q}$ of the arc C as $\mathbf{q}$
   approaches $\mathbf{p}$. A unique tangent exists at every point on the curve.
2. *The osculating circle* is the limit of a circle through $\mathbf{p}$ and two other distinct
   points $\mathbf{q}$ and $\mathbf{r}$ of C as $\mathbf{q}$ and $\mathbf{r}$ approach $\mathbf{p}$.
3. *The principal normal,* $\hat{n}$, is the straight line from $\mathbf{p}$ to the center of the osculating
   circle (the center of the curvature).
4. *The binormal,* $\hat{b}$, is a third line through $\mathbf{p}$ such that the tangent, principal
   normal, and binormal form a set of right-handed rectangular Cartesian axes
   t-n-b.
5. *The moving trihedron* is determined by the system of axes t-n-b and is formed by
   the *osculating plane* and the *normal* and *rectifying planes* normal to the tangent $\hat{t}$
   and principal normal axes $\hat{n}$ at $\mathbf{p}$, respectively.

Path Curvature

The radius of curvature of a space curve is given in terms of the kinematic motion
parameters of the path point $\mathbf{p}$ as

$$\rho_K = \frac{|\dot{\mathbf{p}}|^3}{|\dot{\mathbf{p}} \times \ddot{\mathbf{p}}|} \tag{10.73}$$

and the radius of torsion is found from

$$\rho_T = \frac{|\dot{\mathbf{p}} \times \ddot{\mathbf{p}}|^2}{|\dot{\mathbf{p}} \cdot \ddot{\mathbf{p}} \cdot \dddot{\mathbf{p}}|} \tag{10.74}$$

Therefore, if the motion of any point **p** in a spatial mechanism has been determined by an appropriate kinematic analysis, Eqs. 10.73 and 10.74 may be used to calculate the curvature and torsion of the path of the point. To complete the curvature analysis, we require the directions of the path tangent, principal normal, and binormal. This will allow the determination of the coordinates of the centers of curvature and torsion. The unit direction vectors $\hat{t}$, $\hat{n}$, and $\hat{b}$ are found from

$$\hat{t} = \frac{\dot{\mathbf{p}}}{|\dot{\mathbf{p}}|}$$

$$\hat{b} = \frac{\dot{\mathbf{p}} \times \ddot{\mathbf{p}}}{|\dot{\mathbf{p}} \times \ddot{\mathbf{p}}|}$$

$$\hat{n} = \hat{b} \times \hat{t} \tag{10.75}$$

Example 10-2 Path Curvature Analysis—The RRSS Mechanism. An RRSS mechanism is specified as follows.

The R-R Link	The S-S Link
$\mathbf{a}_0 = (9.960, 10.740, 0.270)$	$\mathbf{b}_0 = (3.550, 2.690, 0.740)$
$\mathbf{u}_0 = (0.985, -0.173, 0.000)$	$\mathbf{b} = (1.400, -3.200, 6.520)$
$\mathbf{a} = (8.879, 4.600, 7.100)$	
$\mathbf{u}_a = (0.650, -0.613, -0.449)$	

Solution Using Phi Matrices   We first form the *phi* matrices $[D']$, $[D'']$, and $[D''']$ from Eqs. 10.28, 10.29, and 10.31.

The first-order phi matrix $[D']$ is formed in terms of the unknown instantaneous rotation axis $\mathbf{u} = (u_x, u_y, u_z)$ and the rate of change of position of the moving revolute point $\mathbf{a}' = (a'_x, a'_y, a'_z)$. Hence, the form of the $[D']$ matrix becomes

$$[D'] = \begin{bmatrix} [R'] & (\mathbf{a}' - [R']\mathbf{a}) \\ 0 & 0 \end{bmatrix} \tag{10.76}$$

We may solve for $\mathbf{u} = (u_x, u_y, u_z)$ and $\mathbf{a}' = (a'_x, a'_y, a'_z)$ from the four constraint equations for the R-R link

$$(\mathbf{u}_0)^T(\mathbf{a}') = 0$$

$$(\mathbf{u}'_a)^T(\mathbf{a} - \mathbf{a}_0) + (\mathbf{u}_a)^T(\mathbf{a}') = 0$$

$$(\mathbf{u}'_a)^T(\mathbf{u}_0) = 0$$

$$[\mathbf{a}' \times \mathbf{u}_a + (\mathbf{a} - \mathbf{a}_0) \times \mathbf{u}'_a]^T(\mathbf{u}_0) = 0 \qquad (10.77)$$

plus the constant-length first-order constraint for the S-S link

$$(\mathbf{b}')^T(\mathbf{b} - \mathbf{b}_0) = 0 \qquad (10.78)$$

and the direction cosine constraint on the unknown rotation axis

$$(\mathbf{u})^T(\mathbf{u}) = 1 \qquad (10.79)$$

In solving Eqs. 10.77, 10.78, and 10.79, note that

$$(\mathbf{u}'_a) = [R'](\mathbf{u}_a) \qquad (10.80)$$

Once u_x, u_y, u_z, a'_x, a'_y, and a'_z are known, we compute the elements of the $[D']$ matrix as

$$[D'] = \begin{bmatrix} 0.0 & -0.4669 & 0.3786 & -2.3152 \\ 0.4669 & 0.0 & -0.7992 & -8.5500 \\ -0.3786 & 0.7992 & 0.0 & -9.6563 \\ 0.0 & 0.0 & 0.0 & 0.0 \end{bmatrix} \qquad (10.81)$$

The $[D'']$ matrix is next calculated in the form

$$[D''] = \begin{bmatrix} [R''] & (\mathbf{a}'' - [R'']\mathbf{a}) \\ 0 & 0 \end{bmatrix} \qquad (10.82)$$

The unknown components of $\mathbf{u}' = (u'_x, u'_y, u'_z)$ and $\mathbf{a}'' = (a''_x, a''_y, a''_z)$ are found from the six second-order constraint equations

$$(\mathbf{u}_0)^T(\mathbf{a}'') = 0$$

$$(\mathbf{u}''_a)^T(\mathbf{a} - \mathbf{a}_0) + 2(\mathbf{u}'_a)^T(\mathbf{a}') + (\mathbf{u}_a)^T(\mathbf{a}'') = 0$$

$$(\mathbf{u}''_a)^T(\mathbf{u}_0) = 0$$

$$[\mathbf{a}'' \times \mathbf{u}_a + 2\mathbf{a}' \times \mathbf{u}'_a + (\mathbf{a} - \mathbf{a}_0) \times \mathbf{u}''_a]^T(\mathbf{u}_0) = 0$$

$$(\mathbf{b}'')^T(\mathbf{b} - \mathbf{b}_0) + (\mathbf{b}')^T(\mathbf{b}') = 0$$

$$(\mathbf{u}')^T(\mathbf{u}) = 0 \qquad (10.83)$$

where, again, the motion of axis $\mathbf{u}_a$ is expressed by

$$(\dot{\mathbf{u}}_a) = [R'](\mathbf{u}_a)$$

and

$$(\mathbf{u}''_a) = [R''](\mathbf{u}_a) \tag{10.84}$$

This leads to

$$[D''] = \begin{bmatrix} -0.3613 & 0.2927 & -1.3189 & 13.8680 \\ 0.3124 & -0.8567 & -0.6190 & 20.5709 \\ 2.0651 & 0.9726 & -0.7820 & -31.4574 \\ 0.0 & 0.0 & 0.0 & 0.0 \end{bmatrix} \tag{10.85}$$

The $[D''']$ matrix is written in the form

$$[D'''] = \begin{bmatrix} [R'''] & (\mathbf{a}''' - [R''']\mathbf{a}) \\ 0 & 0 \end{bmatrix} \tag{10.86}$$

Making use of the results of the two previous calculations for $[D']$ and $[D'']$, we calculate six additional parameters $\mathbf{u}'' = (u''_x, u''_y, u''_z)$ and $\mathbf{a}''' = (a'''_x, a'''_y, a'''_z)$ from the following six third-order constraint equations found by differentiation of Eqs. 10.83.

$$(\mathbf{u}_0)^T(\mathbf{a}''') = 0$$

$$(\mathbf{u}'''_a)^T(\mathbf{a} - \mathbf{a}_0) + 3(\mathbf{u}''_a)^T(\mathbf{a}') + 3(\mathbf{u}'_a)^T(\mathbf{a}'') + (\mathbf{u}_a)^T(\mathbf{a}''') = 0$$

$$(\mathbf{u}'''_a)^T(\mathbf{u}_0) = 0$$

$$[\mathbf{a}''' \times \mathbf{u}_a + 3(\mathbf{a}'' \times \mathbf{u}'_a + \mathbf{a}' \times \mathbf{u}''_a) + (\mathbf{a} - \mathbf{a}_0) \times \mathbf{u}'''_a]^T(\mathbf{u}_0) = 0$$

$$(\mathbf{b}''')^T(\mathbf{b} - \mathbf{b}_0) + 3(\mathbf{b}'')^T(\mathbf{b}') = 0$$

$$(\mathbf{u}'')^T(\mathbf{u}) + (\mathbf{u}')^T(\mathbf{u}') = 0 \tag{10.87}$$

Solution of Eqs. 10.87 leads to

$$[D'''] = \begin{bmatrix} 1.9080 & 6.5807 & -4.7102 & -5.3037 \\ -9.7334 & -1.9218 & -2.8195 & 163.3596 \\ 5.8484 & 0.4609 & 0.0138 & 0.6551 \\ 0.0 & 0.0 & 0.0 & 0.0 \end{bmatrix} \tag{10.88}$$

Once the numerical values of the elements of the $[D']$, $[D'']$, and $[D''']$ matrices are known, we can analyze the higher-order path curvature parameters for an arbitrary point $\mathbf{p}$ moving with the coupler, the R-S link.

Assume a point $\mathbf{p} = (5.800, 0.800, 8.700)$. Then, using Eqs. 10.73 and 10.74, remembering that for the case $\dot{\phi} = 1$, $\dot{\mathbf{p}} = \mathbf{p}'$, $\ddot{\mathbf{p}} = \mathbf{p}''$, and $\dddot{\mathbf{p}} = \mathbf{p}'''$, we may calculate the radius of curvature

$$\rho_K = \frac{|\mathbf{p}'|^3}{|\mathbf{p}' \times \mathbf{p}''|} = 9.681$$

where $\mathbf{p}' = [D']\mathbf{p}$, $\mathbf{p}'' = [D'']\mathbf{p}$, and $\mathbf{p}''' = [D''']\mathbf{p}$.
   The radius of torsion becomes

$$\rho_T = \frac{|\mathbf{p}' \times \mathbf{p}''|^2}{|\mathbf{p}' \cdot \mathbf{p}'' \cdot \mathbf{p}'''|} = -18.680$$

   The direction of the principal normal is calculated from Eqs. 10.75 after which we may locate the center of curvature at point $\mathbf{c}$ where

$$\mathbf{c} = (5.920, 7.183, 1.422)$$

Solution Using Time Derivatives   The same results can be obtained using constraint equations expressed in terms of derivatives taken with time as the independent variable. In this case velocities are calculated using

$$[V] = \begin{bmatrix} [W] & (\dot{\mathbf{a}} - [W]\mathbf{a}) \\ 0 & 0 \end{bmatrix} \tag{10.89}$$

   The velocity matrix $[V]$ for the R-S link has only four unknowns, since we have specified both $\mathbf{a}$ and $\dot{\mathbf{a}}$ in terms of the input angular velocity $\dot{\theta}$. The four unknowns u_x, u_y, u_z, and $\dot{\phi}$ for the coupler motion are found from the solution of the four constraint equations.

$$(\dot{\mathbf{u}}_a)^T(\mathbf{a} - \mathbf{a}_0) + (\mathbf{u}_a)^T(\dot{\mathbf{a}}) = 0$$

$$(\dot{\mathbf{u}}_a)^T(\mathbf{u}_0) = 0$$

$$(\dot{\mathbf{b}})^T(\mathbf{b} - \mathbf{b}_0) = 0$$

$$(\mathbf{u})^T(\mathbf{u}) = 1 \tag{10.90}$$

where

$$(\dot{\mathbf{u}}_a) - [W](\mathbf{u}_a)$$

   We note here that the first of Eqs. 10.77 is automatically satisfied by $(\dot{\mathbf{a}}) = [W_{\theta, \mathbf{u}_0}](\mathbf{a} - \mathbf{a}_0)$, as is also the case for the fourth of Eqs. 10.77.
   Solving Eqs. 10.90 for the elements of the coupler rigid body rotation matrix $[W]$ and then forming the complete velocity matrix $[V]$ from Eq. 10.89, we calculate

$$[V] = \begin{bmatrix} 0.0 & -0.6231 & 0.5053 & -3.0903 \\ 0.6231 & 0.0 & -1.0667 & -11.4122 \\ -0.5053 & 1.0667 & 0.0 & -12.8889 \\ 0.0 & 0.0 & 0.0 & 0.0 \end{bmatrix}$$

The acceleration matrix $[A]$ introduces four additional unknowns $\dot{u}_x$, $\dot{u}_y$, $\dot{u}_z$, and $\ddot{\phi}$ in the angular acceleration matrix $[\dot{W}]$. The acceleration matrix for the R-S link is written in terms of the known $\mathbf{a}$ and $\ddot{\mathbf{a}}$ in the form

$$[A] = \begin{bmatrix} [\dot{W}] & (\ddot{\mathbf{a}} - [\dot{W}]\mathbf{a}) \\ 0 & 0 \end{bmatrix} \tag{10.91}$$

The constraint equations become

$$(\ddot{\mathbf{u}}_a)^T(\mathbf{a} - \mathbf{a}_0) + 2(\dot{\mathbf{u}}_a)^T(\dot{\mathbf{a}}) + (\mathbf{u}_a)^T(\ddot{\mathbf{a}}) = 0$$
$$(\ddot{\mathbf{u}}_a)^T(\mathbf{u}_0) = 0$$
$$(\dot{\mathbf{b}})^T(\mathbf{b} - \mathbf{b}_0) + (\dot{\mathbf{b}})^T(\dot{\mathbf{b}}) = 0$$
$$(\dot{\mathbf{u}})^T(\mathbf{u}) = 0 \tag{10.92}$$

where

$$(\dot{\mathbf{u}}_a) = [W](\mathbf{u}_a)$$
$$(\ddot{\mathbf{u}}_a) = [\dot{W}](\mathbf{u}_a)$$

After solving Eqs. 10.92 for $\dot{u}_x$, $\dot{u}_y$, $\dot{u}_z$, and $\ddot{\phi}$, we calculate the acceleration matrix elements from Eq. 10.91 as

$$[A] = \begin{bmatrix} -0.6437 & 0.1090 & -2.0151 & 22.6611 \\ 0.9691 & -1.5262 & -1.8091 & 29.0936 \\ 3.3445 & 2.4389 & -1.3933 & -64.5769 \\ 0.0 & 0.0 & 0.0 & 0.0 \end{bmatrix}$$

The second acceleration matrix $[J]$ involves the previous eight parameters plus four new parameters $\ddot{u}_x$, $\ddot{u}_y$, $\ddot{u}_z$, and $\dddot{\phi}$. The constraint equations in this case become

$$(\dddot{\mathbf{u}}_a)^T(\mathbf{a} - \mathbf{a}_0) + 3(\ddot{\mathbf{u}}_a)^T(\dot{\mathbf{a}}) + 3(\dot{\mathbf{u}}_a)^T(\ddot{\mathbf{a}}) + (\mathbf{u}_a)^T(\mathbf{a}) = 0$$
$$(\dddot{\mathbf{u}}_a)^T(\mathbf{u}_0) = 0$$
$$(\dddot{\mathbf{b}})^T(\mathbf{b} - \mathbf{b}_0) + 3(\dot{\mathbf{b}})^T(\dot{\mathbf{b}}) = 0$$
$$(\ddot{\mathbf{u}})^T(\mathbf{u}) + (\dot{\mathbf{u}})^T(\dot{\mathbf{u}}) = 0 \tag{10.93}$$

where

$$(\dddot{\mathbf{u}}_a) = [\dddot{W}](u_a), \text{ etc.}$$

After solving Eqs. 10.93 for $\ddot{\mathbf{u}}$ and $\dddot{\phi}$, we form the second acceleration matrix

$$[J] = \begin{bmatrix} [\ddot{W}] & (\dddot{\mathbf{a}} - [\ddot{W}]\mathbf{a}) \\ 0 & 0 \end{bmatrix} \tag{10.94}$$

with elements

$$[J] = \begin{bmatrix} 3.2587 & 12.8162 & -12.7303 & 17.2742 \\ -18.1719 & -7.6011 & -15.5172 & 390.4097 \\ 18.0773 & 11.1595 & 2.7344 & -189.7642 \\ 0.0 & 0.0 & 0.0 & 0.0 \end{bmatrix} \qquad (10.95)$$

The path curvature parameters for the point $\mathbf{p} = (5.800, 0.800, 8.700)$ as used in the previous example can be found by substitution into Eqs. 10.73, 10.74, 10.75 with

$$(\dot{\mathbf{p}}) = [V](\mathbf{p})$$

$$(\ddot{\mathbf{p}}) = [A](\mathbf{p})$$

$$(\dddot{\mathbf{p}}) = [J](\mathbf{p})$$

The results obtained are identical to those obtained with phi matrices.

Comparison of the Two Methods. When the problem is formulated in terms of phi matrices, there are six simultaneous nonlinear equations to be solved for each stage of the problem. In the second case, with specified input crank motion parameters $\dot{\theta}$, $\ddot{\theta}$, and θ, four equations must be solved at each stage. In the second method, after the numerical $[V]$, $[A]$, and $[J]$ matrices are known, we may locate the acceleration pole $\mathbf{q}_0$ from the relationship

$$[A](\mathbf{q}_0) = (0.0)$$

which gives $\mathbf{q}_0 = (11.176, 16.033, 8.543)$.

A point $\mathbf{r}_0$ with zero second acceleration (i.e., a *second acceleration pole*) is found from

$$[J](\mathbf{r}_0) = (0.0)$$

from which $\mathbf{r}_0 = (6.385, 9.816, 12.874)$.

Chapter 11
Dynamics of Mechanisms

11.1 INTRODUCTION

Matrix methods in combination with vector calculus are useful as a systematic method for solving the dynamical equations of motion for constrained rigid body systems. We will be concerned with the dynamics of a system of interconnected rigid bodies with one degree of freedom where the motions of any rigid body of the system can be expressed in terms of the input motion. We will neglect higher-order effect caused by the elasticity of individual members as, for example, the free and forced vibration or dynamic stability of the system.

There are two basic classes of dynamics problems. In the first class the motions of all rigid bodies are specified in terms of the motion of the input member, and we calculate the forces and moments associated with the specified motion. In the second class, or *inverse* dynamics problem, we specify the force system and integrate the nonlinear differential equations of motion to find the motion of the system as a function of time. Examples of both types of problems will be demonstrated.

11.2 DYNAMICS OF THE PLANE FOUR-BAR LINKAGE

As an example of the use of matrix methods for dynamic analysis of plane mechanisms, first consider the plane four-bar mechanism shown in Figure 11.1. The physical properties of the ith link are specified by its mass m_i and moment of inertia about the mass center I_i. The angular position, velocity, and acceleration are specified by θ_i, $\dot{\theta}_i$, and $\ddot{\theta}_i$.

It is assumed that the results of an acceleration analysis are available and the system of internal joint forces $\mathbf{f}_i$ and required driving torque T_0 are to be calculated.

The joint force system is developed as a result of the combined effect of specified external loads acting on the mechanism, and the dynamic force components required to accelerate the links. The sum of the external loads acting on the ith link will be designated as $\sum F_i$ and the sum of the specified external torques as $\sum T_i$. The resultant force acting on the ith link is the sum of the external loads $\sum F_i$ plus the two joint forces $\mathbf{f}_i$ and $\mathbf{f}_{i-1}$. $\mathbf{f}_i$ is described as the *leading* joint force

276

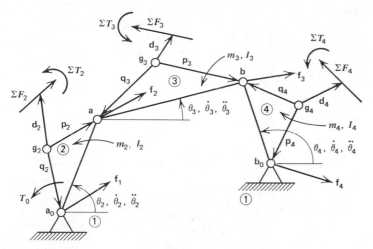

FIGURE 11.1 COORDINATE SYSTEM FOR THE DYNAMIC ANALYSIS OF THE FOUR-BAR LINKAGE.

for a particular member. It has the same index as the member itself and is shown in the positive sense on the free body diagram. The trailing joint force with index $i - 1$ is shown as a negative force vector.

The coordinate system used is shown in Figure 11.1. The corresponding free body diagrams are given in Figure 11.2. To account for the moment of $\sum \mathbf{F}_i$ about $\mathbf{g}_i$ the vector $\mathbf{d}_i$ locates any point on the line of action of $\sum \mathbf{F}_i$.

The equations of motion for the ith member can be written in the form

$$\mathbf{f}_i - \mathbf{f}_{i-1} + \sum \mathbf{F}_i = m_i \ddot{\mathbf{g}}_i \tag{11.1}$$

$$(\mathbf{p}_i \times \mathbf{f}_i) - (\mathbf{q}_i \times \mathbf{f}_{i-1}) + (\mathbf{d}_i \times \sum \mathbf{F}_i) + \sum \mathbf{T}_i = I_i \ddot{\theta}_i \tag{11.2}$$

where

$\mathbf{p}_i$ = the vector from $\mathbf{g}_i$ to the leading joint i

$\mathbf{q}_i$ = the vector from $\mathbf{g}_i$ to the trailing joint $i - 1$

$\mathbf{d}_i$ = the vector from $\mathbf{g}_i$ to one point on $\sum \mathbf{F}_i$

$\sum \mathbf{F}_i$ = the sum of all forces acting on link i other than joint forces $\mathbf{f}_i$ and $\mathbf{f}_{i-1}$

$\sum \mathbf{T}_i$ = the sum of all pure couples acting on link i other than the unknown driving torque T_0 acting on link 2

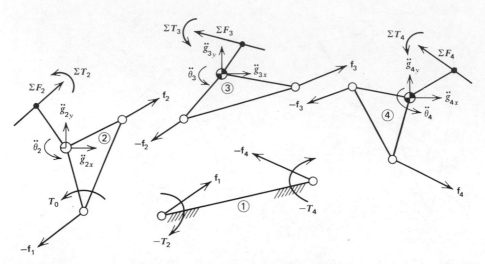

FIGURE 11.2 FREE BODY DIAGRAMS. DYNAMIC ANALYSIS OF THE FOUR-BAR LINKAGE.

Apply Eqs. 11.1 and 11.2 to each link in turn. The vector equations of motion for each link become:

Link 2

$$\mathbf{f}_2 - \mathbf{f}_1 = m_2\, \ddot{\mathbf{g}}_2 - \sum \mathbf{F}_2$$

$$(\mathbf{a} - \mathbf{g}_2) \times \mathbf{f}_2 - (\mathbf{a}_0 - \mathbf{g}_2) \times \mathbf{f}_1 + \mathbf{T}_0 = I_2\, \ddot{\theta}_2 - \sum \mathbf{T}_2 - (\mathbf{d}_2 \times \sum \mathbf{F}_2)$$

Link 3

$$\mathbf{f}_3 - \mathbf{f}_2 = m_3\, \ddot{\mathbf{g}}_3 - \sum \mathbf{F}_3$$

$$(\mathbf{b} - \mathbf{g}_3) \times \mathbf{f}_3 - (\mathbf{a} - \mathbf{g}_3) \times \mathbf{f}_2 = I_3\, \ddot{\theta}_3 - \sum \mathbf{T}_3 - (\mathbf{d}_3 \times \sum \mathbf{F}_3)$$

Link 4

$$\mathbf{f}_4 - \mathbf{f}_3 = m_4\, \ddot{\mathbf{g}}_4 - \sum \mathbf{F}_4$$

$$(\mathbf{b}_0 - \mathbf{g}_4) \times \mathbf{f}_4 - (\mathbf{b} - \mathbf{g}_4) \times \mathbf{f}_3 = I_4\, \ddot{\theta}_4 - \sum \mathbf{T}_4 - (\mathbf{d}_4 \times \sum \mathbf{F}_4) \qquad (11.3)$$

Recalling that $\mathbf{r} \times \mathbf{f} = r_x\, f_y - r_y\, f_x$, Eqs. 11.3 can be expanded as a set of nine scalar equations with nine unknowns $f_{1x}, f_{1y}, f_{2x}, f_{2y}, f_{3x}, f_{3y}, f_{4x}, f_{4y}$, and T_0 (the driving torque acting on link 2).

The nine equations are then conveniently arranged in matrix form as shown in Figure 11.3. The matrix equation in Figure 11.3 is easily solved by Gauss' elimination using a library computer subprogram available in most computer systems. Function LINEQF in program DESIGN in Chapter 12 is an example.

$$
\begin{array}{ccccccccc}
f_{1x} & f_{1y} & f_{2x} & f_{2y} & f_{3x} & f_{3y} & f_{4x} & f_{4y} & T_2
\end{array}
$$

$$
\begin{bmatrix}
-1 & 0 & 1 & 0 & 0 & 0 & 0 & 0 & 0 \\
0 & -1 & 0 & 1 & 0 & 0 & 0 & 0 & 0 \\
q_{2y} & -q_{2x} & -p_{2y} & p_{2x} & 0 & 0 & 0 & 0 & 1 \\
0 & 0 & -1 & 0 & 1 & 0 & 0 & 0 & 0 \\
0 & 0 & 0 & -1 & 0 & 1 & 0 & 0 & 0 \\
0 & 0 & q_{3y} & -q_{3x} & -p_{3y} & p_{3x} & 0 & 0 & 0 \\
0 & 0 & 0 & 0 & -1 & 0 & 1 & 0 & 0 \\
0 & 0 & 0 & 0 & 0 & -1 & 0 & 1 & 0 \\
0 & 0 & 0 & 0 & q_{4y} & -q_{4x} & -p_{4y} & p_{4x} & 0
\end{bmatrix}
\begin{bmatrix}
f_{1x} \\ f_{1y} \\ f_{2x} \\ f_{2y} \\ f_{3x} \\ f_{3y} \\ f_{4x} \\ f_{4y} \\ T_0
\end{bmatrix}
=
\begin{bmatrix}
m_2\ddot{g}_{2x} - \sum F_{2x} \\
m_2\ddot{g}_{2y} - \sum F_{2y} \\
I_2\ddot{\theta}_2 - \sum T_2 - (\mathbf{d}_2 \times \sum \mathbf{F}_2) \\
m_3\ddot{g}_{3x} - \sum F_{3x} \\
m_3\ddot{g}_{3y} - \sum F_{3y} \\
I_3\ddot{\theta}_3 - \sum T_3 - (\mathbf{d}_3 \times \sum \mathbf{F}_3) \\
m_4\ddot{g}_{4x} - \sum F_{4x} \\
m_4\ddot{g}_{4y} - \sum F_{4y} \\
I_4\ddot{\theta}_4 - \sum T_4 - (\mathbf{d}_4 \times \sum \mathbf{F}_4)
\end{bmatrix}
$$

where

$$\mathbf{p}_2 = (\mathbf{a} - \mathbf{g}_2) \qquad \mathbf{p}_3 = (\mathbf{b} - \mathbf{g}_3) \qquad \mathbf{p}_4 = (\mathbf{b}_0 - \mathbf{g}_4)$$

$$\mathbf{q}_2 = (\mathbf{a}_0 - \mathbf{g}_2) \qquad \mathbf{q}_3 = (\mathbf{a} - \mathbf{g}_3) \qquad \mathbf{q}_4 = (\mathbf{b} - \mathbf{g}_4)$$

FIGURE 11.3 MATRIX EQUATION OF MOTION FOR THE FOUR-BAR LINKAGE.

Note that the square matrix on the left side of Eqs. 11.3 in matrix form describes the instantaneous *geometry* of the mechanism. The column matrix on the right side contains the *dynamical* terms. The unknown forces and torques form the *force* vector.

The combined reaction of the two cranks on the frame is often of interest and is called the *shaking force*. Once Eqs. 11.3 have been solved, the shaking force $\mathbf{F}$ is found from

$$\mathbf{F} = \mathbf{f}_1 - \mathbf{f}_4 \tag{11.4}$$

or

$$F_x = f_{1x} - f_{4x}$$
$$F_y = f_{1y} - f_{4y} \tag{11.5}$$

Example 11-1 Dynamic Analysis of a Plane Four-Bar Linkage. A four-bar linkage is specified in its first position by

$$\mathbf{a}_0 = (0.0, 0.0) \qquad \mathbf{a}_1 = (1.0, 0.0)$$
$$\mathbf{b}_0 = (3.0, 0.0) \qquad \mathbf{b}_1 = (3.0, 2.0)$$

with the mass center of each link located at

$$\mathbf{g}_2 = (0.5, 0.0) \qquad \mathbf{g}_3 = (2.0, 1.0) \qquad \mathbf{g}_4 = (3.0, 1.0)$$

and

$$\sum F_2 = \sum F_3 = \sum F_4 = 0.0 \qquad \sum T_2 = \sum T_3 = \sum T_4 = 0.0$$

The physical properties of the links are specified as

$$m_2 = 3.0 \times 10^{-4} \text{ lb-sec}^2/\text{in.} \qquad I_2 = k_2^2 m_2 , \ k_2 = 0.5 \text{ in.}$$

$$m_3 = 6.5 \times 10^{-4} \qquad\qquad\qquad I_3 = k_3^2 m_3 , \ k_3 = 0.8$$

$$m_4 = 5.0 \times 10^{-4} \qquad\qquad\qquad I_4 = k_4^2 m_4 , \ k_4 = 1.0$$

The acceleration analysis of the mechanism for input motion

$$\dot\theta_2 = 200\pi \text{ rad/sec}$$

$$\ddot\theta_2 = 0.0 \text{ rad/sec}^2$$

gave the following results. See subroutine FRBRAN in Chapter 12.

$$\ddot{\mathbf{a}} = (-394,384., 0.0) \text{ ips}^2 \qquad \dot\theta_3 = -314. \text{ rad/sec}$$

$$\ddot{\mathbf{b}} = (-591,576., -197,192.) \qquad \ddot\theta_3 = 0.0 \text{ rad/sec}^2$$

$$\ddot{\mathbf{g}}_2 = (-197,192., 0.0) \qquad\qquad \dot\theta_4 = -314.$$

$$\ddot{\mathbf{g}}_3 = (-492,980., -98,596.) \qquad \ddot\theta_4 = 295,788.$$

$$\ddot{\mathbf{g}}_4 = (-295,788., -98,596.)$$

Solution of Eqs. 11.3 gives

$$\mathbf{f}_1 = (527.3, 340.8) \text{ lb} \qquad T_0 = -340.8 \text{ in-lb}$$

$$\mathbf{f}_2 = (468.3, 340.2) \qquad\qquad \text{Shaking force components, } F_x , F_y$$

$$\mathbf{f}_3 = (147.9, 276.1) \qquad\qquad F_x = 527.5 \text{ lb}$$

$$\mathbf{f}_4 = (0.0, 226.6) \qquad\qquad F_y = 113.4$$

Example 11-2 The Oscillating Slider Mechanism. Consider the mechanism shown in Figure 11.4. From the free body diagrams we may write the equations of motion as

Member 2

$$\mathbf{f}_2 - \mathbf{f}_1 = m_2 \ddot{\mathbf{g}}_2$$

$$T_2 + (\mathbf{p}_2 \times \mathbf{f}_2) - (\mathbf{q}_2 \times \mathbf{f}_1) = I_2 \ddot\theta_2$$

Member 3

$$\mathbf{f}_3 - \mathbf{f}_2 = m_3 \ddot{\mathbf{g}}_3$$

$$((\mathbf{d} + \mathbf{e}_4) \times \mathbf{f}_3) - (\mathbf{q}_3 \times \mathbf{f}_2) = I_3 \ddot\theta_3$$

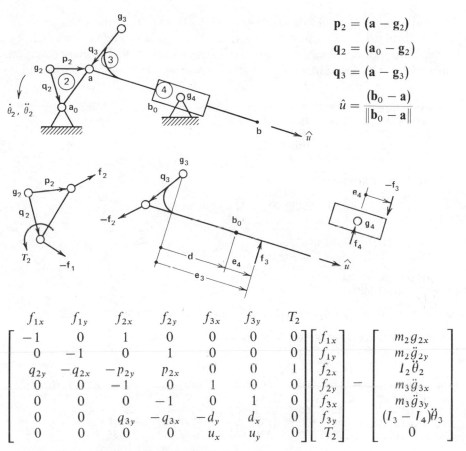

$$\mathbf{p}_2 = (\mathbf{a} - \mathbf{g}_2)$$

$$\mathbf{q}_2 = (\mathbf{a}_0 - \mathbf{g}_2)$$

$$\mathbf{q}_3 = (\mathbf{a} - \mathbf{g}_3)$$

$$\hat{u} = \frac{(\mathbf{b}_0 - \mathbf{a})}{\|\mathbf{b}_0 - \mathbf{a}\|}$$

f_{1x}	f_{1y}	f_{2x}	f_{2y}	f_{3x}	f_{3y}	T_2			
-1	0	1	0	0	0	0	f_{1x}		$m_2 g_{2x}$
0	-1	0	1	0	0	0	f_{1y}		$m_2 \ddot{g}_{2y}$
q_{2y}	$-q_{2x}$	$-p_{2y}$	p_{2x}	0	0	1	f_{2x}		$I_2 \ddot{\theta}_2$
0	0	-1	0	1	0	0	f_{2y}	$-$	$m_3 \ddot{g}_{3x}$
0	0	0	-1	0	1	0	f_{3x}		$m_3 \ddot{g}_{3y}$
0	0	q_{3y}	$-q_{3x}$	$-d_y$	d_x	0	f_{3y}		$(I_3 - I_4)\ddot{\theta}_3$
0	0	0	0	u_x	u_y	0	T_2		0

FIGURE 11.4 MATRIX EQUATION OF MOTION. THE OSCILLATING SLIDER MECHANISM.

Member 4

$$\mathbf{f}_4 - \mathbf{f}_3 = m_4 \ddot{\mathbf{g}}_4$$

$$-\mathbf{e}_4 \times \mathbf{f}_3 - I_4 \ddot{\theta}_4 - I_4 \ddot{\theta}_3$$

To insure that the slider force $\mathbf{f}_3$ is perpendicular to the guide, we must add the constraint equation

$$\mathbf{e}_4 \cdot \mathbf{f}_3 \qquad \text{or} \qquad \mathbf{e}_4 \cdot \mathbf{f}_4 = 0$$

This leads to a set of ten equations in the ten unknown components of $\mathbf{f}_1, \mathbf{f}_2, \mathbf{f}_3$, and $\mathbf{f}_4$ plus the input torque T_2 and length e_4. The direction $\hat{u}$ for $\mathbf{e}_4$ is known.

The solution of Eqs. 11.6 is complicated by the fact that the cross product $e_4 \times f_3$ is involved, leading to a nonlinear set of equations. The vector e_4 can be eliminated by adding the moment equation for member 4 to the moment equation for member 3. This leads to a reduced set of equations. Taking note of the added

```
      GEOMETRY OF THE LINKS
R2=  1.00    S2=  .50    PHI2=   0.
R3=  5.00    S3=  1.00   PHI3=   30.00
R4=  0.      S4=  0.     PHI4=   0.

      PHYSICAL PROPERTIES
M2=   .300E-03   K2=  .500E+00   I2=   .750E-04
M3=   .900E=03   K3=  .800E+00   I3=   .576E-03
M4=   0.         K4=  0.         I4=   0.

      FIXED PIVOT COORDINATES
P(1,1)= 0.     P(1,2)= 0.     P(4,1)= 3.00     P(4,2)= 0.

      INPUT CRANK MOTION PARAMETERS
W2= 628.32 RAD/SEC     A2= 0. RAD/SEC/SEC

RANGE OF INPUT ROTATION=30.00 DEG     INCREMENT DELTH=30.00 DEG

      INPUT CRANK ANGLE, TH2= 30.00 DEG
```

POINT	POSITION		VELOCITY		ACCELERATION	
	X	Y	X	Y	X	Y
A_0	0.	0.	0.	0.	0.	0.
A	.866E+00	.500E+00	-.314E+03	.544E+03	-.342E+06	-.197E+06
B_0	.300E+01	0.	0.	0.	0.	0.
B	.573E+01	-.641E+00	-.553E+03	-.473E+03	-.320E+06	.852E+06
G_2	.433E+00	.250E+00	-.157E+03	.272E+03	-.171E+06	-.987E+05
G_3	.182E+01	.789E+00	-.254E+03	.344E+03	-.443E+06	-.135E+05

```
W3= -.209E+03     W4= -.209E+03     A3= .205E+06     A4= .205E+06

FORCE DISTRIBUTION IN THE MECHANISM

F1X= F21X=    473E+03     F1Y= F21Y=   1404E+03
F2X= F32X=   4219E+03     F2Y= F32Y=   1108E+03
F3X= F43X=  .2310E+02     F3Y= F43Y=  .9859E+02
F4X= F14X=  .2310E+02     F4Y= F14Y=  .9859E+02

DRIVING OR INPUT TORQUE T2= .1150E+03

SHAKING FORCE COMPONENTS SFIX= .4501E+03     SF1Y= .4177E+02

ECCENTRICITY E3=  .133E+01
ECCENTRICITY E4= 0.
```

FIGURE 11.5 DYNAMIC ANALYSIS OF AN OSCILLATING SLIDER MECHANISM.

relations

$$e_3 - e_4 = d = d\hat{u}$$
$$\ddot{\theta}_4 = \ddot{\theta}_3$$

we obtain a set of seven equations.

$$f_{2x} - f_{1x} = m_2 \ddot{g}_{2x}$$

$$f_{2y} - f_{1y} = m_2 \ddot{g}_{2y}$$

$$T_2 + p_{2x} f_{2y} - p_{2yzf_{2x}} - q_{2x} f_{1y} + q_{2y} f_{1x} = I_2 \ddot{\theta}_2$$

$$f_{3x} - f_{2x} = m_3 \ddot{g}_{3x}$$

$$f_{3y} - f_{2y} = m_3 \ddot{g}_{3y}$$

$$d_x f_{3y} - d_y f_{3x} - q_{3x} f_{2y} + q_{3y} f_{2x} = (I_3 + I_4)\ddot{\theta}_3$$

$$u_x f_{3x} + u_y f_{3y} = 0 \qquad\qquad (11.7)$$

The seven equations are arranged in the matrix form shown in Figure 11.4 and solved by Gauss elimination for f_1, f_2, f_3, and T_2. f_4 is then found from the equations for member 4.

Results for a numerical example are given in Figure 11.5.

11.3   DYNAMIC BALANCING OF THE FOUR-BAR LINKAGE

The principle of motion of the mass center in a dynamical system states that "the net force acting on a system of rigid bodies is equal to the product of the total mass of the system times the acceleration of the mass center." If we consider each mass of the four-bar linkage system separately and then sum the forces, we see that forces internal to the system cancel, and we may write

$$\sum F = m\ddot{g} = \sum_{i=1}^{4} m_i \ddot{g}_i \qquad\qquad (11.8)$$

where

$$m = \text{the sum of the masses}$$

$$\ddot{g} = \text{the acceleration of the mass center}$$

The net force $\sum F$ acting on the system of three moving links corresponds to the shaking force F calculated from Eq. 11.4. The shaking force (i.e., the net reaction on the foundation) can be reduced to zero by finding a means to reduce the acceleration of the mass center $\ddot{\bar{g}}$ to zero in all positions of the linkage. This can be accomplished as follows.

We first add a balance weight to the input crank, member 2, such that the center of gravity $\mathbf{g}_2$ is relocated at $\mathbf{a}_0$. A similar procedure relocates the center of gravity of member 4 at the fixed pivot $\mathbf{b}_0$.

These added masses m_{02} and m_{04} must satisfy the relationships*

$$m_{02}\,\mathbf{r}_2 + m_2\,\mathbf{q}_2 = 0$$

$$m_{04}\,\mathbf{r}_4 + m_4\,\mathbf{p}_4 = 0 \tag{11.9}$$

Either the mass to be added or the radius $\mathbf{r}_i$ opposite the center of gravity can be specified arbitrarily and the second quantity calculated to give the required balance moment $m_{0i}\mathbf{r}_i$.

We assume that the center of gravity of the coupler $\mathbf{g}_3$ is located on the line of centers $\overline{\mathbf{ab}}$. In those cases where $\mathbf{g}_3$ is in a more general position, as shown in Figure 11.1, we would first add mass on the opposite side of the line of centers to bring the center of gravity to the line of centers.

To eliminate the unbalance caused by the mass of the coupler, we first calculate a system of two hypothetical point masses m_a and m_b that could replace the combined mass and moment of inertia of the actual coupler, link 3. We assume the center of gravity g_3 is located at distance s from point $\mathbf{a}$ and a distance t from point $\mathbf{b}$. To maintain the same position of the center of gravity with the two equivalent masses,

$$m_a s = m_b t \tag{11.10}$$

and, since

$$m_a + m_b = m_3 \tag{11.11}$$

we have

$$m_a = \frac{t}{s + t} m_3 \tag{11.12}$$

$$m_b = \frac{s}{s + t} m_3 \tag{11.13}$$

Although the equivalent masses m_a and m_b have the same total mass as the actual mass m_3 and the center of gravity is in the same location, it is unlikely that

---

* Note that $\mathbf{q}_2$ in Eq. 11.9 is the vector from $\mathbf{a}_0$ to the center of gravity $\mathbf{g}_2$. The sense of $\mathbf{q}_2$ is opposite to $\mathbf{q}_2$, as used in Eq. 11.2.

their moment of inertia about the mass center is equal to I_3. The moment of inertia of m_a and m_b about $\mathbf{g}_3$ is equal to

$$I'_3 = m_a s^2 + m_b t^2 \tag{11.14}$$

Substituting Eqs. 11.10 and 11.11,

$$I'_3 = \left(\frac{t}{s+t} s^2 + \frac{s}{s+t} t^2 \right) m_3 = stm_3 \tag{11.15}$$

Since the actual moment of inertia $I_3 = k_3^2 m_3$, we see that the equivalent moment of inertia of the coupler is in error by the following amount.

$$\text{Error} = (k_3^2 - st)m_3 \tag{11.16}$$

The error can be appreciable, depending on the mass distribution in the actual coupler.

The balancing is then completed by adding masses to the input and output cranks to balance m_a at pivot $\mathbf{a}$ and m_b at $\mathbf{b}$. Note that this is equivalent to reducing $m_a \mathbf{a}$ and $m_b \mathbf{b}$ to zero; hence the net acceleration of the combined mass center of the three links has been reduced to zero and we would expect perfect dynamic force balancing (i.e., the shaking force is reduced to zero in all positions of the linkage).

It should also be noted that the balancing has been achieved by the addition of mass to the system; therefore, we would expect a completely new system of joint forces and driving torque T_0 after the balancing.

The total balance weight added to the input crank is calculated from

$$\mathbf{e}_2 m_{b2} = -[m_2 \mathbf{q}_2 + m_a(\mathbf{a} - \mathbf{a}_\theta)] \tag{11.17}$$

where the right side is a vector addition.

Similarly, the total balance weight added to the output crank becomes

$$\mathbf{e}_4 m_{b4} = -[m_4 \mathbf{p}_4 + m_b(\mathbf{b} - \mathbf{b}_0)] \tag{11.18}$$

where

$\mathbf{e}_i$ = an assumed radial distance at which the balance mass is added
m_{bi} = balance mass

After the balance weights m_{b2} and m_{b4} have been added, the dynamic analysis can be repeated. In the dynamic analysis of the balanced linkage the *actual* physical properties m_3 and I_3 for the coupler are used in the analysis, since no mass was added to the coupler. Since mass has been added to the input and output cranks, the new locations of their centers of mass must be calculated from

$$\mathbf{q}_2' = \frac{m_2\,\mathbf{q}_2 - m_{b2}\,\mathbf{e}_2}{(m_2 + m_{b2})} \tag{11.19}$$

$$\mathbf{p}_4' = \frac{m_4\,\mathbf{p}_4 - m_{b4}\,\mathbf{e}_4}{(m_4 + m_{b4})} \tag{11.20}$$

The corrected masses become

$$m_2' = m_2 + m_{b2} \tag{11.21}$$

$$m_4' = m_4 + m_{b4} \tag{11.22}$$

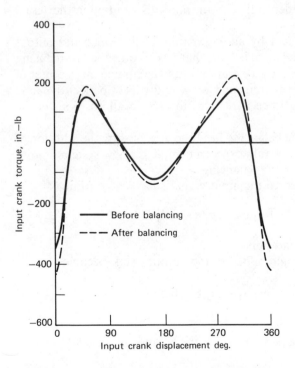

FIGURE 11.6 INPUT CRANK TORQUE VARIATION FOR ONE COMPLETE CYCLE. BEFORE BALANCING FROM EXAMPLE 11.1. AFTER BALANCING FROM EXAMPLE 11.3.

The moments of inertia of members 2 and 4 have increased as a result of adding m_{b2} and m_{b4}. The new moments of inertia are calculated from

$$I_2' = (k_2^2 + q_2^2)m_2 + e_2^2 m_{b2} - (q_2')^2 m_2' \qquad (11.23)$$

$$I_4' = (k_4^2 + p_4^2)m_4 + e_4^2 m_{b4} - (p_4')^2 m_4' \qquad (11.24)$$

The first two terms in Eq. 11.23 are seen to be the new moment of inertia referred to the fixed pivot a_0. The transfer formula then resuls in the third term.

These new physical constants are then used in the dynamic analysis, and the force and torque components are recalculated. As expected, the shaking force is negligible. The input torque, T_0, is of generally increased magnitude, as shown in Figure 11.6. The joint forces are also increased, as shown by the results of Example 11.3.

Example 11-3 Dynamic Balancing of a Four-Bar Linkage. The linkage of Example 11.1 is to be balanced. Assume that $e_2 = e_4 = 0.5$.

$m_a = m_b = \frac{1}{2}m_3 = 3.25 \times 10^{-4}$ lb sec^2/in.

$m_2 = 3.0 \times 10^{-4}$ $\qquad k_2 = 0.5$ in.

$m_4 = 5.0 \times 10^{-4}$ $\qquad k_4 = 1.0$

$m_{b2} = \dfrac{(3.0 \times 10^{-4})(0.5) + (3.25 \times 10^{-4})(1.0)}{0.5} = 9.5 \times 10^{-4}$

$m_{b4} = \dfrac{(5.0 \times 10^{-4})(1.0) + (3.25 \times 10^{-4})(2.0)}{0.5} = 23.0 \times 10^{-4}$

$q_2' = \dfrac{(3.0 \times 10^{-4})(0.5) - (9.5 \times 10^{-4})(0.5)}{(3.0 + 9.5)10^{-4}} = -0.26000$ in.

$p_4' = \dfrac{(5.0 \times 10^{-4})(1.0) - (23. \times 10^{-4})(0.5)}{(5.0 + 23.0)10^{-4}} = -0.23214$ in.

$m_2' = 12.5 \times 10^{-4}$ $\qquad\qquad m_4' = 28.0 \times 10^{-4}$

$g_2' = (-0.26000, 0.0)$ $\qquad g_4' = (3.0, -0.23214)$

$k_2' = 0.492$ in. $\qquad\qquad\quad k_4' = 0.713$ in.

These values gave the results shown in Figures 11.6, 11.7, and 11.8, where forces and torques before and after balancing are compared. The shaking force after balancing was negligible.

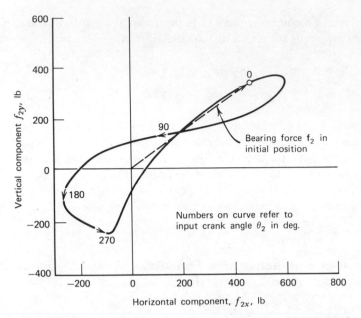

FIGURE 11.7   EXAMPLE 11.1. POLAR DIAGRAM OF BEARING
FORCE AT MOVING PIVOT $A(\mathbf{a})$.

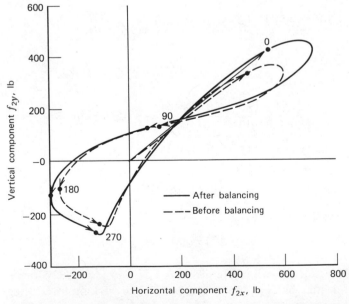

FIGURE 11.8   EXAMPLES 11.1 AND 11.3. COMPARISON OF
BEARING FORCE AT PIVOT A BEFORE AND AFTER BALANCING.

11.4 THE INVERSE DYNAMICS PROBLEM

This class of problem, in which we calculate the time history of the motion as a result of the application of a known system of external forces, can be approached in several ways. We will use a method in which the effect of all masses and forces of the system will be reflected back to the input member. This will result in a nonlinear differential equation that must be integrated numerically.

Reduced Mass or Inertia

The concept of reduced mass is based on the equivalence of kinetic energy in the reduced system and the actual system. As shown in Figure 11.9, we seek an equivalent single mass (or inertia) system that is dynamically equivalent to the actual system in the sense that the response of the hypothetical equivalent single mass system to an input force (or torque) would be identical to the actual multi-link system. The actual system of n members is to be modeled by a single rotating mass of variable moment of inertia I^*. At any instant, assuming the angular velocity of the mass I^* as $\dot{\phi}$, we equate the kinetic energies of the equivalent systems

$$\tfrac{1}{2}I^*\dot{\phi}^2 = \sum_{i=1}^{n} \tfrac{1}{2}m_i(\dot{x}_i^2 + \dot{y}_i^2) + \tfrac{1}{2}I_i\dot{\theta}_i^2 \qquad (11.25)$$

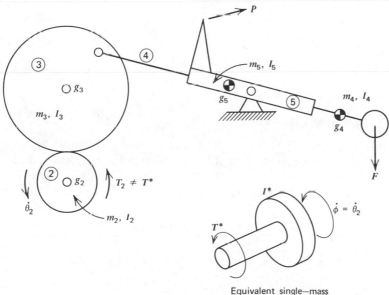

Equivalent single—mass
dynamic system

FIGURE 11.9 PLANE MECHANISM WITH EQUIVALENT REDUCED MASS-TORQUE SYSTEM.

which gives the *reduced moment of inertia* as

$$I^* = \sum_{i=1}^{n} m_i \left(\frac{\dot{x}_i^2 + \dot{y}_i^2}{\dot{\phi}^2} \right) + I_i \left(\frac{\dot{\theta}_i^2}{\dot{\phi}} \right) \tag{11.26}$$

If the equivalent system were assumed as a single mass with linear velocity v the *reduced mass* would be calculated from

$$m^* = \sum_{i=1}^{n} m_i \left(\frac{\dot{x}_i^2 + \dot{y}_i^2}{v^2} \right) + I_i \left(\frac{\dot{\theta}_i}{v} \right)^2 \tag{11.27}$$

Reduced Force or Torque

The equivalent single force or torque acting on the reduced mass or inertia is calculated by equating the rate of doing work or power of the two systems. This leads to

$$\mathbf{F^*} \cdot \mathbf{v} \qquad \text{or} \qquad T^* \cdot \dot{\phi} = \sum_{i=1}^{n} \mathbf{F}_i \cdot \mathbf{v}_i + T_i \cdot \dot{\theta}_i \tag{11.28}$$

which gives

$$F^* = \sum_{i=1}^{n} \frac{F_{xi} \dot{x}_i + F_{yi} \dot{y}_i + T_i \dot{\theta}_i}{v} \tag{11.29}$$

or

$$T^* = \sum_{i=1}^{n} \frac{F_{xi} \dot{x}_i + F_{yi} \dot{y}_i + T_i \dot{\theta}_i}{\dot{\phi}} \tag{11.30}$$

The Work-Energy Method

The work-energy principle offers an alternative method for dynamic analysis when the effect of external forces only need be considered. Internal forces do no work on the complete system, so they need not be included in the analysis. The method is useful as a check on the more involved calculations based on equations of motion.

The work-energy equation for the complete system is given by

$$\int_{0}^{\phi} T^* \, d\phi = \sum_{i=1}^{n} \tfrac{1}{2} m_i (\dot{x}_i^2 + \dot{y}_i^2) + \tfrac{1}{2} I_i \dot{\theta}_i^2 - (KE)_0 \tag{11.31}$$

Note that T^* includes the effect of *all* external forces and moments acting on the system.

Differentiating Eq. 11.31, we have

$$T^* = \frac{1}{\phi} \sum_{i=1}^{n} m_i(\dot{x}_i \ddot{x}_i + \dot{y}_i \ddot{y}_i) + I_i \dot{\theta}_i \ddot{\theta}_i \qquad (11.32)$$

Eq. 11.32 is a very useful tool in dynamic analysis. The terms on the right side are available from the kinematic analysis. Once T^* is known, it may be decomposed using Eq. 11.28 to yield the actual input torque T_0 by subtracting the contributions of other known applied forces and torques.

The Nonlinear Differential Equation of Motion

The work-energy equation is written in the form

$$\int_0^{\phi} T^* \, d\phi = \tfrac{1}{2} I^* \dot{\phi}^2 - (KE)_0 \qquad (11.33)$$

Differentiating with respect to ϕ,

$$T^* = I^* \dot{\phi} \frac{d\dot{\phi}}{d\phi} + \frac{1}{2} \frac{dI^*}{d\phi} \dot{\phi}^2$$

$$= I^* \ddot{\phi} + \frac{1}{2} \frac{dI^*}{d\phi} \dot{\phi}^2$$

$$= I^* \ddot{\phi} + \frac{1}{2} \frac{dI^*}{dt} \dot{\phi} \qquad (11.34)$$

The second term in Eq. 11.34 involves the rate of change of the variable moment of inertia I^* with respect to the position of the mechanism. $dI^*/d\phi$ is a purely geometric quantity. It is convenient to calculate $dI^*/d\phi$ numerically by the method of finite differences using the expression

$$\frac{dI^*}{d\phi} \approx \frac{I^*(\phi + \Delta\phi) - I^*(\phi)}{\Delta\phi} \qquad (11.35)$$

where $I^*(\phi + \Delta\phi)$ is evaluated in terms of the velocity ratios after a small increment $\Delta\phi$ from the current position ϕ.

Solving Eq. 11.34 for $\ddot{\phi}$, we obtain the second-order nonlinear differential equation of motion in the form

$$\ddot{\phi} = \frac{T^* - \dfrac{1}{2} \dfrac{dI^*}{d\phi} \dot{\phi}^2}{I^*} \qquad (11.36)$$

Eq. 11.36 can be solved by any of the well-known numerical integration methods such as the Runge-Kutta or Euler methods. We will use the modified Euler method with predictor-corrector equations.

The Predictor-Corrector Equations

The method is self starting (i.e., at time $t = 0$, $\ddot{\phi}$ can be calculated in terms of the initial values for ϕ and $\dot{\phi}$). The iterative procedure is carried out as follows, where k is the interaction counter per step.

1. Initialize ϕ_k and $\dot{\phi}_k$ at $t = 0$
2. Calculate the initial value of $\ddot{\phi}_k$ from

$$\ddot{\phi}_k = \frac{T_k^* - \frac{1}{2}\left(\frac{dI^*}{d\phi}\right)_k \dot{\phi}_k^2}{I_k^*}$$

3. Predict the value of $\dot{\phi}_{k+1}$ based on constant $\ddot{\phi}_k$ for a time interval Δt.

$$P(\dot{\phi}_{k+1}) = \dot{\phi}_k + \ddot{\phi}_k\, \Delta t$$

4. Predict ϕ_{k+1} using the average of $\dot{\phi}_k$ and $P(\dot{\phi}_{k+1})$.

$$P(\phi_{k+1}) = \phi_k + \frac{(\dot{\phi}_k + P(\dot{\phi}_{k+1}))}{2}\, \Delta t$$

5. Predict $\ddot{\phi}_{k+1}$ in terms of $P(\phi_{k+1})$ and $P(\dot{\phi}_{k+1})$.
6. Correct $\dot{\phi}_{k+1}$ using average of $\ddot{\phi}_k$ and $P(\ddot{\phi}_{k+1})$.

$$C(\dot{\phi}_{k+1}) = \dot{\phi}_k + \frac{\ddot{\phi}_k + P(\ddot{\phi}_{k+1})}{2}\, \Delta t$$

7. Correct ϕ_{k+1} using average of $\dot{\phi}_k$ and $P(\dot{\phi}_{k+1})$.

$$C(\phi_{k+1}) = \phi_k + \frac{\dot{\phi}_k + P(\dot{\phi}_{k+1})}{2}\, \Delta t$$

8. Set $\phi_k = C(\phi_{k+1})$, $\dot{\phi}_k = C(\dot{\phi}_{k+1})$ and repeat steps 2 to 7 until successive values of $C(\phi_{k+1})$ and $C(\dot{\phi}_{k+1})$ agree within a specified tolerance ε.
9. Increment $t = t + \Delta t$, set $\phi_k = C(\phi_{k+1})$ and $\dot{\phi}_k = C(\dot{\phi}_{k+1})$ and return to step 2.

In each of the two classes of mechanism dynamics problems it is essential that an efficient kinematic analysis subprogram be available as required. The vector methods of Chapter 2 can be applied in the analysis of plane mechanisms. The matrix-vector methods of Chapter 4 are useful for spatial mechanisms.

Example 11-4 The Inverse Dynamics Problem. We will consider again the four-bar linkage of Example 11.1. In this case we wish to follow the motion of the linkage caused by the action of a constant input torque $T_2 = 10.0$ in.-lb. A constant load torque $T_4 = -3.0$ in.-lb. acts about the fixed pivot of the output crank at all times. The linkage is assumed to start from rest.

The values of T^*, I^*, and $dI^*/d\theta_2$ required in Eq. 11.36 can be precalculated and stored in the computer memory if desired. Note that these qualities are in terms of velocity ratios. Therefore, the ratios can be calculated in any position from a velocity analysis with the input angular velocity θ_2 arbitrarily set equal to 1.0 rad/sec. In general, however, it is more convenient to form these geometrical factors in terms of the velocity ratios at a given position of the mechanism, since this eliminates the need to interpolate between stored numerical values.

Figure 11.10 shows the variation in T^*, I^*, and $dI^*/d\theta_2$ for the first thirty degrees of input crank motion.

Figure 11.11 gives the dynamic response of the four-bar linkage as a result of the specified input torque T_2 and load torque T_4. Approximately 12 sec of central processor time on a CDC 6400 computer were required to compute fifty points on each of the six curves of Figures 11.10 and 11.11. For each iteration a complete

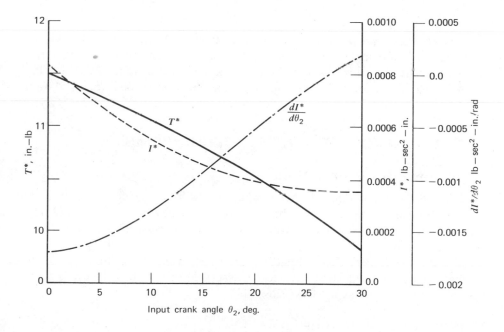

FIGURE 11.10   EXAMPLE 11.4. T^*, I^*, AND $dI^*/d\theta_2$ VERSUS INPUT CRANK ANGLE θ_2.

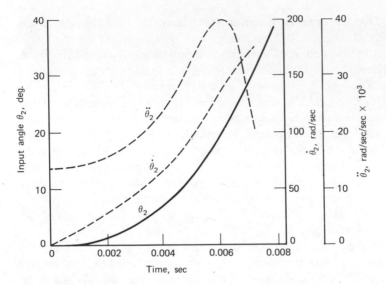

FIGURE 11.11 EXAMPLE 11.4. $\ddot{\theta}_2$, $\dot{\theta}_2$, AND θ_2 VERSUS TIME FOR INPUT CRANK.

velocity analysis must be carried out several times. The solution of this class of problem is a formidable task without the use of a digital computer. The numerical calculations are easily programmed for computation once an appropriate subprogram for the predictor-corrector equations has been developed. (See program PRCR2 in Chapter 12.)

11.5 DYNAMICS OF SPATIAL MECHANISMS

Coordinate Transformations

In describing the dynamics of spatial motion it will be convenient to describe certain vector quantities in a coordinate system that moves with the rigid body. On the other hand, the dynamic equations of motion must be written with an inertial coordinate system used as a reference. An inertial coordinate system is any nonaccelerating system of coordinates, such as one fixed in the earth. These coordinate transformations are conveniently described by adopting the cyclic notation of tensor analysis as used by Pipes [1].

Vector Transformations

The coordinates of a vector ($\mathbf{v}$) in a fixed coordinate system will be denoted as $(\mathbf{v})_x = (v_1, v_2, v_3)_x$ where the subscript x refers to a system of fixed right-handed unit vectors $(\hat{x}) = (\hat{x}_1, \hat{x}_2, \hat{x}_3)$ equivalent to the usual $\hat{i}, \hat{j}, \hat{k}$ system of unit Cartesian vectors.

The coordinates of a vector $(\mathbf{v})$ in one system of coordinates $(\hat{x})$ can be expressed in a second system $(\hat{y})$ by one of the following orthogonal transformations.

$$(\mathbf{v})_x = [R](\mathbf{v})_y = [T_{yx}](\mathbf{v})_y$$
$$(\mathbf{v})_y = [R]^{-1}(\mathbf{v})_x = [T_{xy}](\mathbf{v})_x \qquad (11.37)$$

The rotation matrix $[R]$ would rotate the $(\hat{y})$ system to its current position from an initial position where the axes $(\hat{y})$ were parallel to $(\hat{x})$. The matrix $[T_{xy}]$ is the coordinate transformation matrix that transforms $(\hat{x})$ coordinates into $(\hat{y})$ coordinates and allows any *vector* quantity expressed in fixed axis coordinates to be transformed into equivalent coordinates measured in the moving system; for example

Forces, $(\mathbf{F})_x = [R](\mathbf{F})_y$ or $(\mathbf{F})_y = [T_{xy}](\mathbf{F})_x$

Moments, $(\mathbf{M})_x = [R](\mathbf{M})_y$ or $(\mathbf{M})_y = [T_{xy}](\mathbf{M})_x$

Angular velocities, $(\boldsymbol{\phi})_x = [R](\boldsymbol{\phi})_y$ or $(\boldsymbol{\phi})_y = [T_{xy}](\boldsymbol{\phi})_x$

Direction cosines, $(\mathbf{u})_x = [R](\mathbf{u})_y$ or $(\mathbf{u})_y = [T_{xy}](\mathbf{u})_x$

Example 11-5 Coordinate Transformations. Two coordinate systems are defined as shown in Figure 11.12 such that the moving system $(\hat{y})$ has been rotated 90° about the x_3 axis of the fixed system $(\hat{x})$. Thus,

$$[R] = \begin{bmatrix} \cos(90) & -\sin(90) & 0 \\ \sin(90) & \cos(90) & 0 \\ 0 & 0 & 1 \end{bmatrix} = \begin{bmatrix} 0 & -1 & 0 \\ 1 & 0 & 0 \\ 0 & 0 & 1 \end{bmatrix}$$

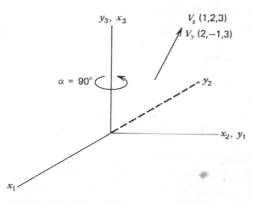

FIGURE 11.12   COORDINATE TRANSFORMATION FOR A FREE VECTOR.

The coordinate transformation matrix $[T]$ becomes

$$[T_{xy}] = \begin{bmatrix} \cos(-90) & -\sin(-90) & 0 \\ \sin(-90) & \cos(-90) & 0 \\ 0 & 0 & 1 \end{bmatrix} = \begin{bmatrix} 0 & 1 & 0 \\ -1 & 0 & 0 \\ 0 & 0 & 1 \end{bmatrix}$$

A vector $(\mathbf{v})_x = (1, 2, 3)$ has coordinates in the $(\hat{y})$ system given by

$$(\mathbf{v})_y = [T_{xy}](\mathbf{v})_x = \begin{bmatrix} 0 & 1 & 0 \\ -1 & 0 & 0 \\ 0 & 0 & 1 \end{bmatrix} \begin{bmatrix} 1 \\ 2 \\ 3 \end{bmatrix} = \begin{bmatrix} 2 \\ -1 \\ 3 \end{bmatrix}$$

Matrix Transformations

As an example of a matrix transformation consider the transformation of the angular velocity matrix $[W]$. Let

$(\hat{y})$ be a system of axes in the moving rigid body.

$(\hat{x})$ be a system of fixed axes.

$(\hat{z})$ be a second system of fixed axes coincident with system $(\hat{y})$.

Transformation of the $(\hat{z})$ system to the $(\hat{x})$ system is described by

$$(\hat{x}) = [T_{zx}](\hat{z})$$

or, in terms of a vector $(\mathbf{v})$ given in $(\hat{z})$ components,

$$(\mathbf{v})_x = [T_{zx}](\mathbf{v})_z$$

Differentiating, remembering that $[\dot{T}_{zx}] = 0.0$,

$$(\dot{\mathbf{v}})_x = [T_{zx}](\dot{\mathbf{v}})_z = [W]_x(\mathbf{v})_x$$

Premultiply by $[T_{zx}]^{-1}$; then

$$(\dot{\mathbf{v}})_z = [T_{zx}]^{-1}[W]_x(\mathbf{v})_x$$
$$= [T_{zx}]^{-1}[W]_x[T_{zx}](\mathbf{v})_z$$

Hence we see that

$$[W]_z = [T_{zx}]^{-1}[W]_x[T_{zx}] \tag{11.38}$$

which is typical of the transformation of any 3×3 matrix from a system $(\hat{x})$ into a system $(\hat{z})$. We also note at this time that the transformation $[T_{yx}]$, when used to transform the angular velocity matrix, is actually with respect to a parallel but *nonrotating* system $(\hat{z})$.

11.6   THE DYNAMICAL EQUATIONS OF MOTION

Figure 11.13 illustrates a rigid body $\sum$ accelerating with respect to a three-dimensional inertial coordinate system under the action of a system of applied forces and moments. The absolute angular motion of the body is described by the angular velocity $\boldsymbol{\phi}$ and angular acceleration $\dot{\boldsymbol{\phi}}$ about an instantaneous rotation axis $(\mathbf{u})_x$. The axis $(\mathbf{u})_x$ is itself rotating with respect to a fixed coordinate system $(\hat{\mathbf{x}})$ at a rate given by $(\dot{\mathbf{u}})_x$ and $(\ddot{\mathbf{u}})_x$. A second coordinate system $(\hat{\mathbf{y}})$, fixed in the moving body, has its origin located at the center of gravity of the moving body and the axes $(\hat{\mathbf{y}})$ are oriented parallel to its principal inertial axes.

At any instant the acceleration of a general point in the body is given by

$$(\ddot{\mathbf{x}}) = [\dot{W}](\mathbf{x} - \mathbf{x}_g) + (\ddot{\mathbf{x}}_g) \tag{11.39}$$

where $[\dot{W}]$ is the angular acceleration matrix.

The equation of motion for an arbitrary single mass particle may be written as

$$(\delta\mathbf{F})_x = \delta m(\ddot{\mathbf{x}})$$

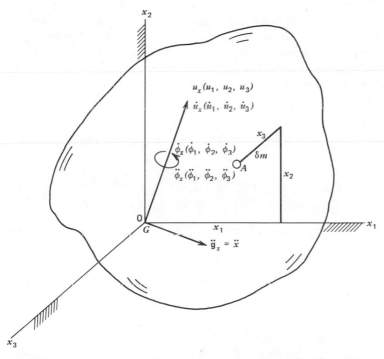

FIGURE 11.13   A RIGID BODY ACCELERATING IN AN INERTIAL COORDINATE SYSTEM.

When summed over the entire rigid body, all internal forces of interaction between particles cancel, and we are left with

$$\left(\sum \mathbf{F}\right)_x = \sum \delta m(\ddot{\mathbf{x}}) \tag{11.40}$$

Substituting Eq. 11.39 into Eq. 11.40, we obtain

$$\left(\sum \mathbf{F}\right)_x = \sum \delta m\{[\dot{W}](\mathbf{x} - \mathbf{x}_g) + (\ddot{\mathbf{x}}_g)\} \tag{11.41}$$

The term $\sum \delta m(\mathbf{x} - \mathbf{x}_g)$ is, by definition of the mass center $(\mathbf{x}_g)$, equal to zero. Hence,

$$\left(\sum \mathbf{F}\right)_x = \sum \delta m(\ddot{\mathbf{x}}_g) = m(\ddot{\mathbf{x}}_g) \tag{11.42}$$

Eq. 11.42 is *Newton's second law of motion*, from which we see that the acceleration of the mass center is always proportional to the sum of *external forces* acting on the rigid body.

Euler's Equations

The angular motion of the rigid body is a function of the net moment or couple acting on the body. In this case it will be found convenient to use the mass center as the moment center (i.e., we assume the origin of both fixed and moving coordinate systems at the mass center). The moment of the elemental force system $\sum (\delta \mathbf{F})_x$ about axes $(\hat{x})$ becomes

$$\left(\sum \mathbf{M}\right)_x = \sum (\mathbf{r})_x \times (\delta \mathbf{F})_x \tag{11.43}$$

The components of the net moment $\left(\sum \mathbf{M}\right)_x$ are found from the vector cross product as

$$\sum M_1 = \sum (x_2 \ \delta F_3 - x_3 \ \delta F_2) = \sum (x_2 \ddot{x}_3 - x_3 \ddot{x}_2) \ \delta m$$
$$\sum M_2 = \sum (x_3 \ \delta F_1 - x_1 \ \delta F_3) = \sum (x_3 \ddot{x}_1 - x_1 \ddot{x}_3) \ \delta m$$
$$\sum M_3 = \sum (x_1 \ \delta F_2 - x_2 \ \delta F_1) = \sum (x_1 \ddot{x}_2 - x_2 \ddot{x}_1) \ \delta m \tag{11.44}$$

From Eq. 3.67, we see that with our choice of coordinate systems $(\hat{x})$,

$$(\ddot{\mathbf{x}}) = [\dot{W}](\mathbf{x}) + (\ddot{\mathbf{x}}_g) \tag{11.45}$$

that is,

$$\ddot{x}_1 = \ddot{a}_{11}x_1 + \ddot{a}_{12}x_2 + \ddot{a}_{13}x_3 + \ddot{x}_{g1}, \text{ etc.}$$

We may also assume, without loss of generality, that instantaneously the nonrotating axes $(\hat{z})$ coincide with a set of principal axes $(\hat{y})$ fixed in the moving body. The moments of inertia of the rigid body expressed in terms of the principal coordinates $(\hat{y})$ are given by

$$I_{11} = \sum (y_2^2 + y_3^2)\, \delta m \qquad I_{12} = \sum y_1 y_2\, \delta m = 0$$
$$I_{22} = \sum (y_1^2 + y_3^2)\, \delta m \qquad I_{23} = \sum y_2 y_3\, \delta m = 0$$
$$I_{33} = \sum (y_1^2 + y_2^2)\, \delta m \qquad I_{13} = \sum y_1 y_3\, \delta m = 0 \qquad (11.46)$$

Substituting Eq. 11.45 into Eq. 11.44, we have

$$\sum M_1 = \sum [x_2(\ddot{a}_{31}x_1 + \ddot{a}_{32}x_2 + \ddot{a}_{33}x_3 + \ddot{x}_{g3})$$
$$- x_3(\ddot{a}_{21}x_1 + \ddot{a}_{22}x_2 + \ddot{a}_{23}x_3 + \ddot{x}_{g2})]\, \delta m \qquad (11.47)$$

which becomes, with $(y) = (x)$,

$$\sum M_1 = \sum (\ddot{a}_{32}y_2^2 - \ddot{a}_{23}y_3^2)\, \delta m \qquad (11.48)$$

where

$$\ddot{a}_{23} = u_2 u_3 \dot{\phi}^2 - \dot{u}_1 \dot{\phi} - u_1 \ddot{\phi} = \dot{\phi}_2 \dot{\phi}_3 - \ddot{\phi}_1$$
$$\ddot{a}_{32} = u_2 u_3 \dot{\phi}^2 + \dot{u}_1 \dot{\phi} + u_1 \ddot{\phi} = \dot{\phi}_2 \dot{\phi}_3 + \ddot{\phi}_1$$

This leads to *Euler's dynamical equations*, referred to the nonrotating system $(\hat{z})$ that instantaneously coincides with $(\hat{y})$.

$$\sum M_1 = I_{11}\ddot{\phi}_1 + \dot{\phi}_2 \dot{\phi}_3 (I_{33} - I_{22})$$
$$\sum M_2 = I_{22}\ddot{\phi}_2 + \dot{\phi}_1 \dot{\phi}_3 (I_{11} - I_{33})$$
$$\sum M_3 = I_{33}\ddot{\phi}_3 + \dot{\phi}_1 \dot{\phi}_2 (I_{22} - I_{11}) \qquad (11.49)$$

Equations of Motion in Matrix Form

Eqs. 11.42 can be written in matrix form as

$$(\mathbf{F})_x = m(\ddot{\mathbf{x}}_g) \qquad (11.50)$$

Eqs. 11.49 become

$$(\mathbf{M})_z = [I]_y(\ddot{\phi})_z + [W]_z [I]_y(\dot{\phi})_z \qquad (11.51)$$

where

$(\mathbf{M})_z$ is the column matrix of the resultant external moments acting on the body expressed with respect to the $(\hat{z})$ system.

$[R]_z$, $[W]_z$, and $[\dot{W}]_z$ are the angular rotation matrices

$(\dot{\phi})_z$ and $(\ddot{\phi})_z$ are the angular motion vectors

The *inertia matrix* $[I]_y$ takes the form

$$[I]_y = \begin{bmatrix} I_{11} & -I_{12} & I_{13} \\ -I_{21} & I_{22} & -I_{23} \\ -I_{31} & -I_{32} & I_{33} \end{bmatrix}_y \tag{11.52}$$

In Eq. 11.51 it will be convenient to express the elements of the inertia matrix in terms of the moving set of principal axes $(\hat{y})$. On the other hand, we may prefer to describe the angular velocities, accelerations, and moments in terms of $(\hat{x})$ or $(\hat{z})$ components in a manner analogous to the force equation (Eq. 11.50). Making the appropriate substitutions,

$$(\dot{\phi})_z = [R]^{-1}(\dot{\phi})_x \qquad [W]_z = [R]^{-1}[W]_x[R]$$

$$[W]_x = [R][W]_z[R]^{-1}$$

$$(\ddot{\phi})_z = [R]^{-1}(\ddot{\phi})_x \qquad (M)_x = [R](M)_z$$

Euler's equation becomes

$$[T_{xz}](M)_x = [I]_y[T_{xz}](\ddot{\phi})_x + [T_{zx}]^{-1}[W]_x[T_{zx}][I]_y[T_{xz}](\dot{\phi})_x$$

which leads to

$$(M)_x = [T_{xz}]^{-1}[I]_y[T_{xz}](\ddot{\phi})_x + [W]_x[T_{xz}]^{-1}[I]_y[T_{xz}](\dot{\phi})_x$$

$$= [I]_x(\ddot{\phi})_x + [W]_x[I]_x(\dot{\phi})_x \tag{11.53}$$

where

$$[T_{xz}] = [T_{xy}] = [R]^{-1}$$

and

$$[I]_x = [T_{xz}]^{-1}[I]_y[T_{xz}] = [R][I]_y[R]^{-1}$$

Example 11-6 Dynamics of a Rotating System. Figure 11.14 shows a uniform bar of cylindrical cross section that is constrained to rotate about a moving axis y_3 with constant angular velocity $(\dot{\phi}_3)_y$. The y_3-axis is in turn rotating about a fixed axis x_3. At the instant shown the moving axis y_3 is assumed to coincide with the x_1-axis and the y_1-axis has rotated an angle α from a position originally coincident with the x_2-axis.

The principal axes of the bar are the axes of symmetry y_1, y_2, and y_3. We note that $(I_2)_y = (I_3)_y$.

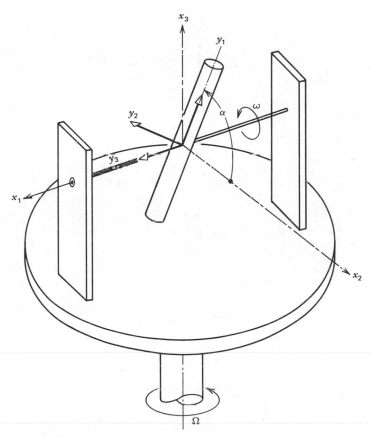

FIGURE 11.14 DYNAMICS OF A ROTATING SYSTEM.

The angular rotation of the bar, referred to the fixed coordinate system, is given by the sum of the rotation $(\dot{\phi}_3)_x$ about the x_3-axis plus the relative rotation $(\dot{\phi}_3)_y = (\dot{\phi}_1)_x$.

$$[R]_x = [R_1]_x[R_3]_x \tag{11.54}$$

Differentiating,

$$[\dot{R}]_x = [\dot{R}_1]_x[R_3]_x + [R_1]_x[\dot{R}_3]_x$$

As the finite rotation angles approach zero,

$$[\dot{R}]_x \rightarrow [W]_x = [W_1]_x + [W_3]_x \tag{11.55}$$

that is,

$$[W]_x = \begin{bmatrix} 0 & 0 & 0 \\ 0 & 0 & -\dot{\phi}_1 \\ 0 & \dot{\phi}_1 & 0 \end{bmatrix} + \begin{bmatrix} 0 & -\dot{\phi}_3 & 0 \\ \dot{\phi}_3 & 0 & 0 \\ 0 & 0 & 0 \end{bmatrix}$$

$$= \begin{bmatrix} 0 & -\dot{\phi}_3 & 0 \\ \dot{\phi}_3 & 0 & -\dot{\phi}_1 \\ 0 & \dot{\phi}_1 & 0 \end{bmatrix}$$

The equivalent angular velocity $(\dot{\phi})_x$ about an axis $(u)_x$ can be found by comparison with the spatial angular velocity matrix (3.60). This leads to

$$(\dot{\phi})_x = \begin{bmatrix} \dot{\phi}_1 \\ 0 \\ \dot{\phi}_3 \end{bmatrix}_x \qquad (u)_x = \begin{bmatrix} u_1 \\ u_2 \\ u_3 \end{bmatrix}_x$$

$$\dot{\phi} = \sqrt{\dot{\phi}_1^2 + \dot{\phi}_3^2}$$

where

$$u_1 = \frac{\dot{\phi}_1}{\sqrt{\dot{\phi}_1^2 + \dot{\phi}_3^2}}$$

$$u_2 = 0$$

$$u_3 = \frac{\dot{\phi}_3}{\sqrt{\dot{\phi}_1^2 + \dot{\phi}_3^2}}$$

The vector $(\dot{u})_x$ is found as follows. Differentiate Eq. 11.54 a second time and again assume that the finite rotation angles approach zero. This gives

$$[\dot{W}]_x = [\dot{W}_1]_x + [\dot{W}_3]_x + 2[\dot{W}_1]_x[\dot{W}_3]_x \tag{11.56}$$

$$= \begin{bmatrix} -\dot{\phi}_3^2 & 0 & 2\dot{\phi}_1\dot{\phi}_3 \\ 0 & -(\dot{\phi}_1^2 + \dot{\phi}_3^2) & 0 \\ 0 & 0 & -\dot{\phi}_1^2 \end{bmatrix}$$

which, when compared with matrix (3.67), leads to

$$2(\dot{u}_3\dot{\phi} + u_3\ddot{\phi}) = 0$$

$$2(\dot{u}_1\dot{\phi} + u_1\ddot{\phi}) = 0$$

$$2(\dot{u}_2\dot{\phi} + u_2\ddot{\phi}) = (2\dot{\phi}_1\dot{\phi}_3)_x$$

from which

$$\dot{u}_1 = -u_1 \frac{\ddot{\phi}}{\dot{\phi}} = -\frac{\phi_1}{\phi_1^2 + \phi_3^2} \ddot{\phi}$$

$$\dot{u}_2 = \frac{\dot{\phi}_1 \phi_3 - u_2 \ddot{\phi}}{\dot{\phi}} = \frac{\phi_1 \phi_3}{\sqrt{\phi_1^2 + \phi_3^2}}$$

$$\dot{u}_3 = -u_3 \frac{\ddot{\phi}}{\dot{\phi}} = -\frac{\phi_3}{\phi_1^2 + \phi_3^2} \ddot{\phi}$$ (11.57)

Since

$$\begin{bmatrix} \ddot{\phi}_1 \\ \ddot{\phi}_2 \\ \ddot{\phi}_3 \end{bmatrix}_x = \begin{bmatrix} \dot{u}_1 \dot{\phi} + u_1 \ddot{\phi} \\ \dot{u}_2 \dot{\phi} + u_2 \ddot{\phi} \\ \dot{u}_3 \dot{\phi} + u_3 \ddot{\phi} \end{bmatrix}_x$$ (11.58)

and

$$\ddot{\phi} = 0 \qquad \dot{u}_1 = \dot{u}_3 = 0$$

we obtain

$$\begin{bmatrix} \ddot{\phi}_1 \\ \ddot{\phi}_2 \\ \ddot{\phi}_3 \end{bmatrix}_x = \begin{bmatrix} 0 \\ \dot{\phi}_1 \phi_3 \\ 0 \end{bmatrix}_x$$ (11.59)

In order to apply Euler's equations we must refer all angular motions to the moving coordinate system using matrix (11.53). This requires both the rotation matrix $[R]$, which describes the rotation of system $(\hat{y})$ from a position originally coincident with $(\hat{x})$, and its inverse $[R]^{-1} = [T_{yx}]$. The rotation matrix

$$[R] = [R_{\alpha, x_1}][R_{90°, x_2}][R_{90°, x_3}]$$

$$= \begin{bmatrix} 0 & 0 & 1 \\ \cos \alpha & -\sin \alpha & 0 \\ \sin \alpha & \cos \alpha & 0 \end{bmatrix}$$ (11.60)

and the corresponding $[T_{yx}]$ matrix becomes

$$[T_{yx}] = [R]^{-1} = \begin{bmatrix} 0 & \cos \alpha & \sin \alpha \\ 0 & -\sin \alpha & \cos \alpha \\ 1 & 0 & 0 \end{bmatrix}$$ (11.61)

Substituting into Eq. 11.53 leads to

$$
\begin{bmatrix} M_1 \\ M_2 \\ M_3 \end{bmatrix}_x = \begin{bmatrix} \frac{1}{2}(I_2 - I_1)\dot{\phi}_3^2 \sin 2\alpha \\ \dot{\phi}_1\dot{\phi}_3[I_2 + (I_1 - I_2)\cos 2\alpha] \\ \dot{\phi}_1\dot{\phi}_3(I_1 - I_2)\sin 2\alpha \end{bmatrix} \tag{11.62}
$$

where $[I] = [I]_y$ and $(\dot{\phi}) = (\dot{\phi})_x$.

Eq. 11.62 then allows the calculation of the torque M_3 required to drive the platform and the torque M_1 required about the axis x_1 in order to maintain the constant angular velocities $(\dot{\phi}_3)_x$ and $(\dot{\phi}_1)_x$. Both of these torque components are seen to be variable functions of the rotor angle α.

Chapter 12

Computer Programs

12.1 INTRODUCTION

The success of numerical methods in kinematics depends to a large extent on the availability of a digital computer. The computer in turn requires a set of instructions in the form of programs written in a language it can understand. The programs listed in this chapter have been coded in the FORTRAN IV language and have been compiled and executed on a CDC 6400 computer using the RUN compiler. Certain statements, particularly WRITE statements, which contain Hollerith text fields between asterisks used as field delimiters, may require modification for use with other compilers.

Contained in the following sections are FORTRAN codes for plane and spatial kinematics, solution of simultaneous linear and nonlinear algebraic equations, and minimization of a function of many variables. One code is given written in BASIC language that will solve either linear or nonlinear algebraic equations.

Detailed user instructions including input format and definition of program variables are included as comments at the beginning of programs DESIGN, PCON, and LSTCON.

12.2 LINKPAC, A SUBROUTINE PACKAGE USEFUL IN PLANE KINEMATICS AND MECHANISMS

The LINKPAC subroutines include FORTRAN codes for all of the plane kinematics included in the vector and matrix expressions from Chapters 2 and 3. Use of the subroutines for position, velocity, or acceleration analysis of three basic elements of plane mechanisms, the two-link dyad, the oscillating slider, and the rotating guide are best shown by example.

As a first example let us use subroutine AGUIDE for the acceleration analysis of an offset slider-crank mechanism, as shown in Figure 12.1. We must recognize that, in this case, the guide is fixed, hence the angle BETA, the angular velocity VBETA, and the angular acceleration ABETA must all be set equal to zero before the call to AGUIDE. The fixed point 4 in Figure 12.1 is identified by the dummy variable N2 in AGUIDE. The moving pivot 2 on the input crank is identified by variable N1 in AGUIDE. The motion of point 2, required as input to AGUIDE, is

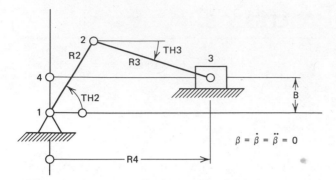

FIGURE 12.1 THE OFFSET SLIDER-CRANK WITH LINKPAC COORDINATE SYSTEM.

calculated by a call to subroutine CRANK. In subroutine CRANK variable N1 identifies the fixed pivot 1 and variable N2 locates the moving pivot 2. The output of subroutine AGUIDE includes the position of the slider, given by variable R4, its velocity VR4, acceleration AR4, and the angular motion of the connecting rod specified by TH3, W3, and A3. The development of the program can be followed by referring to the comments included in program OFFSLD.

Example 12-1 Acceleration Analysis of the Offset Slider-Crank Mechanism.

```
       PROGRAM OFFSLD(INPUT,OUTPUT,TAPE5=INPUT,TAPE6=OUTPUT)
C      ACCELERATION ANALYSIS OF THE OFFSET SLIDER-CRANK MECHANISM
       DIMENSION P(30,2),VP(30,2),AP(30,2)
C
C      INITIALIZE THE MOTION PARAMETERS FOR ALL FIXED PIVOTS
       P(1,1)= P(1,2)= VP(1,1)= VP(1,2)= AP(1,1)= AP(1,2)= 0.0
       P(4,1)= VP(4,1)= VP(4,2)= AP(4,1)= AP(4,2)= 0.0
       BETA= VBETA= ABETA= 0.0
C      READ IN THE GEOMETRICAL PARAMETERS
       READ(5,10) R2,R3,BB
    10 FORMAT(3F10.0)
       P(4,2)= BB
       WRITE(6,11) R2,R3,BB
    11 FORMAT(1H1, 5X,*R2= *,F6.2,2X,*R3= *,F6.2,2X,*BB= *,F6.2)
C      READ IN THE INPUT CRANK MOTION PARAMETERS
       READ(5,15) TH2ZERO, TH2MAX,DELTH, W2, A2
    15 FORMAT(5F10.0)
       WRITE(6,12) TH2ZERO,TH2MAX,W2,A2
    12 FORMAT(/,5X,*TH2O= *,F6.2,* DEG*,3X,*TH2MAX= *,F6.2,* DEG*//
      $ 5X,*INPUT CRANK ANGULAR VELOCITY= *,F6.2,* RAD/SEC*//13X*ANGULAR
      $ ACCELERATION= *,F6.2,* RAD/SEC/SEC*)
C      PRINT HEADINGS FOR THE RESULTS
       WRITE (6,13)
    13 FORMAT(//6X,*R4*9X*VR4*9X*AR4*9X*TH3*10X*W3*10X*A3*)
```

```
C       CONVERT DEGREES TO RADIANS
        CON= ATAN(1.)/45.
        TH20= TH2ZERO*CON
        TH2M= TH2MAX*CON
        DEL= DELTH*CON
C       COMPUTE THE NUMBER OF INCREMENTS FOR THETA
        NPT= INT(TH2MAX/DELTH) + 1
C       SET UP THE DO LOOP FOR ACCELERATION ANALYSIS
        DO 20 J= 1,NPT
        TH2= TH20 + (J-1)*DEL
        TH= TH2/CON
        WRITE(6,14) TH
     14 FORMAT(/5X*TH2= *F6.2* DEGREES*)
C       CALL SUBROUTINE CRANK WITH FIXED REFERENCE POINT 1
C       RETURNS POS, VEL. AND ACC OF POINT 2
        CALL CRANK(1,2,R2,TH2,W2,A2,P,VP,AP)
C       CALL SUBROUTINE AGUIDE WITH FIXED ANGLE BETA AND FIXED REF PT 4
C       RETURNS POS, VEL AND ACC OF POINT 3 , MEMBER 3 AND THE
C       SLIDING MOTION COMPONENTS R4, VR4, AR4
C       SET THE MODE OF ASSEMBLY BEFORE THE CALL TO AGUIDE
        M= +1
        CALL AGJIDE(M,2,4,3,R3,R4,TH3,BETA,P,W3,VBETA,VR4,VP,A3,ABETA,
     $  AR4,AP)
C       PRINT RESULTS FOR THE SLIDER MOTION AND THE ANGULAR MOTION
C       OF THE COUPLER
     20 WRITE(6,30) R4,VR4,AR4,TH3,W3,A3
     30 FORMAT(6E12.3)
        STOP
        END
```

SAMPLE OUTPUT

R2= 1.00   R3= 2.00   BB= 1.00
TH20= 0.00 DEG   TH2MAX= 90.00 DEG

INPUT CRANK ANGULAR VELOCITY= 628.00 RAD/SEC

ANGULAR ACCELERATION= 0.00 RAD/SEC/SEC

R4	VR4	AR4	TH3	W3	A3
TH2= 0.00 DEGREES					
2.732E+00	3.626E+02	−6.980E+05	5.236E 01	3.626E 02	7.590E 04
TH2= 30.00 DEGREES					
2.803E+00	−1.736E+02	−5.554E+05	2.527E−01	−2.809E+02	1.222E+05
TH2= 60.00 DEGREES					
2.496E+00	−5.228E+02	−2.698E+05	6.704E−02	−1.574E+02	1.728E+05
TH2= 90.00 DEGREES					
2.000E+00	−6.280E+02	−2.126E−09	0.	−1.693E−12	1.972E+05

As a second example consider a complete acceleration and force analysis for the plane four-bar linkage with a general system of external forces and moments acting on each link. Joint forces f_1, f_2, f_3, and f_4 are not included in the external force system.

In this case subroutine FRBRAN from the subroutine package is used for the acceleration analysis. Communication between the main program FORCE and subroutine FRBRAN is through the labeled COMMON/FRBR/————, which must be duplicated in the main program.

The dynamic analysis is accomplished by first forming the dynamic equations of motion in the matrix form $[A](f) = (b)$ and then calling a system library function LNEQF, which solves for the unknown force and torque components in (f) by Gauss elimination. Most computer systems will have a similar subprogram available. Function LINEQF as used by SIMEQS in program DESIGN is essentially the same. A BASIC subroutine for the same purpose is included in Section 12.7 beginning at statement 5000.

In the following example a program is given for the solution of Example 11.1. Note that in this case many of the external force system terms are set to zero. The coordinate system using LINKPAC notation is shown in Figure 12.2.

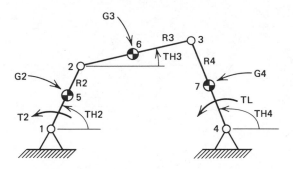

FIGURE 12.2   EXAMPLE 12.2. LINKPAC NOTATION.

Example 12-2 Acceleration and Dynamic Force Anaysis for the Plane Four-Bar Linkage

```
      PROGRAM FORCE(INPUT,OUTPUT,TAPE5=INPUT,TAPE6=OUTPUT)
      COMMON /FBBR/ M,R2,S2,PH2,R3,S3,PH3,R4,S4,PH4,P(30,2),VP(30,2),
     $ AP(30,2),PI(30,2),TH2,TH3,TH4,W2,W3,W4,A2,A3,A4
      DIMENSION A(9,9),B(9),ERASE(9)
      REAL M2,M3,M4,K2,K3,K4,I2,I3,I4
      LOGICAL FORCE
      FORCE=.TRUE.
```

```
C.....READ IN LINK GEOMETRY AND MODE OF ASSEMBLY
   10 READ(5,100) R2,S2,P2,R3,S3,P3,R4,S4,P4,M
      IF(R2.LT.0.0) FORCE=.FALSE.
      IF(EOF,5) 99,11
   11 CONTINUE
      CON= ATAN(1.)/45. $ R2= ABS(R2)
      PH2= P2*CON $ PH3= P3*CON $ PH4= P4*CON
C.....READ IN THE FIXED PIVOT COORDINATES
      READ(5,200) P(1,1),P(1,2),P(4,1),P(4,2)
C.....READ IN THE INPUT CRANK MOTION PARAMETERS
      READ(5,200) THETA,DELTH,W2,A2
      IF(.NOT.FORCE) GO TO 15
C.....READ IN PHYSICAL CONSTANTS FOR THE LINKS
      READ(5,210) M2,M3,M4,K2,K3,K4
      I2= K2**2*M2 $ I3= K3**2*M3 $ I4= K4**2*M4
C.....READ LOAD TORQUE ON THE OUTPUT LINK
      READ(5,210) TL
   15 DEL= DELTH*CON
      NPT= INT(THETA/DELTH) + 1
C.....PRINT HEADINGS FOR RESULTS AND THE PROBLEM DATA
      WRITE(6,500) R2,S2,P2,R3,S3,P3,R4,S4,P4,M2,K2,I2,M3,K3,I3,
     $ M4,K4,I4,P(1,1),P(1,2),P(4,1),P(4,2),W2,A2,THETA,DELTH
C.....INITIALIZE VARIOUS QUANTITIES
      VP(1,1)=VP(1,2)=AP(1,1)=AP(1,2)=VP(4,1)= VP(4,2)=AP(4,1)=AP(4,2)=0.
      DO 400 J= 1,NPT
C.....ACCELERATION ANALYSIS USING SUBROUTINE FRBRAN
      TH2= (J-1)*DEL
      CALL FRBRAN
C.....DYNAMIC FORCE ANALYSIS
      TH2= TH2/CON
      WRITE(6,600) TH2,(((I,P(I,1),P(I,2),VP(I,1),VP(I,2),AP(I,1),
     $ AP(I,2)), I=1,7)
      WRITE(6,650) W3,W4,A3,A4
      IF(.NOT.FORCE) GO TO 400
C.....INITIALIZE A-MATRIX AND RIGHT SIDE VECTOR EQUAL TO 0.0
      DO 20 K= 1,9
      B(K)= 0.0
      DO 20 I= 1,9
   20 A(I,K)- 0.0
C.....FORM MATRIX EQUATION OF MOTION...A-MATRIX
      A(1,1)= -1. $ A(1,3)= 1. $ A(2,2)= -1. $ A(2,4)= 1.
      Q2Y= P(1,2)-P(5,2) $ A(3,1)= Q2Y
      Q2X= P(1,1)-P(5,1) $ A(3.2)= - Q2X
      P2Y= P(2.2)-P(5,2) $ A(3,3)= - P2Y
      P2X= P(2,1)-P(5,1) $ A(3,4)= P2X
      A(3,9)= 1. $ A(4,3)= -1. $ A(4,5)= 1. $ A(5,4)= -1. $ A(5,6)= 1.
      Q3Y= P(2,2)-P(6,2) $ A(6,3)= Q3Y
      Q3X= P(2,1)-P(6,1) $ A(6,4)= - Q3X
      P3Y= P(3,2)-P(6,2) $ A(6,5)= - P3Y
      P3X= P(3,1)-P(6,1) $ A(6,6)= P3X
      A(7,5)= -1. $ A(7,7)= 1. $ A(8,6)= -1. $ A(8,8)= 1.
      Q4Y= P(3,2)-P(7,2) $ A(9,5)= Q4Y
      Q4X= P(3,1)-P(7,1) $ A(9,6)= - Q4X
      P4Y= P(4,2)-P(7,2) $ A(9,7)= - P4Y
      P4X= P(4,1)-P(7,1) $ A(9,8)= P4X
```

```
C.....SPECIFY KNOWN EXTERNAL LOADS AND COUPLES ACTING ON EACH LINK
      SF2X= SF2Y= ST2= 0.0
      SF3X= SF3Y= ST3= 0.0
      SF4X= SF4Y= 0.0
      ST4= TL
C.....RIGHT SIDE B-VECTOR
      DXF2= DXF3= DXF4= 0.0
      B(1)= M2*AP(5,1) - SF2X $ B(2)= M2*AP(5,2) - SF2Y
      B(3)= I2*A2 - DXF2 - ST2 $ B(4)= M3*AP(6,1) - SF3X
      B(5)= M3*AP(6,2) - SF3Y $ B(6)= I3*A3 - DXF3 - ST3
      B(7)= M4*AP(7,1) - SF4X $ B(8)= M4*AP(7,2) - SF4Y
      B(9)= I4*A4 - DXF4 - ST4
C.....SOLVE SET OF LINEAR DYNAMIC EQS OF MOTION BY GAUSS ELIMINATION
C     USING SYSTEM LIBRARY FUNCTION LNEQF
      SCALE= 1.0
      LN= LNEQF(9,9,1,A,B,SCALE,ERASE)
C.....COMPUTE SHAKING FORCE COMPONENTS
      SF1X= B(1) - B(7)
      SF1Y= B(2) - B(8)
C.....WRITE RESULTS FOR JOINT FORCES,INPUT TORQUE AND SHAKING FORCES
      WRITE(6,700) (B(I), I= 1,9),SF1X,SF1Y
  400 CONTINUE
   99 STOP
  100 FORMAT(3(3F10.0/),I2)
  200 FORMAT(4F10.0)
  210 FORMAT(3F10.0)
  500 FORMAT(1H1,15X*ACCELERATION AND FORCE ANALYSIS OF THE FOUR-BAR LIN
     1 KAGE*//10X*GEOMETRY OF THE LINKS*/
     2 5X,*R2= *F6.2,5X,*S2= *F6.2,5X,*PHI2= *F6.2/
     3 5X,*R3= *F6.2,5X,*S3= *F6.2,5X,*PHI3= *F6.2/
     4 5X,*R4= *F6.2,5X,*S4= *F6.2,5X,*PHI4= *F6.2/
     1 //10X*PHYSICAL PROPERTIES*/
     2 5X*M2= *E12.3,5X*K2= *E12.3,5X*I2= *E12.3/
     3 5X*M3= *E12.3,5X*K3= *E12.3,5X*I3= *E12.3/
     4 5X*M4= *E12.3,5X*K4= *E12.3,5X*I4= *E12.3//
     5 10X*FIXED PIVOT COORDINATES*/5X,*P(1,1)= *F6.2,5X,
     6 *P(1,2)= *F6.2,5X,*P(4,1)= *F6.2,5X,*P(4,2)= *F6.2//
     7 10X,*INPUT CRANK MOTION PARAMETERS*/
     8 5X,*W2= *F6.2* RAD/SEC*5X*A2= *F6.2* RAD/SEC/SEC*//
     9 5X*RANGE OF INPUT ROTATION = *F6.2* DEG*5X* INCREMENT DELTH= *
     1 F6.2* DEG*)
  600 FORMAT(1H1,10X*INPUT CRANK ANGLE, TH2= *F6.2* DEG*//* POINT*7X
     1 *POSITION*14X*VELOCITY*12X*ACCELERATION*/* NUMBER*
     2 3(5X*X*10X*Y*5X)//(I6,6E11.3))
  650 FORMAT(/5X*W3= *E10.3,5X*W4= *E10.3,5X*A3= *E10.3,5X*A4= *E10.3)
  700 FORMAT(//10X*FORCE ANALYSIS*//
     1 5X*F1X= F21X= *E12.4,5X,*F1Y= F21Y= *E12.4/
     2 5X,*F2X= F32X= *E12.4,5X,*F2Y= F32Y= *E12.4/
     3 5X,*F3X= F43X= *E12.4,5X,*F3Y= F43Y= *E12.4/
     4 5X,*F4X= F14X= *E12.4,5X*F4Y= F14Y= *E12.4//
     5 5X*DRIVING OR INPUT TORQUE T2= *E12.4//
     6 5X*SHAKING FORCE COMPONENTS SF1X= *E12.4,5X*SF1Y= *E12.4///)
      END
```

ACCELERATION AND FORCE ANALYSIS OF THE FOUR-BAR LINKAGE

GEOMETRY OF THE LINKS
R2= 1.00   S2=   .50   PHI2= 0.
R3= 2.83   S3= 1.41   PHI3= 0.
R4= 2.00   S4= 1.00   PHI4= 0.

PHYSICAL PROPERTIES
M2= .300E−03   K2= .500E+00   I2= .750E−04
M3= .650E 03   K3= .800E+00   I3− .416E−03
M4− .500E−03   K4= .100E+01   I4= .500E−03

FIXED PIVOT COORDINATES
P(1,1)= 0.   P(1,2)= 0.   P(4,1)= 3.00   P(4,2)= 0.

INPUT CRANK MOTION PARAMETERS
W2= 628.00 RAD/SEC   A2=   0.   RAD/SEC/SEC
RANGE OF INPUT ROTATION= 30.00 DEG   INCREMENT DELTH= 30.00 DEG

INPUT CRANK ANGLE, TH2= 0. DEG

POINT	POSITION		VELOCITY		ACCELERATION	
NUMBER	X	Y	X	Y	X	Y
1	0.	0.	0.	0.	0.	0.
2	.100E+01	0.	0.	.628E+03	−.394E+06	0.
3	.300E+01	.200E+01	.628E+03	.190E+00	−.592E+06	−.197E+06
4	.300E+01	0.	0.	0.	0.	0.
5	.500E+00	0.	0.	.314E+03	−.197E+06	0.
6	.200E+01	.100E+01	.314E+03	.314E+03	−.493E+06	−.987E+05
7	.300E+01	.100E+01	.314E+03	.948E−01	−.296E+06	−.987E+05

W3= −.314E+03   W4= −.314E+03   A3= −.893E+02   A4= .296E+06

FORCE ANALYSIS

F1X= F21X=   .5273E+03   F1Y= F21Y=   .3401E+03
F2X= F32X=   .4681E+03   F2Y= F32Y=   .3401E+03
F3X= F43X=   .1478E+03   F3Y= F43Y=   .2759E+03
F4X= F14X= −.6943E−01   F4Y= F14Y=   .2266E+03

DRIVING OR INPUT TORQUE T2= −.3401E+03

SHAKING FORCE COMPONENTS SF1X=   .5274E+03   SF1Y=   .1135E+03

```
        INPUT CRANK ANGLE, TH2= 30.00 DEG

   POINT           POSITION                  VELOCITY                ACCELERATION
  NUMBER       X            Y            X            Y            X            Y
       1    0.           0.           0.           0.           0.           0.
       2    .866E+00     .500E+00    -.314E+03     .544E+03    -.342E+06    -.197E+06
       3    .328E+01     .198E+01     .222E+02    -.309E+01    -.693E+06     .960E+05
       4    .300E+01    0.           0.           0.           0.           0.
       5    .433E+00     .250E+00    -.157E+03     .272E+03    -.171E+06    -.986E+05
       6    .207E+01     .124E+01    -.146E+03     .270E+03    -.517E+06    -.506E+05
       7    .314E+01     .990E+00     .111E+02    -.154E+01    -.346E+06     .480E+05
```

W3= −.227E+03   W4= −.112E+02   A3= .153E+06   A4= .350E+06

FORCE ANALYSIS

```
F1X= F21X=  .5977E+03   F1Y= F21Y=  .3051E+03
F2X= F32X=  .5465E+03   F2Y= F32Y=  .2755E+03
F3X= F43X=  .2103E+03   F3Y= F43Y=  .2426E+03
F4X= F14X=  .3705E+02   F4Y= F14Y=  .2666E+03
```

DRIVING OR INPUT TORQUE T2= .3463E+02

SHAKING FORCE COMPONENTS SF1X= .5606E+03   SF1Y= .3846F+02

12.3 DESIGN—SOLUTION OF SETS OF SIMULTANEOUS NONLINEAR ALGEBRAIC EQUATIONS BY THE NEWTON-RAPHSON METHOD

Program DESIGN is a driver for subroutine SIMEQS, which carries out the Newton-Raphson iterative procedure. The required partial derivatives $\partial f_i / \partial Z_j$ are formed numerically by function GRAD, so expressions for partial derivatives in analytical form are not required.

The equations to be solved are supplied by the user in function YCOMP (Z, I) where I is the index of a function $y_i = f_i(Z)$, $i = 1, n$. Each $f_i(Z)$ approaches zero at a solution.

Program DESIGN provides for the solution of independent sets of equations in the design variables (Z) with the results for each independent set in sequence available to the next set of equations.

Program DESIGN also provides for the generation of a family of solutions in terms of one variable design parameter through use of the special variable V. Variable V is automatically incremented by an amount ΔV after each convergence to a solution. This allows any variable $Z(J)$ to be incremented by use of a dummy

function in YCOMP in the form YCOMP = Z(J) − V. At convergence this function, along with all other functions YCOMP, must approach zero. This forces the
variable Z(J) to assume the current value of V.

An important feature of subroutine SIMEQS is the inclusion of the *damping*
feature. As each correction vector (ΔZ) is calculated a test is made of the possible
new estimate of a solution at point $(Z + \Delta Z)$.

If the norm of the functions vector increases, as indicated by

$$\sum_{i=1}^{n} |f_i(Z + \Delta Z)| > \sum_{i=1}^{n} |f_i(Z)|$$

then it is assumed that the corrections ΔZ would not lead to a region of a solution
and a smaller correction vector is tried. On the first pass for any iteration that fails
the NORM test, the correction vector (ΔZ) is normalized about its maximum
element. On subsequent tries the correction vector (ΔZ) is replaced by $(\Delta Z) =$
$(\Delta Z/5)$ for a maximum of ten trys. It is obvious that this procedure can lead to
very small correction vector elements, and it has been found a most effective
procedure in overcoming the basic instability of the Newton-Raphson procedure
with poor initial guesses.

A complete set of user instructions for program DESIGN is included as comments in the program listing as given in the appendix.

Example 12-3 Use of program DESIGN. As a first example consider the
use of program DESIGN for a simple three-variable problem. The problem is to
find the common intersections of the sphere

$$x^2 + y^2 + z^2 = 1$$

the plane

$$x + y + z = 1$$

and the parabolic surface

$$x = y^2$$

With the initial guess vector $(0.562, 0.750, -0.312)$ there is rapid convergence to
a solution at $(0.56984, 0.75488, -0.32472)$.

With initial guesses $(0.0, 0.0, 0.0)$ we see that the first correction vector at
iteration 1 is normalized once then damping occurs on the second try at iteration
1 and the first try at iteration 2. Iterations 3 to 8 continue without damping and
lead to a second solution at $(0.0, 0.0, 1.0)$.

YCOMP   FORTRAN COMPILATION   RUN 2.3C0-75274   27 JUL 76 15 : 38 : 09

```
              FUNCTION YCOMP(Z,J)
000005        DIMENSION Z(3)
000005        GO TO (1,2,3) J
000013      1 YCOMP= Z(1)**2 + Z(2)**2 + Z(3)**2 - 1.0
000020        RETURN
000021      2 YCOMP= Z(1) - Z(2)**2
000023        RETURN
000024      3 YCOMP= Z(1) + Z(2) + Z(3) - 1.0
000030        RETURN
000031        END
```

```
******************************** NEW DATA *********************************

NTEQ NSEQ NINC NFEI NITR NSET(I), I= 1,5                  V        DELV
  -3    1    1    1   20    3   0    0    0    0    0.       .1000000E+01

INITIAL GUESSES FOR VARIABLES Z(I), I= 1,NTEQ
  .5620000E+00   .7500000E+00   -.3120000E+00

AT ITERATION 1
CORRECTNS     .7908060E-02     .4936239E-02    -.1284430E-01
VARIABLES     .5699081E+00     .7549362E+00    -.3248443E+00
FUNCTIONS    -.2477408E-03     .2066427E-04    -.3552714E-13

AT ITERATION 2
CORRECTNS    -.6777016E-04    -.5854153E-04     .1263117E-03
VARIABLES     .5698403E+00     .7548777E+00    -.3247180E+00
FUNCTIONS    -.6576079E-07     .4762223E-07   -0.

AT ITERATION 3
CORRECTNS     .7226033E-09    -.3104886E-07     .3032626E-07
VARIABLES     .5698403E+00     .7548777E+00    -.3247180E+00
FUNCTIONS    -.1318057E-10     .2344080E-10     .7105427E-14

CONVERGED IN 3 ITERATIONS
ENDING Z WITH V=   0.
  .5698403E+00   .7548777E+00   -.3247180E+00
```

```
INITIAL GUESSES FOR VARIABLES Z(I), I= 1,NTEQ
0.              0.              0.

AT ITERATION 1
CORRECTNS    -.2305612E+19    -.2305612E+23     .2305843E+23
VARIABLES    -.2305612E+19    -.2305612E+23     .2305843E+23
FUNCTIONS    -.1063276E+46     .5315849E+45     .1000000E+01
```

```
AT ITERATION 1                                                          G
CORRECTNS   -.9999000E-04   -.9999000E+00   .1000000E+01    Z
VARIABLES   -.9999000E-04   -.9999000E+00   .1000000E+01    ₫
FUNCTIONS   -.9998000E+00    .9999000E+00   .1000000E+01    ≥
                                                            Q

AT ITERATION 1
CORRECTNS   -.1999800E-04   -.1999800E+00   .2000000E+00
VARIABLES   -.1999800E-04   -.1999800E+00   .2000000E+00
FUNCTIONS    .9200080E+00    .4001200E-01   .1000000E+01

AT ITERATION 2
CORRECTNS    .3746976E+00   -.8372162E+00   .1462519E+01
VARIABLES    .3746776E+00   -.1037196E+01   .1662519E+01
FUNCTIONS   -.2980127E+01    .7010984E+00   .1449507E-11    G
                                                            Z
AT ITERATION 2                                              ₫
CORRECTNS    .7493951E-01   -.1674432E+00   .2925037E+00    ≥
VARIABLES    .7491951E-01   -.3674232E+00   .4925037E+00    Q
FUNCTIONS    .6168273E+00    .6008032E-01   .8000000E+01



AT ITERATION 7
CORRECTNS    .1491299E-04    .5980251E-05   -.2089324E-04
VARIABLES    .5622618E-09   -.1031178E-07    .1000000E+01
FUNCTIONS   -.1949904E-07   -.5622517E-09  -0.

AT ITERATION 8
CORRECTNS    .5012313E-09    .1030637E-07   -.9745135E-08
VARIABLES    .1030530E-11   -.5413969E-11    .1000000E+01
FUNCTIONS   -.8768097E-11   -.1030530E-11  -11.

CONVERGED IN 8 ITERATIONS
ENDING Z WITH V=  0.
  .1030530E-11  -.5413969E-11   .1000000E+01
```

Example 12-4 Center and Circle Points for Four-Position Rigid Body Guidance. A function YCOMP is given for the solution of Example 6-7. In this case a family of solutions is given for a series of specified values for the x-coordinate of the center point $\mathbf{a}_0$. A new series of points could be found by changing statement 25 to any of the following

$$YCOMP = V - A0Y$$

$$YCOMP = V - A1X$$

or

$$YCOMP = V - A1Y$$

This allows the user to generate the complete center and circle point curves in several parts.

```
      FUNCTION YCOMP(X,II)
C.....CENTER AND CIRCLE POINT COORDINATES FOR FOUR-POSITION
C     RIGID BODY GUIDANCE
      DIMENSION X(4),XX(4),YY(4),THETA(4),DM(2,3,4),AX(4),AY(4),TH(4)
      COMMON /YCOMP/ ONCE,NOSE,V
      LOGICAL ONCE
      IF(ONCE) GO TO 15
C.....READ IN POSITION DATA FOR THE RIGID BODY
      READ(5,5) ((XX(I),YY(I),THETA(I)), I= 1,4)
    5 FORMAT(3F10.0)
C.....CALCULATE AND STORE DISPLACEMENT MATRIX ELEMENTS
      CON= ATAN(1.)/45.
      DO 10 J= 2,4
      TH(J)= (THETA(J)-THETA(1))*CON
      CT= COS(TH(J))
      ST= SIN(TH(J))
      DM(1,1,J)= CT
      DM(1,2,J)= -ST
      DM(1,3,J)= XX(J)  - XX(1)*DM(1,1,J)  - YY(1)*DM(1,2,J)
      DM(2,1,J)= ST
      DM(2,2,J)= CT
   10 DM(2,3,J)= YY(J)  - XX(1)*DM(2,1,J)  - YY(1)*DM(2,2,J)
      ONCE= .TRUE.
   15 CONTINUE
C.....DEFINE CONSTANT LENGTH DESIGN EQUATIONS IN TERMS OF A
C     CENTER POINT (A0X,A0Y) AND A CIRCLE POINT AX(1),AY(1)
      A0X= X(1)
      A0Y= X(2)
      AX(1)= X(3)
      AY(1)= X(4)
C.....CALCULATE AX(J) AND AY(J) IN TERMS OF AX(1) AND AY(1)
      DO 20 J= 2,4
      AX(J)= AX(1)*DM(1,1,J) + AY(1)*DM(1,2,J) + DM(1,3,J)
   20 AY(J)= AX(1)*DM(2,1,J) + AY(1)*DM(2,2,J) + DM(2,3,J)
C.....CALCULATE FUNCTION OF (X) WITH INDEX II
      IF(II.GT.3) GO TO 25
      JJ= II + 1
      YCOMP= (AX(JJ)-A0X)**2 + (AY(JJ)-A0Y)**2 - (AX(1)-A0X)**2 -
     1 (AY(1)-A0Y)**2
      RETURN
C.....CHANGE NEXT CARD TO INCREMENT A VARIABLE OTHER THAN A0X
   25 YCOMP= V - A0X
      RETURN
      END
```

```
******************************* NEW DATA *************************************
NTEQ NSEQ NINC NFEI NITR NSET(I), I= 1,5                    V              DELV
  4    1    5    1   20    4   0    0    0   0    -.1000000E+01    .1000000E+00

INITIAL GUESSES FOR VARIABLES Z(1), I= 1,NTEQ
 -.1000000E+01   .1000000E+01   -.5000000E+00   .2500000E+01


CONVERGED IN   3 ITERATIONS
ENDING Z WITH  V=    -1.000
 -.1000000E+01   .9940841E+00   -.8209411E+00   .2602202E+01


CONVERGED IN   4 ITERATIONS
ENDING Z WITH  V=     -.900
 -.9000000E+00   .1098663E+01   -.7605226E+00   .2627618E+01


CONVERGED IN   4 ITERATIONS
ENDING Z WITH  V=     -.800
 -.8000000E+00   .1197322E+01   -.6982746E+00   .2650673E+01


CONVERGED IN   4 ITERATIONS
ENDING Z WITH  V=     -.700
 -.7000000E+00   .1289902E+01   -.6343184E+00   .2671265E+01


CONVERGED IN   4 ITERATIONS
ENDING Z WITH  V=     -.600
 -.6000000E+00   .1376845E+01   -.5687676E+00   .2689310E+01


CONVERGED IN   4 ITERATIONS
ENDING Z WITH  V=     -.500
 -.5000000E+00   .1458188E+01   -.5017265E+00   .2704725E+01
```

Example 12-5 Displacement Anaysis of the RRSC Mechanism Using SPAPAC Subroutines in YCOMP. In this case the equations of constraint for the RRSC mechanism (Eqs. 4.109 and 4.110) are solved for a specified input crank motion θ. The variable THETA in YCOMP is incremented using the variable V. The second YCOMP is introduced as a means of calculating the output rotation ϕ in terms of the position of point (**b**). In this case the primary variables are the relative rotation about moving axis ($\mathbf{u}_a$) and the sliding motion component s in direction ($\mathbf{u}_b$). Eqs. 4.109 and 4.110 are included as the third and fourth YCOMP. With the use of the SPAPAC subroutines the development of the computer program is not difficult. Note that each problem variable has been defined in terms of DESIGN variables (**Z**) before writing additional statements.

This makes interpretation of the program much easier. SPAPAC subroutine argument lists are written in exactly the same order as the corresponding matrix-vector equation where appropriate. For example,

$$\text{CALL ROTATE (AJ,A0,RM,A1,A0,J)}$$

is equivalent to the solution of

$$(\mathbf{a}_j - \mathbf{a}_0) = [R_{1j}](\mathbf{a}_1 - \mathbf{a}_0)$$

The constraint equations are computed as functions that return the current residual (i.e., error). For example, the constant length and plane constraint functions become

$$\text{DLENGTH} = (\mathbf{b}_j - \mathbf{b}_0')^T(\mathbf{b}_j - \mathbf{b}_0') - (\mathbf{b}_1 - \mathbf{b}_0)^T(\mathbf{b}_1 - \mathbf{b}_0) \to 0$$

and

$$\text{PLANE} = (\mathbf{u}_b)^T(\mathbf{b} - \mathbf{b}_0') \to 0$$

```
      FUNCTION YCOMP(Z,JJ)
C.....KINEMATIC ANALYSIS OF THE RRSC SPATIAL MECHANISM. NOTE THAT THIS
C     IS BASICALLY A TWO VARIABLE PROBLEM. THE FIRST TWO YCOMP
C     EXPRESSIONS ARE CONCERNED WITH 1) INCREMENTING THE INPUT ANGLE
C     AND 2) COMPUTING THE OUTPUT ROTATION ANGLE IN TERMS OF AN
C     ASSUMED KNOWN POSITION FOR POINT BJ.
C     PRIMARY UNKNOWNS ARE THE RELATIVE ROTATION ANGLE ALPHA AND THE
C     LINEAR DISPLACEMENT S.
      DIMENSION Z(4),U0(3),UA1(3),UAJ(3),U(3),A0(3),A1(3),AJ(3),B0(3),
     1 B0P(3),B1(3),B1P(3),BJ(3),BJP(3),RM(3,3,2)
      COMMON /YCOMP/ ONCE,NOSE,V
      LOGICAL ONCE
      IF(ONCE) GO TO 10
      READ(5,1000) U0,UA1,U,A1,A0,B1,B0
 1000 FORMAT(3F10.0)
      WRITE(6,2000)  U0,UA1,U,A1,A0,B1,B0
 2000 FORMAT(//10X*U0=*3F8.5/10X*UA1=*3F8.5/10X* U=*3F8.5/10X*A1=*3E12.4
     1 /10X*A0=*3E12.4/10X*B1=*3E12.4/10X*B0=*3E12.4//
     2 4X*Z(1)= THETA*,4X*Z(2)= ALPHA*,6X*Z(3)= PHI*,8X*Z(4)= S*//)
      ONCE= .TRUE.
   10 THETA= Z(1)*.0174533
      ALPHA= Z(2)*.0174533
      PHI= Z(3)*.0174533
      S= Z(4)
      CALL RMAXIS(U0,THETA,RM,2)
```

```
      CALL  ROTVEC(UAJ,RM,UA1,2)
      CALL  ROTATE(B1P,A0,RM,B1,A0,2)
      CALL  ROTATE(AJ,A0,RM,A1,A0,2)
      CALL  RMAXIS(UAJ,ALPHA,RM,2)
      CALL  ROTATE(BJ,AJ,RM,B1P,AJ,2)
      DO 20 I= 1,3
   20 BOP(I)= BO(I)  +  S*U(I)
      GO  TO  (1,2,3,4)  JJ
    1 YCOMP= Z(1)  -  V
      RETURN
    2 YCOMP=  PHI  -  ATAN2(BJ(3),BJ(2))
      RETURN
    3 YCOMP=  DLENGTH(BJ,BOP,B1,BO)
      RETURN
    4 YCOMP=  PLANE(U,BJ,BOP)
      RETURN
      END
```

```
***************************** NEW DATA ********************************

NTEQ NSEQ NINC NFEI NITR NSET(I), I= 1,5                    V        DELV
  4    1   10    1   100    4    0    0    0    0   0.           .1000000E+02

INITIAL GUESSES FOR VARIABLES Z(I), I= 1,NTEQ
0.              0.              0.              0.

        U0= 0.          0.  1.00000
       UA1=  .60000      0.   .86670
        U= 1.00000      0.   0.
       A1= 0.               .1000E+01  0.
       A0= 0.              0.          0.
       B1=   .3000E+01      .1000E+01  0.
       B0=   .3000E+01    0.           0.

    Z(1)= THETA    Z(2)= ALPHA    Z(3)= PHI    Z(4)= S

CONVERGED IN  0 ITERATIONS
ENDING Z WITH V=    0.
0.              0.              0.              0.

CONVERGED IN  5 ITERATIONS
ENDING Z WITH V=    10.000
 .1000000E+02  -.1122800E+02  .1425828E+01  -.1737217E+00

CONVERGED IN  5 ITERATIONS
ENDING Z WITH V=    20.000
 .2000000E+02  -.2197565E+02  .5420112E+01  -.3437796E+00

CONVERGED IN  4 ITERATIONS
ENDING Z WITH V=    30.000
 .3000000E+02  -.3252237E+02  .1176346E+02  -.5085492E+00
```

```
CONVERGED IN   5 ITERATIONS
ENDING Z WITH  V=      40.000
  .4000000E+02   -.4328380E+02   .2071117E+02   -.6676587E+00

CONVERGED IN   5 ITERATIONS
ENDING Z WITH  V=      50.000
  .5000000E+02   -.5515938E+02   .3387186E+02   -.8229529E+00

NO CONVERGENCE FOR SET NUMBER 1

DAMPING LIMIT EXCEEDED AT ITERATION 9
CURRENT Z
  .6000000E+02   -.7241960E+02   .6092864E+02   -.9924226E+00

NO CONVERGENCE FOR SET NUMBER 1

DAMPING LIMIT EXCEEDED AT ITERATION 1
CURRENT Z
  .6000000E+02   -.7241960E+02   .6092864E+02   -.9924226E+00
```

Example 12-6 Spatial Three-Position Guidance. The following YCOMP coding provides for the synthesis of a variety of two-joint cranks for three position spatial rigid body guidance using SPAPAC subroutines to form the constraint equations. This code was used to generate the solutions for the R-R, S-S, C-S, or R-C links as given for Example 6-10.

```
      FUNCTION YCOMP(X,J)
C
C.....THREE FINITELY SEPARATED POSITIONS ... RIGID BODY GUIDANCE
C
      COMMON /YCOMP/ ONCE,NOSE,V
      COMMON /PRNTR/ PRNT
      DATA DAT / .FALSE. /
      LOGICAL ONCE,PRNT,DAT
      DIMENSION X(12),P12(3),P13(3),U12(3),U13(3),DM(4,4,3),RM(3,3,3)
     1 ,A0(3),U0(3),A1(3),A2(3),A3(3).U1(3),U2(3),U3(3)
      DIMENSION A2P(3),A3P(3),A02P(3).A03P(3)
      CON= ATAN(1.)/45.
      IF(DAT) GO TO 5
      WRITE(6,1)
    1 FORMAT(1H1,10X,35H THREE FINITELY SEPARATED POSITIONS,
     1 25H ... RIGID BODY GUIDENCE /)
      PRNT= .FALSE.
      READ(5,120) P12,U12,S12,PHI12,P13,U13,S13,PHI13
      WRITE(6,130)  (P12(I),I=1,3),(U12(I),I=1,3),S12,PHI12,
     1 (P13(I),I=1,3),(U13(I),I=1,3),S13,PHI13
  120 FORMAT(8F10.5)
  130 FORMAT(//10X,5(2H* ),18H DISPLACEMENT DATA ,5(2H* ) //
```

```
      2 5X*P12=*3F10.3/5X*U12=*3F10.6/5X*S12=*F10.3/5X*PHI12=*F10.3/
      3 5X*P13=*3F10.3/5X*U13=*3F10.6/5X*S13=*F10.3/5X*PHI13=*F10.3)
        PH12= PHI12*CON
        PH13= PHI13*CON
        CALL DMSCREW(P12,U12,S12,PH12,RM,DM,2)
        CALL DMSCREW(P13,U13,S13,PH13,RM,DM,3)
        DAT= .TRUE.
      5 CONTINUE
        IF(ONCE) GO TO 6
        READ(5,2) IPROB
      2 FORMAT(I2)
        IF(IPROB.LT.0) PRNT= .TRUE.
        IPROB= IABS(IPROB)
C       IF IPROB= 1 ... R-R LINK SYNTHESIS
C                 2 ... S-S
C                 3 ... C-S
C                 4 ... R-C
      6 GO TO(1000,2000,3000,4000) IPROB
C
   1000 CONTINUE
        IF(ONCE) GO TO 1002
        WRITE(6,1001) IPROB
   1001 FORMAT(//10X,7H IPROB=,I2//
      $ 51H R-R LINK SYNTHESIS - 12 VARIABLES - NONE SPECIFIED //)
        ONCE= .TRUE.
   1002 CONTINUE
        A0(1)= X(1) $ A0(2)= X(2) $ A0(3)= X(3)
        A1(1)= X(4) $ A1(2)= X(5) $ A1(3)= X(6)
        U0(1)= X(7) $ U0(2)= X(8) $ U0(3)= X(9)
        U1(1)= X(10) $ U1(2)= X(11) $ U1(3)= X(12)
        CALL DISP(A2,DM,A1,2)
        CALL DISP(A3,DM,A1,3)
        CALL ROTVEC(U2,RM,U1,2)
        CALL ROTVEC(U3,RM,U1,3)
        GO TO (11,12,13,14,15,16,17,18,19,20,21,22) J
C.....THE PLANE EQUATIONS
     11 YCOMP= PLANE(U0,A1,A0)
        RETURN
     12 YCOMP- PLANE(U0,A2,A0)
        RETURN
     13 YCOMP= PLANE(U0,A3,A0)
        RETURN
     14 YCOMP= PLANE(U1,A1,A0)
        RETURN
     15 YCOMP= PLANE(U2,A2,A0)
        RETURN
     16 YCOMP= PLANE(U3,A3,A0)
        RETURN
C.....THE DIRECTION COSINE EQUATIONS
     17 YCOMP= DIRCOS(U0)
        RETURN
     18 YCOMP= DIRCOS(U1)
        RETURN
C.....THE CONSTANT TWIST ANGLE EQUATIONS
```

```
   19 YCOMP= DTWIST(U2,U1,U0)
      RETURN
   20 YCOMP= DTWIST(U3,U1,U0)
      RETURN
C.....THE CONSTANT MOMENT EQUATIONS
   21 YCOMP= DMOMENT(U0,U1,U2,A0,A1,A2)
      RETURN
   22 YCOMP= DMOMENT(U0,U1,U3,A0,A1,A3)
      RETURN
C
 2000 CONTINUE
      IF(ONCE) GO TO 2003
      READ(5,2001) A0,A1(3)
 2003 FORMAT(8F10.3)
      WRITE(6,2002) IPROB,A0,A1(3)
 2002 FORMAT(//10X,7H IPROB=,I2//
     $ 10X,54H S-S LINK SYNTHESIS - 2 VARIABLES - A0,A1(3) SPECIFIED /
     $ 4H A0=,3F8.3,10X,7H A1(3)=,F8.3//)
      ONCE= .TRUE.
 2003 CONTINUE
      A1(1)= X(1) $ A1(2)= X(2)
      CALL DISP(A2,DM,A1,2)
      CALL DISP(A3,DM,A1,3)
      GO TO(201,202) J
C.....CONSTANT LENGTH CONSTRAINT ONLY
201   YCOMP= DLENGTH(A2,A0,A1,A0)
      RETURN
202   YCOMP= DLENGTH(A3, A0,A1,A0)
      RETURN
C
 3000 CONTINUE
      IF (ONCE) GO TO 3003
      READ(5,2001) A0,U0(1),U0(2)
      WRITE(6,3002) IPROB,A0,U0(1),U0(2)
 3002 FORMAT(//10X,7H IPROB=,I2,//,10X,
     $ 60H C-S LINK SYNTHESIS - 6 VARIABLES - A0,U0(1),U0(2) SPECIFIED
     $ /10X,4H A0=,3F8.3,2X,7H U0(1)=,F8.3,2X,7H U0(2)=,F8.3 //)
      ONCE= .TRUE.
 3003 CONTINUE
      U0(3)= X(1)
      A1(1)= X(2) $ A1(2)= X(3) $ A1(3)= X(4)
      S12= X(5) $ S13= X(6)
      CALL DISP(A2,DM,A1,2)
      CALL DISP(A3,DM,A1,3)
      DO 3001 I= 1,3
      A02P(I)= A0(I) + S12*U0(I)
 3001 A03P(I)= A0(I) + S13*U0(I)
      GO TO (301,302,303,304,305,306) J
C.....THE CONSTANT LENGTH EQUATIONS
  301 YCOMP= DLENGTH(A2,A02P,A1,A0)
      RETURN
  302 YCOMP= DLENGTH(A3,A03P,A1,A0)
      RETURN
C.....THE PLANE EQUATIONS
```

```
303     YCOMP= PLANE(U0,A1,A0)
        RETURN
304     YCOMP= PLANE(U0,A2,A02P)
        RETURN
305     YCOMP= PLANE(U0,A3,A03P)
        RETURN
C.....THE DIRECTION COSINE EQUATION
306     YCOMP= DIRCOS(U0)
        RETURN
C
 4000 CONTINUE
        IF (ONCE) GO TO 4003
        READ(5,2001) U0(1),U0(2)
        WRITE(6,4002) IPROB,U0(1),U0(2)
 4002 FORMAT(//10X,7H IPROB=,I2//
      $ 10X,58H R-C LINK SYNTHESIS - 12 VARIABLES - U0(1),U0(2) SPECIFIED
      $ /, 10X,7H U0(1)=,F8.3,2X,7H U0(2)=,F8.3 //)
        ONCE= .TRUE.
 4003 CONTINUE
        A0(1)= X(1) $ A0(2)= X(2) $ A0(3)= X(3)
        U0(3)= X(4)
        A1(1)= X(5) $ A1(2)= X(6) $ A1(3)= X(7)
        U1(1)= X(8) $ U1(2)= X(9) $ U1(3)= X(10)
        S12= X(11) $ S13= X(12)
        CALL DISP(A2,DM,A1,2)
        CALL DISP(A3,DM,A1,3)
        CALL ROTVEC(U2,RM,U1,2)
        CALL ROTVEC(U3,RM,U1,3)
        DO 4001 I= 1,3
        A2P(I)= A2(I) - S12*U2(I)
 4001 A3P(I)= A3(I) - S13*U3(I)
        GO TO (401,402,403,404,405,406,407,408,409,410,411,412) J
C.....THE PLANE EQUATIONS
401     YCOMP= PLANE(U0,A1,A0)
        RETURN
402     YCOMP= PLANE(U0,A2P,A0)
        RETURN
403     YCOMP= PLANE(U0,A3P,A0)
        RETURN
404     YCOMP= PLANE(U1,A1,A0)
        RETURN
405     YCOMP= PLANE(U2,A2P,A0)
        RETURN
406     YCOMP= PLANE(U3,A3P,A0)
        RETURN
C.....THE DIRECTION COSINE EQUATIONS
407     YCOMP= DIRCOS(U0)
        RETURN
408     YCOMP= DIRCOS(U1)
        RETURN
C.....THE CONSTANT TWIST EQUATIONS
409     YCOMP= DTWIST(U2,U1,U0)
        RETURN
410     YCOMP= DTWIST(U3,U1,U0)
```

```
      RETURN                      .
C.....THE CONSTANT MOMENT EQUATIONS
411   YCOMP= DMOMENT(U0,U1,U2,A0,A1,A2P)
      RETURN
412   YCOMP= DMOMENT(U0,U1,U3,A0,A1,A3P)
      RETURN
      END
```

********************************** NEW DATA **********************************

```
NTEG NSEQ NINC NFEI NITR NSET(I), I= 1,5                         V              DELV
 -12    1    0    0   100   12    -0    -0    -0    -0    -0.            -0.
```

INITIAL GUESSES FOR VARIABLES Z(I), I= 1,NTEQ
```
 -.5000000E+00     .2000000E+01   -.1500000E+01   .1500000E+01   0.
 -.7500000E+00    -.4000000E+00   -.6000000E+00   .7000000E+00   .5000000E+00
 0.                .1000000E+01
```

THREE FINITELY SEPARATED POSITIONS ... RIGID BODY GUIDANCE

***** DISPLACEMENT DATA *****

```
      P12=        .941        .521     -.419
      U12=     .360000     .096500   .928000
      S12=       1.384
      PHI12=   133.200
      P13=       -.610        .837      .123
      U13=     .145700    -.039000   .988600
      S13=       1.899
      PHI13=    70.620
```

SCREW MATRIX DM12

```
 -.4662298E+00   -.6179617E+00    .6331190E+00   .2466281E+01
  .7350041E+00   -.6688602E+00   -.1115741E+00   .2650848E+00
  .4924280E+00    .4132833E+00    .7661579E+00   .5076414E+01
 0.               0.               0.             .1000000E+01
```

SCREW MATRIX DM13

```
  .3460161E+00   -.9363812E+00    .5945208E-01   .6541046E+00
  .9287878E+00    .3328481E+00   -.1632059E+00   .1071321E+01
  .1330325E+00    .1116829E+00    .9848526E+00   .1867013E+01
 0.               0.               0.             .1000000E+01
```

IPROB= 1

R–R LINK SYNTHESIS – 12 VARIABLES – NONE SPECIFIED

CONVERGED IN 5 ITERATIONS
ENDING Z
```
 -.4448413E+01    .3849848E+01   -.3588110E+01   -.1060904E+01   -.3334244E+00
 -.5212973E+01   -.4008629E+00   -.5951430E+00    .6965011E+00    .4782305E+00
  .4661706E-01    .6769962E+00
```


******************************* NEW DATA ***********************************

```
NTEQ NSEQ NINC NFEI NITR NSET(I), I= 1,5                          V              DELV
  -2    1    0    0   100    2    -0    -0    -0    -0    -0.        -0.
```

INITIAL GUESSES FOR VARIABLES Z(I), I= 1,NTEQ
.1000000E+01   .1000000E+01

1PROB= 2

S–S LINK SYNTHESIS – 2 VARIABLES – A0,A1(3) SPECIFIED
A0=   1.000   0.   0.   A1(3)=   0.

CONVERGED IN 3 ITERATIONS
ENDING Z
 -.9069445E+00   .5569138E-01

******************************* NEW DATA ***********************************

```
NTEQ NSEQ NINC NFEI NITR NSET(I), I= 1,5                          V              DELV
  -6    1    0    0   100    6    -0    -0    -0    -0    -0.        -0.
```

INITIAL GUESSES FOR VARIABLES Z(I), I= 1,NTEQ
```
.1000000E+01    .1000000E+01    .1000000E+01    0.                      .1000000E+01
.2000000E+01
```

IPROB= 3

C-S LINK SYNTHESIS - 6 VARIABLES - A0,U0(1),U0(2) SPECIFIED
A0=   0.   0.   0.   U0(1)=   -.935   U0(2)=   .041

CONVERGED IN 8 ITERATIONS
ENDING Z
   .3522581E+00   .4179008E+00   .2937692E+01   .7664780E+00   -.5505268E-01
   .2936367E+01

*********************************** NEW DATA ***********************************

NTEQ NSEQ NINC NFEI NITR NSET(I), I= 1,5   V   DELV
  -12   1   0   0   100   12   -0   -0   -0   -0   -0.   -0.

INITIAL GUESSES FOR VARIABLES Z(I), I= 1,NTEQ
  -.3000000E+00   .4000000E+02   -.6000000E+00   .4000000E+00   .4000000E+01
   .1000000E+01   -.2000000E+01   .5000000E+00   -.3000000E+00   .8000000E+00
   .1000000E+01   .2000000E+01

IPROB= 4

R-C LINK SYNTHESIS - 12 VARIABLES - U0(1),U0(2) SPECIFIED
U0(1)= -.935 U0(2)= .041

CONVERGED IN 5 ITERATIONS
ENDING Z
   .5937989E+00   .4318059E+02   -.6297499E+01   .3522581E+00   -.5086994E+01
   .6716943E+01   -.1712162E+02   .4836969E+00   -.3174733E+00   .8156274E+00
  -.3312619E+02   -.2059514E+02

12.4   PCON—CONSTRAINED MINIMIZATION USING POWELL'S DIRECT SEARCH-CONJUGATE DIRECTIONS METHOD AND THE SUMT PROCEDURE

Program PCON is a driver for subroutine POWELL, which carries out the direct search procedure. The nonderivative search is based on quadratic extrapolation, and partial derivatives are not required.

Subroutine RESTNT forms the modified objective function with penalty functions by calling a user-supplied function FFUN. The penalty functions are formed by functions GIN (internal penalty), GEX (external penalty), or H (equality penalty) which, in turn, call user-supplied subroutines GFUN and HFUN. GFUN returns the current values for inequality constraint functions $g_i(\mathbf{X})$ $i = 1$, mg and HFUN returns the equality constraint functions $h_j(\mathbf{X})$ $j = 1$, mh.

PCON automatically normalizes the constraint functions and sets a predetermined ratio of penalty to objective function at the beginning of the problem. This procedure is designed to prevent the dominance of an individual or group of constraint functions over the objective function at the start of the search. These ratios GFAC and HFAC are supplied by the user. The rate of change for the sequential multiplier r in the SUMT procedure is controlled by a user supplied factor FAC. The value of r is calculated by one of the following expressions, with IR automatically incremented by 1 after each successful subproblem convergence.

For an internal inequality penalty function,

$$r = FAC ** (1 - IR)$$

For an external inequality penalty or the equality penalty function,

$$r = FAC ** (IR - 1)$$

For normal execution a value of FAC = 10. is suggested, which results in either a sequence

$$r = 1., .1, .01, .001$$

or

$$r = 1., 10., 100., 1000$$

depending on the type of penalty function.

Additional weighting factors WTG, and WTH are computed in PCON to form the desired starting ratios

$$GFAC = WTG * GEX(X)/FFUN(X)$$

$$HFAC = WTH * H(X)/FFUN(X)$$

See subroutine RESTNT for additional detail.

Example 12-7 The Rosenbrock Test Function With Added Inequality Constraints. The Rosenbrock test function exhibits a long narrow curved valley and is widely used as an example that causes difficulty with steepest descent search techniques. The problem has been further complicated in this example by the added inequality constraints included in subroutine GFUN, which force variable x_1 to lie within the range

$$0.0 \leq x_1 \leq 0.75$$

```
      FUNCTION FFUN(X)
      DIMENSION X(2)
C
C.....ROSENBROCK TEST FUNCTION WITH ADDED INEQUALITY CONSTRAINT
C
      FFUN= 100.*(X(2)-X(1)**2)**2 + (1.-X(1))**2
      RETURN
      END
C

      SUBROUTINE GFUN(G,X)
      DIMENSION G(1),X(2)
      G(1)= - X(1)
      G(2)= X(1) - .75
      RETURN
      END
C
```

****************************** NEW DATA ******************************

N	MAXIT	ISW	IEPS	IPRNT	MG	MH	NRUN	ISRCH	REUSE
2	100	0	0	0	2	0	10	0	0

ESCALE	EPS1	FAC	GFAC	HFAC
.1000E+04	.1000E-03	.1000E+02	.1000E+01	.1000E+01

INITIAL GUESSES (X(I),I=1,N)

  .200000E+01   .100000E+01

*********************** START PCON WITH IR= 1 ***********************

*********************** INITIAL CONDITIONS ***********************

F-INITIAL=   .901000E+03

VARIABLES ARE
 .200000E+01   .100000E+01

INEQUALITY CONSTRAINTS
 −.200000E+01   .125000E+01

INITIAL GUESSES ARE NOT FEASIBLE
BEGIN SEARCH FOR A FEASIBLE BASE POINT

  NO UNIQUE MINIMUM
OBJECTIVE FUNCTION MAY BE INDEPENDENT OF AT LEAST ONE VARIABLE

      TIME USED FOR THIS MINIMIZATION .04000 SECONDS

************************* INITIAL CONDITIONS **************************

F-INITIAL= .375710E+02

VARIABLES ARE
 .740000E+00   .116000E+01

INEQUALITY CONSTRAINTS
 −.740000E+00   −.100000E−01

INITIAL GUESSES ARE FEASIBLE
BEGIN SEARCH WITH INTERIOR PENALTY FUNCTION

IR= 1 ITERATIONS= 4 FUNC EVAL= 50 SUBTOTAL= 50
U-INITIAL=   .7514195E+02   U-FINAL=   .2320220E+02
F-INITIAL=   .3757098E+02   F-FINAL=   .1071721E+00
  G-INT= −.1229408E+01   G-EXT=   .4619096E−10   H= 0.

VARIABLES
 .672629E+00   .452429E+00

INEQUALITY CONSTRAINTS
 −.672629E+00   −.773715E−01

      TIME USED FOR THIS MINIMIZATION .07300 SECONDS

********************* START PCON WITH IR= 4 ************************

IR= 4 ITERATIONS= 0 FUNC EVAL= 16 SUBTOTAL= 135
U-INITIAL=   .6484860E−01   U-FINAL=   .6484860E−01
F-INITIAL=   .6478382E−01   F-FINAL=   .6478382E−01
  G-INT= −.2000000E+01   G-EXT=   .1462006E+01   H= 0.

VARIABLES
.745473E+00   .555730E+00

INEQUALITY CONSTRAINTS
−.745473E+00   −.452666E−02

TIME USED FOR THIS MINIMIZATION .02000 SECONDS

******************** CONSTRAINED MINIMUM FOUND ********************

RATIO CHANGE/CURRENT LESS THAN .100E−03 FOR ALL X(I)

TOTAL FUNC EVAL = 135

TOTAL TIME USED .21700 SECONDS

Example 12-8 Optimal Synthesis of a Combined Path Generation and Plane Rigid Body Guidance Mechanism (Example 9-2). The data of Example 9-2 has been used but with addition of our inequality constraints such that

$$0.5 \le a_{1x} \le 1.0$$

$$0.75 \le a_{1y} \le 1.0$$

and the equality constraint that specifies that

$$b_{1x} = b_{1y}$$

An option in PCON is the use of a subroutine MORPRNT with which the user can organize calculations or special output format at the end of normal execution. An example of the use of MORPRNT for the analysis of the optimal solution is included. Note the use of LINKPAC subroutines for the error analysis in this case.

```
      FUNCTION FFUN(X)
      DIMENSION X(10),CTH(9),STH(9),CA(9),SA(9)
      COMMON /PATH/ PX(9),PY(9),QX(9),QY(9),ALPHA(9),THETA(9)
C
C.....OPTIMAL SYNTHESIS OF A COMBINED RIGID BODY GUIDANCE AND
C     PATH GENERATION LINKAGE ... EXAMPLE 9.2
C
```

```
      LOGICAL ONCE
      DATA ONCE /.FALSE./
      IF(ONCE) GO TO 10
      DATA THETA / 0.0,-40.,-80.,-120.,-160.,-200.,-240.,-280.,-320. /
      DATA ALPHA(1),ALPHA(2),ALPHA(3),ALPHA(4),ALPHA(5)/
     1 0.0,12.,21.,22.,15./
      DATA A0X,A0Y,B0X,B0Y/0.0,0.0,4.0,0.0/
      DATA QX / 1.00,1.00,.60,.10.-.30,-.50.-.40,-.10,.60 /
      DATA QY / 2.60,2.10,1.50,1.00,1.10,1.50,2.00,2.50,2.88 /
      CON= ATAN(1.)/45.
      DO 5 J= 1,9
      CTH(J)= COS(THETA(J)*CON)
      STH(J)= SIN(THETA(J)*CON)
    5 CONTINUE
      ONCE= .TRUE.
   10 A1X= X(1) $ A1Y= X(2) $ B1X= X(3) $ B1Y= X(4)
      P1X= X(5) $ P1Y= X(6)
      ALPHA(6)= X(7) $ ALPHA(7)= X(8) $ ALPHA(8)= X(9) $ ALPHA(9)= X(10)
      FFUN= 0.0
      DO 15 J= 1,9
      CA= COS(ALPHA(J)*CON) $ SA= SIN(ALPHA(J)*CON)
      AX= (A1X-A0X)*CTH(J) - (A1Y-A0Y)*STH(J) + A0X
      AY= (A1X-A0X)*STH(J) + (A1Y-A0Y)*CTH(J) + A0Y
      BX= (B1X-A1X)*CA - (B1Y-A1Y)*SA + AX
      BY= (B1X-A1X)*SA + (B1Y-A1Y)*CA + AY
      PX(J)= (P1X-A1X)*CA - (P1Y-A1Y)*SA + AX
      PY(J)= (P1X-A1X)*SA + (P1Y-A1Y)*CA + AY
      F1= (QX(J)-PX(J))*(QX(J)-PX(J)) + (QY(J)-PY(J))*(QY(J)-PY(J))
      F2= (BX-B0X)*(BX-B0X) + (BY-B0Y)*(BY-B0Y) - (B1X-B0X)*(B1X-B0X)
     1 - (B1Y-B0Y)*(B1Y-B0Y)
   15 FFUN= FFUN + F1**2 + F2**2
      RETURN
      END
C

      SUBROUTINE GFUN(G,X)
      DIMENSION G(1),X(1)
C.....RESTRICT LOCATION OF MOVING PIVOT A SUCH THAT
C     (0.5).LE.A1X.LE.(1.0) AND (0.75).LE.A1Y.LE.(1.0)
      G(1)= 0.5 - X(1)
      G(2)= X(1) - 1.0
      G(3)- 0.75 - X(2)
      G(4)= X(2) - 1.0
      RETURN
      END


      SUBROUTINE HFUN(H,X)
      DIMENSION H(1),X(1)
C.....RESTRICT LOCATION OF MOVING PIVOT B SUCH THAT B1X= B1Y
      H(1)= X(3) - X(4)
      RETURN
      END
C
```

```
      FUNCTION RMAG(N1,N2,P)
C.....COMPUTES DISTANCE BETWEEN TWO POINTS N1 AND N2
      DIMENSION P(30,2)
      RMAG= SQRT((P(N1,1)-P(N2,1))**2 + (P(N1,2)-P(N2,2))**2)
      RETURN
      END
C


      SUBROUTINE MORPRINT(X,U,F,G,H)
      DIMENSION X(10),G(1),H(1)
      DIMENSION P1(30,2),P(30,2),ERROR(3,9)
      COMMON /PATH/ PX(9),PY(9),QX(9),QY(9),ALPHA(9),THETA(9)
      WRITE(6,50) (X(I), I=1,6)
   50 FORMAT(1H1,10X,*OPTIMAL SYNTHESIS OF A COMBINED PATH GENERATION*/
     1 11X,*AND RIGID BODY GUIDANCE MECHANISM*///
     2 * A0= (      0.0,     0.0)*10X*B0= (     4.0,     0.0)*
     3 /* A1= (*F10.5*,*F10.5*)*10X*B1= (*F10.5*,*F10.5*)*/
     4 * P1= (*F10.5*,*F10.5*)*//
     5 5X*SPECIFIED PATH POINTS*11X*PATH ERROR*6X*ANGULAR
     6 ERROR*/
     7 55X*RADIANS*)
      PI(1,1)= P(1,1)= 0.0 $ PI(1,2)= P(1,2)= 0.0
      PI(4,1)= P(4,1)= 4.0 $ PI(4,2)= P(4,2)= 0.0
      PI(2,1)= P(2,1)= X(1) $ PI(2,2)= P(2,2)= X(2)
      PI(3,1)= P(3,1)= X(3) $ PI(3,2)= P(3,2)= X(4)
      PI(5,1)= P(5,1)= X(5) $ PI(5,2)= P(5,2)= X(6)
      R2= RMAG(1,2,P) $ R3= RMiG(2,3,P) $ R4= RMAG(3,4,P)
      R5= RMAG(2,5P) $ R6= RMAG(3,5,P)
      DO 200 I= 1,9
      TH2= THETA(I)*.0174533
      CALL DISP(1,2,TH2,P,P1)
      CALL PDYAD(1,2,4,3,R3,R4,TH3,TH4,P)
      CALL PDYAD(1,2,3,5,R5,R6,TH5,TH6,P)
      ERROR(1,I)= QX(I)-P(5,1)
      ERROR(2,I)= QY(I)-P(5,2)
      ALPHA(I)= ALPHA(I)*.0174533
  200 ERROR(3,1)=ALPHA(I)-TH3+ATAN2((PI(3,2)-PI(2,2)),(PI(3,1)-PI(2,1)))
      WRITE(6,100) ((QX(I),QY(I),ERROR(1,I),ERROR(2,I),ERROR(3,I)), I=1,9)
  100 FORMAT(5F12.5)
      RETURN
      END
```

```
*****************************  NEW DATA  *****************************
```

N	MAXIT	ISW	IEPS	IPRNT	MG	MH	NRUN	ISRCH	REUSE
-10	100	0	0	0	4	1	-10	0	0

ESCALE	EPSI	FAC	GFAC	HFAC
.1000E+04	.1000E-02	.1000E+02	.5000E+01	.5000E+01

INITIAL GUESSES (X(I),I=1,N)

```
 .550000E+00  .700000E+00   .260000E+01   .470000E+01   .100000E+01
 .250000E+01  .660000E+01  -.120000E+01  -.600000E+01  -.600000E+01
      VARIABLE  SCALING  OPTION  ADOPTED
```

*********************** START PCON WITH IR= 1 ***********************

*********************** INITIAL CONDITIONS ************************

F-INITIAL=   .257487E+00

VARIABLES ARE
```
 .550000E+00  .700000E+00   .260000E+01   .470000E+01   .100000E+01
 .250000E+01  .660000E+01  -.120000E+01  -.600000E+01  -.600000E+01
```

INEQUALITY CONSTRAINTS
```
 -.500000E-01  -.450000E+00  .500000E-01  -.300000E+00
```

EQUALITY CONSTRAINTS
```
 -.210000E+01
```

INITIAL GUESSES ARE NOT FEASIBLE
BEGIN SEARCH FOR A FEASIBLE BASE POINT

   NO UNIQUE MINIMUM
OBJECTIVE FUNCTION MAY BE INDEPENDENT OF AT LEAST ONE VARIABLE

        TIME USED FOR THIS MINIMIZATION .24300 SECONDS

*********************** START PCON WITH IR= 1 ***********************

*********************** INITIAL CONDITIONS ************************

Γ-INITIAL=   .679055E+03

VARIABLES ARE
```
 .605000E+00   .850000E+00   .676000E+01   .470000E+01   .100000E+01
 .250000E+01   .660000E+01  -.120000E+01  -.600000E+01  -.600000E+01
```

INEQUALITY CONSTRAINTS
```
 -.105000E+00  -.395000E+00  -.100000E+00  -.150000E+00
```

EQUALITY CONSTRAINTS
```
 .206000E+01
```

INITIAL GUESSES ARE FEASIBLE
BEGIN SEARCH WITH INTERIOR PENALTY FUNCTION

```
*********************** START PCON WITH IR= 5 ***********************


IR=  5  ITERATIONS=   0  FUNC EVAL=   54  SUBTOTAL=   1028
U-INITIAL=    .7150026E+00   U-FINAL=    .7150026E+00
F-INITIAL=    .7014215E+00   F-FINAL=    .7014215E+00
   G-INT=  -.4000000E+01    G-EXT=  0.   H=  .5048710E-20
```

VARIABLES
```
   .523089E+00    .756361E+00    .283500E+01    .283500E+01    .909253E+00
   .265020E+01    .336532E+01   -.592871E+01   -.976761E+01   -.772976E+01
```

INEQUALITY CONSTRAINTS
```
  -.230892E-01   -.476911E+00   -.636076E-02   -.243639E+00
```

EQUALITY CONSTRAINTS
```
   .710543E-13
```

TIME USED FOR THIS MINIMIZATION .41600 SECONDS


```
******************** CONSTRAINED MINIMUM FOUND ********************
```

RATIO CHANGE/CURRENT LESS THAN .100E-02 FOR ALL X(I)

TOTAL FUNC EVAL = 1028

TOTAL TIME USED 8.44100 SECONDS

OPTIMAL SYNTHESIS OF A COMBINED PATH GENERATION
AND RIGID BODY GUIDANCE MECHANISM


```
A0= (     0.0,      0.0)   B0= (     4.0,     .0.)
A1= (  .52309,   .75636)   B1= ( 2.83500,  2.83500)
P1= (  .90925,  2.65020)
```

SPECIFIED PATH POINTS		PATH ERROR		ANGULAR ERROR RADIANS
1.00000	2.60000	.09075	-.05020	-.00000
1.00000	2.10000	.14600	-.07570	-.00872
.60000	1.50000	.15543	-.00902	-.03840
.10000	1.00000	.11237	-.05853	-.02873
-.30000	1.10000	.05082	.06045	-.00042
-.50000	1.50000	-.02411	.11860	.00001
-.40000	2.00000	-.06316	.08135	.00001
-.10000	2.50000	-.14790	.05267	.00004
.60000	2.80000	.04809	.05966	-.00000

12.5 LSTCON—CONSTRAINED MINIMIZATION USING POWELL'S LEAST SQUARES METHOD

Program LSTCON is a driver for subroutine POWEL, which carries out the nonderivative least squares search. LSTCON forms the modified objective function in a manner similar to program PCON. See the instructions in the program listing for additional detail.

Example 12-9 Solution of Three Nonlinear Algebraic Equations with Constraints. Solve the system of equations

$$x_1^2 + x_2^2 + x_3^2 = 1.0$$

$$x_1 - x_2^2 = 0.0$$

$$x_1 + x_2 + x_3 = 1.0$$

with the regional (inequality) constraints

$$x_1 > 0.5$$

$$x_2 > 0.7$$

$$x_3 > 0.9$$

and the functional (equality) constraint

$$x_1 = x_2$$

```
SUBROUTINE FFUN(F,X)
DIMENSION F(1),X(1)
F(1)= X(1)**2 + X(2)**2 + X(3)**2 - 1.0
F(2)= X(1) - X(2)**2
F(3)= X(1) + X(2) + X(3) - 1.0
RETURN
END


SUBROUTINE GFUN(G,X)
DIMENSION G(1),X(1)
G(1)= .5 - X(1)
G(2)= .7 - X(2)
G(3)= .9 - X(3)
RETURN
END


SUBROUTINE HFUN(H,X)
DIMENSION H(1),X(1)
H(1)= X(1) - X(2)
RETURN
END
```

```
  DATA CARD ONE
N MAXTRY MAXFUN    MF   MG   MH   ISRCH  IPRINT
3    20    1000    3    3    1      0       0
  DATA CARD TWO
  ESCALE      TOL      FAC      GFAC      HFAC
.1000E+03 .1000E-02 .1000E+03 .1000E+01 .1000E+01
  DATA CARD(S) THREE PLUS
  INITIAL GUESSES X(I), I= 1,N
.10000E+01     .10000E+01      .10000E+01
```

          .....BEGIN LEAST SQUARES MINIMIZATION.....

INEQUALITY WEIGHTING FACTORS

```
     WG( 1)=      .200E+01
     WG( 2)=      .333E+01
     WG( 3)=      .100E+02
     WTG=      .229E+02
```

EQUALITY WEIGHTING FACTORS

```
     WH( 1)=      .100E+07
     WTH=              R
```

              ...SEARCH FOR A FEASIBLE POINT...

****************** SUBMINIMUM FOUND FOR IR= 1 ********************
ITERATIONS= 3 FUNC EVAL= 16 SUM SQS OF RESIDUALS= .667E-24

```
  VARIABLES ARE
.60000E+00   .80000E+00   .10000E+01

  FUNCTIONS
.67909E-12   .28193E-12   .35548E-12

  INEQUALITY CONSTRAINTS
.12301E-12   .30642E-13   .12879E-13

  EQUALITY CONSTRAINTS
.10000E-05
```

            ...SEARCH FOR A FEASIBLE POINT COMPLETED...

              ...INTERIOR PENALTY FUNCTION...

****************** SUBMINIMUM FOUND FOR IR= 1 ********************
ITERATIONS= 12 FUNC EVAL= 104 SUM SQS OF RESIDUALS= .259E+04

VARIABLES ARE
.31056E+01   .31066E+01   .33970E+01

FUNCTIONS
.29835E+02   −.65454E+01   .86092E+01

INEQUALITY CONSTRAINTS
−.26056E+01   −.24066E+01   −.24970E+01

EQUALITY CONSTRAINTS
−.10281E−02

****************** SUBMINIMUM FOUND FOR IR= 8 ************************
ITERATIONS=   4   FUNC EVAL=   92   SUM SQS OF RESIDUALS= .239E+01

VARIABLES ARE
.70297E+00   .70297E+00   .90020E+00

FUNCTIONS
.79870E+00   .20880E+00   .13061E+01

INEQUALITY CONSTRAINTS
−.20297E+00   −.29750E−02   −.19836E−03

EQUALITY CONSTRAINTS
.42633E−13

***************** CONSTRAINED MINIMUM FOUND ***********************

RATIO DELX/X LESS THAN E(I) FOR ALL VARIABLES X(I)

TOTAL ITERATIONS= 65
TOTAL FUNC EVAL= 694
SUM OF THE SQUARES OF THE RESIDUALS= .23884E+01
FINAL VALUES FOR THE VARIABLES
.70297E+00   .70297E+00   .90020E+00

FUNCTIONS
.79870E+00   .20880E+00   .13061E+01

INEQUALITY CONSTRAINTS
−.20297E+00   −.29750E−02   −.19836E−03

EQUALITY CONSTRAINTS
.42633E−13

12.6 NONLIN—SOLUTION OF SIMULTANEOUS NONLINEAR ALGEBRAIC EQUATIONS WITH RANDOM NUMBER INITIAL GUESSES

Program NONLIN is a driver for subroutine SIMEQ, another somewhat simpler version of SIMEQS as used in DESIGN. SIMEQ is useful separately when the user wishes to write another main program to control input and output for special problems.

A specified number of initial guesses are generated randomly for each variable Z(J) within a range given by specified values ZMIN(J) and ZMAX(J). Each new initial guess vector is supplied to subroutine SIMEQ. A new solution that differs from a previously found solution by an assigned tolerance is printed and stored. A message giving the total number of solutions found is printed at the end of execution.

The equations to be solved are contained in a function YCOMP and called in a manner identical to that used in DESIGN with the exception that variable V is not incremented and must be defined. This makes program NONLIN useful in situations where it is difficult to estimate initial guesses intuitively.

Example 12-10 Solution of a Set of Nonlinear Algebraic Equations with Random Initial Guesses. The set of equations to be solved is

$$x_1 + 2x_2 - x_3 = 2$$

$$x_1^2 - x_2^2 + x_3^2 = 6$$

$$x_1 + x_2^2 + x_3 = 8$$

with one obvious solution at $x_1 = 1$, $x_2 = 2$, and $x_3 = 3$. Random initial guesses were generated over the range

$$-10 \leq x_1 \leq 10$$

$$-10 \leq x_2 \leq 10$$

$$-10 \leq x_3 \leq 10$$

giving a second solution at $x_1 = -4.13$, $x_2 = 3.38$, and $x_3 = 0.65$.

```
FUNCTION YCOMP(X,J)
DIMENSION X(1)
GO TO(1,2,3) J
1 YCOMP= X(1) + 2.*X(2) - X(3) - 2.
RETURN
```

```
2 YCOMP=  X(1)**2 - X(2)**2 + X(3)**2 - 6.
  RETURN
3 YCOMP=  X(1) + X(2)**2 + X(3) - 8.
  RETURN
  END
```

```
  ...SOLUTION NUMBER 1...
  INITIAL GUESSES
   -.86431E+01  -.20265E+01    .81176E+01

  SOLUTION
    .10000E+01   .20000E+01    .30000E+01

   ...SOLUTION NUMBER 2...
  INITIAL GUESSES
   -.64398E+01   .62150E+01    .75283E+01

  SOLUTION
   -.41312E+01   .33889E+01    .64662E+00


ITERATIONS EXCEEDED 10 PLUS NUMBER OF EQUATIONS

CURRENT Z
  .47655E+01  -.29236E+01  -.30816E+01


DAMPING EXCEEDS 10 AT ITERATION 11

CURRENT Z
  .44510E+01  -.25596E+01  -.26683E+01

ITERATIONS EXCEEDED 10 PLUS NUMBER OF EQUATIONS

CURRENT Z
  .48816E+01  -.30607E+01  -.32397E+01


DAMPING EXCEEDS 10 AT ITERATION 6

CURRENT Z
  .44720E+01  -.25849E+01  -.26977E+01


ITERATIONS EXCEEDED 10 PLUS NUMBER OF EQUATIONS

CURRENT Z
  .32497E+01  -.25158E+01  -.21112E+01


2 SOLUTIONS FOUND FOR 10 INITIAL GUESSES
```

12.7   PROGRAM IN BASIC LANGUAGE FOR THE SOLUTION OF SIMULTANEOUS NONLINEAR AND LINEAR EQUATIONS

In cases where it is desired to solve constraint equations using the interactive BASIC programming language, program SIMEQ may be helpful.

Normal execution is initiated by the BASIC command RUN. The program will then ask for data and proceed to solve the set of equations contained in statements beginning with 1000. The example given solves the set of three equations from Example 12.3.

A second use would be for the solution of a set of linear equations in an interactive mode. A second program that begins at statement 6000 is included. Type GO TO 6000 to start the linear equation solver. Sample teletype output for both nonlinear and linear equations are given below. Note that damping is included in the nonlinear equation solver, which is similar to SIMEQS.

Examples of use of BASIC programs:

```
RUN
NUMBER OF VARIABLES? 3
CONVERGENCE TOLERANCE? .0001        1? 1? 1
ITERATION LIMIT? 100                VALUE OF NORM
INITIAL GUESSES                      4
? 10? 10? 10                         2.6405E+14
VALUE OF NORM                       DAMPING   1
 418                                 61.8746
 2.92906E+16                        DAMPING   1
DAMPING   1                          1.5578
 274.708                             .355368
 70.0944                             2.73836E-2
 19.1829                             2.17966E-4
 12.3688                             1.23037E-8
 2.9663                             CONVERGED IN 7 ITERATIONS
 6.97815
DAMPING   1                         ANSWERS ARE
 .636866                            -1.23036E-8
 .334678                             2.21742E-7
 2.76053E-2                          1
 3.23935E-4                         INITIAL GUESSES
 4.65664E-10                         ?
 4.65209E-13
CONVERGED IN 13 ITERATIONS

ANSWERS ARE
-4.62685E-13
 5.02337E-8
 1
INITIAL GUESSES
?
```

0? 1? 0   GO TO 6000
VALUE OF NORM   NUMBER OF EQUATIONS? 5
  1   COEFFICIENT MATRIX BY ROWS IS
  1.99986   ? 1? 0? 3? 0? 5
DAMPING   1   ? 2? 3? 0? 0? 3
DAMPING   2   ? 1? 2? 3? 4? 5
  .326073   ? 0? 4? 0? 0? 1
  2.65414E−2   ? 1? 2? 0? 4? 3
  2.38895E−4   RIGHT SIDE VECTOR
  1.19209E−7   ? 1? 2? 3? 4? 5
CONVERGED IN 5 ITERATIONS   ANSWERS ARE
                                   2.6
ANSWERS ARE   .533335
  .56984   −1.91111
  .754878   .233333
  −.324718   1.86667
INITIAL GUESSES
?   NUMBER OF EQUATIONS?

12.8   PRCR2, A PREDICTOR-CORRECTOR METHOD FOR THE SOLUTION OF A SECOND-ORDER NONLINEAR DIFFERENTIAL EQUATION

The FORTRAN coding follows the algorithm outlined in Section 11-4 and is listed in Appendix 8.

Example 12.11 Function DERIV2 for the inverse dynamics problem, Example 11.4.

```
      FUNCTION DERIV2(T,PHI,PHIDOT)
      LOGICAL ONCE,EOF
      DATA ONCE /.FALSE./
      COMMON /MOR/ TSTAR, ISTAR, DIDPHI
      COMMON /FRBR/ M,R2,S2,PH2,R3,S3,PH3,R4,S4,PH4,P(30,2),VP(30,2),
     $ AP(30,2),PI(30,2),TH2,TH3,TH4,W2,W3,W4,A2,A3,A4
      REAL M2,M3,M4,K2,K3,K4,I2,I3,I4,ISTAR,ISTARP
C READ LINKAGE GEOMETRY AND MODE OF ASSEMBLY
      IF(ONCE) GO TO 10
      READ(5,100) R2,S2,P2,R3,S3,P3,R4,S4,P4,M
  100 FORMAT(3(3F10.0/),I2)
      IF(EOF(5)) STOP
      CON= ATAN(1.)/45.
      PH2= P2*CON $ PH3= P3*CON $ PH4= P4*CON
C READ FIXED PIVOT COORDINATES
      READ(5,200) P(1,1),P(1,2),P(4,1),P(4,2)
  200 FORMAT(4F10.0)
```

```
      VP(1,1)= VP(1,2)= VP(4,1)= VP(4,2)= 0.0
      AP(1,1)= AP(1,2)= AP(4,1)= AP(4,2)= 0.0
C READ PHYSICAL CONSTANTS FOR THE LINKS
      READ(5,300) M2,M3,M4,K2,K3,K4
  300 FORMAT(3F10.0)
      I2= M2*K2**2 $ I3= M3*K3**2 $ I4= M4*K4**2
C READ INPUT AND OUTPUT CONSTANT TORQUE VALUES
      READ(5,300) T2,T4
      ONCE= .TRUE.
   10 CONTINUE
      TH2= PHI
C VEL RATIOS ARE REQD. SET W2 GT 0.0 TO AVOID DIV BY ZERO.
      W2= PHIDOT
      IF(ABS(W2).LT.1.E-6) W2= 1.E-6
      A2= 0.0
C NOTE THAT FRBRAN IS USED FOR CONVENIENCE ONLY-VELOCITY ANALYSIS
C ONLY IS REQUIRED-COMPUTE THE REDUCED TORQUE TSTAR
      CALL FRBRAN
      TSTAR= T2 + T4*W4/W2
C COMPUTE THE REDUCED INERTIA ISTAR
      ISTAR= (M2*(VP(5,1)**2+VP(5,2)**2) + M3*(VP(6,1)**2+VP(6,2)**2)
     $ + M4*(VP(7,1)**2+VP(7,2)**2))/W2**2 + I2 + I3*(W3/W2)**2
     $ + I4*(W4/W2)**2
C COMPUTE RATE OF CHANGE DISTAR/DPHI
      TH2= TH2 + .00001
      CALL FRBRAN
      ISTARP=(M2*(VP(5,1)**2+VP(5,2)**2) + M3*(VP(6,1)**2+VP(6,2)**2)
     $ + M4*(VP(7,1)**2+VP(7,2)**2))/W2**2 + I2 + I3*(W3/W2)**2
     $ + I4*(W4/W2)**2
      DIDPHI= (ISTARP - ISTAR) / .00001
C COMPUTE DERIV2= PHI DOUBLE DOT
      DERIV2= (TSTAR - 0.5*DIDPHI*W2**2)/ISTAR
      RETURN
      END
```

PREDICTOR-CORRECTOR SOLUTION FOR
A GENERAL NON-LINEAR SECOND-ORDER
DIFFERENTIAL EQUATION D/DX(DY/DX)=F(X,Y,DY/DX)

XO= 0.   YO= 0.   (DY/DX)= 0.

DELX= .100E-03 IPRINT= 10 XMAX= .100E-01 TEST= .100E-06

..........SOLUTION..........

X	Y	DY	DDY	
0.	0.	0.	.13870E+05	
.10000E-02	.69680E-02	.14000E+02	.14261E+05	2
.20000E-02	.28273E-01	.28822E+02	.15554E+05	2
.30000E-02	.65242E-01	.45544E+02	.18161E+05	2

.40000E−02	.12055E+00	.65862E+02	.22930E+05	2
.50000E−02	.19908E+00	.92533E+02	.31019E+05	2
.60000E−02	.30881E+00	.12850E+03	.40277E+05	2
.70000E−02	.45730E+00	.16708E+03	.31498E+05	2
.80000E−02	.63526E+00	.18407E+03	.35108E+04	2
.90000E−02	.81868E+00	.18126E+03	−.59495E+04	2
.10000E−01	.99700E+00	.17563E+03	−.46508E+04	2
.10100E−01	.10145E+01	.17518E+03	−.43783E+04	2

Appendixes

Appendix 1. LINKPAC

Appendix 2. SPAPAC

Appendix 3. DESIGN

Appendix 4. NONLIN

Appendix 5. PCON

Appendix 6. LSTCON

Appendix 7. SIMEQ

Appendix 8. PRCR2

Appendix 1.
LINKPAC

LINKPAC KINEMATIC ANALYSIS SUBROUTINES

1. CRANK(N1, N2, R, THETA, W, A, P, VP, AP)

2. CRANK2(N1, N2, N3, R, S, PHI, THETA, W, A, P, VP, AP)

3. MOTION(N1, N2, N3, R, S, PHI, THETA, W, A, PI, P, VP, AP)

4. OFFSLD
   COMMON/OFF/M, R2, S2, PH2, R3, S3, PH3, P(30, 2), VP(30, 2), AP(30, 2),
      TH2, TH3, W2, W3, A2, A3, R4, VR4, AR4, BETA, VBETA, ABETA

5. FRBRAN
   COMMON/FRBR/M, R2, S2, PH2, R3, S3, PH3, R4, S4, PH4, P(30, 2),
      VP(30, 2), AP(30, 2), PI(30, 2), TH2, TH3, TH4, W2, W3, A2, A3, A4

6. OSCSLD
   COMMON/OSC/M, R2, S2, PH2, R3, S3, PH3, R4, S4, PH4, P(30, 2),
      VP(30, 2), AP(30, 2), PI(30, 2), TH2, TH3, TH4, W2, W3, W4, A2, A3, A4, E

7. GEOM(N1, N2, N3, R, S, PHI, PI)

8. POS(N1, N2, N3, R, S, PHI, THETA, P)

9. DISP(N1, N2, THETA, P, PI)

10. VEL(N1, N2, W, P, VP)

11. ACC(N1, N2, W, A, P, VP, AP)

12. PDYAD(M, N1, N2, N3, R1, R2, TH1, TH2, P)

13. VDYAD(M, N1, N2, N3, R1, R2, TH1, TH2, P, W1, W2, VP)

14. ADYAD(M, N1, N2, N3, R1, R2, TH1, TH2, P, W1, W2, VP, A1, A2, AP)

15. POSC(M, N1, N2, N3, E, R2, R3, THETA, P)

16. VOSC(M, N1, N2, N3, E, R2, R3, THETA, P, W2, VR2, VP)

17. AOSC(M, N1, N2, N3, E, R2, R3, THETA, P, W2, VR2, VP, A2, AR2, AP)

18. PGUIDE(M, N1, N2, N3, R1, R2, TH1, BETA, P)

19. VGUIDE(M, N1, N2, N3, R1, R2, TH1, BETA, P, W1, VBETA, VR2, VP)

20. AGUIDE(M, N1, N2, N3, R1, R2, TH1, BETA, P, W1, VBETA, VR2, VP,
    A1, ABETA, AR2, AP)

LINKPAC INPUT-OUTPUT SUBROUTINES

1. RDACC
   COMMON/FRBR/M, R2, S2, PH2, R3, S3, PH3, R4, S4, PH4, P(30, 2),
   VP(30, 2) AP(30, 2), PI(30, 2), TH2, TH3, TH4, W2, W3, W4, A2, A3, A4
   COMMON/RD/NTH, NPTS, TH20, DELTH
   CALLS SUBROUTINES RDGEOM, RDFIXED, RDINPUT TO READ DATA FOR
   ACCELERATION ANALYSIS FROM A STANDARDIZED INPUT FORMAT.

2. RDFORCE
   COMMON/FRBR/- - - - (DASHES INDICATE ARGUMENT LIST GIVEN PREVIOUSLY)
   COMMON/RD/- - - -
   COMMON/PH/M2, M3, M4, K2, K3, K4, I2, I3, I4
   CALLS SUBROUTINES RDGEOM, ROFIXED, RDINPUT, RDPHYS TO READ DATA FOR
   FORCE-TORQUE ANALYSIS FROM A STANDARDIZED INPUT FORMAT.

3. WRACC
   COMMON/FRBR/- - - -
   COMMON/RD/- - - -
   PRINTS RESULTS OF ACCELERATION ANALYSIS IN A STANDARDIZED
   OUTPUT FORMAT.

4. WRFORCE
   COMMON/FRBR/- - - -
   COMMON/FT/F1X, F1Y, F2X, F2Y, F3X, F3Y, F4X, F4Y, SF1X, SF1Y, T2, F(30, 2)
   PRINTS RESULTS OF FORCE-TORQUE ANALYSIS IN A STANDARDIZED
   OUTPUT FORMAT.

5. RDGEOM
   COMMON/FRBR/- - - -
   READS R, S, PHI FOR LINKS 2, 3, 4 IN FORMAT(3(3F10.0/)) AND MODE OF
   ASSEMBLY OF THE LINKAGE IN FORMAT(I2).

6. RDFIXED
   COMMON/FRBR/- - - -
   READS COORDINATES FOR FIXED PIVOTS 1 AND 4 IN FORMAT(4F10.00) AND
   SETS VEL AND ACC TO ZERO.

7. RDINPUT
   COMMON/FRBR/- - - -
   COMMON/RD/- - - -
   READS INPUT CRANK MOTION PARAMETERS NTH, TH20(DEG.), DEL, W2, A2
   IN FORMAT(I5, 4F10.0).

8. RDPHYS
   COMMON/PH/M2, M3, M4, K2, K3, K4, I2, I3, I4
   COMMON/EXLDS/- - - -
   READS PHYSICAL PROPERTIES OF LINKS AS REAL VARIABLES
   M2, M3, M4, K2, K3, K4 IN FORMAT(3F10.0) AND COMPUTES I2, I3, I4.

LINKPAC DYNAMIC ANALYSIS SUBROUTINES

1. FCRANK(N1, N2, N3, J1, K1, K2, A, P, AP, F, DT)

2. FDYAD(N1, N2, N3, N4, N5, J1, J2, K1, K2, K3, A1, A2, P, AP, F)

3. FOSC(N1, N2, N3, N4, N5, J1, J2, K1, K2, K3, D1, ALF1, R2, D2,
   ALF2, A2, P, AP, F, EC1, EC2)

4. FGUIDE(N1, N2, N3, N4, N5, J1, J2, K1, K2, K3, BETA, ABETA, A1,
   P, AP, F, EC2)

5. FSPRING(N1, N2, N3, N4, J1, J2, K, L0, LC, L, P)

6. FDAMPER(N1, N2, N3, N4, J1, J2, C, P, VP)

   THE FIRST SIX SUBROUTINES REQUIRE THE LABELLED COMMON/EXLDS/
   IN THE CALLING PROGRAM.

   COMMON/EXLDS/XM(15), XI(15), DXF(15), ST(15), SF(15,2)

7. FUNCTION DCROSSF(N1, N2, SF2, SFY, P)

```
      SUBROUTINE FRBRAN
C.... ACCELERATION ANALYSIS OF THE FOUR-BAR LINKAGE
      COMMON /FRBR/ M,R2,S2,PH2,R3,S3,PH3,R4,S4,PH4,P(30,2),VP(30,2),
     1 AP(30,2),PI(30,2),TH2,TH3,TH4,W2,W3,W4,A2,A3,A4
C     COMPUTE MOTION OF POINTS ON THE INPUT CRANK
      CALL GEOM(1,2,5,R2,S2,PH2,PI)
      CALL DISP(1,2,TH2,P,PI)
      CALL DISP(1,5,TH2,P,PI)
      CALL ACC(1,2,W2,A2,P,VP,AP)
      CALL ACC(1,5,W2,A2,P,VP,AP)
C     ACCELERATION ANALYSIS OF THE BASIC DYAD
      CALL ADYAD(M,2,4,3,R3,R4,TH3,TH4,P,W3,W4,VP,A3,A4,AP)
C     COMPUTE MOTION OF POINTS 6 AND 7
      CALL GEOM(2,3,6,R3,S3,PH3,PI)
      CALL GEOM(4,3,7,R4,S4,PH4,PI)
      CALL DISP(2,6,TH3,P,PI)
      CALL DISP(4,7,TH4,P,PI)
      CALL ACC(2,6,W3,A3,P,VP,AP)
      CALL ACC(4,7,W4,A4,P,VP,AP)
      RETURN
      END
C
```

```
      SUBROUTINE OFFSLD
C.... ACCELERATION ANALYSIS OF THE OFFSET SLIDER-CRANK MECHANISM
      COMMON/OFF/M,R2,S2,PH2,R3,S3,PH3,P(30,2),VP(30,2),AP(30,2),
     1 TH2,TH3,W2,W3,A2,A3,R4,VR4,AR4,BETA,VBETA,ABETA
C     COMPUTE MOTION OF POINT  2
      CALL CRANK(1,2,R2,TH2,W2,A2,P,VP,AP)
C     SET THE SLIDE DIRECTION
      BETA=VBETA=ABETA=0.0
C     ACCELERATION OF THE SLIDER AS A SPECIAL CASE OF A ROTATING GUIDE
      CALL AGUIDE(M,2,4,3,R3,R4,TH3,BETA,P,W3,VBETA,VR4,VP,A3,ABETA,
     1 AR4,AP)
C     COMPUTE MOTION OF POINTS 5 AND 6
      CALL POS(1,2,5,R2,S2,PH2,TH2,P)
      CALL POS(2,3,6,R3,S3,PH3,TH3,P)
      CALL ACC(1,5,W2,A2,P,VP,AP)
      CALL ACC(2,6,W3,A3,P,VP,AP)
      RETURN
      END
C
      SUBROUTINE DONKEY
C.... ACCELERATION CF THE DONKEY MECHANISM
      COMMON /DONK/ M,R2,S2,PH2,R3,S3,PH3,R4,S4,PH4,P(30,2),VP(30,2),
     1 AP(30,2),PI(30,2),TH2,TH3,TH4,W2,W3,W4,A2,A3,A4,E
C     COMPUTE MOTION OF POINT  2
      CALL CRANK(1,2,R2,TH2,W2,A2,P,VP,AP)
C     COMPUTE MOTION OF POINT  4
      E=0.0
      CALL AOSC(M,2,3,4,E,R3,R4,TH3,P,W3,VR3,VP,A3,AR3,AP)
C     COMPUTE MOTION OF POINTS 5 AND 6
      CALL POS(1,2,5,R2,S2,PH2,TH2,P)
      CALL POS(2,3,6,R3,S3,PH3,TH3,P)
      CALL ACC(1,5,W2,A2,P,VP,AP)
      CALL ACC(2,6,W3,A3,P,VP,AP)
      RETURN
      END
C
      SUBROUTINE GEOM(N1,N2,N3,R,S,PHI,PI)
C.... CALCULATES INITIAL COORDINATES PI IN REFERENCE POSITION
C     FOR THREE POINTS WITH INDICES N1,N2,N3 ON A LINK WITH
C     GEOMETRY PARAMETERS R, S, AND PHI.
      DIMENSION PI(30,2)
      PI(N1,1)= PI(N1,2)= PI(N2,2)= 0.0
      PI(N2,1)= R
      PI(N3,1)= S*COS(PHI)
      PI(N3,2)= S*SIN(PHI)
      RETURN
      END
C
      SUBROUTINE POS(N1,N2,N3,R,S,PHI,THETA,P)
C.... GIVEN POSITION FOR POINT N1 PLUS PARAMETERS R,S, AND PHI
C     AND POSITION ANGLE THETA, RETURNS POSITION COORDINATES FOR POINTS
C     N2 AND N3.
      DIMENSION P(30,2)
      CT= COS(THETA)
      ST= SIN(THETA)
      CP= COS(PHI)
      SP= SIN(PHI)
      P(N2,1)= P(N1,1) + R*CT
      P(N2,2)= P(N1,2) + R*ST
      P(N3,1)= P(N1,1) + S*CP*CT - S*SP*ST
      P(N3,2)= P(N1,2) + S*CP*ST + S*SP*CT
      RETURN
      END
C
      SUBROUTINE DISP(N1,N2,THETA,P,PI)
C.... SOLVES ROTATION MATRIX EQUATION P(N2)= R*(PI(N2)-PI(N1)) + P(N1)
      DIMENSION P(30,2),PI(30,2)
      C= COS(THETA)
```

```
      S= SIN(THETA)
      RX= PI(N2,1) - PI(N1,1)
      RY= PI(N2,2) - PI(N1,2)
      P(N2,1)= P(N1,1) + RX*C - RY*S
      P(N2,2)= P(N1,2) + RX*S + RY*C
      RETURN
      END
C
      SUBROUTINE VEL(N1,N2,W,P,VP)
C.... SOLVES VELOCITY MATRIX EQUATION VP(N2)= W*(P(N2)-P(N1)) + VP(N1)
      DIMENSION P(30,2),VP(30,2),VREL(2)
      VP(N2,1)= VP(N1,1) - W*(P(N2,2)-P(N1,2))
      VP(N2,2)= VP(N1,2) + W*(P(N2,1)-P(N1,1))
      RETURN
      END
C
      SUBROUTINE ACC(N1,N2,W,A,P,VP,AP)
C.... SOLVES ACCEL MATRIX EQUATION AP(N2)= WDOT*(P(N2)-P(N1)) + AP(N1)
      DIMENSION P(30,2),VP(30,2),AP(30,2)
      RX= P(N2,1)-P(N1,1)
      RY= P(N2,2)-P(N1,2)
      CALL VEL(N1,N2,W,P,VP)
      AP(N2,1)= AP(N1,1) - W*W*RX - A*RY
      AP(N2,2)= AP(N1,2) + A*RX - W*W*RY
      RETURN
      END
C
      SUBROUTINE CRANK(N1,N2,R,THETA,W,A,P,VP,AP)
C.... POSITION VELOCITY AND ACCELERATION ANALYSIS FOR THE MCVING PIVOT
C     ON A ROTATING CRANK WITH ONE FIXED POINT
      DIMENSION P(30,2),VP(30,2),AP(30,2)
      C= COS(THETA)
      S= SIN(THETA)
      VP(N1,1)=VP(N1,2)=AP(N1,1)=AP(N1,2)=0.
      RX= R*C
      RY= R*S
      P(N2,1)= P(N1,1) + RX
      P(N2,2)= P(N1,2) + RY
      VP(N2,1)= - RY*W
      VP(N2,2)=   RX*W
      AP(N2,1)= -RY*A - RX*W*W
      AP(N2,2)=   RX*A - RY*W*W
      RETURN
      END
C
      SUBROUTINE CRANK2(N1,N2,N3,R,S,PHI,THETA,W,A,P,VP,AP)
C.... POSITION VELOCITY AND ACCELERATION ANALYSIS FOR TWO MCVING
C     POINTS ON A ROTATING CRANK WITH ONE FIXED PIVOT
      DIMENSION P(30,2),VP(30,2),AP(30,2)
      VP(N1,1)=VP(N1,2)=AP(N1,1)=AP(N1,2)=0.
      RX= R*COS(THETA)
      RY= R*SIN(THETA)
      SX= S*COS(THETA+PHI)
      SY= S*SIN(THETA+PHI)
      P(N2,1)= P(N1,1) + RX
      P(N2,2)= P(N1,2) + RY
      P(N3,1)= P(N1,1) + SX
      P(N3,2)= P(N1,2) + SY
      VP(N2,1)= -RY*W
      VP(N2,2)   RX*W
      VP(N3,1)= -SY*W
      VP(N3,2)   SX*W
      AP(N2,1)= -RY*A - RX*W*W
      AP(N2,2)=  RX*A - RY*W*W
      AP(N3,1)= -SY*A - SX*W*W
      AP(N3,2)=  SX*A - SY*W*W
      RETURN
      END
```

```
C
      SUBROUTINE MOTION(N1,N2,N3,R,S,PHI,THETA,W,A,PI,P,VP,AP)
C.... COMPUTES NEW POSITION VELOCITY AND ACCELERATION FOR TWO
C     SPECIFIED POINTS WITH INDICES N2 AND N3 IN TERMS OF THE MOTION
C     OF A REFERENCE POINT WITH INDEX N1. THE RIGID BODY IS DESCRIBED BY
C     PARAMETERS R,S, AND PHI. MOTION PARAMETERS ARE THETA,W, AND A.
      DIMENSION PI(30,2),P(30,2),VP(30,2),AP(30,2)
      CALL GEOM(N1,N2,N3,R,S,PHI,PI)
      CALL DISP(N1,N3,THETA,P,PI)
      CALL ACC(N1,N3,W,A,P,VP,AP)
      RETURN
      END
C
      SUBROUTINE PDYAD(M,N1,N2,N3,R1,R2,TH1,TH2,P)
C.... POSITION ANALYSIS OF THE TWO LINK DYAD
      DIMENSION P(30,2)
      LOGICAL PRNT
      PRNT= .TRUE.
      IF(N1.LT.0) PRNT= .FALSE.
      N1= IABS(N1)
      DELX= P(N2,1)-P(N1,1)
      IF(ABS(DELX).LE.1.E-10) DELX= 1.E-10
      DELY= P(N2,2)-P(N1,2)
      PHI= ATAN2(DELY,DELX)
      SSQ=(P(N2,1)-P(N1,1))**2+(P(N2,2)-P(N1,2))**2
      S=SQRT(SSQ)
      TEST=S-(R1+R2)
      IF(TEST)40,40,500
   40 TEST= ABS(R1-R2) - S
      IF(TEST)50,50,500
   50 CONTINUE
      COSIN=(R1**2+SSQ-R2**2)/(2.*R1*S)
      ALPHA=ATAN2(SQRT(1.-COSIN**2),COSIN)
      IF(M)200,100,100
  100 THETA=PHI+ALPHA
      GO TO 300
  200 THETA=PHI-ALPHA
  300 CONTINUE
      P(N3,1)=P(N1,1)+R1*COS(THETA)
      P(N3,2)=P(N1,2)+R1*SIN(THETA)
      TH1= ATAN2((P(N3,2)-P(N1,2)),(P(N3,1)-P(N1,1)))
      TH2= ATAN2((P(N3,2)-P(N2,2)),(P(N3,1)-P(N2,1)))
      RETURN
  500 IF(PRNT) WRITE(6,600)
  600 FORMAT(* DYAD CANNOT BE ASSEMBLED *)
      RETURN
      END
C
      SUBROUTINE VDYAD(M,N1,N2,N3,R1,R2,TH1,TH2,P,W1,W2,VP)
C.... POSITION AND VELOCITY ANALYSIS OF THE TWO LINK DYAD
      DIMENSION P(30,2),VP(30,2)
      CALL PDYAD(M,N1,N2,N3,R1,R2,TH1,TH2,P)
      R2X= P(N3,1)-P(N2,1)
      R2Y= P(N3,2)-P(N2,2)
      A1= (VP(N2,1)-VP(N1,1))*R2X
      A2= (VP(N2,2)-VP(N1,2))*R2Y
      R1X= P(N3,1)-P(N1,1)
      R1Y= P(N3,2)-P(N1,2)
      DET= R1Y*R2X - R1X*R2Y
      B1= (VP(N2,2)-VP(N1,2))*R1Y
      B2= (VP(N2,1)-VP(N1,1))*R1X
      W1= -(A1+A2)/DET
      W2= -(B1+B2)/DET
      VP(N3,1)= VP(N1,1) - W1*R1Y
      VP(N3,2)= VP(N1,2) + W1*R1X
      RETURN
      END
C
```

```
      SUBROUTINE ADYAD(M,N1,N2,N3,R1,R2,TH1,TH2,P,W1,W2,VP,A1,A2,AP)
C.... POSITION VELOCITY AND ACCELERATION ANALYSIS OF THE TWC LINK DYAD
      DIMENSION P(30,2),VP(30,2),AP(30,2)
      CALL VDYAD(M,N1,N2,N3,R1,R2,TH1,TH2,P,W1,W2,VP)
      R1X= P(N3,1)-P(N1,1)
      R1Y= P(N3,2)-P(N1,2)
      R2X= P(N3,1)-P(N2,1)
      R2Y= P(N3,2)-P(N2,2)
      DET= R1Y*R2X - R2Y*R1X
      E= AP(N2,1) - AP(N1,1) + W1**2*R1X - W2**2*R2X
      F= AP(N2,2) - AP(N1,2) + W1**2*R1Y - W2**2*R2Y
      A1= -(E*R2X + F*R2Y)/DET
      A2= -(F*R1Y + E*R1X)/DET
      AP(N3,1)= AP(N1,1) - W1**2*R1X - A1*R1Y
      AP(N3,2)= AP(N1,2) + A1*R1X - W1**2*R1Y
      RETURN
      END
C
      SUBROUTINE POSC(M,N1,N2,N3,E,R2,R3,THETA,P)
C.... POSITION ANALYSIS OF THE OSCILLATING SLIDER
      DIMENSION P(30,2)
      LOGICAL PRNT
      PRNT= .TRUE.
      IF(N1.LT.0) PRNT= .FALSE.
      N1= IABS(N1)
      TEST=((P(N2,1)-P(N1,1))**2 + (P(N2,2)-P(N1,2))**2 - E**2)
      IF(TEST) 500,50,50
   50 R2= SQRT(TEST)
      ALPHA= ATAN2((P(N2,2)-P(N1,2)),(P(N2,1)-P(N1,1)))
      BETA= ATAN(E/R2)
      IF(M) 200,100,100
  100 THETA= ALPHA + BETA
      GO TO 300
  200 THETA= ALPHA - BETA
  300 CONTINUE
      P(N3,1)= P(N2,1) + (R3-R2)*COS(THETA)
      P(N3,2)= P(N2,2) + (R3-R2)*SIN(THETA)
      RETURN
  500 IF(PRNT) WRITE(6,600)
  600 FORMAT(* OSCILLATING SLIDER CANNOT BE ASSEMBLED*/)
      RETURN
      END
C
      SUBROUTINE VOSC(M,N1,N2,N3,E,R2,R3,THETA,P,W2,VR2,VP)
C.... POSITION AND VELOCITY ANALYSIS OF THE OSCILLATING SLIDER
      DIMENSION P(30,2),VP(30,2)
      CALL POSC(M,N1,N2,N3,E,R2,R3,THETA,P)
      C= COS(THETA)
      S= SIN(THETA)
      SX= R2*C + E*S
      SY= R2*S - E*C
      W2= ((VP(N2,1)-VP(N1,1))*S - (VP(N2,2)-VP(N1,2))*C)/(-SX*C-SY*S)
      VR2=(-(VP(N2,2)-VP(N1,2))*SY-(VP(N2,1)-VP(N1,1))*SX)/(-SY*S-SX*C)
      VP(N3,1)= VP(N1,1) - W2*(R3*S-E*C)
      VP(N3,2)= VP(N1,2) + W2*(R3*C+E*S)
      RETURN
      END
C
      SUBROUTINE AOSC(M,N1,N2,N3,E,R2,R3,THETA,P,W2,VR2,VP,A2,AR2,AP)
C.... POSITION VELOCITY AND ACCELERATION ANALYSIS OF THE CSC SLIDER
      DIMENSION P(30,2),VP(30,2),AP(30,2)
      CALL VOSC(M,N1,N2,N3,E,R2,R3,THETA,P,W2,VR2,VP)
      C= COS(THETA)
      S= SIN(THETA)
      SX= R2*C + E*S
      SY= R2*S - E*C
      E2= (AP(N2,1)-AP(N1,1)) + W2**2*SX + 2.*W2*VR2*S
      F2= (AP(N2,2)-AP(N1,2)) + W2**2*SY - 2.*W2*VR2*C
```

```
      A2= (F2*C-E2*S)/(SX*C+SY*S)
      AR2= (E2*SX+F2*SY)/(SX*C+SY*S)
      R3X= R3*C+E*S
      R3Y= R3*S-E*C
      AP(N3,1)= AP(N1,1) - W2**2*R3X - A2*R3Y
      AP(N3,2)= AP(N1,2) - W2**2*R3Y + A2*R3X
      RETURN
      END
C
      SUBROUTINE PGUIDE(M,N1,N2,N3,R1,R2,TH1,BETA,P)
C.... POSITION ANALYSIS OF THE ROTATING GUIDE
      DIMENSION P(30,2)
      LOGICAL PRNT
      PRNT= .TRUE.
      IF(N1.LT.0) PRNT= .FALSE.
      N1= IABS(N1)
      SSQ= (P(N1,1)-P(N2,1))**2 + (P(N1,2)-P(N2,2))**2
      E= 2.*((P(N2,1)-P(N1,1))*COS(BETA) + (P(N2,2)-P(N1,2))*SIN(BETA))
      F= SSQ - R1**2
      TEST= E**2 - 4.*F
      IF(TEST) 500,50,50
   50 SQROOT= SQRT(TEST)
      MODE= M
      RSQ= R1*R1
      IF(RSQ.GE.SSQ) MODE= + 1
      IF(MODE) 200,100,100
  100 R2=ABS(-E+SQROOT)/2.
      GO TO 300
  200 R2=ABS(-E-SQROOT)/2.
  300 CONTINUE
      P(N3,1)= P(N2,1) + R2*COS(BETA)
      P(N3,2)= P(N2,2) + R2*SIN(BETA)
      TH1= ATAN2((P(N3,2)-P(N1,2)),(P(N3,1)-P(N1,1)))
      RETURN
  500 IF(PRNT) WRITE(6,600)
  600 FORMAT(/* ROTATING GUIDE CANNOT BE ASSEMBLED*/)
      RETURN
      END
C
      SUBROUTINE VGUIDE(M,N1,N2,N3,R1,R2,TH1,BETA,P,W1,VBETA,VR2,VP)
C.... POSITION AND VELOCITY ANALYSIS OF THE ROTATING GUIDE
      DIMENSION P(30,2),VP(30,2)
      CALL PGUIDE(M,N1,N2,N3,R1,R2,TH1,BETA,P)
      CB= COS(BETA)
      SB= SIN(BETA)
      CT= COS(TH1)
      ST= SIN(TH1)
      E1= (VP(N2,1)-VP(N1,1)) - R2*VBETA*SB
      F1= (VP(N2,2)-VP(N1,2)) + R2*VBETA*CB
      DET= ST*SB + CT*CB
      W1= (F1*CB - E1*SB)/(R1*DET)
      VR2= -(E1*CT + F1*ST)/DET
      VP(N3,1)= VP(N1,1) - R1*W1*ST
      VP(N3,2)= VP(N1,2) + R1*W1*CT
      RETURN
      END
C
      SUBROUTINE AGUIDE(M,N1,N2,N3,R1,R2,TH1,BETA,P,
C.... POSITION VELOCITY AND ACCELERATION ANALYSIS OF THE ROTATING GUIDE
     1 W1,VBETA,VR2,VP,A1,ABETA,AR2,AP)
      DIMENSION P(30,2),VP(30,2),AP(30,2)
      CALL VGUIDE(M,N1,N2,N3,R1,R2,TH1,BETA,P,W1,VBETA,VR2,VP)
      CB= COS(BETA)
      SB= SIN(BETA)
      CT= COS(TH1)
      ST= SIN(TH1)
      E2= AP(N2,1)-AP(N1,1) + W1**2*R1*CT - ABETA*R2*SB
     1 - VBETA**2*R2*CB - 2.*VBETA*VR2*SB
```

```
        F2= AP(N2,2)-AP(N1,2) + W1**2*R1*ST + ABETA*R2*CB
      2 - VBETA**2*R2*SB + 2.*VBETA*VR2*CB
        DET= ST*SB + CT*CB
        A1= (F2*CB - E2*SB)/(R1*DET)
        AR2= - (E2*CT + F2*ST)/DET
        AP(N3,1)= AP(N1,1) - R1*A1*ST - R1*W1**2*CT
        AP(N3,2)= AP(N1,2) + R1*A1*CT - R1*W1**2*ST
        RETURN
        END
C
        SUBROUTINE DGEAR(M,N1,N2,RHO1,RHO3,D1,D2,D3,TH20,P)
C.... DISPLACEMENT ANALYSIS FOR A TWO-GEAR SET WITH MOVING CENTERS AT POINTS
C     N1 AND N2. TH20 IS INITIAL ARM POSITION. D1 IS ANG DISP FOR GEAR1.
C     D2 IS ANG DISP FOR THE ARM R2. D3 IS THE COMPUTED ANG DISP GEAR3.
        DIMENSION P(30,2)
        IF(M.LT.0) GO TO 10
        R2= RHO1 + RHO3
        D3= - (RHO1/RHO3)*D1 + ((RHO1+RHO3)/RHO3)*D2
        R2X= R2*COS(TH20)
        R2Y= R2*SIN(TH20)
        P(N2,1)= P(N1,1) + R2X*COS(D2) - R2Y*SIN(D2)
        P(N2,2)= P(N1,2) + R2X*SIN(D2) + R2Y*COS(D2)
        RETURN
     10 CONTINUE
        R2= RHO1 - RHO3
        D3= (RHO1/RHO3)*D1 - ((RHO1-RHO3)/RHO3)*D2
        R2X= R2*COS(TH20)
        R2Y= R2*SIN(TH20)
        P(N2,1)= P(N1,1) + R2X*COS(D2) - R2Y*SIN(D2)
        P(N2,2)= P(N1,2) + R2X*SIN(D2) + R2Y*COS(D2)
        RETURN
        END
C
        SUBROUTINE VGEAR(M,N1,N2,RHO1,RHO3,D1,D2,D3,TH20,P,W1,W2,W3,VP)
C.... VELOCITY ANALYSIS OF THE TWO-GEAR SET, W3 IS ANG VEL GEAR3.
C     W1 AND W2 ARE SPECIFIED.
        DIMENSION P(30,2),VP(30,2)
        CALL DGEAR(M,N1,N2,RHO1,RHO3,D1,D2,D3,TH20,P)
        IF(M.LT.0) GO TO 10
        W3= - (RHO1/RHO3)*W1 + ((RHO1+RHO3)/RHO3)*W2
        GO TO 20
     10 W3= (RHO1/RHO3)*W1 + ((RHO1-RHO3)/RHO3)*W2
     20 CALL VEL(N1,N2,W2,P,VP)
        RETURN
        END
C
        SUBROUTINE AGEAR(M,N1,N2,RHO1,RHO3,D1,D2,D3,TH20,P,W1,W2,W3,VP,
C.... ACCELERATION ANALYSIS OF THE TWO-GEAR SET. A3 IS ANG ACC OF GEAR3.
C     A1 AND A2 ARE SPECIFIED.
      1 A1,A2,A3,AP)
        DIMENSION P(30,2),VP(30,2),AP(30,2)
        CALL VGEAR(M,N1,N2,RHO1,RHO3,D1,D2,D3,TH20,P,W1,W2,W3,VP)
        IF(M.LT.0) GO TO 10
        A3= - (RHO1/RHO3)*A1 + ((RHO1+RHO3)/RHO3)*A2
        GO TO 20
     10 A3= (RHO1/RHO3)*A1 - ((RHO1-RHO3)/RHO3)*A2
     20 CALL ACC(N1,N2,W2,A2,P,VP,AP)
        RETURN
        END
C
        SUBROUTINE FCRANK(N1,N2,N3,J1,K1,K2,A,P,AP,F,DT)
C.... FORCE ANALYSIS FOR THE INPUT CRANK.
        COMMON/EXLDS/ XM(15),XI(15),DXF(15),ST(15),SF(15,2)
        DIMENSION P(30,2),AP(30,2),F(30,2)
        DIMENSION AA(3,3),B(3),ERASE(3)
        AA(1,2)=AA(1,3)=AA(2,1)=AA(2,3)=0.
        AA(1,1)=-1.   $  AA(2,2)=-1.
        AA(3,3)=1.
```

```
      QX= P(N1,1) - P(N3,1)   $  AA(3,2)= -QX
      QY= P(N1,2) - P(N3,2)   $  AA(3,1)=  QY
      PX= P(N2,1) - P(N3,1)
      PY= P(N2,2) - P(N3,2)
      B(1)= XM(J1)*AP(N3,1) - SF(J1,1) - F(K2,1)
      B(2)= XM(J1)*AP(N3,2) - SF(J1,2) - F(K2,2)
      B(3)= XI(J1)*A - ST(J1) - DXF(J1) - PX*F(K2,2) + PY*F(K2,1)
      SCALE=1.
      IF(LNEQF(3,3,1,AA,B,SCALE,ERASE).EQ.2) GOTO 98
      F(K1,1)=B(1)
      F(K1,2)=B(2)
      DT=       B(3)
      RETURN
   98 WRITE(6,60)
   60 FORMAT(/* ERROR HALT - THE COEFFICIENT MATRIX A IN FCRANK IS SINGU
     1LAR*/* GAUSSIAN ELIMINATION DOES NOT CONVERGE*)
      STOP
      END
C
      SUBROUTINE FDYAD(N1,N2,N3,N4,N5,J1,J2,K1,K2,K3,A1,A2,P,AP,F)
C.... FORCE ANALYSIS FOR THE TWO-LINK DYAD.
      COMMON/EXLDS/ XM(15),XI(15),DXF(15),ST(15),SF(15,2)
      DIMENSION P(30,2),AP(30,2),F(30,2)
      DIMENSION A(6,6),B(6),ERASE(6)
      DO 10 K=1,6
      DO 10 KK=1,6
   10 A(K,KK)=0.
      A(1,1)=-1. $ A(2,2)=-1. $ A(4,3)=-1. $ A(5,4)=-1.
      A(1,3)= 1. $ A(2,4)= 1. $ A(4,5)= 1. $ A(5,6)= 1.
      P1X= P(N3,1) - P(N4,1)   $   A(3,4)= P1X
      P1Y= P(N3,2) - P(N4,2)   $   A(3,3)=-P1Y
      Q1X= P(N1,1) - P(N4,1)   $   A(3,2)=-Q1X
      Q1Y= P(N1,2) - P(N4,2)   $   A(3,1)= Q1Y
      P2X= P(N2,1) - P(N5,1)   $   A(6,6)= P2X
      P2Y= P(N2,2) - P(N5,2)   $   A(6,5)=-P2Y
      Q2X= P(N3,1) - P(N5,1)   $   A(6,4)=-Q2X
      Q2Y= P(N3,2) - P(N5,2)   $   A(6,3)= Q2Y
      B(1)= XM(J1)*AP(N4,1) - SF(J1,1)
      B(2)= XM(J1)*AP(N4,2) - SF(J1,2)
      B(3)= XI(J1)*A1 - DXF(J1) - ST(J1)
      B(4)= XM(J2)*AP(N5,1) - SF(J2,1)
      B(5)= XM(J2)*AP(N5,2) - SF(J2,2)
      B(6)= XI(J2)*A2 - DXF(J2) - ST(J2)
      SCALE=1.
      IF(LNEQF(6,6,1,A,B,SCALE,ERASE).EQ.2) GOTO 98
      F(K1,1)=B(1)
      F(K1,2)=B(2)
      F(K3,1)=B(3)
      F(K3,2)=B(4)
      F(K2,1)=B(5)
      F(K2,2)=B(6)
      RETURN
   98 WRITE(6,60)
   60 FORMAT(/* ERROR HALT - THE COEFFICIENT MATRIX A IN FDYAD IS SINGUL
     1AR*/* GAUSSIAN ELIMINATION DOES NOT CONVERGE*)
      STOP
      END
C
      SUBROUTINE FOSC(N1,N2,N3,N4,N5,J1,J2,K1,K2,K3,
     1 D1,ALF1,R2,D2,ALF2,A2,P,AP,F,EC1,EC2)
C.... FORCE ANALYSIS FOR THE OSC SLIDER DYAD.
      COMMON/EXLDS/ XM(15),XI(15),DXF(15),ST(15),SF(15,2)
      DIMENSION P(30,2),AP(30,2),F(30,2)
      DIMENSION A(6,6),B(6),ERASE(6)
      TOL= 1.E-05
      SCALE=1.
      DO 10 K=1,6
      DO 10 KK=1,6
```

```
    10 A(K,KK)=0.
       A(1,1)=A(2,2)=A(4,3)=A(5,4)= -1.
       A(1,3)=A(2,4)=A(4,5)=A(5,6)=  1.
       P2X= P(N2,1) - P(N5,1)   $   A(3,6)= P2X
       P2Y= P(N2,2) - P(N5,2)   $   A(3,5)=-P2Y
       Q1X= P(N1,1) - P(N4,1)   $   A(3,2)=-Q1X
       Q1Y= P(N1,2) - P(N4,2)   $   A(3,1)= Q1Y
       DX= P(N3,1) - P(N2,1)
       DY= P(N3,2) - P(N2,2)
       DMAG= SQRT(DX*DX+DY*DY)
       UX= DX/DMAG   $   A(6,3)= UX
       UY= DY/DMAG   $   A(6,4)= UY
       AMAG= R2 - D1*COS(ALF1) + D2*COS(ALF2)
       AX= UX*AMAG   $   A(3,4)= AX
       AY= UY*AMAG   $   A(3,3)=-AY
       B(1)= XM(J1)*AP(N4,1) - SF(J1,1)
       B(2)= XM(J1)*AP(N4,2) - SF(J1,2)
       B(3)= (XI(J1) + XI(J2))*A2 - DXF(J1) - ST(J1) - DXF(J2) - ST(J2)
       B(4)= XM(J2)*AP(N2,1) - SF(J2,1)
       B(5)= XM(J2)*AP(N2,2) - SF(J2,2)
       B(6)=0.
       IF(LNEQF(6,6,1,A,B,SCALE,ERASE),EQ.2) GOTO 98
       F(K1,1)=B(1)
       F(K1,2)=B(2)
       F(K2,1)=B(3)
       F(K2,2)=B(4)
       F(K3,1)=B(5)
       F(K3,2)=B(6)
       TCON= XI(J2)*A2 - P2X*F(K3,2) + P2Y*F(K3,1) - DXF(J2) - ST(J2)
       DET= -(F(K2,1)*F(K2,1) + F(K2,2)*F(K2,2))
       IF(ABS(DET).LT.TOL) DET=TOL
       E2X= TCON*F(K2,2)/DET
       E2Y=-TCON*F(K2,1)/DET
       EC2= SQRT(E2X*E2X+E2Y*E2Y)
       TEST= E2X*UX + E2Y*UY
       IF(TEST.LT.0.) EC2= -EC2
       EC1= AMAG + EC2
       RETURN
    98 WRITE(6,60)
    60 FORMAT(/* ERROR HALT - THE COEFFICIENT MATRIX A IN FOSC IS SINGULA
      1R*/* GAUSSIAN ELIMINATION DOES NOT CONVERGE*)
       STOP
       END
C
       SUBROUTINE FGUIDE(N1,N2,N3,N4,N5,J1,J2,K1,K2,K3,
      1 BETA,ABETA,A1,P,AP,F,EC2)
C.... FORCE ANALYSIS FOR THE ROTATING GUIDE DYAD.
       COMMON/EXLDS/ XM(15),XI(15),DXF(15),ST(15),SF(15,2)
       DIMENSION P(30,2),AP(30,2),F(30,2)
       DIMENSION A(6,6),B(6),ERASE(6)
       SCALE=1.
       TOL= 1.E-05
       DO 10 K=1,6
       DO 10 KK=1,6
    10 A(K,KK)=0.
       A(1,1)=A(2,2)=A(4,3)=A(5,4)= -1.
       A(1,3)=A(2,4)=A(4,5)=A(5,6)=  1.
       Q1X= P(N1,1) - P(N4,1)   $   A(3,2)= -Q1X
       Q1Y= P(N1,2) - P(N4,2)   $   A(3,1)=  Q1Y
       P1X= P(N3,1) - P(N4,1)   $   A(3,4)=  P1X
       P1Y= P(N3,2) - P(N4,2)   $   A(3,3)= -P1Y
       Q2X= P(N3,1) - P(N5,1)
       Q2Y= P(N3,2) - P(N5,2)
       UX= COS(BETA)            $   A(6,5)=  UX
       UY= SIN(BETA)            $   A(6,6)=  UY
       B(1)= XM(J1)*AP(N4,1) - SF(J1,1)
       B(2)= XM(J1)*AP(N4,2) - SF(J1,2)
       B(3)= XI(J1)*A1 - ST(J1) - DXF(J1)
```

```
      B(4)= XM(J2)*AP(N5,1) - SF(J2,1)
      B(5)= XM(J2)*AP(N5,2) - SF(J2,2)
      B(6)=0.
      IF(LNEQF(6,6,1,A,B,SCALE,ERASE).EQ.2) GOTO 98
      F(K1,1)=B(1)
      F(K1,2)=B(2)
      F(K2,1)=B(3)
      F(K2,2)=B(4)
      F(K3,1)=B(5)
      F(K3,2)=B(6)
      TCON= XI(J2)*ABETA + Q2X*F(K2,2) - Q2Y*F(K2,1) - DXF(J2) - ST(J2)
      DET= F(K3,1)*F(K3,1) + F(K3,2)*F(K3,2)
      IF(ABS(DET).LT.TOL) DET=TOL
      E2X=  TCON*F(K3,2)/DET
      E2Y= -TCON*F(K3,1)/DET
      EC2= SQRT(E2X*E2X+E2Y*E2Y)
      TEST= E2X*UX + E2Y*UY
      IF(TEST.LT.0.) EC2= -EC2
      RETURN
   98 WRITE(6,60)
   60 FORMAT(/* ERROR HALT - THE COEFFICIENT MATRIX A IN FGLIDE IS SINGU
     1LAR*/* GAUSSIAN ELIMINATION DOES NOT CONVERGE*)
      STOP
      END
C
      SUBROUTINE FSPRING(N1,N2,N3,N4,J1,J2,K,L0,LC,L,P)
C.... COMPUTES SPRING FORCE BETWEEN TWO GIVEN POINTS.
      COMMON/EXLDS/ XM(15),XI(15),DXF(15),ST(15),SF(15,2)
      DIMENSION P(30,2)
      REAL K,L0,L,LC
      X= P(N2,1) - P(N1,1)
      Y= P(N2,2) - P(N1,2)
      L= SQRT(X*X+Y*Y)
      IF(L.LT.LC) GOTO 10
      UX= X/L
      UY= Y/L
      RF= K*(L-L0)
      RFX= RF*UX
      RFY= RF*UY
      SF(J1,1)=  RFX + SF(J1,1)
      SF(J1,2)=  RFY + SF(J1,2)
      SF(J2,1)= -RFX + SF(J2,1)
      SF(J2,2)= -RFY + SF(J2,2)
      DXF(J1)= DXF(J1) + DCROSSF(N3,N1, RFX, RFY,P)
      DXF(J2)= DXF(J2) + DCROSSF(N4,N2,-RFX,-RFY,P)
      RETURN
   10 WRITE(6,60) N1,N2
      STOP
   60 FORMAT(/* THE SPRING BETWEEN PIVOTS*I3* AND*I3* IS LESS THAN ITS S
     1PECIFIED COMPRESSED LENGTH*/)
      END
C
      SUBROUTINE FDAMPER(N1,N2,N3,N4,J1,J2,C,P,VP)
C.... COMPUTES VISCOUS DAMPER FORCE BETWEEN TWO GIVEN POINTS.
      COMMON/EXLDS/ XM(15),XI(15),DXF(15),ST(15),SF(15,2)
      DIMENSION P(30,2),VP(30,2)
      TOL= 1.E-05
      X= P(N2,1) - P(N1,1)
      Y= P(N2,2) - P(N1,2)
      D= SQRT(X*X+Y*Y)
      IF(D.LT.TOL) GOTO 10
      UX= X/D
      UY= Y/D
      VX= VP(N2,1) - VP(N1,1)
      VY= VP(N2,2) - VP(N1,2)
      RF= C*(VX*UX + VY*UY)
      RFX= RF*UX
      RFY= RF*UY
```

```
      SF(J1,1)=  RFX + SF(J1,1)
      SF(J1,2)=  RFY + SF(J1,2)
      SF(J2,1)= -RFX + SF(J2,1)
      SF(J2,2)= -RFY + SF(J2,2)
      DXF(J1)= DXF(J1) + DCROSSF(N3,N1, RFX, RFY,P)
      DXF(J2)= DXF(J2) + DCROSSF(N4,N2,-RFX,-RFY,P)
      RETURN
   10 WRITE(6,60) N1,N2
      RETURN
   60 FORMAT(/* THE DAMPER PIVOTS*I3* AND*I3* ARE COINCIDENT*/* THE DIRE
     1CTION OF A CORRESPONDING DAMPER FORCE IS UNDEFINED*/)
      END
      FUNCTION DCROSSF(N1,N2,SFX,SFY,P)
      DIMENSION P(30,2)
      DX= P(N2,1) - P(N1,1)
      DY= P(N2,2) - P(N1,2)
      DCROSSF= DX*SFY - DY*SFX
      RETURN
      END


      SUBROUTINE RDACC
C.... READ DATA FOR ACCELERATION ANALYSIS
      COMMON /FRBR/ M,R2,S2,PH2,R3,S3,PH3,R4,S4,PH4,P(30,2),VP(30,2),
     1 AP(30,2),PI(30,2),TH2,TH3,TH4,W2,W3,W4,A2,A3,A4
      COMMON /RD/ NTH,NPTS,TH2Ø,DELTH
      WRITE(6,100)
  100 FORMAT(1H1,15X,22(2H* )/10X*KINEMATIC ANALYSIS FOUR-LINK*
     1 * MECHANISMS*/16X,22(2H* ))
      CALL RDGEOM
      CALL RDFIXED
      CALL RDINPUT
      RETURN
      END
C
      SUBROUTINE RDFORCE
C.... READ DATA FOR FORCE-TORQUE ANALYSIS
      COMMON /FRBR/ M,R2,S2,PH2,R3,S3,PH3,R4,S4,PH4,P(30,2),VP(30,2),
     1 AP(30,2),PI(30,2),TH2,TH3,TH4,W2,W3,W4,A2,A3,A4
      COMMON /RD/ NTH,NPTS,TH2Ø,DELTH
      COMMON /PH/ M2,M3,M4,K2,K3,K4,I2,I3,I4
      REAL M2,M3,M4,K2,K3,K4,I2,I3,I4
      WRITE(6,100)
  100 FORMAT(1H1,15X,22(2H* )/16X*FORCE-TORQUE ANALYSIS FOUR-LINK*
     1 * MECHANISMS*/16X,22(2H* ))
      CALL RDGEOM
      CALL RDFIXED
      CALL RDINPUT
      CALL RDPHYS
      RETURN
      END
C
      SUBROUTINE WRACC
C.... PRINT RESULTS FOR ACCELERATION ANALYSIS
      COMMON /FRBR/ M,R2,S2,PH2,R3,S3,PH3,R4,S4,PH4,P(30,2),VP(30,2),
     1 AP(30,2),PI(30,2),TH2,TH3,TH4,W2,W3,W4,A2,A3,A4
      COMMON /RD/ NTH,NPTS,TH2Ø,DELTH
      CON= ATAN(1.)/45.
      THET2= TH2/CON $ THET3= TH3/CON $ THET4= TH4/CON
      WRITE(6,100) THET2,THET2,THET3,THET4,W2,W3,W4,A2,A3,A4
      WRITE(6,200)
      DO 400 J= 1,NPTS
  400 WRITE(6,300) J,P(J,1),P(J,2),VP(J,1),VP(J,2),
     1 AP(J,1),AP(J,2)
  100 FORMAT(//1X,5(2H* )*  ACCELERATION RESULTS FOR THETA2= *F6.2* DEG*
     1 2X,5(2H* )//5X*ANGULAR POSITION*
     2 /5X*TH2= *E11.4,3X,*TH3= *E11.4,3X,*TH4= *E11.4* DEGREES*
     3 //5X*ANGULAR VELOCITY*
     4 /5X* W2= *E11.4,3X,* W3= *E11.4,3X,* W4= *E11.4* RAD/SEC*
```

```
      5 //5X*ANGULAR ACCELERATION*
      6 /5X* A2= *E11.4,3X,* A3= *E11.4,3X,* A4= *E11.4* RAD/SEC/SEC*)
  200 FORMAT(//2X*POINT*,7X,*POSITION*14X*VELOCITY*12X
      1 *ACCELERATION*/* NUMBER*3(5X*X*10X*Y*5X))
  300 FORMAT(I6,6E11.3)
      RETURN
      END
C
      SUBROUTINE WRFORCE
C.... PRINT RESULTS FOR FORCE-TORQUE ANALYSIS
      COMMON /FRBR/ M,R2,S2,PH2,R3,S3,PH3,R4,S4,PH4,P(30,2),VP(30,2),
      1 AP(30,2),PI(30,2),TH2,TH3,TH4,W2,W3,W4,A2,A3,A4
      COMMON /FT/ F1X,F1Y,F2X,F2Y,F3X,F3Y,F4X,F4Y,SF1X,SF1Y,T2,F(30,2)
      CON= ATAN(1.)/45. $ THET2= TH2/CON
      F1X= F(1,1) $ F1Y= F(1,2) $ F2X= F(2,1) $ F2Y= F(2,2)
      F3X= F(3,1) $ F3Y= F(3,2) $ F4X= F(4,1) $ F4Y= F(4,2)
      WRITE(6,100) THET2,F1X,F1Y,F2X,F2Y,F3X,F3Y,F4X,F4Y,SF1X,SF1Y,T2
  100 FORMAT(//1X,5(2H* )* FORCE-TORQUE RESULTS FOR THETA2= *F6.2* DEG*
      1    2X,5(2H* )//5X*F1X=F21X= *E12.4,3X,*F1Y=F21Y= *E12.4
      2 /5X*F2X=F32X= *E12.4,3X,*F2Y=F32Y= *E12.4
      3 /5X*F3X=F43X= *E12.4,3X,*F3Y=F43Y= *E12.4
      4 /5X*F4X=F14X= *E12.4,3X,*F4Y=F14Y= *E12.4
      5 //5X*SHAKING FORCE COMPONENTS*/5X*SF1X= *E12.4,3X,*SF1Y= *
      6 E12.4//5X*INPUT TORQUE ON LINK 2 = *E12.4)
      RETURN
      END
C
      SUBROUTINE RDGEOM
C.... READ GEOMETRY FOR LINKS NUMBER 2, 3, 4
      COMMON /FRBR/ M,R2,S2,PH2,R3,S3,PH3,R4,S4,PH4,P(30,2),VP(30,2),
      1 AP(30,2),PI(30,2),TH2,TH3,TH4,W2,W3,W4,A2,A3,A4
      CON= ATAN(1.)/45.
      READ(5,100) R2,S2,P2,R3,S3,P3,R4,S4,P4
  100 FORMAT(3F10.0)
      PH2= P2*CON $ PH3= P3*CON $ PH4=P4*CON
C     READ MODE OF ASSEMBLY FOR THE MECHANISM
      READ(5,200) M
  200 FORMAT(I2)
C     WRITE GEOMETRY DATA
      WRITE(6,300) R2,S2,P2,R3,S3,P3,R4,S4,P4,M
  300 FORMAT(//10X*LINK GEOMETRY DATA*
      1 /5X*R2= *F7.3,5X,*S2= *F7.3,5X,*PHI2= *F7.3* DEG*
      2 /5X*R3= *F7.3,5X,*S3= *F7.3,5X,*PHI3= *F7.3
      3 /5X*R4= *F7.3,5X,*S4= *F7.3,5X,*PHI4= *F7.3
      4 /5X*MODE FACTOR M= *I2)
      RETURN
      END
C
      SUBROUTINE RDFIXED
C.... READ FIXED PIVOT COORDINATES FOR POINTS 1 AND 4
      COMMON /FRBR/ M,R2,S2,PH2,R3,S3,PH3,R4,S4,PH4,P(30,2),VP(30,2),
      1 AP(30,2),PI(30,2),TH2,TH3,TH4,W2,W3,W4,A2,A3,A4
      READ(5,100) P(1,1),P(1,2),P(4,1),P(4,2)
  100 FORMAT(4F10.0)
C     SET VEL AND ACC OF POINTS 1 AND 4 = ZERO
      VP(1,1)= VP(1,2)= VP(4,1)= VP(4,2)= 0.0
      AP(1,1)= AP(1,2)= AP(4,1)= AP(4,2)= 0.0
      WRITE(6,200) P(1,1),P(1,2),P(4,1),P(4,2)
  200 FORMAT(//10X*FIXED PIVOT COORDINATES*
      1 /5X*P(1,1)= *F7.3,3X,*P(1,2)= *F7.3,3X,*P(4,1)= *F7.3,
      2 3X,*P(4,2)= *F7.3)
      RETURN
      END
C
      SUBROUTINE RDINPUT
C.... READ INPUT CRANK MOTION PARAMETERS
      COMMON /FRBR/ M,R2,S2,PH2,R3,S3,PH3,R4,S4,PH4,P(30,2),VP(30,2),
      1 AP(30,2),PI(30,2),TH2,TH3,TH4,W2,W3,W4,A2,A3,A4
```

```
      COMMON /RD/ NTH,NPTS,TH2Ø,DELTH
      READ(5,100) NTH,TH2ZERO,DEL,W2,A2
  100 FORMAT(I5,4F10.0)
      CON= ATAN(1.)/45.
      TH2Ø= TH2ZERO*CON $ DELTH= DEL*CON
      WRITE(6,2ØØ) NTH,TH2ZERO,DEL,W2,A2
  200 FORMAT(//10X*INPUT CRANK MOTION PARAMETERS*
     1 /5X*NTH= *I3,3X,*TH2Ø= *F7.3,3X,*DELTH= *F7.3* DEGREES*
     2 /5X* W2= *F7.3* RAD/SEC*,3X,*A2= *F7.3* RAD/SEC/SEC*)
      RETURN
      END
C
      SUBROUTINE RDPHYS
C.... READ MASS AND RADIUS OF GYRATION FOR LINKS 2 3 4
      COMMON /PH/ M2,M3,M4,K2,K3,K4,I2,I3,I4
      COMMON /EXLDS/ XM(15),XI(15),DXF(15),ST(15),SF(15,2)
      REAL M2,M3,M4,K2,K3,K4,I2,I3,I4
      READ(5,100) M2,M3,M4,K2,K3,K4
  100 FORMAT(3F10.0)
      I2= M2*K2*K2 $ I3=M3*K3*K3 $ I4= M4*K4*K4
      XM(2)= M2 $ XM(3)= M3 $ XM(4)= M4
      XI(2)= I2 $ XI(3)= I3 $ XI(4)= I4
      WRITE(6,2ØØ) M2,M3,M4,K2,K3,K4,I2,I3,I4
  200 FORMAT(//10X*PHYSICAL PROPERTIES OF THE LINKS*
     1 /5X*M2= *E12.4,3X,*M3= *E12.4,3X,*M4= *E12.4
     2 /5X*K2= *E12.4,3X,*K3= *E12.4,3X,*K4= *E12.4
     3 /5X*I2= *E12.4,3X,*I3= *E12.4,3X,*I4= *E12.4)
      RETURN
      END
C
```

Appendix 2.
SPAPAC

SPATIAL KINEMATICS SUBPROGRAMS

SPAPAC
1. FUNCTION DOT(V1,V2)
2. SUBROUTINE CROSS (VX,VI,V2)
3. FUNCTION DOTU2P(U,P1,P2)
4. FUNCTION DIRCOS(U)
5. FUNCTION DLENGTH(PJ,QJ,P1,Q1)
6. FUNCTION VLENGTH(VPJ,PJ,P0)
7. FUNCTION ALENGTH(APJ,VPJ,PJ,P0)
8. FUNCTION DTWIST(UJ,U1,U0)
9. FUNCTION VTWIST(VUJ,U0)
10. FUNCTION ATWIST(AUJ,U0)
11. FUNCTION PLANE(U,PJ,P0)
12. FUNCTION VPLANE(VUJ,UJ,VPJ,PJ,P0)
13. FUNCTION APLANE(AUJ,VUJ,UJ,APJ,VPJ,PJ,P0)
14. FUNCTION DMOMENT(U0,U1,UJ,A0,A1,AJ)
15. FUNCTION VMOMENT(U0,UJ,VUJ,A0,AJ,VAJ)
16. FUNCTION AMOMENT(U0,UJ,VUJ,AUJ,A0,AJ,VAJ,AAJ)
17. SUBROUTINE ROTATE(PJ,QJ,RM,P1,Q1,J)
18. SUBROUTINE DISP(PJ,DM,P1,J)
19. SUBROUTINE ROTVEC(V2,RM,V1,J)
20. SUBROUTINE PMTX(U,PM)
21. SUBROUTINE QMTX(U, QM)
22. SUBROUTINE QIMTX(U,QM,QIM)
23. SUBROUTINE MTXVEC(TEMP,A,VEC)
24. SUBROUTINE RMAXIS(U,PHI,RM,J)
25. SUBROUTINE RMABG(ALPHA,BETA,GAMMA,RM,J)
26. SUBROUTINE RMEULER(PSI,THETA,PHI,RM,J)
27. SUBROUTINE DMABG(ALPHA,BETA,GAMMA,RM,DM,J,PJ,P1)
28. SUBROUTINE DMAXIS(U,PHI,RM,DM,J,PJ,P1)
29. SUBROUTINE DMSCREW(P,U,S,PHI,RM,DM,J)
30. SUBROUTINE DMEULER(PSI,THETA,PHI,RM,DM,J,PJ,P1)
31. SUBROUTINE WMTX(U,VPHI,WM,J)
32. SUBROUTINE WDOT(U,VU,VPHI,APHI,WD,J)
33. SUBROUTINE ROTVEL(VP,VQ,WM,P,Q,J)
34. SUBROUTINE ROTACC(AP,AQ,WD,P,Q,J)
35. SUBROUTINE COMPERP(P1,P2,U1,U2,Q1,Q2,D1,D2)
36. LOGICAL FUNCTION EFG(E,F,G,A1,A2,A)

```
     LOGICAL FUNCTION EFG(E,F,G,A1,A2,A)
C   SOLVES E*COS(A)+F*SIN(A)+G=0. AS A QUADRATIC IN TAN(A/2.)
```

```
C   RETURNS TWO SOLUTIONS FOR A1 AND A2 AND THE VALUE CLOSEST
C   TO THE PREVIOUS VALUE OF A AS THE NEW VALUE OF A
C   IF EFG = .FALSE. THERE IS NO REAL SOLUTION
      EFG= .TRUE.
      TEMP= E*E + F*F - G*G
      IF(TEMP.LT.0.0) EFG= .FALSE.
      IF(.NOT.EFG) GO TO 10
      TEMP= SQRT(TEMP)
      GE= G-E
      IF(ABS(GE).LT.1.E-10) GE= 1.E-10
      A1= 2.*ATAN((-F-TEMP)/GE)
      A2= 2.*ATAN((-F+TEMP)/GE)
      T1= ABS(A1-A)
      T2= ABS(A2-A)
      A= A1
      IF(T2.LT.T1) A= A2
      RETURN
   10 WRITE(6,100)
  100 FORMAT(5X,28H NO SOLUTION IN FUNCTION EFG )
      RETURN
      END
C
      FUNCTION DOT(V1,V2)
C.... COMPUTES VECTOR DOT PRODUCT V1*V2
      DIMENSION V1(3),V2(3)
      DOT= 0.0
      DO 10 I= 1,3
   10 DOT= DOT + V1(I)*V2(I)
      RETURN
      END
C
      SUBROUTINE CROSS(VX,V1,V2)
C.... COMPUTES VECTOR CROSS PRODUCT VX= V1 X V2
      DIMENSION V1(3),V2(3),VX(3)
      VX(1)= V1(2)*V2(3) - V1(3)*V2(2)
      VX(2)= V1(3)*V2(1) - V1(1)*V2(3)
      VX(3)= V1(1)*V2(2) - V1(2)*V2(1)
      RETURN
      END
C
      FUNCTION DOTU2P(U,P1,P2)
C.... COMPUTES DOT PRODUCT U*(P1-P2)
      DIMENSION U(3),P1(3),P2(3)
      DOTU2P= 0.0
      DO 10 I= 1,3
   10 DOTU2P= DOTU2P + U(I)*(P1(I)-P2(I))
      RETURN
      END
C
      FUNCTION DIRCOS(U)
C.... COMPUTES U*U - 1. = 0.
      DIMENSION U(3)
      DIRCOS= U(1)*U(1) + U(2)*U(2) + U(3)*U(3) - 1.0
      RETURN
      END
C
      FUNCTION DLENGTH(PJ,QJ,P1,Q1)
C.... CONSTANT LENGTH - DISPLACEMENT CONSTRAINT
C     COMPUTES (PJ-QJ)*(PJ-QJ) - (P1-Q1)*(P1-Q1) = 0.
      DIMENSION PJ(3),QJ(3),P1(3),Q1(3)
      DLENGTH= 0.0
      DO 10 I= 1,3
   10 DLENGTH= DLENGTH + (PJ(I)-QJ(I))**2 - (P1(I)-Q1(I))**2
      RETURN
      END
C
      FUNCTION VLENGTH(VPJ,PJ,P0)
C.... CONSTANT LENGTH - VELOCITY CONSTRAINT
```

```
C       COMPUTES VPJ*(PJ-P0) = 0.
        DIMENSION VPJ(3),PJ(3),P0(3)
        VLENGTH= 0.0
        DO 10 I= 1,3
   10 VLENGTH= VLENGTH + VPJ(I)*(PJ(I)-P0(I))
        RETURN
        END
C
        FUNCTION ALENGTH(APJ,VPJ,PJ,P0)
C.... CONSTANT LENGTH - ACCELERATION CONSTRAINT
C       COMPUTES APJ*(PJ-P0) + VPJ*VPJ = 0.
        DIMENSION APJ(3),VPJ(3),PJ(3),P0(3)
        ALENGTH= 0.0
        DO 10 I= 1,3
   10 ALENGTH= ALENGTH + APJ(I)*(PJ(I)-P0(I)) + VPJ(I)*VPJ(I)
        RETURN
        END
C
        FUNCTION DTWIST(UJ,U1,U0)
C.... CONSTANT TWIST - DISPLACEMENT CONSTRAINT
C       COMPUTES UJ*U0 - U1*U0 = 0.
        DIMENSION UJ(3),U1(3),U0(3)
        DTWIST= 0.0
        DO 10 I= 1,3
   10 DTWIST= DTWIST + UJ(I)*U0(I) - U1(I)*U0(I)
        RETURN
        END
C
        FUNCTION VTWIST(VUJ,U0)
C.... CONSTANT TWIST - VELOCITY CONSTRAINT
C       COMPUTES VUJ*L0 = 0.
        DIMENSION VUJ(3),U0(3)
        VTWIST= 0.0
        DO 10 I= 1,3
   10 VTWIST= VTWIST + VUJ(I)*U0(I)
        RETURN
        END
C
        FUNCTION ATWIST(AUJ,U0)
C.... CONSTANT TWIST - ACCELERATION CONSTRAINT
C       COMPUTES AUJ*U0 = 0.
        DIMENSION AUJ(3),U0(3)
        ATWIST= 0.0
        DO 10 I= 1,3
   10 ATWIST= ATWIST + AUJ(I)*U0(I)
        RETURN
        END
C
        FUNCTION PLANE(U,PJ,P0)
C.... PLANE EQUATION
C       COMPUTES U*(PJ-P0) = 0.
        DIMENSION U(3),PJ(3),P0(3)
        PLANE= 0.0
        DO 10 I= 1,3
   10 PLANE= PLANE + U(I)*(PJ(I)-P0(I))
        RETURN
        END
C
C
        FUNCTION VPLANE(VUJ,UJ,VPJ,PJ,P0)
C.... PLANE EQUATION - VELOCITY CONSTRAINT
C       ASSUMES FIXED AXIS U0
C       COMPUTES VUJ*(PJ-P0) + UJ*VPJ = 0.
        DIMENSION VUJ(3),UJ(3),VPJ(3),PJ(3),P0(3)
        VPLANE= 0.0
        DO 10 I= 1,3
   10 VPLANE= VPLANE + VUJ(I)*(PJ(I)-P0(I)) + UJ(I)*VPJ(I)
```

```
      RETURN
      END
C
      FUNCTION APLANE(AUJ,VUJ,UJ,APJ,VPJ,PJ,P0)
C.... PLANE EQUATION - ACCELERATION CONSTRAINT
C     ASSUMES FIXED AXIS UJ
C     COMPUTES AUJ*(PJ-P0) + 2.*VUJ*VPJ + UJ*APJ = 0.
      DIMENSION AUJ(3),VUJ(3),UJ(3),APJ(3),VPJ(3),PJ(3),P0(3)
      APLANE= 0.0
      DO 10 I= 1,3
   10 APLANE= APLANE + AUJ(I)*(PJ(I)-P0(I)) + 2.*(VUJ(I)*VPJ(I)) +
     1 UJ(I)*APJ(I)
      RETURN
      END
C
      FUNCTION DMOMENT(U0,U1,UJ,A0,A1,AJ)
C.... COMPUTES U0*((AJ-A0) X UJ) - U0*((A1-A0) X U1) = 0.
      DIMENSION U0(3),U1(3),UJ(3),A0(3),A1(3),AJ(3),T1(3),T2(3),T3(3),
     1 T4(3)
      DO 10 I= 1,3
      T1(I)= A1(I) - A0(I)
   10 T2(I)= AJ(I) - A0(I)
      CALL CROSS(T3,T1,U1)
      CALL CROSS(T4,T2,UJ)
      DMOMENT= DOT(U0,T4) - DOT(U0,T3)
      RETURN
      END
C
     1 T4(3)
      DO 10 I= 1,3
   10 T1(I)= AJ(I) - A0(I)
      CALL CROSS(T2,T1,VUJ)
      CALL CROSS(T3,VAJ,UJ)
      VMOMENT= DOT(U0,T3) + DOT(U0,T2)
      RETURN
      END
C
      FUNCTION AMOMENT(U0,UJ,VUJ,AUJ,A0,AJ,VAJ,AAJ)
C.... COMPUTES U0*((AAJ X UJ) + 2.*(VAJ X VUJ) + (AJ-A0) X AUJ)= 0.
      DIMENSION U0(3),UJ(3),VUJ(3),AUJ(3),A0(3),AJ(3),VAJ(3),AAJ(3),
     1 T1(3),T2(3),T3(3),T4(3),T5(3)
      DO 10 I= 1,3

      VAJ(I)= 2.*VAJ(I)
   10 T1(I)= AJ(I) - A0(I)
      CALL CROSS(T3,AAJ,UJ)
      CALL CROSS(T4,VAJ,VUJ)
      CALL CROSS(T5,T1,AUJ)
      AMOMENT= DOT(U0,T3) + DOT(U0,T4) + DOT(U0,T5)
      RETURN
      END
C
      SUBROUTINE ROTATE(PJ,QJ,RM,P1,Q1,J)
C.... ROTATES A VECTOR (P1-Q1) TO (PJ-QJ)
C     COMPUTES (PJ-QJ)= (RM)*(P1-Q1)
C     COMPUTES PJ GIVEN P1,Q1,QJ,RM
      DIMENSION PJ(3),QJ(3),P1(3),Q1(3),RM(3,3,J)
      DO 10 I= 1,3
   10 PJ(I)= RM(I,1,J)*(P1(1)-Q1(1)) + RM(I,2,J)*(P1(2)-Q1(2))
     1 + RM(I,3,J)*(P1(3)-Q1(3)) + QJ(I)
      RETURN
      END
C
      SUBROUTINE DISP(PJ,DM,P1,J)
C.... COMPUTES PJ= (DM)*P1
      DIMENSION PJ(3),DM(4,4,J),P1(3)
      DO 10 I= 1,3
```

```
   10 PJ(I)= DM(I,1,J)*P1(1)+DM(I,2,J)*P1(2)+DM(I,3,J)*P1(3)+DM(I,4,J)
      RETURN
      END
C
      SUBROUTINE ROTVEC(V2,RM,V1,J)
      DIMENSION V2(3),RM(3,3,J),V1(3)
C.... COMPUTES V2= (RM)*V1
      DO 10 I= 1,3
      V2(I)= 0.0
      DO 10 K= 1,3
   10 V2(I)= RM(I,K,J)*V1(K) + V2(I)
      RETURN
      END
C
      SUBROUTINE PMTX(U,PM)
C.... COMPUTES ELEMENTS OF THE P-MATRIX (EQ. 3.57)
      DIMENSION U(3),PM(3,3)
      PM(1,1)= 0.0 $ PM(1,2)= -U(3) $ PM(1,3)= U(2)
      PM(2,1)= U(3) $ PM(2,2)= 0.0 $ PM(2,3)= -U(1)
      PM(3,1)= -U(2) $ PM(3,2)= U(1) $ PM(3,3)= 0.0
      RETURN
      END
C
      SUBROUTINE QMTX(U,QM)
C.... COMPUTES ELEMENTS OF THE Q-MATRIX  ( EQ. 3.57)
      DIMENSION U(3),QM(3,3)
      QM(1,1)= U(1)*U(1) $ QM(1,2)= U(1)*U(2) $ QM(1,3)= U(1)*U(3)
      QM(2,1)= U(2)*U(1) $ QM(2,2)= U(2)*U(2) $ QM(2,3)= U(2)*U(3)
      QM(3,1)= U(3)*U(1) $ QM(3,2)= U(3)*U(2) $ QM(3,3)= U(3)*U(3)
      RETURN
      END
C
      SUBROUTINE QIMTX(U,QM,QIM)
C.... COMPUTES ELEMENTS OF THE ( I-Q ) MATRIX  (EQ. 3.57)
      DIMENSION U(3),QM(3,3),QIM(3,3)
      DO 10 I= 1,3
      DO 10 J= 1,3
   10 QIM(I,J)= - QM(I,J)
      QIM(1,1)= 1. + QIM(1,1)
      QIM(2,2)= 1. + QIM(2,2)
      QIM(3,3)= 1. + QIM(3,3)
      RETURN
      END
C
      SUBROUTINE MTXVEC(TEMP,A,VEC)
C.... RETURNS PRODUCT OF MATRIX A TIMES VECTOR VEC IN TEMP
      DIMENSION A(3,3),VEC(3),TEMP(3)
      DO 10 I= 1,3
      TEMP(I)= 0.0
      DO 10 J= 1,3
   10 TEMP(I)= TEMP(I) + A(I,J)*VEC(J)
      RETURN
      END
C
      SUBROUTINE DM3PT(A1,B1,C1,AJ,BJ,CJ,DM,J)
C.... COMPUTE DISPLACEMENT MATRIX FROM MOTION OF THREE POINTS
      DIMENSION A1(3),B1(3),C1(3),AJ(3),BJ(3),CJ(3),D1(3),DC(3),
     $ DM(4,4,2),UAB1(3),UABJ(3),RM(3,3,2),P(4,4),Q(4,4),L(4,4),ERASE(4)
      COMMON /PRNTR/ PRNT
      LOGICAL PRNT
C COMPUTE DIRECTIONS FOR VECTORS B1-A1 AND BJ-AJ
      AB1= SQRT((B1(1)-A1(1))*(B1(1)-A1(1)) +
     $ (B1(2)-A1(2))*(B1(2)-A1(2)) + (B1(3)-A1(3))*(B1(3)-A1(3)))
      ABJ= SQRT((BJ(1)-AJ(1))*(BJ(1)-AJ(1)) +
     $ (BJ(2)-AJ(2))*(BJ(2)-AJ(2)) + (BJ(3)-AJ(3))*(BJ(3)-AJ(3)))
      UAB1(1)= (B1(1)-A1(1))/AB1
      UAB1(2)= (B1(2)-A1(2))/AB1
      UAB1(3)= (B1(3)-A1(3))/AB1
```

```
      UABJ(1)= (BJ(1)-AJ(1))/ABJ
      UABJ(2)= (BJ(2)-AJ(2))/ABJ
      UABJ(3)= (BJ(3)-AJ(3))/ABJ
C   COMPUTE COORDINATES FOR A FOURTH POINT BY ROTATION OF C ABOUT AB
      CON= ATAN(1.)/45.
      PHI= 90.*CON
      CALL RMAXIS(UAB1,PHI,RM,2)
      CALL ROTATE(D1,A1,RM,C1,A1,2)
      CALL RMAXIS(UABJ,PHI,RM,2)
      CALL ROTATE(DJ,AJ,RM,CJ,AJ,2)
C   FILL LEFT AND RIGHT SIDE MATRICES FOR LNEQF
      P(1,1)= AJ(1) $ P(1,2)= BJ(1) $ P(1,3)= CJ(1) $ P(1,4)= DJ(1)
      P(2,1)= AJ(2) $ P(2,2)= BJ(2) $ P(2,3)= CJ(2) $ P(2,4)= DJ(2)
      P(3,1)= AJ(3) $ P(3,2)= DJ(3) $ P(3,3)= CJ(3) $ P(3,4)= DJ(3)
      P(4,1)= P(4,2)= P(4,3)= P(4,4) = 1.0
      Q(1,1)= A1(1) $ Q(1,2)= B1(1) $ Q(1,3)= C1(1) $ Q(1,4)= D1(1)
      Q(2,1)= A1(2) $ Q(2,2)= B1(2) $ Q(2,3)= C1(2) $ Q(2,4)= D1(2)
      Q(3,1)= A1(3) $ Q(3,2)= B1(3) $ Q(3,3)= C1(3) $ Q(3,4)= D1(3)
      Q(4,1)= Q(4,2)= Q(4,3)= Q(4,4)= 1.0
      U(1,1)= U(2,2)= U(3,3)= U(4,4)= 1.0
      U(1,2)= U(1,3)= U(1,4)= U(2,1)= U(2,3)= U(2,4)= 0.0
      U(3,1)= U(3,2)= U(3,4)= U(4,1)= U(4,2)= U(4,3)= 0.0
C   INVERT RIGHT SIDE MATRIX - INVERSE STORED IN U
      NDIM= 4
      SCALE= 1.0
      N= NB= 4
      IF(LNEQF(NDIM,N,NB,Q,U,SCALE,ERASE) .EQ. 1 ) GO TO 20
      GO TO 30
C   MULTIPLY THE LEFT SIDE TIMES THE RIGHT SIDE INVERSE
   20 DO 25 I= 1,4
      DO 25 JJ= 1,4
      DM(I,JJ,J)= 0.0
      DO 25 K= 1,4
   25 DM(I,JJ,J)= DM(I,JJ,J) + P(I,K)*U(K,JJ)
      IF(PRNT) WRITE(6,50) J,((DM(I,K,J), K= 1,4), I= 1,4)
      RETURN
   30 WRITE(6,40)
      RETURN
   40 FORMAT(* NO INVERSE - THE MATRIX IS SINGULAR*)
   50 FORMAT(//10X,17H FROM DM3PT   DM 1,I1,//(4E15.5))
      END
C
      SUBROUTINE RMAXIS(U,PHI,RM,J)
C.... COMPUTES RM(U,PHI,J)
C      COMPUTES ROTATION MATRIX ELEMENTS IN TERMS OF
C      ROTATION ANGLE PHI ABOUT AN AXIS U
      DIMENSION RM(3,3,J), U(3)
      COMMON /PRNTR/ PRNT
      LOGICAL PRNT
      C= COS(PHI)
      S= SIN(PHI)
      V= 1. - C
      RM(1,1,J)= U(1)*U(1)*V + C
      RM(1,2,J)= U(1)*U(2)*V - U(3)*S
      RM(1,3,J)= U(1)*U(3)*V + U(2)*S
      RM(2,1,J)= U(1)*U(2)*V + U(3)*S
      RM(2,2,J)= U(2)*U(2)*V + C
      RM(2,3,J)= U(2)*U(3)*V - U(1)*S
      RM(3,1,J)= U(1)*U(3)*V - U(2)*S
      RM(3,2,J)= U(2)*U(3)*V + U(1)*S
      RM(3,3,J)= U(3)*U(3)*V + C

      IF(.NOT.PRNT) GO TO 99
      WRITE(6,100) J,((RM(I,K,J), K= 1,3), I= 1,3)
  100 FORMAT(//10X*AXIS ROTATION MATRIX RM1*I1//(3E15.7))
   99 RETURN
      END
C
```

```
      SUBROUTINE RMABG(A,B,G,RM,J)
C.... COMPUTES RM(ALPHA,BETA,GAMMA,J)
C     COMPUTES ROTATION MATRIX ELEMENTS IN TERMS OF PITCH YAW ROLL
C     ANGLES ALPHA BETA GAMMA
      DIMENSION RM(3,3,J)
      COMMON /PRNTR/ PRNT
      LOGICAL PRNT
      CA= COS(A)
      SA= SIN(A)
      CB= COS(B)
      SB= SIN(B)
      CG= COS(G)
      SG= SIN(G)
      RM(1,1,J)= CA*CB
      RM(1,2,J)= -SA*CB
      RM(1,3,J)= SB
      RM(2,1,J)= SA*CG + CA*SB*SG
      RM(2,2,J)= CA*CG - SA*SB*SG
      RM(2,3,J)= -CB*SG
      RM(3,1,J)= SA*SG - CA*SB*CG
      RM(3,2,J)= CA*SG + SA*SB*CG
      RM(3,3,J)= CB*CG
      IF(.NOT.PRNT) GO TO 99
      WRITE(6,100) J,((RM(I,K,J), K= 1,3), I= 1,3)
  100 FORMAT(//10X*ABG ROTATION MATRIX RM1*I1//(3E15.7))
   99 RETURN
      END
C
      SUBROUTINE RMEULER(PSI,THETA,PHI,RM,J)
C.... COMPUTES RM(PSI,THETA,PHI)
C     COMPUTES ROTATION MATRIX ELEMENTS IN TERMS OF EULER ANGLES
      DIMENSION RM(3,3,J)
      COMMON /PRNTR/ PRNT
      LOGICAL PRNT
      CPSI= COS(PSI)
      SPSI= SIN(PSI)
      CTH= COS(THETA)
      STH= SIN(THETA)
      CPHI= COS(PHI)
      SPHI= SIN(PHI)
      RM(1,1,J)= CPSI*CPHI - SPSI*CTH*SPHI
      RM(1,2,J)= - CPSI*SPHI - SPSI*CTH*CPHI
      RM(1,3,J)= SPSI*STH
      RM(2,1,J)= SPSI*CTH + CPSI*CTH*SPHI
      RM(2,2,J)= - SPSI*STH + CPSI*CTH*CPHI
      RM(2,3,J)= - CPSI*STH
      RM(3,1,J)= STH*SPHI
      RM(3,2,J) = STH*CPHI
      RM(3,3,J)= CTH
      IF(.NOT.PRNT) GO TO 99
      WRITE(6,100) J, ((RM(I,K,J), K= 1,3), I= 1,3)
  100 FORMAT(//10X*EULER ROTATION MATRIX RM1*I1//(3E15.7))
   99 RETURN
      END
C
      SUBROUTINE DMABG(A,B,G,RM,DM,J,PJ,P1)
C.... COMPUTES DM(U,PHI) FROM RM(U,PHI)
      DIMENSION RM(3,3,J), DM(4,4,J), PJ(3), P1(3)
      COMMON /PRNTR/ PRNT
      LOGICAL PRNT
      CALL RMABG(A,B,G,RM,J)
      DO 10 I= 1,3
      DO 10 K= 1,3
   10 DM(I,K,J)= RM(I,K,J)
      DM(4,4,J)= 1.0
      DO 20 I= 1,3
      DM(4,I,J)= 0.0
   20 DM(I,4,J)= PJ(I) - DM(I,1,J)*P1(1) - DM(I,2,J)*P1(2) - DM(I,3,J)*
```

```
      1 P1(3)
        IF(.NOT.PRNT) GO TO 99
        WRITE(6,100) J,((DM(I,K,J), K= 1,4), I= 1,4)
  100 FORMAT(//10X*DISPLACEMENT MATRIX DM1*I1//(4E15.7))
   99 RETURN
        END
C
        SUBROUTINE DMAXIS(U,PHI,RM,DM,J,PJ,P1)
C.... COMPUTES DM(U,PHI) FROM RM(U,PHI)
        DIMENSION U(3),RM(3,3,J),DM(4,4,J),PJ(3),P1(3)
        COMMON /PRNTR/ PRNT
        LOGICAL PRNT
        CALL RMAXIS(U,PHI,RM,J)
        DO 10 I= 1,3
        DO 10 K= 1,3
   10 DM(I,K,J)= RM(I,K,J)
        DM(4,4,J)= 1.0
        DO 20 I= 1,3
        DM(4,I,J)= 0.0
   20 DM(I,4,J)= PJ(I) - DM(I,1,J)*P1(1) - DM(I,2,J)*P1(2) - DM(I,3,J)*
      1 P1(3)
        IF(.NOT.PRNT) GO TO 99
        WRITE(6,100) J,((DM(I,K,J), K= 1,4), I= 1,4)
  100 FORMAT(//10X*DISPLACEMENT MATRIX DM1*I1//(4E15.7))
   99 RETURN
        END
C
        SUBROUTINE DMSCREW(P,U,S,PHI,RM,DM,J)
C.... COMPUTES FINITE SCREW MATRIX DM(P,U,S,PHI)
        DIMENSION P(3),PP(3),U(3),RM(3,3,J),DM(4,4,J)
        COMMON /PRNTR/ PRNT
        LOGICAL PRNT
        DO 10 I= 1,3
   10 PP(I)= P(I) + S*U(I)
        CALL DMAXIS(U,PHI,RM,DM,J,PP,P)
  100 FORMAT(//10X*SCREW MATRIX DM1*I1//(4E15.7))
        WRITE(6,100) J,((DM(I,K,J), K=1,4), I=1,4)
        IF(.NOT.PRNT) GO TO 99
   99 RETURN
        END
C
        SUBROUTINE DMEULER(PSI,THETA,PHI,RM,DM,J,PJ,P1)
C.... COMPUTES DM(PSI,THETA,PHI) FROM RM(PSI,THETA,PHI)
        DIMENSION  RM(3,3,J),DM(4,4,J)
        COMMON /PRNTR/ PRNT
        LOGICAL PRNT
        CALL RMEULER(PSI,THETA,PHI,RM,J)
        DO 10 I= 1,3
        DO 10 K= 1,3
   10 DM(I,K,J)= RM(I,K,J)
        DM(4,4,J)= 1.0
        DO 20 I= 1,3
        DM(4,I,J)= 0.0
   20 DM(I,4,J)= PJ(I) - DM(I,1,J)*P1(1) - DM(I,2,J)*P1(2) - DM(I,3,J)*
      1 P1(3)
        IF(.NOT.PRNT) GO TO 99
        WRITE(6,100) J,((DM(I,K,J), K= 1,4), I= 1,4)
  100 FORMAT(//10X*DISPLACEMENT MATRIX DM1*I1//(4E15.7))
   99 RETURN
        END
C
        SUBROUTINE WMTX(U,VPHI,WM,J)
C.... COMPUTES ANGULAR VELOCITY MATRIX (WM)
        DIMENSION WM(3,3,J),U(3)
        COMMON /PRNTR/ PRNT
        LOGICAL PRNT
        WM(1,1,J)= WM(2,2,J)= WM(3,3,J)= 0.0
        WM(1,2,J)= - U(3)*VPHI
```

```
      WM(1,3,J)= U(2)*VPHI
      WM(2,1,J)= -WM(1,2,J)
      WM(2,3,J)= - U(1)*VPHI
      WM(3,1,J)= - WM(1,3,J)
      WM(3,2,J)= - WM(2,3,J)
      IF(.NCT.PRNT) GO TO 99
      WRITE(6,100) J,((WM(I,K,J), K= 1,3), I= 1,3)
  100 FORMAT(//10X*ANGULAR VELOCITY MATRIX WM*I1//(3E15.7))
   99 RETURN
      END
C
      SUBROUTINE WDOT(U,VU,VPHI,APHI,WD,J)
C.... COMPUTES ANGULAR ACCELERATION MATRIX (WD)
      DIMENSION WD(3,3,J),U(3), VU(3)
      COMMON /PRNTR/ PRNT
      LOGICAL PRNT
      WD(1,1,J)= (U(1)*U(1) - 1.)*VPHI*VPHI
      WD(1,2,J)= U(1)*U(2)*VPHI*VPHI - VU(3)*VPHI - U(3)*APHI
      WD(1,3,J)= U(3)*U(1)*VPHI*VPHI + VU(2)*VPHI + U(2)*APHI
      WD(2,1,J)= U(1)*U(2)*VPHI*VPHI + VU(3)*VPHI + U(3)*APHI
      WD(2,2,J)= (U(2)*U(2) - 1.)*VPHI*VPHI
      WD(2,3,J)= U(2)*U(3)*VPHI*VPHI - VU(1)*VPHI - U(1)*APHI
      WD(3,1,J)= U(3)*U(1)*VPHI*VPHI - VU(2)*VPHI - U(2)*APHI
      WD(3,2,J)= U(3)*U(2)*VPHI*VPHI - VU(1)*VPHI + U(1)*APHI
      WD(3,3,J)= (U(3)*U(3) - 1.)*VPHI*VPHI
      IF(.NOT.PRNT) GO TO 99
      WRITE(6,100) J,((WD(I,K,J), K= 1,3), I= 1,3)
  100 FORMAT(//10X*ANGULAR ACCELERATION MATRIX WD*I1//(3E15.7))
   99 RETURN
      END
C
      SUBROUTINE ROTVEL(VP,VQ,WM,P,Q,J)
C.... COMPUTES (VP-VQ)= (WM)*(P-Q)
      DIMENSION VP(3),VQ(3),WM(3,3,J),P(3),Q(3)
      CALL ROTATE(VP,VQ,WM,P,Q,J)
      RETURN
      END
C
      SUBROUTINE ROTACC(AP,AQ,WD,P,Q,J)
C.... COMPUTES (AP-AQ)= (WD)*(P-Q)
      DIMENSION AP(3),AQ(3),WD(3,3,J),P(3),Q(3)
      CALL ROTATE(AP,AQ,WD,P,Q,J)
      RETURN
      END
C
      SUBROUTINE COMPERP(P1,P2,U1,U2,Q1,Q2,D1,D2)
C.... FINDS TWO POINTS ON THE COMMON PERPENDICULAR BETWEEN TWO LINES
C     DEFINED BY POINTS P1 AND P2  WITH DIRECTIONS U1 AND U2,
C     DISTANCES D1 AND D2 ARE MEASURED FROM POINTS P1 AND P2,
      DIMENSION P1(3),P2(3),U1(3),U2(3),Q1(3),Q2(3)
      T1= DOTU2P(U2,P1,P2)
      T2= DOT(U1,U2)
      T3= DOTU2P(U1,P1,P2)
      D1= (T1*T2 - T3)/(1. - T2*T2)
      D2= (T1 - T3*T2)/(1. - T2*T2)
      DO 10 I= 1,3
      Q1(I)= P1(I) + D1*U1(I)
   10 Q2(I)= P2(I) + D2*U2(I)
      RETURN
      END
```

Appendix 3.

DESIGN

```
      PROGRAM DESIGN(INPUT,OUTPUT,TAPE5=INPUT,TAPE6=OUTPUT)
      COMMON /SIMEQS/ CONVRG,LIM,NORM,PRINT,NITR,ITNS,DAMP
      COMMON /YCOMP/ ONCE,NOSE,V
      LOGICAL CONVRG,LIM,NORM,PRINT,INCMT,ONCE,EOF,DAMP
      DIMENSION Z(30),A(30,30),B(30),C(30),D(30),E(30),AA(30,30),BB(30),
     1          F(30),NSET(5)
C
C     *********************USER INSTRUCTIONS***************************
C
C     PROGRAM DESIGN WILL SOLVE UP TO 5 SETS OF NONLINEAR ALGEBRAIC
C     EQUATIONS IN SEQUENCE WITH THE RESULTS FOR EACH SET AVAILABLE TO
C     THE FOLLOWING SETS.
C
C     A SELECTED VARIABLE Z(I) MAY BE INCREMENTED BY DELV AND THE
C     SOLUTION REPEATED NINC TIMES.  THIS ALLOWS THE GENERATION OF
C     CURVES REPRESENTING FAMILILIES OF SOLUTIONS AS A FUNCTION OF ONE
C     VARIABLE PARAMETER Z(I).
C
C     THE USER MUST SUPPLY THE EQUATIONS TO BE SOLVED IN THE FORM F(Z)=
C     WHERE EACH FUNCTION F(Z) IS EVALUATED IN A USER SUPPLIED FUNCTION
C     YCOMP.   SEE THE SAMPLE PROBLEMS FOR YCOMP EXAMPLES.
C     ********************* INPUT FORMAT ****************************
C
C     READ(5,2) NTEQ,NSEQ,NINC,NFEI,NITR,(NSET(I),I=1,5),L,DELV
C     5 FORMAT(10I5,2F10.0)
C     READ(5,6) (Z(I),  I=1,NTEQ)
C     6 FORMAT(8F10.0)
C
C     NTEQ= TOTAL NUMBER OF EQUATIONS = TOTAL NUMBER OF VARIABLES
C           IF NTEQ IS SET NEGATIVE PRNT IS SET .TRUE. AND SIMEQS WILL
C           PRINT AFTER EACH ITERATION.
C     NSEQ= NUMBER OF SETS INTO WHICH THE NTEQ EQUATIONS ARE DIVIDED
C     NINC= NUMBER OF TIMES THE VARIABLE V IS TO BE INCREMENTED
C     NFEI= IDENTIFIER OF THE FIRST SET WHICH CONTAINS A VARIABLE BEING
C           INCREMENTED.
C     NITR= ITERATION LIMIT ( SUGGEST LET NITR= 10 + NTEQ )
C     NSET(I)= NUMBER OF EQUATIONS IN EACH SET(I)
C     U= ORIGINAL VALUE OF V. V IS RESET TO U FOR EACH NEW INITIAL GUESS
C     DELV= INCREMENT FOR THE VARIABLE V
C     (Z(I), I=1,NTEQ) = INITIAL VARIABLES VECTOR
C
C     A VARIABLE Z(I) IS INCREMENTED IN FUNCTION YCOMP BY ADDING A DUMMY
C     FUNCTION IN THE FORM   YCOMP= Z(I) - V
C                     WHERE V IS TRANSFERRED IN COMMON /YCOMP/
C
C     WHEN THIS DUMMY FUNCTION APPROACHES ZERO THE VALUE OF Z(I) IS
C     INCREMENTED TO THE CURRENT VALUE OF V.
C
C     MULTIPLE INITIAL GUESSES MAY BE SUPPLIED. THE PROGRAM RETURNS TO T
C     SECOND READ STATEMENT AFTER EACH SOLUTION IS FOUND.
C     TO SEND THE PROGRAM BACK TO THE FIRST READ STATEMENT INSERT A
C     DATA CARD WITH ANY INTEGER PUNCHED IN COLUMN 1 .      THIS WILL
C     RESTART THE PROGRAM AND READ IN A COMPLETE NEW SET OF PARAMETERS.
C     INCLUDE BLANK DATA CARDS IF FORMER NTEQ WAS GREATER THAN 8
C
```

371

```
C      ********************* OUTPUT FORMAT *****************************
C      THE DATA CARD IMAGES ARE PRINTED FIRST.
C      DEPENDING ON THE STATE OF LOGICAL VARIABLES CONVRG,LIM,DAMP,NORM,
C      PRINT ... MESSAGES ARE PRINTED BY BOTH PROGRAM DESIGN AND SUBROUTI
C      SIMEQS.
C
C      IF(PRINT) .. SIMEQS WILL PRINT AFTER EACH ITERATION
C      IF(CONVRG) .. MESSAGE .. CONVERGED IN XX ITERATIONS
C         WHICH INDICATES DELZ/Z LESS THAN 1.E-07 FOR ALL VARIABLES.
C      IF(.NOT.CONVRG.AND.NORM) .. MESSAGE .. NORM DECREASED
C         WHICH INDICATES SOME PROGRESS TOWARD A SOLUTION
C      IF(.NOT.CONVRG.AND.DAMP) .. MESSAGE .. DAMPING LIMIT EXCEEDED
C      IF(.NOT.CONVRG.AND.LIM) .. MESSAGE .. ITERATION LIMIT EXCEEDED
C      WHEN THE GAUSS ELIMINATION IN FUNCTION LINEQF FAILS .. MESSAGE ..
C      LINEQF FAILED AT ITERATION .. THE CURRENT MATRICES B(I) AND A(I,J)
C
C      FINALLY THE CURRENT OR ENDING VALUES FOR THE Z VECTOR ARE PRINTED.
C
C      CCCCCCCCCCCCCCCCCCCCCCCCCCCCCCCCCCCCCCCCCCCCCCCCCCCCCCCCCCCCCCCCCC
     5 READ (5,50) NTEQ,NSEQ,NINC,NFEI,NITR,(NSET(I),I=1,5),L,DELV
       IF (EOF(5)) GO TO 45
       WRITE (6,55)
       WRITE (6,60) NTEQ,NSEQ,NINC,NFEI,NITR,(NSET(I),I=1,5),U,DELV
       PRINT=.FALSE.
       INCMT=.FALSE.
       ONCE=.FALSE.
       IF (NINC.GT.0) INCMT=.TRUE.
       IF (NTEQ) 10,45,15
    10 PRINT=.TRUE.
       NTEQ=-NTEQ
    15 READ (5,65) (Z(I),I=1,NTEQ)
       IF (Z(1).GE.1.E6) GO TO 5
       IF (EOF(5)) GO TO 45
       WRITE (6,70) (Z(I),I=1,NTEQ)
       V=U
       ICNT=0
       IN=1
       JJ=0
    20 DO 25 NOSE=IN,NSEQ
       NBR=NSET(NOSE)
       II=JJ+1
       JJ=NSET(NOSE)+JJ
       CALL SIMEQS (NBR,II,JJ,Z,A,B,C,D,E,AA,BB,F)
       IF (.NOT.CONVRG) WRITE (6,75) NOSE
       IF (.NOT.CONVRG.AND.LIM) WRITE (6,80) ITNS,(Z(I),I=II,JJ)
       IF (.NOT.CONVRG.AND.DAMP) WRITE (6,85) ITNS,(Z(I),I=II,JJ)
       IF (CONVRG.AND.INCMT) WRITE (6,90) ITNS,V,(Z(I),I=II,JJ)
       IF (.NOT.INCMT.AND.CONVRG) WRITE (6,95) ITNS,(Z(I),I=II,JJ)
       IF (.NOT.CONVRG.AND.NORM) WRITE (6,100) ITNS,(Z(I),I=II,JJ)
    25 CONTINUE
       IF (NSEQ.EQ.1.OR.NFEI.EQ.1) GO TO 35
       IN=NFEI
       NN=NFEI-1
       JJ=0
       DO 30 K=1,NN
    30 JJ=JJ+NSET(K)
       GO TO 40
    35 IN=1
       JJ=0
    40 V=V+DELV
       ICNT=ICNT+1
       IF (NINC.EQ.1) GO TO 15
       IF (ICNT.LE.NINC) GO TO 20
       GO TO 15
    45 STOP
C
    50 FORMAT(10I5,2F10.0)
    55 FORMAT(1H1,30(1H*)*   NEW DATA  *,30(1H*)//* NTEQ NSEQ NINC NFEI NIT
```

```
      1R NSET(I), I= 1,5*15X   *     V           DELV*)
   60 FORMAT(10I5,2E15.7)
   65 FORMAT(8F10.0)
   70 FORMAT(47H0 INITIAL GUESSES FOR VARIABLES Z(I), I= 1,NTEQ  /
      1 (5E15.7))
   75 FORMAT(31H0NO CONVERGENCE FCR SET NUMBER ,I1)
   80 FORMAT(38H0ITERATION LIMIT EXCEEDED AT ITERATION ,I3/10H CURRENT Z
      1/(5E15.7))
   85 FORMAT(36H0DAMPING LIMIT EXCEEDED AT ITERATION ,I3/
      1 10H CURRENT Z/(5E15.7))
   90 FORMAT(14H0CONVERGED IN ,I3,11H ITERATIONS/
      1 17H ENDING Z WITH V=,F10.3/(5E15.7))
   95 FORMAT(14H0CONVERGED IN ,I3,11H ITERATIONS/
      1 9H ENDING Z/(5E15.7))
  100 FORMAT(28H0NORM DECREASED AT ITERATION ,I3/
      1 10H0CURRENT Z/(5E15.7))
      END
      SUBROUTINE SIMEQS (NBR,II,JJ,Z,A,B,C,D,E,AA,BB,F)
C     ... SOLVES SIMULTANEOUS NONLINEAR EQUATIONS BY NEWTON,S METHOD.
      LOGICAL CONVRG,LIM,NORM,PRINT,DAMP
      COMMON /SIMEQS/ CONVRG,LIM,NORM,PRINT,NITR,ITNS,DAMP
      DIMENSION Z(NBR), A(NBR,NBR), B(NBR), C(NBR), D(NBR), E(NBR),
      1                 AA(NBR,NBR),BB(NBR)
C     ... IF CONVRG IS TRUE, CORRECTION VECTOR IS .LT. 1.0E-7 OF PRESEN
C     ... VARIABLE VECTOR, SO CONVERGENCE HAS OCCURRED.  IF LIM IS TRUE
C     ... THE ITERATION LIMIT HAS BEEN EXCEEDED.  IF NORM IS TRUE, THE
C     ... NORM OF THE FUNCTION VECTOR HAS DECREASED, INDICATING THAT SO
C     ... PROGRESS HAS BEEN MADE TOWARD A SOLUTION.  IF PRINT IS TRUE,
C     ... ITERATIONS ARE PRINTED.
      NORM=.FALSE.
      CONVRG=.FALSE.
      LIM=.FALSE.
      DAMP=.FALSE.
      MARK=0
      ITNS=0
      GMIN=0.0
      DO 5 I=II,JJ
      D(I)=-GRAD(Z,I,0,NBR,F)
    5 GMIN=GMIN+ABS(D(I))
      IF (GMIN.LT.1.0E-7) GO TO 80
C     ... COMPUTE PARTIAL DERIVS.
   10 DO 15 I=II,JJ
      B(I)=D(I)
      DO 15 J=II,JJ
   15 A(I,J)=GRAD(Z,I,J,NBR,F)
      IF (NBR.EQ.1) 20,25
   20 B(I)=B(I)/A(I,I)
      GO TO 40
C     ... USE LINEQF TO SOLVE FOR CORRECTION VECTOR.
   25 NN=0
      DO 30 I=II,JJ
      NN=1+NN
      BB(NN)=B(I)
      MM=0
      DO 30 J=II,JJ
      MM=1+MM
   30 AA(NN,MM)=A(I,J)
      DET=1.0
      IF (LINEQF(NBR,NBR,AA,BB,DET,E).NE.1) GO TO 90
      NN=0
      DO 35 I=II,JJ
      NN=1+NN
   35 B(I)=BB(NN)
   40 ITNS=ITNS+1
C     ... SEE IF CORRECTION TO Z DECREASE NORM OF FUNCTICN VECTOR.
   45 DO 50 I=II,JJ
   50 C(I)=Z(I)+B(I)
      GNORM=0.0
```

```
      DO 55 I=II,JJ
      D(I)=-GRAD(C,I,0,NBR,F)
  55 GNORM=GNORM+ABS(D(I))
      IF (PRINT) GO TO 115
      IF (GNORM.LE.1.0E-7) GO TO 65
  60 IF (GNORM.GT.GMIN) GO TO 120
  65 CONTINUE
      NORM=.TRUE.
      GMIN=GNORM
      NDAMP=0
      DO 70 I=II,JJ
  70 Z(I)=C(I)
C     ... CONVERGENCE TEST.
      DO 75 I=II,JJ
      TEST=ABS(B(I))
      IF (ABS(Z(I)).GT.1.0) TEST=ABS(B(I)/Z(I))
      IF (TEST.GT.1.0E-7) GO TO 85
  75 CONTINUE
  80 CONVRG=.TRUE.
      RETURN
  85 IF (ITNS.LE.NITR) GO TO 10
      LIM=.TRUE.
      RETURN
C     ... FOR FAILURE OF LINEQF.  TRY SCALING MATRICES.
  90 IF (MARK.EQ.1) RETURN
      MARK=1
      WRITE (6,140) ITNS
      DO 95 I=II,JJ
  95 WRITE (6,145) I,B(I),(A(I,J),J=II,JJ)
      DO 105 I=II,JJ
      SCALE=0.0
      DO 100 J=II,JJ
      AIJ=ABS(A(I,J))
 100 IF (AIJ.GT.SCALE) SCALE=AIJ
      IF (SCALE.LT.1.0E-10) SCALE=1.0
      B(I)=B(I)/SCALE
      DO 105 J=II,JJ
 105 A(I,J)=A(I,J)/SCALE
      WRITE (6,150)
      DO 110 I=II,JJ
 110 WRITE (6,145) I,B(I),(A(I,J),J=II,JJ)
      GO TO 25
 115 WRITE (6,155) ITNS,(B(I),I=II,JJ)
      WRITE (6,160) (C(I),I=II,JJ)
      WRITE (6,165) (D(I),I=II,JJ)
      GO TO 60
C     ..40 IF NO DECREASE IN NORM OF FUNCTION VECTOR TRY DAMPING I.E.
C     ... REDUCE CORRECTION VECTOR IN CASE PROBLEM IS DUE TO INSTABILIT
 120 NORM=.FALSE.
      WRITE(6,170)
      NDAMP=NDAMP+1
      IF (NDAMP.GT.10) DAMP=.TRUE.
      IF (DAMP) RETURN
      SCALE=5.0
      IF (NDAMP.GT.1) GO TO 130
      DO 125 I=II,JJ
      BI=ABS(B(I))
 125 IF (BI.GT.SCALE) SCALE=BI
 130 DO 135 I=II,JJ
 135 B(I)=B(I)/SCALE
      GO TO 45
C
 140  FORMAT (27H0LINEQF FAILED AT ITERATION I2,31H.  THE MATRICES B(I)
     1 AND A(I,J))
 145 FORMAT (I4, E15.7, 5X, 5E15.7/(24X, 5E15.7))
 150 FORMAT (15H WERE SCALED TO  )
 155 FORMAT(13H AT ITERATION I2/11H CORRECTNS /(5X,5E15.7))
 160 FORMAT(11H VARIABLES /(5X,5E15.7))
```

```
  165 FORMAT(11H FUNCTIONS /(5X,5E15.7))
  170 FORMAT(8H DAMPING )
      END
      FUNCTION GRAD (Z, I, J, NBR, F)
C     ...  APPROXIMATES PARTIAL DERIV OF ITH YCOMP WRT JTH Z.  IF J=0,
C     ...  RETURNS VALUE OF ITH YCOMP.
      DIMENSION Z(NBR),F(NBR)
      IF (J.NE.0) GO TO 5
      F(I)=YCOMP(Z,I)
      GRAD=F(I)
      RETURN
    5 Y1=F(I)
      TEMP=Z(J)
      H=0.001*ABS(TEMP)
      IF (H.LT.0.0001) H=0.0001
      Z(J)=TEMP+H
      Y2=YCOMP(Z,I)
      Z(J)=TEMP
      GRAD=(Y2-Y1)/H
      RETURN
      END
      FUNCTION LINEQF(M,N,A,B,DTRMNT,Z
      REAL   A(M,M), B(M), Z(M)
C     .. SOLVES SIMULTANEOUS LINEAR EQUATIONS BY GAUSSIAN REDUCTION.
C     .. FORTRAN IV EQUIVALENT OF LNEQS.
      EPS=1.0E-30
      NM1=N-1
      DO 40 J=1,NM1
      J1=J+1
      LMAX=J
      RMAX=ABS(A(J,J))
      DO 5 K=J1,N
      RNEXT=ABS(A(K,J))
      IF (RMAX.GE.RNEXT) GO TO 5
      RMAX=RNEXT
      LMAX=K
    5 CONTINUE
      IF (LMAX.NE.J) GO TO 10
      IF (ABS(A(J,J)).LT.EPS) GO TO 70
      GO TO 20
   10 DO 15 L=J,N
      W=A(J,L)
      A(J,L)=A(LMAX,L)
      A(LMAX,L)=W
   15 CONTINUE
      W=B(J)
      B(J)=B(LMAX)
      B(LMAX)=W
      DTRMNT=-DTRMNT
   20 Z(J)=1./A(J,J)
      DO 35 K=J1,N
      IF (A(K,J)) 25,35,25
   25 W=-Z(J)*A(K,J)
      DO 30 L=J1,N
      A(K,L)=W*A(J,L)+A(K,L)
   30 CONTINUE
      B(K)=W*B(J)+B(K)
   35 CONTINUE
   40 CONTINUE
      IF (ABS(A(N,N)).LT.1.0E-30) GO TO 70
      Z(N)=1./A(N,N)
      B(N)=Z(N)*B(N)
      DO 50 K=1,NM1
      J=N-K
      J1=J+1
      W=0.
      DO 45 I=J1,N
      W=A(J,I)*B(I)+W
```

```
   45 CONTINUE
      B(J)=(B(J)-W)*Z(J)
   50 CONTINUE
      IF (DTRMNT.EQ.0.0) GO TO 60
      DO 55 J=1,N
   55 DTRMNT=DTRMNT*A(J,J)
   60 LINEQF=1
   65 CONTINUE
      RETURN
C     ONTINUE HERE FOR SINGULAR OR NEAR-SINGULAR CASE...................
   70 LINEQF=2
      DTRMNT=0.
      GO TO 65
      END
```

Appendix 4.

NONLIN

```
        PROGRAM NONLIN(INPUT,OUTPUT,TAPE5=INPUT,TAPE6=OUTPUT)
        COMMON /KNOW/ CONVRG,LIM,NORM,PRINT,NITR,ITNS,N DAMP
        COMMON NOSE,V,ONCE
        COMMON /PRNTR/ PRNT
        LOGICAL CONVRG,LIM,NORM,PRINT,PRNT,INCMT,ONCE,EOF
        DIMENSION Z(30),A(30,30),B(30),C(30),D(30),E(30)
        DIMENSION ZMAX(30),ZMIN(30),ZI(30),TEMP(30,100)
C
CCCCCCCCCCCCCCCCCCCCCCCCCCCCCCCCCCCCCCCCCCCCCCCCCCCCCCCCCCCCCCCCCCCCCCCC
C                                                                      C
C       SOLVES A MAXIMUM OF THIRTY SIMULTANEOUS NONLINEAR EQUATIONS     C
C       WITH RANDOM INITIAL GUESSES.                                   C
C                                                                      C
C       SUPPLY NEQ AS A NEGATIVE INTEGER IF YOU WISH TO PRINT EACH      C
C       ITERATION IN SIMEQ. INITIAL GUESSES ARE GENERATED AUTOMATICALLY C
C       BY A RANDOM NUMBER GENERATOR.                                  C
C                                                                      C
C       NEQ= NUMBER OF EQUATIONS= NUMBER OF VARIABLES                   C
C       NSOL= TOTAL NUMBER OF RANDOM INITIAL GUESSES TO BE TRIED        C
C       ZMIN(I)= MINIMUM VALUE OF EACH VARIABLE IN THE SEARCH SPACE     C
C       ZMAX(I)= MAXIMUM VALUE                                         C
C                                                                      C
CCCCCCCCCCCCCCCCCCCCCCCCCCCCCCCCCCCCCCCCCCCCCCCCCCCCCCCCCCCCCCCCCCCCCCCC
C
      1 WRITE(6,5)
      5 FORMAT(1H1)
        ONCE= .FALSE.
        PRINT=.FALSE.
        ICOUNT=1
        READ(5,10) NEQ,NSOL
        IF(EOF,5) 100,11
     10 FORMAT(2I3)
     11 IF(NEQ.LT.0) PRINT= .TRUE.
        IF(NEQ.GT.30) GO TO 75
        NEQ= IABS(NEQ)
        READ (5,15) ((ZMIN(N),ZMAX(N)), N= 1,NEQ)
     15 FORMAT(2F10.5)
        DO 99 I= 1,NSOL
        DO 12 K= 1,NEQ
C
CCCCCCCCCCCCCCCCCCCCCCCCCCCCCCCCCCCCCCCCCCCCCCCCCCCCCCCCCCCCCCCCCCCCCCCC
C                                                                      C
C       THE RANDOM NUMBER FUNCTION MAY REQUIRE REPLACEMENT             C
C                                                                      C
CCCCCCCCCCCCCCCCCCCCCCCCCCCCCCCCCCCCCCCCCCCCCCCCCCCCCCCCCCCCCCCCCCCCCCCC
C
        Z(K)= ZMIN(K) + (ZMAX(K) - ZMIN(K))*RANF(0)
        ZI(K)= Z(K)
     12 CONTINUE
        CALL SIMEQ(NEQ,Z,A,B,C,D,E)
        N=0
        IF(CONVRG) GO TO 30
        IF(LIM) GO TO 55
        IF(N DAMP.GE.10) GO TO 65
        GO TO 99
     30 DO 31 II=1,NEQ
```

377

```
31      TEMP(II,ICOUNT)=Z(II)
        IF(ICOUNT.EQ.1) GO TO 40
        DO 33 J= 1, ICOUNT
        IF(J.EQ.1) GO TO 33
        DO 32  M=1,NEQ
        JJ= J-1
        TEST= ABS(Z(M)-TEMP(M,JJ))
        IF(TEST.LT. 0.00001) N= N+1
        IF(N.EQ.NEQ) GO TO 50
32      CONTINUE
33      CONTINUE
40      N= 0
        DO 41 M= 1,NEQ
        TEST= ABS(Z(M) - ZI(M))
        IF(TEST.LT.0.0001) N= N + 1
41      IF(N.EQ.NEQ) GO TO 61
        GO TO 43
61      CONVRG= .FALSE.
        WRITE(6,62)
62      FORMAT(//* NO PROGRESS FROM INITIAL GUESSES*)
        GO TO 99
43      CONTINUE
        WRITE(6,45) ICOUNT,(ZI(N), N=1,NEQ)
45      FORMAT(//* ... SOLUTION NUMBER *I2* ...*/
       1 * INITIAL GUESSES*/(5E14.5))
        WRITE(6,46) (Z(N), N= 1,NEQ)
46      FORMAT(/* SOLUTION*/(5E14.5))
        ICOUNT= ICOUNT + 1
50      CONVRG= .FALSE.
        GO TO 99
55      WRITE(6,60) (Z(N), N= 1,NEQ)
        LIM=.FALSE.
60      FORMAT(//* ITERATIONS EXCEEDED 10 PLUS NUMBER OF EQUATIONS*
       1 //* CURRENT Z*/(5E14.5))
        GO TO 99
65      WRITE(6,68) ITNS,(Z(N), N= 1,NEQ)
68      FORMAT(//* DAMPING EXCEEDS 10 AT ITERATION *I3//
       1 * CURRENT Z*/(5E14.5))
99      CONTINUE
        IKOUNT= ICOUNT - 1
        WRITE(6,70) IKOUNT,NSOL
70      FORMAT(///I2* SOLUTIONS FOUND FOR*I3* INITIAL GUESSES*)
        GO TO 1
75      WRITE(6,80)
80      FORMAT(//20X*... NUMBER OF EQUATIONS EXCEEDS 30 ...*)
  100   GO TO 1
        END


        SUBROUTINE SIMEQ(NBR, Z, A, B, C, D, E)
C...    SOLVES SIMULTANEOUS NONLINEAR EQUATIONS BY NEWTON≠S METHOD.
        COMMON /KNOW/ CONVRG,LIM,NORM,PRINT,NITR,ITNS,N DAMP
        LOGICAL CONVRG,LIM,NORM,PRINT,EOF
        DIMENSION Z(NBR), A(NBR,NBR), B(NBR), C(NBR), D(NBR), E(NBR)
C...    IF CONVRG IS TRUE, CORRECTION VECTOR IS .LT. 1.0E-7 OF PRESENT
C...    VARIABLE VECTOR, SO CONVERGENCE HAS OCCURRED.  IF LIM IS TRUE,
C...    THE ITERATION LIMIT HAS BEEN EXCEEDED.  IF NORM IS TRUE, THE
C...    NORM OF THE FUNCTION VECTOR HAS DECREASED, INDICATING THAT SOME
C...    PROGRESS HAS BEEN MADE TOWARD A SOLUTION.  IF PRINT IS TRUE,
C...    ITERATIONS ARE PRINTED.
        LIMIT = 10 + NBR
        ITNS = 0
        N DAMP = 0
        GMIN = 0.0
        DO 100 I=1,NBR
        D(I) = - GRAD (Z, I, 0, NBR)
  100   GMIN = GMIN + ABS(D(I))
        IF(GMIN.LT.1.E-7) GO TO 500
C...    COMPUTE PARTIAL DERIVS.
```

```
   200   DO 205 I=1,NBR
         B(I) = D(I)
         DO 205 J=1,NBR
   205   A(I,J) = GRAD (Z, I, J, NBR)
C...   USE LINEQF TO SOLVE FOR CORRECTION VECTOR.
         DET = 1.0
         IF (LINEQF(NBR, NBR, A, B, DET, E) .NE. 1) GO TO 700
   210   ITNS = ITNS + 1
C...   SEE IF ADDING CORRECTION VECTOR TO Z WILL DECREASE NORM OF
C...   FUNCTION VECTOR.
   300   DO 305 I=1,NBR
   305   C(I) = Z(I) + B(I)
         GNORM = 0.0
         DO 310 I=1,NBR
         D(I) = - GRAD (C, I, 0, NBR)
   310   GNORM = GNORM + ABS(D(I))
         IF (PRINT) GO TO 800
         IF(GNORM.LE.1.0E-7) GO TO 316
   315   IF (GNORM .GT. GMIN) GO TO 900
   316   CONTINUE
         NORM = .TRUE.
         GMIN = GNORM
         N DAMP = 0
         DO 400 I=1,NBR
   400   Z(I) = C(I)
C...   CONVERGENCE TEST.
         DO 500 I=1,NBR
         TEST = ABS(B(I))
         IF(ABS(Z(I)).GE. 1.0)  TEST= ABS(B(I)/Z(I))
         IF (TEST .GT. 1.0E-7) GO TO 600
   500   CONTINUE
         CONVRG = .TRUE.
         RETURN
C...   SEE IF ITERATION LIMIT HAS BEEN EXCEEDED.
   600   IF (ITNS .LE. LIMIT) GO TO 200
         LIM = .TRUE.
         RETURN
C...   FOR FAILURE OF LINEQF,  TRY SCALING MATRICES IN CASE FAILURE
C...   WAS DUE TO UNDER- OR OVER-FLOW.
   700   DO 705 I=1,NBR
         B(I) = D(I)
         DO 705 J=1,NBR
   705   A(I,J) = GRAD (Z, I, J, NBR)
         PRINT 10, ITNS
    10   FORMAT (27H0LINEQF FAILED AT ITERATION I2,15H.   THE MATRICES )
         DO 710 I=1,NBR
   710   PRINT 20, I, B(I), (A(I,J), J=1,NBR)
    20   FORMAT (I4, E15.7, 5X, 5E15.7/(24X, 5E15.7))
         DO 720 I=1,NBR
         SCALE = 0.0
         DO 715 J=1,NBR
         AIJ = ABS(A(I,J))
   715   IF (AIJ .GT. SCALE) SCALE = AIJ
         IF (SCALE .LT. 1.0E-10) SCALE = 1.0
         B(I) = B(I)/SCALE
         DO 720 J=1,NBR
   720   A(I,J) = A(I,J)/SCALE
         PRINT 30
    30   FORMAT (15H WERE SCALED TO  )
         DO 725 I=1,NBR
   725   PRINT 20, I, B(I), (A(I,J), J=1,NBR)
         IF (LINEQF (NBR, NBR, A, B, DET, E) .EQ. 1) GO TO 210
         RETURN

C...   FOR PRINTING ITERATIONS.
   800   PRINT 40, ITNS, (B(I), I=1,NBR)
    40   FORMAT (13H0AT ITERATION I2/11H CORRECTNS 5E15.7/(11X,5E15.7))
         PRINT 50, (C(I), I=1,NBR)
```

```
   50   FORMAT (11H VARIABLES   5E15.7/(11X, 5E15.7))
        PRINT 60, (D(I), I=1,NBR)
   60   FORMAT (11H FUNCTIONS   5E15.7/(11X, 5E15.7))
        GO TO 315
C...  FOR NO DECREASE IN NORM OF FUNCTION VECTOR.  TRY REDUCING
C...  (DAMPING) CORRECTION VECTOR IN CASE PROBLEM IS DUE TO INSTABILITY.
  900   N DAMP = N DAMP + 1
        IF (N DAMP .GT. 10) RETURN
        NORM= .FALSE.
        SCALE = 5.0
  910   DO 915 I=1,NBR
  915   B(I) = B(I)/SCALE
        GO TO 300
        END
```

```
        PROGRAM     PCON(INPUT,OUTPUT,TAPE5=INPUT,TAPE6=OUTPUT)
CDC RUN COMPILER C W RADCLIFFE MECH ENG U C BERKELEY UPDATED APRIL 21, 1977
C
C       PROGRAM PCON SOLVES THE PROBLEM OF FINDING THE CONSTRAINED MINIMUM
C       OF A FUNCTION OF UP TO 30 VARIABLES USING A MODIFICATION OF
C       POWELL.S DIRECT NUMERICAL SEARCH - CONJUGATE DIRECTIONS METHOD.
C
C       REFERENCE ...
C       POWELL, M J D , AN EFFICIENT METHOD FOR FINDING THE MINIMUM OF A
C       FUNCTION WITHOUT CALCULATING DERIVATIVES, COMPUTER JOURNAL, VOL7,
C       64,P155
C
C       THE MODIFICATION COMBINES POWELL≠S UNCONSTRAINED METHOD WITH A
C       SEQUENTIAL  UNCONSTRAINED MINIMIZATION TECHNIQUE(SIMILAR TO SUMT).
C       REFERENCE ...
C       FIACCO AND MCCORMICK, NONLINEAR PROGRAMMING - SEQUENTIAL
C       UNCONSTRAINED MINIMIZATION TECHNIQUE, JOHN WILEY, 1968,
C
C       A FEATURE OF PCON IS THE AUTOMATIC GENERATION OF WEIGHTING
C       FACTORS FOR THE INEQUALITY CONSTRAINT FUNCTIONS G(I)=≤ 0 AND THE
C       EQUALITY CONSTRAINT FUNCTIONS H(I)= 0 SUCH THAT ALL CONSTRAINTS
C       ARE NORMALIZED AT THE BASE POINT. THIS TENDS TO REDUCE THE CHANCE
C       THAT ONE OR MORE OF THE CONSTRAINTS WILL DOMINATE THE PENALTY
C       FUNCTION.
C
C       THREE CONSTANTS FAC, GFAC, HFAC PROVIDE CONTROL OVER THE
C       START OF THE SEARCH FOR A MINIMUM. THE MODIFIED OBJECTIVE FUNCTION
C       TAKES ONE OF THE FOLLOWING FORMS (SEE SUBROUTINE RESTRT).
C       VALU= FFUN(X) - WTG*FAC**(1-IR)*GIN(X) + WTH*FAC**(IR-1)*H(X)
C       OR
C       VALU= FFUN(X) + WTG*FAC**(IR-1)*GEX(X) + WTH*FAC**(IR-1)*H(X)
C       WHERE
C       FAC**(1-IR)= INTERIOR WEIGHTING FACTOR WHICH DECREASES AS IR IS
C        INCREASED
C       FAC**(IR-1)= EXTERIOR WEIGHTING FACTOR HICH INCREASES AS IR IS
C       INCREASED,
C       GIN(X)=SUM(1,/G(I))    I=1,MG
C       GEX(X)=SUM(G(I)**2)    I=1,MG
C       H(X)=SUM(H(I)**2)      I=1,MH
C       GFAC= WTG*GEX(X)/FFUN(X)     GFAK= 1/GFAC = WTG*GIN(X)/FFUN(X)
C       HFAC= WTH*H(X)/FFUN(X)
C
C       THE USER MUST SUPPLY THE FOLLOWING SUBPROGRAMS
C       WHERE X IS THE VECTOR OF PROBLEM VARIABLES
C
C       FUNCTION FFUN(X)
C       RETURNS THE VALU OF THE UNCONSTRAINED OBJECTIVE FUNCTION
C
C       SUBROUTINE GFUN(G,X)
C       RETURNS THE CURRENT INEQUALITY CONSTRAINT FUNCTIONS G(I), I= 1,MG
C
C       SUBROUTINE HFUN(H,X)
C       RETURNS THE CURRENT EQUALITY CONSTRAINT FUNCTIONS H(I), I= 1,MH
C
C       SUBROUTINE MORPRNT(X,F,FC,GG,HH)
```

```
C     PROVIDES OPTIONAL OUTPUT PRINT FORMAT AT END OF EXECUTION
C
C     THE CONSTRAINED OBJECTIVE FUNCTION IS CALCULATED BY SUBROUTINE
C     RESTNT(IR,VALU), WHERE IR IS THE SEQUENTIAL EXPONENTIAL FACTOR.
C
C     INPUT DATA FORMAT
C
C     FIRST CARD ... READ N,MAXIT,ISW,IEPS,IPRNT,MG,MH,NRUN,ISRCH,REUSE
C                    FORMAT(10I5)
C
C     DATA CARD ONE CAN BE BLANK AFTER ISW FOR UNCONSTRAINED MINIMUM.
C
C               N= NUMBER OF VARIABLES
C               MAXIT= ITERATION LIMIT
C                 ISW=0 ... WRITE FINAL RESULTS ONLY
C                    =1 ... WRITE STATUS AFTER EACH ITERATIOIN
C                    =2 ... WRITE STATUS AFTER EACH SEARCH ALONG A LINE
C                    =3 ... WRITE STATUS AFTER EACH TEN ITERATIONS
C                 IEPS=0 ... SET ALL EPS(I)= EPS1
C                    =1 ... READ IN VALUES FOR ALL EPS(I)
C                 IPRNT= 0 ... SUPPRESS PRINT OF WEIGHTING FACTORS
C                      = 1 ... PRINT WEIGHTING FACTORS
C                 MG= NUMBER OF INEQUALITY CONSTRAINT FUNCTIONS
C                 MH= NUMBER OF EQUALITY CONSTRAINT FUNCTIONS
C                 NRUN=NUMBER OF TIMES IR IS TO BE INCREMENTED BY 1 DURING
C                      THE SEQUENTIAL MINIMIZATIONS.
C                 ISRCH=0 ... IF INITIAL GUESSES ARE NOT FEASIBLE SEARCH
C                      FOR A FEASIBLE POINT BY MINIMIZATION OF
C                      SUM((G(I) + 100.*EPS1)**2)
C                    =1 ... FORCE USE OF EXTERIOR PENALTY FUNCTION
C                    =2 ... USE INTERIOR PENALTY IF INITIAL GUESSES ARE
C                      ... FEASIBLE OTHERWISE USE EXTERIOR PENALTY.
C                 REUSE ... ALLOWS RESULTS OF A PREVIOUS PROBLEM TO BE
C                      USED  AS INITIAL GUESSES.  X(1) THROUGH X(REUSE) WILL
C                      BE SET TO THEIR PREVIOUS VALUES.  THE USER THEN
C              SPECIFIES X(REUSE+1) THROUGH X(N) FOR THE CURRENT PROBLEM.
C                      SET REUSE= 0 FOR NORMAL EXECUTION.
C
C                      SET LOGICAL VARIABLES SCAL, TRY, AND MORE=.TRUE.
C                      BY READING IN CERTAIN DATA AS NEGATIVE INTEGERS
C                      WHICH ARE RESET POSITIVE BEFORE CALCULATIONS BEGIN.
C                      IF(N.LT.0) SCAL= .TRUE.
C                      IF(MAXIT.LT.0) TRY= .TRUE.
C                      IF(NRUN.LT.0) MORE= .TRUE.
C                      IF SCAL= .TRUE.
C                      THE  ..AUTOMATIC SCALING OPTION .. IS ADOPTED.
C                      IN THIS CASE ALL VARIABLES ARE NORMALIZED ABOUT
C                      THEIR INITIAL VALUES AT THE BEGINNING OF A SEARCH
C                      FOR EACH NEW VALUE OF IR.
C                      IF TRY= .TRUE. THE TRY AGAIN OPTION IS ADOPTED WITH
C                      REDUCTION OF EPS1 TO .01*EPS1 AFTER A CONSTRAINED
C                      MINIMUM HAS BEEN FOUND.
C                      IF MORE= .TRUE. SUBROUTINE MORPRNT(X,F,FC,GG,HH) WILL
C                      BE CALLED AT THE END OF NORMAL EXECUTION.
C
C     SECOND CARD ... READ ESCALE,EPS1,FAC,GFAC,HFAC
C                    FORMAT(5F10.0)
C                    ESCALE ... MAX MOVE FOR X(I)= ESCALE*EPS(I)
C                    EPS1 ... IF(IEPS.EQ.0) EPS(I)= EPS1 FOR I= 1,N
C                    FAC= EXPONENTIAL FACTOR IN PENALTY FUNCTIONS
C                         SET BETWEEN 10. AND 4. ... 10. IS NORMAL
C                    GFAC= DESIRED RATIO GEX(X)/FFUN(X) AT START
C                         SET BETWEEN 1. AND 5. ... TRY 1. FIRST
C                    HFAC= DESIRED RATIO H(X)/FFUN(X) AT START
C                         SET BETWEEN 1. AND 5. ... TRY 1. FIRST
C
C     THIRD CARD ... READ X(I) I= 1,N  (INITIAL GUESSES FOR X VECTOR)
C                    FORMAT(8F10.0)
```

```
C
C      FOURTH CARD ... READ EPS(I), I=1,N  (ONLY IF IEPS= 1)
C                      FORMAT(8F10.0)
C
C DESCRIPTION OF USE OF CONSTANTS L1 THROUGH L6
C   IF L1=0 RESET ISRCH
C   IF L2 OR L5=0 MESSAGES ARE PRINTED REGARDING FEASIBILITY
C   IF L3=0 SCALE INEQUALITY CONSTRAINTS
C   IF L3=0 COMPUTE WEIGHTING FACTORS WTG AND WTH
C   IF L4=0 SCALE EQUALITY CONSTRAINTS
C   IF L6=0 PRINT MESSAGE REGARDING EQUALITY WEIGHTING FACTORS
C
C      ........... END OF COMMENTS .................................C
C
       DIMENSION S(30)
       COMMON /POW/ X(30),Y(30),N,EPS(30),ESCALE,F,ISW
       COMMON/SUCC/NMOVE,NOLUCK,NFCC
       COMMON/POWDER/ICON,MAXIT,W(1100)
       COMMON /PEN/ JPEN,L1,L2,L3,L4,L5,L6,GEXP,GINP,HP,ISRCH,DEL(30)
      1 ,GG(30),HH(30)
       COMMON /GH/ MG,MH,FAC,GFAC,HFAC,WTG,WTH,GFAK,IPRNT,SX(30),SCAL
       COMMON /SAB/ FINTIAL
       COMMON /FTEST/ TEST
       COMMON /JSR/ JSRCH
       LOGICAL CONVRG,CONVRJ,TEST,TRY,SCAL,MORE
       INTEGER REUSE
C
    5 NRUNC=0
       IR=1
       TTUSED=0.
       NFTOT=0
       ICOUNT=0
       JPEN=L1=L2=L3=L4=L5=L6=0
       NFCC=0
       CONVRJ=.FALSE.
       TRY=.FALSE.
       SCAL=.FALSE.
       TEST=.TRUE.
       MORE=.FALSE.
       WRITE (6,240)
       READ (5,165) N,MAXIT,ISW,IEPS,IPRNT,MG,MH,NRUN,ISRCH,REUSE
       IF (EOF,5) 150,10
   10 CONTINUE
       IF (ISRCH.GT.2) ISRCH=0
       JSRCH=ISRCH
       READ (5,170) ESCALE,EPS1,FAC,GFAC,HFAC
       GFAK=1./GFAC
       WRITE (6,180) N,MAXIT,ISW,IEPS,IPRNT,MG,MH,NRUN,ISRCH,REUSE
       WRITE (6,185) ESCALE,EPS1,FAC,GFAC,HFAC
       IF (N.LT.0) SCAL=.TRUE.
       N=IABS(N)
       IF (MAXIT.LT.0) TRY=.TRUE.
       MAXIT=IABS(MAXIT)
       IF (NRUN.LT.0) MORE=.TRUE.
       NRUN=IABS(NRUN)
       ICON=1
       IF (REUSE.GE.N) GO TO 25
       M=REUSE+1
       WRITE (6,245)
       READ (5,190) (X(J),J=M,N)
       WRITE (6,195) (X(J),J=M,N)
       IF (TRY) WRITE (6,155)
       IF (SCAL) WRITE (6,160)
       DO 15 K=1,N
   15 S(K)=X(K)
       IF(MG.EQ.0) GO TO 25
       DO 20 K=1,MG
   20 DEL(K)=100.*EPS1
```

```
   25 IF (IEPS.GT.0) GO TO 35
      DO 30 J=1,N
   30 EPS(J)=EPS1
      GO TO 40
   35 READ (5,190) (EPS(J),J=1,N)
      WRITE (6,190) (EPS(J),J=1,N)
   40 L1=L3=L4=0
   45 IF (MG.EQ.0.AND.MH.EQ.0) GO TO 50
      WRITE (6,175) IR
   50 CALL SECOND (TI)
      FS=FFUN(X)
      IF (IR.GT.1) GO TO 60
      WRITE (6,200) FS,(X(I),I=1,N)
      IF (MG.EQ.0.AND.MH.EQ.0) GO TO 60
      IF (MG.EQ.0) GO TO 55
      CALL GFUN (GG,X)
      WRITE (6,230) (GG(I),I=1,MG)
   55 IF (MH.EQ.0) GO TO 60
      CALL HFUN (HH,X)
      WRITE (6,235) (HH(I),I=1,MH)
   60 CONTINUE
C  IF SCAL IS .TRUE. ALL VARIABLES ARE NORMALIZED ABOUT THEIR INITIAL
C  GUESSES BEFORE BEGINNING SEARCH FOR A MINIMUM.
      IF (SCAL) CALL SCALE (X,SX,N)
      CALL POWELL (IR)
      IF (SCAL) CALL DESCALE (X,SX,N)
      FC=FFUN(X)
      GN=GIN(X)
      GX=GEX(X)
      HE=H(X)
      CALL SECOND (TF)
      TUSED=TF-TI
      IF (MG.EQ.0.AND.MH.EQ.0) GO TO 65
      IF (NOLUCK.EQ.1) GO TO 90
   65 GO TO (70,75,80,85), NOLUCK
   70 WRITE (6,205)
      GO TO 90
   75 WRITE (6,210)
      IF (CONVRJ) GO TO 150
      GO TO 90
   80 WRITE (6,215)
      IF (CONVRJ) GO TO 150
      GO TO 90
   85 WRITE (6,220)
   90 CONTINUE
      IF(MG.EQ.0.AND.MH.EQ.0) WRITE(6,275) NMOVE,NFCC,FS,FC,(X(I)
     1,I=1,N)
      IF (MG.EQ.0.AND.MH.EQ.0.AND.MORE) CALL MORPRNT (X,F,FC,GG,HH)
C  FC= CURRENT OBJECTIVE FUNCTION WITHOUT PENALTY TERMS
C  F= MODIFIED OBJECTIVE FUNCTION WITH PENALTY TERMS
      IF (MG.EQ.0.AND.MH.EQ.0) GO TO 5
      IF (CONVRJ) GO TO 100
      IF (ISRCH.NE.0) GO TO 95
      IF (.NOT.TEST.AND.ISRCH.EQ.0) IR=IR-1
      ISRCH=2
      TEST= .TRUE.
      JSRCH= ISRCH
      NRUNC=NRUNC-1
      L1=L3=L4=L6=0
C  .....WHEN CONVRJ= .TRUE. SEARCH FOR FEASIBLE POINT IS FINISHED
   95 CONVRJ=.TRUE.
  100 IF (MG.NE.0) CALL GFUN (GG,X)
      IF (MH.NE.0) CALL HFUN (HH,X)
      IF (NOLUCK.NE.1) GO TO 110
      NFTOT=NFTOT+NFCC
      WRITE (6,225) IR,NMOVE,NFCC,NFTOT,FINTIAL,F,FS,FC,GN,GX,HE,(X(I),I
     1=1,N)
      IF (MG.EQ.0) GO TO 105
```

```
      WRITE (6,230) (GG(I),I=1,MG)
  105 IF (MH.EQ.0) GO TO 110
      WRITE (6,235) (HH(I),I=1,MH)
  110 CONTINUE
      CONVRG=.TRUE.
      DO 115 K=1,N
      TST=100.
      IF (X(K).EQ.0.0) GO TO 115
      TST=ABS((S(K)-X(K))/X(K))
C     .....WHEN CONVRG= .TRUE. A CONSTRAINED MINIMUM HAS BEEN FOUND
  115 IF (TST.GT.EPS1) CONVRG=.FALSE.
      DO 120 K=1,N
  120 S(K)=X(K)
      WRITE (6,250) TUSED
      TTUSED=TTUSED+TUSED
      NRUNC=NRUNC+1
      NFLIM=N*1000
      IF (NFCC.LT.NFLIM.AND.NRUNC.LT.NRUN) GO TO 125
      GO TO 145
  125 IR=IR+1
      IF (CONVRG) GO TO 130
      GO TO 40
  130 IF (MG.EQ.0.AND.MH.EQ.0) GO TO 135
      WRITE (6,255) EPS1
  135 WRITE (6,260) NFTOT
      WRITE (6,265) TTUSED
      IF (ICOUNT.EQ.2) GO TO 150
      IF (.NOT.TRY) GO TO 145
      IR=IR-1
      ICOUNT=ICOUNT+1
      DO 140 K=1,N
  140 EPS(K)=.01*EPS(K)
      EPS1=.01*EPS1 $ ESCALE= 100.*ESCALE
      WRITE (6,270) EPS1
  145 CONTINUE
      IF (MORE) CALL MORPRNT (X,F,FC,GG,HH)
      IF (NFCC.GE.NFLIM.OR.NRUNC.GE.NRUN) GO TO 5
      IF (.NOT.TRY) GO TO 5
      GO TO 45
  150 STOP
C
  155 FORMAT(5X,*TRY AGAIN OPTION ADOPTED*)
  160 FORMAT(5X,*VARIABLE SCALING OPTION ADOPTED*)
  165 FORMAT (10I5)
  170 FORMAT(5F10.0)
  175 FORMAT(/25(1H*)* START PCON WITH IR= *I3,25(1H*)/)
  180 FORMAT(5X*N*1X*MAXIT*3X*ISW*2X*IEPS*1X*IPRNT*4X*MG*4X*MH*2X*NRUN*
     1 1X*ISRCH*1X*REUSE*/(10I6))
  185 FORMAT(/7X*ESCALE*9X*EPS1*10X*FAC*9X*GFAC*9X*HFAC*/(5E13.4))
  190 FORMAT(8F10.0)
  195 FORMAT(5E14.6)
  200 FORMAT(25(1H*)* INITIAL CONDITIONS *25(1H*)//* F-INITIAL= *E14.6//
     1 * VARIABLES ARE*/(5E14.6))
  205 FORMAT(///20(1H*),* UNCONSTRAINED MINIMUM FOUND *,20(1H*)///)
  210 FORMAT(1H ,* NO UNIQUE MINIMUM*/* OBJECTIVE FUNCTION MAY BE INDEPE
     1NDENT OF AT LEAST ONE VARIABLE*)
  215 FORMAT(* CONVERGENCE CANNOT BE ACHIEVED FOR CURRENT EPS*)
  220 FORMAT(1H ,*MAXIMUM NUMBER OF ITERATIONS.  FINAL VALUES.*)
  225 FORMAT(//2X*IR= *I2,2X*ITERATIONS= *I3,2X*FUNC EVAL= *I4,2X,
     1 *SUBTOTAL= *I4/2X*U-INITIAL=*E15.7,2X*U-FINAL=*E15.7/
     2 2X*F-INITIAL=*E15.7,2X*F-FINAL=*E15.7/
     3 6X*G-INT=*E15.7,4X*G-EXT=*E15.7,6X*H=*E15.7//
     4 2X*VARIABLES*/(5E14.6))
  230 FORMAT(/* INEQUALITY CONSTRAINTS*/(5E14.6))
  235 FORMAT(/* EQUALITY CONSTRAINTS*/(5E14.6))
  240 FORMAT(1H1,30(1H*),10H NEW DATA ,30(1H*)//)
  245 FORMAT(//* INITIAL GUESSES (X(I),I=1,N)*/)
  250 FORMAT(1H0,8X,*TIME USED FOR THIS MINIMIZATION*,F9.5,* SECONDS*//)
```

```
  255 FORMAT(///21(1H*)* CONSTRAINED MINIMUM FOUND *21(1H*)//
     1  *   RATIO CHANGE/CURRENT LESS THAN* E10.3* FOR ALL X(I)*)
  260 FORMAT(//* TOTAL FUNC EVAL = *I4)
  265 FORMAT(//* TOTAL TIME USED *,F9.5,* SECONDS *)
  270 FORMAT(* TRY AGAIN WITH EPS1=*E12.3//)
  275 FORMAT(//2X*CONVERGED IN *I3* ITERATIONS AND *I4* FUNCTIONAL *
     1  *EVALUATIONS*/2X*F-INITIAL=*E15.7 ,5X*F-CURRENT=*E15.7/
     2  2X*VARIABLES ARE*/(5E14.6))
      END

      SUBROUTINE PWRITE (I)
C     WRITES SYSTEM STATUS.
      COMMON /POW/ X(30),Y(30),N,EPS(30),ESCALE,F,ISW
      DIMENSION MESSAGE(9)
      DATA MESSAGE/10HF(INITIAL),10HF(CURRENT),8HF(FINAL)/
      NPBY2=(N+1)/2
      NBY2=N/2
      WRITE (6,20) MESSAGE(I),F
      IF (N.EQ.1) GO TO 10
      DO 5 J=1,NBY2
      K=J+NPBY2
      WRITE (6,15) J,X(J),K,X(K)
    5 CONTINUE
      IF (NPBY2.EQ.NBY2) RETURN
 . 10 J=NBY2+1
      WRITE (6,15) J,X(J)
      RETURN
C
   15 FORMAT (I10,E20.8,10X,I10,E20.8)
   20 FORMAT(1H ,20X,A10,2H =,E15.8//,9X,1HJ,11X,4HX(J),24X,1HJ,11X,
     14HX(J)/)
      END

      SUBROUTINE POWELL (LI)
C
C     THE ORIGINAL POWELL PROGRAM WITH MODIFICATIONS TO ALLOW SEQUENTIAL
C     UNCONSTRAINED  MINIMIZATION WITH PENALTY FUNCTIONS.
C
      COMMON /POW/ X(30),Y(30),N,E(30),ESCALE,F,IWRITE
      COMMON /SUCC/ ITERC,NOLUCK,NFCC
      COMMON /POWDER/ ICON,MAXIT,W(1100)
      COMMON /SAB/ PINTIAL
      COMMON /PEN/ JPEN,L1,L2,L3,L4,L5,L6,GEXP,GINP,HP,ISRCH,DEL(30)
     1 ,GG(30),HH(30)
      COMMON /GH/ MG,MH,FAC,GFAC,HFAC,WTG,WTH,GFAK,IPRNT,SX(30),SCAL
      LOGICAL SCAL
      NOLUCK=1
      DDMAG=0.1*ESCALE
      SCER=0.05/ESCALE
      JJ=N*N+N
      JJJ=JJ+N
      K=N+1
      NFCC=1
      ITER=0
      IND=1
      INN=1
      DO 20 I=1,N
      DO 15 J=1,N
      W(K)=0.
      IF (I-J) 10,5,10
    5 W(K)=ABS(E(I))
   10 K=K+1
   15 CONTINUE
   20 CONTINUE
      ITERC=0
      ISGRAD=2
      CALL RESTNT (LI,F)
```

```
      PINTIAL=F
      FKEEP=ABS(F)+ABS(F)
   25 ITONE=1
      FP=F
      SUM=0.
      IXP=JJ
      DO 30 I=1,N
      IXP=IXP+1
      W(IXP)=X(I)
   30 CONTINUE
      IDIRN=N+1
      ILINE=1
      KLINE=1
   35 DMAX=W(ILINE)
      DACC=DMAX*SCER
      DMAG=AMIN1(DDMAG,0.1*DMAX)
      DMAG=AMAX1(DMAG,20.*DACC)
      DDMAX=10.*DMAG
      GO TO (40,40,155), ITONE
   40 DL=0.
      D=DMAG
      FPREV=F
      IS=5
      FA=F
      DA=DL
   45 DD=D-DL
      DL=D
   50 K=IDIRN
      DO 55 I=1,N
      X(I)=X(I)+DD*W(K)
      K=K+1
   55 CONTINUE
      CALL RESTNT (LI,F)
      NFCC=NFCC+1
      NFLIM=N*1000
      IF (NFCC.GT.NFLIM) WRITE (6,435)
      IF (NFCC.GT.NFLIM) RETURN
      GO TO (160,145,135,120,60,265), IS
   60 IF (F-FA) 80,65,85
   65 IF (ABS(D)-DMAX) 70,70,75
   70 D=D+D
      GO TO 45
   75 NOLUCK=2
      GO TO 410
   80 FB=F
      DB=D
      GO TO 90
   85 FB=FA
      DB=DA
      FA=F
      DA=D
   90 GO TO (100,95), ISGRAD
   95 D=DB+DB-DA
      IS=1
      GO TO 45
  100 D=0.5*(DA+DB-(FA-FB)/(DA-DB))
      IS=4
      IF ((DA-D)*(D-DB)) 105,45,45
  105 IS=1
      IF (ABS(D-DB)-DDMAX) 45,45,110
  110 D=DB+SIGN(DDMAX,DB-DA)
      IS=1
      DDMAX=DDMAX+DDMAX
      DDMAG=DDMAG+DDMAG
      IF (DDMAG.GE.1.0E+60) DDMAG=1.0E+60
      IF (DDMAX-DMAX) 45,45,115
  115 DDMAX=DMAX
```

```
          GO TO 45
    120 IF (F-FA) 125,95,95
    125 FC=FB
          DC=DB
    130 FB=F
          DB=D
          GO TO 165
    135 IF (F-FB) 125,125,140
    140 FA=F
          DA=D
          GO TO 165
    145 IF (F-FB) 150,160,160
    150 FA=FB
          DA=DB
          GO TO 130
    155 DL=1.
          DDMAX=5.
          FA=FP
          DA=-1.
          FB=FHOLD
          DB=0.
          D=1.
    160 FC=F
          DC=D
    165 A=(DB-DC)*(FA-FC)
          B=(DC-DA)*(FB-FC)
          IF ((A+B)*(DA-DC)) 170,170,175
    170 FA=FB
          DA=DB
          FB=FC
          DB=DC
          GO TO 110
    175 D=0.5*(A*(DB+DC)+B*(DA+DC))/(A+B)
          DI=DB
          FI=FB
          IF (FB-FC) 185,185,180
    180 DI=DC
          FI=FC
    185 GO TO (195,195,190), ITONE
    190 ITONE=2
          GO TO 205
    195 IF (ABS(D-DI)-DACC) 225,225,200
    200 IF (ABS(D-DI)-0.03*ABS(D)) 225,225,205
    205 IF ((DA-DC)*(DC-D)) 215,210,210
    210 FA=FB
          DA=DB
          FB=FC
          DB=DC
          GO TO 105
    215 IS=2
          IF ((DB-D)*(D-DC)) 220,45,45
    220 IS=3
          GO TO 45
    225 F=FI
          D=DI-DL
          DD=SQRT((DC-DB)*(DC-DA)*(DA-DB)/(A+B'))
          DO 230 I=1,N
          X(I)=X(I)+D*W(IDIRN)
          W(IDIRN)=DD*W(IDIRN)
          IDIRN=IDIRN+1
    230 CONTINUE
          FCUR=FFUN(Y)
          IF (IWRITE.EQ.2) WRITE (6,420) ITERC,F,FCUR,(X(I),I=1,N)
          W(ILINE)=W(ILINE)/DD
          ILINE=ILINE+1
          GO TO (235,335), ITONE
    235 IF (FPREV-F-SUM) 245,240,240
```

```
240 SUM=FPREV-F
    JIL=ILINE
245 IF (IDIRN-JJ) 35,35,250
250 GO TO (255,340), IND
255 FHOLD=F
    IS=6
    IXP=JJ
    DO 260 I=1,N
    IXP=IXP+1
    W(IXP)=X(I)-W(IXP)
260 CONTINUE
    DD=1.
    GO TO 50
265 GO TO (270,280), IND
270 IF (FP-F) 320,320,275
275 D=2.*(FP*F-2.*FHOLD)/(FP-F)**2
    IF (D*(FP-FHOLD-SUM)**2-SUM) 280,320,320
280 J=JIL*N+1
    IF (J-JJ) 285,285,300
285 DO 290 I=J,JJ
    K=I-N
    W(K)=W(I)
290 CONTINUE
    DO 295 I=JIL,N
    W(I-1)=W(I)
295 CONTINUE
300 IDIRN=IDIRN-N
    ITONE=3
    K=IDIRN
    IXP=JJ
    AAA=0.
    DO 315 I=1,N
    IXP=IXP+1
    W(K)=W(IXP)
    IF (AAA-ABS(W(K)/E(I))) 305,310,310
305 AAA=ABS(W(K)/E(I))
310 K=K+1
315 CONTINUE
    DDMAG=1.
    W(N)=ESCALE/AAA
    ILINE=N
    GO TO 35
320 IXP=JJ
    AAA=0.
    F=FHOLD
    DO 330 I=1,N
    IXP=IXP+1
    X(I)=X(I)-W(IXP)
    IF (AAA*ABS(E(I))-ABS(W(IXP))) 325,330,330
325 AAA=ABS(W(IXP)/E(I))
330 CONTINUE
    GO TO 340
335 AAA=AAA*(1.+DI)
    GO TO (340,405), IND
340 CONTINUE
    KLINE=KLINE+1
    GO TO (345,360), IND
345 IF (AAA-0.1) 410,410,350
350 IF (F-FP) 365,355,355
355 NOLUCK=3
    GO TO 410
360 IND=1
365 DDMAG=0.4*SQRT(FP-F)
    IF (DDMAG.GE.1.0E+60) DDMAG=1.0E+60
    ISGRAD=1
    ITERC=ITERC+1
    ITER=ITER+1
```

```
      IF (ITER.EQ.10.AND.IWRITE.EQ.3) GO TO 370
      IF (IWRITE.EQ.1) GO TO 370
      GO TO 385
  370 FCUR=FFUN(Y)
      WRITE (6,420) ITERC,F,FCUR,(X(I),I=1,N)
      IF (MG.EQ.0) GO TO 375
      CALL GFUN (GG,X)
      WRITE (6,425) (GG(I),I=1,MG)
  375 IF (MH.EQ.0) GO TO 380
      CALL HFUN (HH,X)
      WRITE (6,430) (HH(I),I=1,MH)
  380 ITER=0
  385 CONTINUE
      IF (ITERC-MAXIT) 25,25,390
  390 NOLUCK=4
      IF (F-FKEEP) 410,410,395
  395 F=FKEEP
      DO 400 I=1,N
      JJJ=JJJ+1
      X(I)=W(JJJ)
  400 CONTINUE
      GO TO 410
  405 IF (AAA-0.1) 410,410,415
  410 EF=F
      RETURN
  415 INN=1
      GO TO 365
C
  420 FORMAT(//* ITERATIONS=*I4* U-CURRENT=*E15.7* F-CURRENT=*E15.7/
     1 * VARIABLES ARE*/(5E14.6))
  425 FORMAT(* INEQUALITY CONSTRAINTS*/(5E14.6))
  430 FORMAT(* EQUALITY CONSTRAINTS*/(5E14.6))
  435 FORMAT(* ... NUMBER OF FUNCTIONAL EVALUATIONS EXCEEDS*/
     1 /*    N TIMES 1000*)
      END


      SUBROUTINE RESTNT (IR,VALU)
      COMMON /POW/ X(30),Y(30),N,EPS(30),ESCALE,F,ISW
      COMMON /PEN/ JPEN,L1,L2,L3,L4,L5,L6,GEXP,GINP,HP,ISRCH,DELL(30),
     1 GG(30),HH(30)
      COMMON /GH/ MG,MH,FAC,GFAC,HFAC,WTG,WTH,GFAK,IPRNT,SX(30),SCAL
      COMMON /FTEST/ TEST
      COMMON /JSR/ JSRCH
      LOGICAL TEST,SCAL
      DO 10 I=1,N
      IF (SCAL) GO TO 5
      Y(I)=X(I)
      GO TO 10
    5 Y(I)=SX(I)*X(I)
   10 CONTINUE
      IF (MG.EQ.0.AND.MH.EQ.0) GO TO 60
      IF (L1.EQ.1) GO TO 25
      TRY=FFUN(Y)
      IF (MG.EQ.0) GO TO 15
      TST=GEX(Y)
   15 IF (MH.EQ.0) GO TO 20
      TST=H(Y)
   20 L1=1
      J=2
      IF (.NOT.TEST.AND.JSRCH.EQ.0) J=3
      IF (TEST.AND.JSRCH.EQ.2) J= 1
      IF (TEST.AND.JSRCH.EQ.0) J=1
   25 CONTINUE
      IF (MG.EQ.0) GO TO 65
      GO TO (30,40,50), J
   30 IF (L2.EQ.1) GO TO 35
      L2=1
```

```
      WRITE (6,70)
   35 GINP=GIN(Y)
      HP=H(Y)
      VALU=FFUN(Y)-WTG*FAC**(1-IR)*GINP+WTH*FAC**(IR-1)*HP
      RETURN
   40 IF (L2.EQ.1) GO TO 45
      L2=1
      IF (.NOT.TEST) WRITE (6,75)
      IF (TEST) WRITE (6,85)
   45 GEXP=GEX(Y)
      HP=H(Y)
      VALU=FFUN(Y)+WTG*FAC**(IR-1)*GEXP+WTH*FAC**(IR-1)*HP
      RETURN
   50 IF (L5.EQ.1) GO TO 55
      L5=1
      WRITE (6,80)
   55 GEXP=GEX(Y)
      VALU=GEXP
      RETURN
   60 VALU=FFUN(Y)
      RETURN
   65 HP=H(Y)
      VALU=FFUN(Y)+WTH*FAC**(IR-1)*HP
      RETURN
C
   70 FORMAT(/* INITIAL GUESSES ARE FEASIBLE*/* BEGIN SEARCH*
     1 * WITH INTERIOR PENALTY FUNCTION*/)
   75 FORMAT(/* INITIAL GUESSES ARE NOT FEASIBLE*/* BEGIN SEARCH*
     1 * WITH EXTERIOR PENALTY FUNCTION*/)
   80 FORMAT(/* INITIAL GUESSES ARE NOT FEASIBLE*/
     1 * BEGIN SEARCH FOR A FEASIBLE BASE POINT*/)
   85 FORMAT(/* INTIAL GUESSES ARE FEASIBLE*/
     1 * BUT FORCE SEARCH WITH EXTERIOR PENALTY FUNCTION*/)
      END

      FUNCTION   GIN(X)
      DIMENSION X(1)
      COMMON /WW/ WG(30),WH(30)
      COMMON /PEN/ JPEN,L1,L2,L3,L4,L5,L6,GEXP,GINP,HP,ISRCH,DELL(30),
     1 GG(30),HH(30)
      COMMON /GH/ MG,MH,FAC,GFAC,HFAC,WTG,WTH,GFAK,IPRNT,SX(30),SCAL
      COMMON /JSR/ JSRCH
      LOGICAL SCAL
      IF (MG.EQ.0) GO TO 20
      GIN=0.0
      CALL GFUN (GG,X)
      DO 15 L=1,MG
      GG(L)=WG(L)*GG(L)
      IF (GG(L).LT.0.0) GO TO 10
      IF (GG(L).GT.0.0) GO TO 5
      IF (GG(L).EQ.0.0) GG(L)=-1.E-6
      GO TO 10
    5 GG(L)=-1./(1.E+8*GG(L)**2)
   10 GIN=GIN+1.0/GG(L)
   15 CONTINUE
      RETURN
   20 GIN=0.0
      RETURN
      END

      FUNCTION   GEX(X)
      DIMENSION X(1)
      COMMON /WW/ WG(30),WH(30)
      COMMON /PEN/ JPEN,L1,L2,L3,L4,L5,L6,GEXP,GINP,HP,ISRCH,DELL(30),
     1 GG(30),HH(30)
      COMMON /GH/ MG,MH,FAC,GFAC,HFAC,WTG,WTH,GFAK,IPRNT,SX(30),SCAL
      COMMON /FTEST/ TEST
```

```
      COMMON /JSR/ JSRCH
      LOGICAL SCAL
      LOGICAL TEST
      IF (MG.EQ.0) GO TO 45
      IF (L3.EQ.1.AND.JSRCH.EQ.0) GO TO 35
      IF (L3.EQ.1) GO TO 25
C     .... TEST IF INITIAL GUESSES ARE FEASIBLE
      TEST= .TRUE.
      CALL GFUN (GG,X)
      IF (IPRNT.EQ.1) WRITE (6,50)
      DO 5 L=1,MG
      IF (GG(L).GT.0.0) TEST=.FALSE.
C     .... SCALE INEQUALITY CONSTRAINTS ON FIRST PASS
      IF (ABS(GG(L)).LE.1.E-6) GG(L)=-1.E-6
      WG(L)=ABS(1.0/GG(L))
    5 IF (IPRNT.EQ.1) WRITE (6,55) L,WG(L)
C     ....WEIGHTING FACTOR FOR EXTERIOR PENALTY FUNCTION
      IF (JSRCH.NE.0.AND.L6.EQ.1) GO TO 20
      IF (TEST) GO TO 15
      GEX=0.0
      CALL GFUN (GG,X)
      DO 10 L=1,MG
   10 GEX=GEX+(WG(L)*GG(L))**2
      WTG=ABS(GFAC*FFUN(X)/GEX)
      IF (IPRNT.EQ.1) WRITE (6,60) WTG,GFAC
      IF (MH.EQ.0) L6=1
      GO TO 20
C     .... WEIGHTING FACTOR FOR INTERIOR PENALTY FUNCTION
   15 WTG=ABS(GFAK*FFUN(X)/GIN(X))
      IF (IPRNT.EQ.1) WRITE (6,65) WTG,GFAK
      IF (MH.EQ.0) L6=1
   20 CONTINUE
      L3=1
C     .... COMPUTE EXTERIOR PENALTY FUNCTION
   25 GEX=0.0
      CALL GFUN (GG,X)
      DO 30 L=1,MG
      IF (GG(L).LE.0.0) GG(L)=0.0
   30 GEX=GEX+(WG(L)*GG(L))**2
      RETURN
C     .... COMPUTE SPECIAL PENALTY FUNCTION USED IN SEARCH FOR FEASIBLE
   35 GEX=0.0
      TEMP=FFUN(X)
      CALL GFUN (GG,X)
      DO 40 K=1,MG
      GG(K)=GG(K)+DELL(K)
      IF (GG(K).LT.0.0) GG(K)=1.0E-6*GG(K)
   40 GEX=GEX+(WG(K)*GG(K))**2
      RETURN
   45 GEX=0.0
      WTG=0.0
      RETURN
C
   50 FORMAT(//*        INEQUALITY WEIGHTING FACTORS*)
   55 FORMAT(* WG(*,I2,*)=*E15.5)
   60 FORMAT(*   WTG=*E15.5,2X,*GFAC=*E15.5)
   65 FORMAT(* WTG=*E15.5,2X,*GFAK=*E15.5)
      END
      FUNCTION    H(X)
      COMMON /WW/ WG(30),WH(30)
      COMMON /GH/ MG,MH,FAC,GFAC,HFAC,WTG,WTH,GFAK,IPRNT,SX(30),SCAL
      COMMON /PEN/ JPEN,L1,L2,L3,L4,L5,L6,GEXP,GINP,HP,ISRCH,DELL(30),
     1 GG(30),HH(30)
      COMMON /JSR/ JSRCH
      LOGICAL SCAL
      IF (MH.EQ.0) GO TO 25
      IF (L4.EQ.1) GO TO 15
```

```
      IF (IPRNT.EQ.1) WRITE (6,35)
      L4=1
      CALL HFUN (HH,X)
      DO 5 J=1,MH
C     .... SCALE EQUALITY CONSTRAINTS ON FIRST PASS
      IF (ABS(HH(J)).LE.1.E-3) HH(J)=1.E-3
      WH(J)=ABS(1./HH(J))
      IF (IPRNT.EQ.1) WRITE (6,30) J,WH(J)
    5 CONTINUE
      IF (JSRCH.NE.0.AND.L6.EQ.1) GO TO 15
C     .... WEIGHTING FACTOR FOR EQUALITY PENALTY FUNCTION
      H=0.0
      CALL HFUN (HH,X)
      DO 10 J=1,MH
      HH(J)=WH(J)*HH(J)
   10 H=H+HH(J)*HH(J)
      WTH=ABS(HFAC*FFUN(X)/H)
      IF (IPRNT.EQ.1) WRITE (6,40) WTH,HFAC
      L6=1
   15 CONTINUE
      H=0.0
      CALL HFUN (HH,X)
C     .... COMPUTE EQUALITY PENALTY FUNCTION H(X)
      DO 20 J=1,MH
      HH(J)=WH(J)*HH(J)
   20 H=H+HH(J)*HH(J)
      RETURN
   25 H=0.0
      WTH=0.0
      RETURN
C
   30 FORMAT(* WH(*I2*)=*E15.5)
   35 FORMAT(//*     EQUALITY WEIGHTING FACTORS*)
   40 FORMAT(*  WTH=*E15.5,*  HFAC= *E14.6)
      END

      SUBROUTINE SCALE (X,SX,N)
      DIMENSION X(N),SX(N)
      DO 5 I=1,N
      SX(I)=ABS(X(I))
      IF (ABS(SX(I)).LT.1.E-5) SX(I)=1.E-5
    5 X(I)=X(I)/SX(I)
      RETURN
      END
      SUBROUTINE DESCALE (X,SX,N)
      DIMENSION X(N),SX(N)
      DO 5 I=1,N
    5 X(I)=SX(I)*X(I)
      RETURN
      END
```

Appendix 6.

LSTCON

```
      PROGRAM LSTCON(INPUT,OUTPUT,TAPE5=INPUT,TAPE6=OUTPUT)
CDC VERSION   C w RADCLIFFE MECH ENG DEPT U C BERKELEY  NOVEMBER 1976
C     CCCCCCCCCCCCCCCCCCCCCCCCCCCCCCCCCCCCCCCCCCCCCCCCCCCCCCCCCCCCCCCCCCC
C                                                                       C
C          ****************USER INSTRUCTIONS**************              C
C                                                                       C
C     PROGRAM LSTCON SOLVES THE GENERAL NONLINEAR PROGRAMMING PROBLEM   C
C     FINDING THE CONSTRAINED MINIMUM OF THE SUM OF THE SQUARES OF M    C
C     FUNCTIONS OF N VARIABLES USING A MODIFICATION OF POWELLS LEAST    C
C     SQUARES METHOD.                                                   C
C                                                                       C
C     REFERENCE                                                         C
C     POWELL M J D , A METHOD FOR MINIMIZING A SUM OF SQUARES OF NON    C
C     LINEAR FUNCTIONS WITHOUT CALCULATING DERIVATIVES, THE COMPUTER    C
C     JOURNAL, JAN 1965, PP303 - 30 7.                                  C
C                                                                       C
C     THE MODIFICATION COMBINES POWELLS UNCONSTRAINED METHOD WITH A     C
C     TECHNIQUE SIMILAR TO SUMT.                                        C
C                                                                       C
C     REFERENCE                                                         C
C     FIACCO AND MCCORMICK, NONLINEAR PROGRAMMING - SEQUENTIAL UNCON-   C
C     CONSTRAINED MINIMIZATION TECHNIQUE, JOHN WILEY, 1968.             C
C                                                                       C
C     A FEATURE OF LSTCON IS THE AUTOMATIC SCALING OF THE CONSTRAINT    C
C     FUNCTIONS SUCH THAT ALL ARE NORMALIZED= 1. AT THE START OF A      C
C     SEARCH BY CALCULATION OF WEIGHTING FACTORS. THIS TENDS TO REDUCE  C
C     THE CHANCE THAT ONE OR MORE OF THE CONSTRAINTS WILL DOMINATE      C
C     THE OBJECTIVE FUNCTION.                                           C
C                                                                       C
C     A SECOND FEATURE CONTROLS THE RATIO OF PENALTY TERMS TO THE UN-   C
C     CONSTRAINED OBJECTIVE FUNCTION AT THE START OF THE SEARCH. THE    C
C     MODIFIED OBJECTIVE FUNCTION IS THE SUM OF THREE TERMS             C
C                                                                       C
C        U(X)= U1(X) + U2(X) + U3(X)                                    C
C                                                                       C
C     WHERE U1(X)= UNCONSTRAINED SUM OF THE SQUARES OF F(I), I= 1,MF    C
C           U2(X)= PENALTY FUNCTION FORMED AS SUM OF SQUARES OF FUNCTION C
C                  FH(I)= SQRT(WTH*FAC**(IR-1)/MH)*WH(I)*H(I)           C
C           U3(X)= SUM OF SQUARES OF FUNCTIONS FG(I) IN EITHER THE      C
C     INTERIOR FORM                                                     C
C           FG(I)= SQRT(WTG*FAC**(1-IR)/MG)*WG(I)/G(I)                  C
C     OR THE EXTERIOR FORM                                              C
C           FG(I)= SQRT(WTG*FAC**(IR-1)/MG)*WG(I)*G(I))                 C
C                                                                       C
C     THE USER MUST SUPPLY THE FOLLOWING SUBROUTINES                    C
C     WHERE X IS THE VECTOR OF THE PROBLEM VARIABLES                    C
C                                                                       C
C     SUBROUTINE FFUN(F,X)                                              C
C     RETURNS CURRENT VALUES OF RESIDUAL FUNCTIONS F(I), I= 1,MF        C
C                                                                       C
C     SUBROUTINE GFUN(G,X)                                              C
C     RETURNS CURRENT VALUES OF INEQUALITY CONSTRAINTS G(I), I= 1,MG    C
C                                                                       C
C     SUBROUTINE HFUN(H,X)                                              C
C     RETURNS CURRENT VALUES OF THE EQUALITY CONSTRAINTS H(I), I= 1,MH C
```

```
C                                                                    C
C          ***********************INPUT DATA FORMAT*****************  C
C                                                                    C
C     FIRST CARD ... READ N,MAXTRY,MAXFUN,MF,MG,MH,ISRCH,IPRINT      C
C     FORMAT(8I5)                                                    C
C     IF N IS READ IN AS A NEGATIVE INTEGER MORE IS SET= .TRUE.      C
C     AND SUBROUTINE MORPRNT(N,X,FF,MF,MG,MH) IS CALLED AT END OF    C
C     NORMAL EXECUTION.                                              C
C                                                                    C
C        N= NUMBER OF VARIABLES IN VECTOR X                          C
C        MAXTRY= NUMBER OF SUB MINIMIZATIONS= INCREMENTS IN IR       C
C        MAXFUN= MAXIMUM NUMBER OF FUNCTIONAL EVALUATIONS            C
C        IPRINT= 0 PRINT FINAL RESULTS ONLY                          C
C              = 1 PRINT AFTER EACH ITERATION                        C
C              = 2 PRINT EACH TENTH ITERATION                        C
C        MF= NUMBER OF FUNCTIONS F(I)                                C
C        MG= NUMBER OF INEQUALITY CONSTRAINTS G(I)                   C
C        MH= NUMBER OF EQUALITY CONSTRAINTS H(I)                     C
C        ISRCH= 0 CHECK FEASIBILITY ... IF NONFEASIBLE SEARCH FOR A  C
C                 FEASIBLE POINT ... THEN SET ISRCH= 2 AND CONTINUE  C
C        ISRCH= 1 USE EXTERIOR PENALTY FUNCTION                      C
C              = 2 USE INTERIOR PENALTY WITH FEASIBLE START          C
C                 OR EXTERIOR PENALTY WITH NONFEASIBLE START         C
C        IR= INTEGER WHICH IS INCREMENTED AUTOMATICALLY TO SET       C
C        DECREASING VALUES FOR R= FAC**(1-IR)                        C
C                                                                    C
C     SECOND CARD ... READ ESCALE, TOL, FAC, GFAC, HFAC              C
C     FORMAT(5F10.0)                                                 C
C                                                                    C
C        ESCALE ... MAX MOVE FOR X(I)= ESCALE*TOL                    C
C        TOL ... CONVERGENCE WHEN ALL DELX(I)/X(I) LESS THAN TOL     C
C        FAC= FACTOR USED IN R= FAC*(IR-1)                           C
C        GFAC= DESIRED RATIO OF INEQUALITY PENALTY TO THE OBJECTIVE  C
C              FUNCTION AT THE START OF THE SEARCH                   C
C        HFAC= DESIRED RATIO OF THE EQUALITY PENALTY TO THE OBJECTIVE C
C              FUNCTION AT THE START OF THE SEARCH                   C
C     IF TOL IS READ AS NEGATIVE NUMBER , ADDITIONAL DATA CARDS WITH C
C     INDIVIDUAL VALUES FOR E(I) FOR EACH VARIABLE MUST FOLLOW DATA  C
C     CARDS FOR THE INITIAL GUESSES.                                 C
C                                                                    C
C     THIRD CARD(S) ... READ (X(I), I= 1,N)                          C
C     FORMAT(5F10.0)       (INITIAL GUESSES FOR THE UNKNOWN VARIABLES) C
C                                                                    C
C          ********************END OF COMMENTS********************   C
C                                                                    C
CCCCCCCCCCCCCCCCCCCCCCCCCCCCCCCCCCCCCCCCCCCCCCCCCCCCCCCCCCCCCCCCCCCCCCC
C                                                                    C
      DIMENSION X(30),E(30),S(30),F(200),G(30),H(30)
      COMMON /LSTCOM/ IR,MF,MG,MH,ISRCH,FAC,GFAC,HFAC,
     1 ONCE,TOL,TOL2,MCT,G,H
      COMMON/CONTRL/ABSACC,RELACC,XTRY
      COMMON /WORK/ W(1000),IT
      LOGICAL EOF , MORE
    5 ITRY=0
      MORE=.FALSE.
      MCTOT=0
      ITCTOT=0
      ONCE=0.0
      WRITE (6,45)
      READ (5,50) N,MAXTRY,MAXFUN,MF,MG,MH,ISRCH,IPRINT
      IF (EOF(5)) GO TO 40
      IF (N.LT.0) MORE=.TRUE.
      WRITE (6,55) N,MAXTRY,MAXFUN,MF,MG,MH,ISRCH,IPRINT
      N=IABS(N)
      IR=IRR=1
      M=MF+MG+MH
      WRITE (6,60)
      READ (5,65) ESCALE,TOL,FAC,GFAC,HFAC
```

```
      wRITE (6,70) ESCALE,TOL,FAC,GFAC,HFAC
      DO 10 I=1,N
   10 E(I)=TOL
      wRITE (6,75)
      READ (5,80) (X(I),I=1,N)
      WRITE (6,85) (X(I),I=1,N)
      IF (TOL.LT.0.0) READ (5,80) (E(I),I=1,N)
      IF (TOL.LT.0.0) WRITE (6,90) (E(I),I=1,N)
      TOL=ABS(TOL)
      IF (MORE) WRITE (6,95)
      WRITE (6,100)
   15 ITRY=ITRY+1
      DO 20 I=1,N
   20 S(I)=X(I)
      IF (ITRY.GT.MAXTRY) GO TO 5
      IF (ISRCH.EQ.0) M=MG
      CALL POWEL (M,N,F,X,E,ESCALE,IPRINT,MAXFUN,ITC,MC,FF)
      wRITE (6,105) IR,ITC,MCT,FF,(X(I),I=1,N)
      wRITE (6,120) (F(I),I=1,MF)
      IF (MG.GT.0) WRITE (6,125) (G(I),I=1,MG)
      IF (MH.GT.0) WRITE (6,130) (H(I),I=1,MH)
      ITCTOT=ITCTOT+ITC
      MCTOT=MCTOT+MCT
      IF (ISRCH.NE.0) GO TO 30
      IR=IRR
      ITRY=ITRY-1
      ISRCH=2
      M=MF+MG+MH
      ONCE=0.0
      DO 25 I=1,N
   25 E(I)=0.1*E(I)
      wRITE (6,110)
      GO TO 15
   30 IR=IR+1
      IF (M.EQ.MF) GO TO 40
      DO 35 I=1,N
      TEST=ABS((X(I)-S(I))/S(I))
      IF (TEST.GT.TOL) GO TO 15
   35 CONTINUE
      WRITE (6,115) ITCTOT,MCTOT,FF,(X(I),I=1,N)
      wRITE (6,120) (F(I),I=1,MF)
      IF (MG.GT.0) WRITE (6,125) (G(I),I=1,MG)
      IF (MH.GT.0) wRITE (6,130) (H(I),I=1,MH)
      IF (MCRE) CALL MORPRNT (N,X,F,FF,MF,MG,MH)
      GO TO 5
   40 CONTINUE
      STOP
C
   45 FORMAT(1H1,//5X*DATA CARD ONE*/
     1 6X*N*1X*MAXTRY*1X*MAXFUN*5X*MF*5X*MG*5X*MH*2X*ISRCH*1X*IPRINT*)
   50 FORMAT(8I5)
   55 FORMAT(8I7)
   60 FORMAT(5X*DATA CARD TWO*/
     1 *     ESCALE          TOL          FAC         GFAC         HFAC*)
   65 FORMAT(5F10.0)
   70 FORMAT(5E10.4)
   75 FORMAT(5X*DATA CARD(S) THREE PLUS*/
     15X*INITIAL GUESSES X(I), I= 1,N*)
   80 FORMAT(5F10.0)
   85 FORMAT(5E14.5)
   90 FORMAT(5X*TOL VALUES E(I), I= 1,N*/(5E14.5))
   95 FORMAT(//10X*... USER MUST SUPPLY SUBROUTINE MORPRNT(N,X,FF,MF,MG,
     1MH) ...*)
  100 FORMAT(//,15X *.....BEGIN LEAST SQUARES MINIMIZATION.....*//)
  105 FORMAT(//20(1H*)* SUBMINIMUM FOUND FOR IR*I3,1X,20(1H*)/
     1 2X*ITERATIONS= *I3,2X*FUNC EVAL= *I6,2X*SUM SQS OF RESIDUALS= *
     2 E9.3//5X*VARIABLES ARE*/(5E13.5))
  110 FORMAT(//*... SEARCH FOR A FEASIBLE POINT COMPLETED ...*//)
```

```
115 FORMAT(1H1,20(1H*)* CONSTRAINED MINIMUM FOUND *,20(1H*)//
   1 5X*RATIO DELX/X LESS THAN E(I) FOR ALL VARIABLES X(I)*//
   2 5X*TOTAL ITERATIONS= *I3/5X*TOTAL FUNC EVAL= *I6/
   3 5X*SUM OF THE SQUARES OF THE RESIDUALS= *E13.5/
   4 5X*FINAL VALUES FOR THE VARIABLES*/(5E13.5))
120 FORMAT(/5X*FUNCTIONS*/(5E13.5))
125 FORMAT(/5X*INEQUALITY CONSTRAINTS*/(5E13.5))
130 FORMAT(/5X*EQUALITY CONSTRAINTS*/(5E13.5))
    END

    SUBROUTINE POWEL (M,N,F,X,E,ESCALE,IPRINT,MAXFUN,ITC,MC,FF)
    DIMENSION X(30),E(30),S(30),F(200),G(30),H(30)
    COMMON /WORK/ W(1000),IT
    COMMON / CONTRL / ABSACC,RELACC,XTRY
    COMMON /LSTCOM/ IR,MF,MG,MH,ISRCH,FAC,GFAC,HFAC,
   1 ONCE,TOL,TOL2,MCT,G,H
    IPK=-1
    IT=0
    MCT=0
    MPLUSN=M+N
    KST=N+MPLUSN
    NPLUS=N+1
    KINV=NPLUS*(MPLUSN+1)
    KSTORE=KINV-MPLUSN-1
    CALL CALFUN (M,N,F,X,E)
    NN=N+N
    K=NN
    DO 5 I=1,M
    K=K+1
    W(K)=F(I)
  5 CONTINUE
    IINV=2
    K=KST
    I=1
 10 X(I)=X(I)+E(I)
    CALL CALFUN (M,N,F,X,E)
    X(I)=X(I)-E(I)
    DO 15 J=1,N
    K=K+1
    W(K)=0.0
 15 W(J)=0.0
    SUM=0.0
    KK=NN
    DO 20 J=1,M
    KK=KK+1
    F(J)=F(J)-W(KK)
    SUM=SUM+F(J)*F(J)
 20 CONTINUE
    IF (SUM) 25,25,35
 25 WRITE (6,415) I
    DO 30 J=1,M
    NN=NN+1
    F(J)=W(NN)
 30 CONTINUE
    GO TO 150
 35 SUM=1.0/SQRT(SUM)
    J=K-N+I
    W(J)=E(I)*SUM
    DO 45 J=1,M
    K=K+1
    W(K)=F(J)*SUM
    KK=NN+J
    DO 40 II=1,I
    KK=KK+MPLUSN
    W(II)=W(II)+W(KK)*W(K)
 40 CONTINUE
 45 CONTINUE
    ILESS=I-1
```

```
       IGAMAX=N+I-1
       INCINV=N-ILESS
       INCINP=INCINV+1
       IF (ILESS) 50,50,55
   50  W(KINV)=1.0
       GO TO 95
   55  B=1.0
       DO 60 J=NPLUS,IGAMAX
       W(J)=0.0
   60  CONTINUE
       KK=KINV
       DO 80 II=1,ILESS
       IIP=II+N
       W(IIP)=W(IIP)+W(KK)*W(II)
       JL=II+1
       IF (JL-ILESS) 65,65,75
   65  DO 70 JJ=JL,ILESS
       KK=KK+1
       JJP=JJ+N
       W(IIP)=W(IIP)+W(KK)*W(JJ)
       W(JJP)=W(JJP)+W(KK)*W(II)
   70  CONTINUE
   75  B=B-W(II)*W(IIP)
       KK=KK+INCINP
   80  CONTINUE
       B=1.0/B
       KK=KINV
       DO 90 II=NPLUS,IGAMAX
       BB=-B*W(II)
       DO 85 JJ=II,IGAMAX
       W(KK)=W(KK)-BB*W(JJ)
       KK=KK+1
   85  CONTINUE
       W(KK)=BB
       KK=KK+INCINV
   90  CONTINUE
       W(KK)=B
   95  GO TO (115,100), IINV
  100  I=I+1
       IF (I-N) 10,10,105
  105  IINV=1
       FF=0.0
       KL=NN
       DO 110 I=1,M
       KL=KL+1
       F(I)=W(KL)
       FF=FF+F(I)*F(I)
  110  CONTINUE
C      --PRINTING ROUTINES
       ICONT=1
       ISS=1
       MC=N+1
       IPP=IABS(IPRINT)*(IABS(IPRINT)-1)
       ITC=0
       IPS=1
       IPC=0
  115  IPC=IPC-IABS(IPRINT)
       IPK=IPK+1
       MCT=MCT+MC
       IF (IPC) 120,140,140
  120  CONTINUE
       IF (IPRINT.EQ.2.AND.IPK.NE.10) GO TO 130
       WRITE (6,420) ITC,MC,FF
       WRITE (6,425) (X(I),I=1,N)
       IF (IPRINT) 135,125,125
  125  WRITE (6,430) (F(I),I=1,M)
       IPK=0
  130  CONTINUE
```

```
135 IPC=IPP
    GO TO (140,155), IPS
140 GO TO (165,145), ICONT
145 IF (CHANGE-1.0) 150,150,160
150 IF (IPRINT.EQ.0) GO TO 155
    IPK=10
    IF (IT.EQ.3) WRITE (6,435) TOL
    IF (IT.EQ.4) WRITE (6,440) MAXFUN
    IPS=2
    GO TO 120
155 ITC=ITC+1
    RETURN
160 ICONT=1
165 ITC=ITC+1
    K=N
    KK=KST
    DO 175 I=1,N
    K=K+1
    W(K)=0.0
    KK=KK+N
    W(I)=0.0
    DO 170 J=1,M
    KK=KK+1
    W(I)=W(I)+W(KK)*F(J)
170 CONTINUE
175 CONTINUE
    DM=0.0
    K=KINV
    DO 200 II=1,N
    IIP=II+N
    W(IIP)=W(IIP)+W(K)*W(II)
    JL=II+1
    IF (JL-N) 180,180,190
180 DO 185 JJ=JL,N
    JJP=JJ+N
    K=K+1
    W(IIP)=W(IIP)+W(K)*W(JJ)
    W(JJP)=W(JJP)+W(K)*W(II)
185 CONTINUE
    K=K+1
190 IF (DM-ABS(W(II)*W(IIP))) 195,200,200
195 DM=ABS(W(II)*W(IIP))
    KL=II
200 CONTINUE
    II=N+MPLUSN*KL
    CHANGE=0.0
    DO 215 I=1,N
    JL=N+I
    W(I)=0.0
    DO 205 J=NPLUS,NN
    JL=JL+MPLUSN
    W(I)=W(I)+W(J)*W(JL)
205 CONTINUE
    II=II+1
    W(II)=W(JL)
    W(JL)=X(I)
    IF (ABS(E(I)*CHANGE)-ABS(W(I))) 210,210,215
210 CHANGE=ABS(W(I)/E(I))
215 CONTINUE
    DO 220 I=1,M
    II=II+1
    JL=JL+1
    W(II)=W(JL)
    W(JL)=F(I)
220 CONTINUE
    FC=FF
    ABSACC=0.1/CHANGE
    IT=3
```

```
        XC=0.0
        XL=0.0
        IS=3
        XSTEP=-AMIN1(XTRY,ESCALE/CHANGE)
        IF (CHANGE-1.0) 225,225,230
225 ICONT=2
230 CALL MINUM (IT,XC,FC,XSTEP,MAXFUN,MC)
        GO TO (235,320,320,320), IT
235 MC=MC+1
        IF (MC-MAXFUN) 245,245,240
240 WRITE (6,445) MAXFUN
        ISS=2
        GO TO 320
245 XL=XC-XL
        DO 250 J=1,N
        X(J)=X(J)+XL*W(J)
250 CONTINUE
        XL=XC
        CALL CALFUN (M,N,F,X,E)
        FC=0.0
        DO 255 J=1,M
        FC=FC+F(J)*F(J)
255 CONTINUE
        GO TO (275,275,260), IS
260 K=N
        IF (FC-FF) 265,230,270
265 IS=2
        FMIN=FC
        FSEC=FF
        GO TO 305
270 IS=1
        FMIN=FF
        FSEC=FC
        GO TO 305
275 IF (FC-FSEC) 280,230,230
280 K=KSTORE
        GO TO (285,290), IS
285 K=N
290 IF (FC-FMIN) 300,230,295
295 FSEC=FC
        GO TO 305
300 IS=3-IS
        FSEC=FMIN
        FMIN=FC
305 DO 310 J=1,N
        K=K+1
        W(K)=X(J)
310 CONTINUE
        DO 315 J=1,M
        K=K+1
        W(K)=F(J)
315 CONTINUE
        GO TO 230
320 K=KSTORE
        KK=N
        GO TO (330,325,330), IS
325 K=N
        KK=KSTORE
330 SUM=0.0
        DM=0.0
        JJ=KSTORE
        DO 335 J=1,N
        K=K+1
        KK=KK+1
        JJ=JJ+1
        X(J)=W(K)
        W(JJ)=W(K)-W(KK)
335 CONTINUE
```

```
      DO 340 J=1,M
      K=K+1
      KK=KK+1
      JJ=JJ+1
      F(J)=W(K)
      W(JJ)=W(K)-W(KK)
      SUM=SUM+W(JJ)*W(JJ)
      DM=DM+F(J)*W(JJ)
340   CONTINUE
      GO TO (345,150), ISS
345   J=KINV
      KK=NPLUS-KL
      DO 350 I=1,KL
      K=J+KL-I
      J=K+KK
      W(I)=W(K)
      W(K)=W(J-1)
350   CONTINUE
      IF (KL-N) 355,365,365
355   KL=KL+1
      JJ=K
      DO 360 I=KL,N
      K=K+1
      J=J+NPLUS-I
      W(I)=W(K)
      W(K)=W(J-1)
360   CONTINUE
      W(JJ)=W(K)
      B=1.0/W(KL-1)
      W(KL-1)=W(N)
      GO TO 370
365   B=1.0/W(N)
370   K=KINV
      DO 380 I=1,ILESS
      BB=B*W(I)
      DO 375 J=I,ILESS
      W(K)=W(K)-BB*W(J)
      K=K+1
375   CONTINUE
      K=K+1
380   CONTINUE
      IF (FMIN-FF) 390,385,385
385   CHANGE=0.0
      GO TO 395
390   FF=FMIN
      CHANGE=ABS(XC)*CHANGE
395   XL=-DM/FMIN
      SUM=1.0/SQRT(SUM+DM*XL)
      K=KSTORE
      DO 400 I=1,N
      K=K+1
      W(K)=SUM*W(K)
      W(I)=0.0
400   CONTINUE
      DO 410 I=1,M
      K=K+1
      W(K)=SUM*(W(K)+XL*F(I))
      KK=NN+I
      DO 405 J=1,N
      KK=KK+MPLUSN
      W(J)=W(J)+W(KK)*W(K)
405   CONTINUE
410   CONTINUE
      GO TO 55
C
415   FORMAT(*0   E(*,I3,*) UNREASONABLY SMALL.*)
420   FORMAT(/5X*CURRENT ITERATION=*I4,2X*FUNC EVAL= *I5,2X
     1 *SUM SQS OF RESIDUALS= *E12.3)
```

```
425 FORMAT(5X*CURRENT VARIABLES*/(5E13.5))
430 FORMAT(5X*CURRENT RESIDUALS*/(5E13.5))
435 FORMAT(2X*ROUND OFF ERROR MAY HAVE MADE IT IMPOSSIBLE TO ACHIEVE*
   1 * CONVERGENCE WITH TOL= *E12.3)
440 FORMAT(/I6* EVALUATIONS OF U(X) MADE - NO MINIMUM FOUND*)
445 FORMAT(*0    NUMBER OF CALLS TO CALFUN ALLOWED WAS LIMITED TO *,I6,
   1* TIMES.*)
    END

    SUBROUTINE MINUM (ITEST,X,F,XSTEP,MAXFUN,MC)
    COMMON/CONTRL/ABSACC,RELACC,XTRY
    GO TO (35,5,5), ITEST
  5 IS=6-ITEST
    ITEST=1
    IINC=1
    XINC=XSTEP+XSTEP
    MC=IS-3
    IF (MC) 45,45,30
 10 MC=MC+1
    IF (MAXFUN-MC) 15,30,30
 15 ITEST=4
 20 X=DB
    F=FB
    IF (FB-FC) 30,30,25
 25 X=DC
    F=FC
 30 RETURN
 35 GO TO (85,75,50,40), IS
 40 IS=3
 45 DC=X
    FC=F
    X=X+XSTEP
    GO TO 10
 50 IF (FC-F) 60,55,65
 55 X=X+XINC
    XINC=XINC+XINC
    GO TO 10
 60 DB=X
    FB=F
    XINC=-XINC
    GO TO 70
 65 DB=DC
    FB=FC
    DC=X
    FC=F
 70 X=DC+DC-DB
    IS=2
    GO TO 10
 75 DA=DB
    DB=DC
    FA=FB
    FB=FC
 80 DC=X
    FC=F
    GO TO 135
 85 IF (FB-FC) 105,90,90
 90 IF (F-FB) 95,80,80
 95 FA=FB
    DA=DB
100 FB=F
    DB=X
    GO TO 135
105 IF (FA-FC) 115,115,110
110 XINC=FA
    FA=FC
    FC=XINC
    XINC=DA
    DA=DC
```

```
      DC=XINC
115 XINC=DC
      IF ((C-DB)*(D-DC)) 80,120,120
120 IF (F-FA) 125,130,130
125 FC=FB
      DC=DB
      GO TO 100
130 FA=F
      DA=X
135 IF (FB-FC) 140,140,145
140 IINC=2
      XINC=DC
      IF (FB-FC) 145,200,145
145 D=(FA-FB)/(DA-DB)-(FA-FC)/(DA-DC)
      IF (D*(DB-DC)) 180,180,150
150 D=0.5*(DB+DC-(FB-FC)/D)
      IF (ABS(D-X)-ABS(ABSACC)) 160,160,155
155 IF (ABS(D-X)-ABS(D*RELACC)) 160,160,165
160 ITEST=2
      GO TO 20
165 IS=1
      X=D
      IF ((DA-DC)*(DC-D)) 10,205,170
170 IS=2
      GO TO (175,190), IINC
175 IF (ABS(XINC)-ABS(DC-D)) 185,10,10
180 IS=2
      GO TO (185,195), IINC
185 X=DC
      GO TO 55
190 IF (ABS(XINC-X)-ABS(X-DC)) 195,195,10
195 X=0.5*(XINC+DC)
      IF ((XINC-X)*(X-DC)) 205,205,10
200 X=0.5*(DB+DC)
      IF ((DB-X)*(X-DC)) 205,205,10
205 ITEST=3
      GO TO 20
      END

      BLOCKDATA
      DATA ABSACC,RELACC,XTRY / 0.1,0.1,0.5 /
      COMMON / CONTRL / ABSACC,RELACC,XTRY
      END

      SUBROUTINE CALFUN (M,N,F,X,E)
      COMMON /LSTCOM/ IR,MF,MG,MH,ISRCH,FAC,GFAC,HFAC,
     1 ONCE,TOL,TOL2,MCT,G,H
      DIMENSION X(1),F(1),E(1)
      DIMENSION G(30),H(30),DEL(30),WG(30),WH(30)
C     IF ISRCH= 0 SEARCH FOR A FEASIBLE STARTING VECTOR THEN SET
C                   ISRCH = 2 AND CONTINUE.
C               = 1 FORCE USE OF AN EXTERIOR PENALTY FUNCTION
C               = 2 TEST FOR A FEASIBLE INITIAL GUESS VECTOR.
C                   IF FEASIBLE SET ISRCH= 2    OTHERWISE SET ISRCH= 1 .
      IF (ONCE.EQ.1.0) GO TO 70
      ONCE=1.0
      FSUM=0.0
      GSUM=0.0
      HSUM=0.0
      CALL FFUN (F,X)
      IF(MG.EQ.0.AND.MH.EQ.0) RETURN
      IF (MG.NE.0) CALL GFUN (G,X)
      IF (MH.NE.0) CALL HFUN (H,X)
      DO 5 I=1,MF
    5 FSUM=FSUM+F(I)*F(I)
      IF (MG.EQ.0) GO TO 15
      WRITE (6,130)
      DO 10 I=1,MG
```

```
          GSUM=GSUM+G(I)*G(I)
          IF (ABS(G(I)).LE.1.E-6) G(I)=1.E-6
          WG(I)=ABS(1./G(I))
   10     WRITE (6,135) I,WG(I)
          WTG=FSUM*GFAC/GSUM
          WRITE (6,140) WTG
   15     IF (MH.EQ.0) GO TO 25
          WRITE (6,145)
          DO 20 I=1,MH
          HSUM=HSUM+H(I)*H(I)
          IF (ABS(H(I)).LE.1.E-6) H(I)=1.E-6
          WH(I)=ABS(1./H(I))
   20     WRITE (6,150) I,WH(I)
          WTH=FSUM*HFAC/HSUM
          WRITE (6,155) WTH
   25     IF (MG.EQ.0) GO TO 75
          IF (ISRCH.EQ.0) GO TO 35
          IF (ISRCH.EQ.1) GO TO 45
          DO 30 I=1,MG
          IF (ISRCH.EQ.2.AND.G(I).LT.0.0) GO TO 30
          ISRCH=0
          GO TO 55
   30     CONTINUE
          GO TO 50
   35     DO 40 I=1,MG
   40     DEL(I)= 100.*TOL
          GO TO 55
   45     WRITE (6,160)
          GO TO 65
   50     WRITE (6,165)
          GO TO 65
   55     WRITE (6,170)
          DO 60 I=1,N
   60     E(I)=10.*E(I)
   65     CONTINUE
   70     IF (ISRCH.EQ.0) GO TO 120
   75     CALL FFUN (F,X)
          IF (MG.EQ.0) GO TO 80
          CALL GFUN (G,X)
   80     IF (MH.EQ.0) GO TO 85
          CALL HFUN (H,X)
   85     IF (MG.EQ.0) GO TO 105
          DO 100 I=1,MG
          IG=I+MF
          GO TO (90,95), ISRCH
   90     IF (G(I).LE.0.0) G(I)=0.0
          F(IG)= SQRT(WTG*FAC**(IR-1))*WG(I)*G(I)
          GO TO 100
   95     IF(G(I).GE.0.0) G(I)= -1./(1.E4*(1.+GG(I)))
          F(IG)= SQRT(WTG*FAC**(1-IR))*WG(I)*(1./G(I))
  100     CONTINUE
  105     IF (MH.EQ.0) GO TO 115
          DO 110 I=1,MH
          IH=I+MG+MF
          F(IH)= SQRT(WTH*FAC**(IR-1))*WH(I)*H(I)
  110     CONTINUE
  115     RETURN
  120     M=MG
          CALL GFUN (G,X)
          DO 125 I=1,MG
          G(I)=G(I)+DEL(I)
          IF (G(I).LT.0.0) G(I)=1000.*G(I)
  125     F(I)=SQRT(WTG*FAC**(IR-1)/MG)*WG(I)*G(I)
          RETURN
C
  130     FORMAT(/* INEQUALITY WEIGHTING FACTORS*/)
  135     FORMAT(5X*WG(*I2*)= *E12.3)
```

```
140 FORMAT(5X*WTG= *E12.3)
145 FORMAT(/* EQUALITY WEIGHTING FACTORS*/)
150 FORMAT(5X*WH(*I2*)= *E12.3)
155 FORMAT(5X*WTH= *E12.3)
160 FORMAT(/15X* ... EXTERIOR PENALTY FUNCTION ...*/)
165 FORMAT(/15X* ... INTERIOR PENALTY FUNCTION ...*/)
170 FORMAT(/15X* ... SEARCH FOR A FEASIBLE POINT ...*/)
    END
```

Appendix 7.

PRCR 2

```
      PROGRAM PRCR2(INPUT,OUTPUT,TAPE5=INPUT,TAPE6=OUTPUT)
      LOGICAL MORE
      MORE= .FALSE.
C PRCR2 WILL PRINT AFTER EACH IPRINT STEPS. IF IPRINT IS READ
C IN AS A NEGATIVE NUMBER A USER SUPPLIED SUBROUTINE MORPRNT WILL
C BE CALLED AFTER EACH PRINTING.


C        READ INPUT DATA


      READ 50, X0,Y0,DY0,DELX,XMAX,TEST
   50 FORMAT(6F10,0)
      READ 100, IPRINT
  100 FORMAT(I3)
      IF(IPRINT.LT.0) MORE= .TRUE.
      IPRINT= IABS(IPRINT)


C        PRINT INPUT DATA

      PRINT 101, X0,Y0,DY0,DELX,IPRINT,XMAX,TEST
  101 FORMAT(1H1,*     PREDICTOR-CORRECTOR SOLUTION FOR*/
     1 *      A GENERAL NON-LINEAR SECOND-ORDER*/
     2 *      DIFFERENTIAL EQUATION D/DX(DY/DX)=F(X,Y,DY/DX)*///
     3 *     X0=*E10.3*   Y0=*E10.3*   (DY/DX)=*E10.3//
     4 *   DELX=*E10.3*  IPRINT=*I3*   XMAX=*E10.3*   TEST=*
     5 E10.3///* ......... S O L U T I O N .........*//
     6 *       X         Y          DY          DDY*////)

C        COMPUTE SECOND DERIVATIVE

      J= 0
      X=X0
      Y=Y0
      IP= 0
      DY=DY0
      DDY= DERIV2(X,Y,DY)
      PRINT 102, X,Y,DY,DDY
      IF(MORE) CALL MORPRNT(X,Y,DY)
   15 DDY=DERIV2(X,Y,DY)

C        PREDICT DY(I+1) AND Y(I+1)
```

406

```
      J=1
      PDY=DY+DDY*DELX
      PY=Y+0.5*(DY+PDY)*DELX

C        COMPUTE CORRECTED VALUES

  11  CDDY=DERIV2(X+DELX),PY,PDY)
      CDY=DY+0.5*(DDY+CDDY)*DELX
      CY=Y+0.5*(DY+CDY)*DELX

C        TEST CONVERGENCE

      IF(ABS(CY).LE.1.0) GO TO 20
      ERROR=(CY PY)/CY
      GO TO 21
  20  ERROR=(CY PY)
  21  IF(ABS(ERROR).LE.TEST) GO TO 10
      PY=CY
      PDY=CDY
      J=J+1
      GO TO 11

C        STEP FORWARD

  10  Y=CY
      DY=CDY
      DDY=CDDY
      X=X+DELX

C        PRINT TEST

      IP=IP+ 1
      IF(IP.EQ.IPRINT) GO TO 12
      GO TO 13
  12  PRINT 102, X,Y,DY,DDY,J
      IF(MORE) CALL MORPRNT(X,Y,DY)
      IP= 0

C        TEST X=XMAX

  13  IF(X.GE.XMAX) GO TO 14
      GO TO 15
  14  PRINT 102, X,Y,DY,DDY,J
 102  FORMAT(4E15.5,15)
      STOP
      END
```

Appendix 8.
SIMEQ

```
LIST
1    REM   ...   SOLUTION OF NONLINEAR OR LINEAR EQUATIONS
2    REM          TYPE RUN TO SOLVE A SET OF EQUATIONS IN SEC 1000
3    REM          TYPE GOTO 6000 TO SOLVE A SET OF LINEAR EQUATIONS
4    DIM X[10],D[10],F[10],P[10,10],A[10,11]
5    PRINT "NUMBER OF VARIABLES";
6    INPUT M
7    PRINT
8    PRINT "CONVERGENCE TOLERANCE";
9    INPUT E
10   PRINT
11   PRINT "ITERATION LIMIT";
12   INPUT L
13   PRINT
14   PRINT "INITIAL GUESSES";
15   PRINT
20   REM   ...INITIAL GUESSES
23   FOR J=1 TO M
25      INPUT X[J]
30   NEXT J
31   PRINT
32   IF E > .00001 GOTO 36
33   LET E=.0001
34   PRINT
36   LET K1=0
38   PRINT "VALUE OF NORM";
39   PRINT
42   GOSUB   1000
43   GOSUB   3000
44   PRINT T
45   LET S=T
50   FOR I=1 TO M
52      LET A[I,M+1]=-F[I]
55   NEXT I
58   GOSUB 2000
62   FOR I=1 TO M
63      FOR J=1 TO M
64         LET A[I,J]=P[I,J]
65      NEXT J
66   NEXT I
67   LET K1=K1+1
69   LET K= 0
85   GOSUB 5000
95   FOR J=1 TO M
```

```
1000     LET  X[J]=X[J]+D[J]
105   NEXT J
106   GOSUB   1000
107   GOSUB   3000
108   PRINT T
111   REM      ...TEST CONVERGENCE
115   FOR J=1 TO M
116     IF ABS X[J] <1 GOTO 119
117     LET T9= ABS (D[J]/X[J])
118     GOTO 122
119     LET T9= ABS (D[J])
122     IF T9 >E GOTO 145
125   NEXT J
130   PRINT "CONVERGED IN"; "ITERATIONS"
133   PRINT
134   PRINT "ANSWERS ARE";
135   PRINT
136   FOR J=1 TO M
137      PRINT X[J]
138   NEXT J
140   GOTO 14
141   REM      ...DAMPING ... DIVIDE STEP BY 5 IF NORM INCREASES
142   REM
145   IF K1 >L GOTO 200
150   IF ABS T > ABS S GOTO 155
151   LET S=T
152   GOTO 50
155   LET S=T
156   LET K=K+1
157   PRINT "DAMPING";K
158   IF K >1 GOTO 161
159   GOSUB   400
160   GOTO 176
161   FOR J=1 TO M
165     LET T8=D[J]
166     LET D[J]=D[J/5
170     LET X[J]=X[J]-T8+D[J]
175   NEXT J
176   GOSUB   1000
177   GOSUB   3000
180   IF K >10 GOTO 190
186   GOTO 145
190   PRINT " DAMPING LIMIT EXCEEDED"
195   GOTO 14
200   PRINT " ITERATION LIMIT EXCEEDED"
205   GOTO 14
400   REM  ON THE FIRST DAMPING CYCLE
401   REM  NORMALIZE D(J) ABOUT THE MAX STEP
405   LET D1=0
410   FOR J=1 TO M
415     IF ABS D[J] >D1 GOTO 425
420     GOTO 435
425     LET D1= ABS (D[J])
435   NEXT J
440   FOR J=1 TO M
```

```
445    LET T8=D[J]
450    LET D[J]=D[J]/D1
455    LET X[J]=X[J]-T8+D[J]
460   NEXT J
1000  REM     ...INSERT NEW EQUATIONS HERE
1001  REM
1010  LET F[1]=X[1]*X[1]+X[2]*X[2]+X[3]*X[3]-1
1020  LET F[2]=X[1]-X[2]*X[2]
1030  LET F[3]=X[1]+X[2]+X[3]-1
1031  RETURN
2000  REM     ...COMPUTE PARTIAL DERIVATIVE MATRIX P(I,J)
2003  FOR I=1 TO M
2005    FOR J=1 TO M
2010      LET Z=X[J]
2015      IF X[J]<1 GOTO 2030
2020      LET Y=.001*X[J]
2025      GOTO 2035
2030      LET Y=.001
2035      LET X[J]=X[J]+Y
2040      GOSUB 1000
2045      LET G=F[I]
2050      LET X[J]=Z
2055      GOSUB 1000
2060      LET P[I,J]=(G-F[I])/Y
2065    NEXT J
2070  NEXT I
2075  RETURN
3000  REM     ...COMPUTE NORM = SUM F(I) I=1 TO M
3003  LET T= 0
3005  FOR I=1 TO M
3010    LET T=T+ ABS (F[I])
3015  NEXT I
3020  RETURN
5000  REM     ... SOLVE LINEAR EQUATIONS
5003  LET N=M
5005  LET N1=N+1
5035  FOR I=2 TO N
5040    FOR J=I TO N
5045      IF ABS A[I-1,I-1]>1E-20 GOTO 5100
5050      LET M1=I-1
5055      FOR M3=M1 TO M
5060        IF A[M3,M1]= 0 GOTO 5090
5065        FOR M2=M1 TO N1
5070          LET S=A[M3,M2]
5075          LET A[M3,M2]=A[M1,M2]
5080          LET A[M1,M2]=S
5085        NEXT M2
5090      NEXT M3
5095      PRINT "A MATRIX IS SINGULAR"
5096      GOTO 14
5100      LET R=A[J,I-1]/A[I-1,I-1]
5105      FOR K4=1 TO N1
5110        LET A[J,K4]=A[J,K4]-R*A[I-1,K4]
5115      NEXT K4
5116    NEXT J
```

```
5117   NEXT I
5120   FOR I=2 TO N
5125     LET K5=N-I+2
5130     LET R=A[K5,N1]/A[K5,K5]
5135     FOR J=I TO N
5140       LET L5=N-J+1
5145       LET A[L5,N1]=A[L5,N1]-R*A[L5,K5]
5150     NEXT J
5155   NEXT I
5160   FOR I=1 TO N
5165     LET D[I]=A[I,N1]/A[I,I]
5170   NEXT I
5180   RETURN
6000   REM     ... TEST OF LINEQS ...
6010   REM
6021   DIM A[10,11],D[10],X[10],B[10]
6022   PRINT
6030   PRINT "NUMBER OF EQUATIONS";
6040   INPUT M
6041   PRINT
6050   PRINT "COEFFICIENT MATRIX BY ROWS IS";
6051   PRINT
6060   FOR I=1 TO M
6070     FOR J=1 TO M
6080       INPUT A[I,J]
6090     NEXT J
6091     PRINT
6100   NEXT I
6110   PRINT "RIGHT SIDE VECTOR";
6111   PRINT
6120   FOR J=1 TO M
6130     INPUT A[J,M+1]
6140   NEXT J
6141   PRINT
6150   GOSUB 5000
6160   PRINT "ANSWERS ARE";
6170   PRINT
6180   FOR J=1 TO M
6190     PRINT D[J]
6200   NEXT J
6300   GOTO 6000
```

References

CHAPTER 1

1. Reuleaux, F., *Theoretische Kinematik*, Friedrich Vieweg und Sohn, Brunswick, Germany, 1875. Translation by A. B. W. Kennedy, *Reuleaux, Kinematics of Machinery*, Macmillan, London, 1876. Reprinted Dover Publications, New York, 1963.
2. Rothbart, H. A., *Cams*, Wiley, New York, 1956.
3. Mabie, H. H., and Ocvirk, F. W., *Mechanisms and Dynamics of Machinery*, Second Edition, Wiley, New York, 1963.
4. Hall, A. S., Jr., *Kinematics and Linkage Design*, Prentice-Hall, Englewood Cliffs, N.J., 1961.
5. Hartenberg, R. S., and Denavit, J., *Kinematic Synthesis of Linkages*, McGraw-Hill, New York, 1964.
6. Beyer, R., *Kinematische Getriebsynthese*, Springer-Verlag, Berlin, 1953. Translated by H. Kuenzel, *The Kinematic Synthesis of Mechanisms*, McGraw-Hill, New York, 1963.
7. Freudenstein, F., "Approximate Synthesis of Four-bar Linkages," *Trans. ASME*, Vol. 77, No. 6, pp. 853–861, 1955.
8. Denavit, J., and Hartenberg, R. S., "A Kinematic Notation for Lower-Pair Mechanisms Based on Matrices," *Trans. ASME, J. Appl. Mech.*, Ser. E, Vol. 22, June 1955.
9. Sandor, G., "On the Loop Equations in Kinematics," Proceedings of the Seventh Conference on Mechanisms, Penton Publishing, Cleveland, 1962.
10. Yang, A. T., and Freudenstein, F., "Application of Dual-Number and Quaternion Algebra to the Analysis of Spatial Mechanisms," *Trans. ASME, J. Appl. Mech.*, Ser. E, Vol. 86, pp. 300–308, 1964.
11. Roth, B., "Finite Position Theory Applied to Mechanism Synthesis," *Trans. ASME, J. Appl. Mech.*, Ser. E, Vol. 89, pp. 599–605, 1967.
12. Suh, C. H., "Design of Space Mechanisms for Rigid Body Guidance," *Trans. ASME, J. Eng. Ind.*, Vol. 90, Series B, No. 3, pp. 499–507, August 1968.

CHAPTER 2

1. Raven, F. H., "Velocity and Acceleration Analysis of Plane and Space Mechanisms by Means of Independent-Position Equations," *Trans. ASME, J. Appl. Mech.*, Vol. 25, No. 1, pp. 1–6, March 1958.

CHAPTER 3

1. Denavit, J., and Hartenberg, R. S., "A Kinematic Notation for Lower-Pair Mechanisms Based on Matrices," *Trans. ASME, J. Appl. Mech.*, Vol. 22, No. 2, pp. 215–221, June 1955.

CHAPTER 4

1. Gupta, V. K., "Kinematic Analysis of Plane and Spatial Mechanisms," *Trans. ASME, J. Eng. Ind.*, Vol. 95, Series B, No. 2, pp. 481–486, May 1973.
2. Suh, C. H., "Design of Space Mechanisms for Rigid Body Guidance," *Trans. ASME, J. Eng. Ind.*, Vol. 90, Series B, No. 3, pp. 499–507, August 1968.
3. Suh, C. H., "Differential Displacement Matrix and Generation of Screw Axes Surfaces in Kinematics," *Trans. ASME, J. Eng. Ind.*, Vol. 93, Series B, No. 1, pp. 1–10, February 1971.

CHAPTER 5

1. Grubler, M., *Getreibelehre*, Springer-Verlag, Berlin, 1917.
2. Kutzbach, K., "Bernegliche Verbindunger. Vortrag, gehalten auf der Tagung fur Maschinen Elemente," *Z. VDI.*, Bd. 77, pp. 1168, 1933.
3. Kutzbach, K., "Quer—und Winkelbewegliche Gleichganggelenke fur Wellenleitunger," *Z. VDI.*, Bd. 81, pp. 889–892, 1937.
4. Bennett, G. T., "A New Mechanism," *Engineering*, Vol. 76, pp. 777–778, 1903.
5. Goldberg, M., "New Five-Bar and Six-Bar Linkage in Three Dimensions," *Trans. ASME*, Vol. 65, pp. 649–661, 1943.
6. Bricard, R., "Lecons de Cinematique," Bd. 11, Paris, pp. 7–12, 1927.
7. Sarrus, P. T., "Note sur la transformation des mouvements rectilignes alternatifs en mouvements circulaires et reciproquement," C.R.—Acad. Sci., Paris, Bd. 36, pp. 1036–1038, 1853.
8. Franke, R., "Von Aufbau der Getriebe," *Deutscher Ingenieur*, Dusseldorf, pp. 97–106, 1951.
9. Artobolevski, I. I., *Teoria Mehanizmov i Masin, Gosudarstv.* Izdatl Tehn—Teori. Lit, Moscow, 1953.
10. Dobrovol'ski, W. W., *Teoria Mehanizmov*, Maschgis, Moscow, 1953.
11. Harrisberger, L., and Soni, A. H., "A Survey of Three-Dimensional Mechanisms with One General Constraint," ASME paper No. 66-Mech-44, 1966.
12. Soni, A. H., and Harrisberger, L., "Existence Criteria of Mechanisms," ASME paper No. 68-Mech-33, 1968.
13. Waldron, K. J., "Application of the Theory of Screw Axes to Linkages which Disobey the Kutzbach—Grubler Constraint Criterion," ASME paper No. 66-MECH-36, 1966.
14. Grashof, F., *Theoretische Maschinenlehrc*, Vol. 2, Voss, Hamburg, 1883.
15. Hartenberg, R. S., and Denavit, J., *Kinematic Synthesis of Linkages*, McGraw-Hill, New York, 1964.
16. Chace, M. A., "Vector Analysis of Linkages" *Trans. ASME, J. Eng. Ind.*, Vol. 85, Series B, No. 3, pp. 289–297, August 1963.
17. Denavit, J., and Hartenberg, R. S., "A Kinematic Notation for Lower-Pair Mechanisms Based on Matrices," *Trans. ASME, J. Appl. Mech.*, Vol. 22, No. 2, pp. 215–221, June 1955.
18. Suh, C. H., "Design of Space Mechanisms for Rigid Body Guidance," *Trans. ASME, J. Eng. Ind.*, Vol. 90, Series B, No. 3, pp. 499–507, August 1968.
19. Skreiner, M., "Methods to Identify the Mobility Regions of a Spatial Four-Link Mechanism," *J. Mechanisms*, Vol. 2, pp. 415–427, 1967.
20. Ogawa, K., Funabashi, H., and Hayakawa, O., "On the Rotational Conditions of the Spatial Four-bar Mechanisms," *Bull. JSME*, Vol. 11, No. 43, pp. 180–188, 1968.
21. Hunt, K. H., "Screw Axes and Mobility in Spatial Mechanisms via the Linear Complex," *J. Mechanisms*, Vol. 3, pp. 307–327, 1967.
22. Jenkins, E. M., Jr., Crossley, F. R. E., and Hunt, K. H., "Gross Motion Attributes of Certain Spatial Mechanisms," *Trans. ASME, J. Eng. Ind.*, Vol. 91, Series B, No. 1, pp. 83–90, February 1969.
23. Gupta, V. K., and Radcliffe, C. W., "Mobility Analysis of plane and Spatial Mechanisms," *Trans. ASME, J. Eng. Ind.*, Vol. 93, Series B, No. 1, pp. 125–130, February 1971.

CHAPTER 6

1. Wilson, J. T., "Analytical kinematic synthesis by finite displacements," *J. Eng. Ind., Trans. ASME*, 87B, pp. 161–169, 1965.

2. Suh, C. H., and Radcliffe, C. W., "Synthesis of plane mechanisms with use of the displacement matrix," *ASME, J. Eng. Ind.*, Ser. B, Vol. 89, No. 2, pp. 206–214, 1967.
3. Suh, C. H., and Radcliffe, C. W., "Synthesis of spherical mechanisms with use of the displacement matrix," *ASME, J. Eng. Ind.*, Ser. B, Vol. 89, No. 2, pp. 215–222, 1967.
4. Suh, C. H., "Design of space mechanisms for rigid body guidance," *J. Eng. Ind., Trans. ASME*, 89B, pp. 215–222, 1967.
5. Roth, B., "On the screw axes and other special lines associated with spatial displacements of a rigid body," *J. Eng. Ind., Trans. ASME*, Series B, 89, pp. 102–110, 1967.
6. Roth, B., "Finite position theory applied to mechanism synthesis," *J. Appl. Mech., Trans. ASME*, Series E, 34, pp. 599–605, 1967.
7. Roth, B., "The design of binary cranks with revolute, cylindric and prismatic joints," *J. of Mech.*, Vol. 3, No. 2, pp. 61–72, Summer 1968.
8. Chen, P., and Roth, B., "Design equations for the finitely and infinitesimally separated position synthesis of binary links and combined link chains," *J. Eng. Ind., Trans. ASME*, 91B, pp. 209–219, 1969.
9. Sandor, G. N., and Bisshopp, K. E., "On a general quaternion-operator method of spatial kinematic synthesis by means of a stretch rotation tensor," *J. Eng. Ind., Trans. ASME*, 91B, pp. 115–121, 1969.
10. Suh, C. H., "On the duality in the existence of R-R links for three positions," *J. Eng. Ind., Trans. ASME*, 91B, pp. 129–134, 1969.
11. Tsai, L. W., and Roth, B., "Design of dyads with helical, cylindrical, spherical, revolute, and prismatic joints, *mechanism and machine theory*," Vol. 7, No. 1, pp. 85–102, Spring 1972.
12. Tsai, L. W., and Roth, B., "A note on the design of revolute-revolute cranks, *mechanism and machine theory*" Vol. 8, No. 1, pp. 23–31, Spring 1973.

CHAPTER 7

1. Freudenstein, F., "Approximate synthesis of four-bar linkages," *Trans. ASME*, Vol. 77, pp. 853–861, August 1955.
2. Freudenstein, F., "Structural error analysis in plane kinematic synthesis," *ASME, J. Eng. Ind.*, Ser. E, Vol. 81, pp. 15–22, February 1959.
3. Freudenstein, F., "Four-bar function generators," *Trans. 5th Conf. Mech.*, Penton Publ. Co., Cleveland, pp. 104–107, 1958.

CHAPTER 8

1. Freudenstein, F., and Sandor, G. N., "Synthesis of path-generating mechanisms by means of a programmed digital computer," *Trans. ASME*, Vol. 81B, pp. 159–168, 1959.
2. Roth, B., and Freudenstein, F., "Synthesis of path-generating mechanisms by numerical methods," *Trans. ASME*, Vol. 85B, pp. 298–306, 1963.

CHAPTER 9

1. Freudenstein, F., "Four-Bar Function Generators," *Trans. The Fifth Conf. on Mechanisms*, Penton Pub., Cleveland, pp. 104–107, 1958.
2. Himmelblau, D. M., *Applied Nonlinear Programming*, McGraw-Hill, New York, 1972.

3. Powell, M. J. D., "An efficient method for finding the minimum of a function of several variables without calculating derivatives," *Computer J*. 7, pp. 155–162, 1964.
4. Powell, M. J. D., "A method for minimizing a sum of squares of non-linear functions without calculating derivatives," *Computer J*. 7, pp. 303–307, 1965.
5. Carroll, C. W., *Operations Res*., Vol. 9, No. 169, 1961; Ph.D. dissertation, Institute of Paper Chemistry, Appleton, Wis., 1959.
6. Fiacco, A. V., and McCormick, G. P., *Nonlinear Programming: Sequential Unconstrained Minimization Techniques*, Wiley, New York, 1968.
7. Gustavson, R. E., "Computer-Aided Design of Window Regulator Mechanisms," ASME paper 66-Mech-41, 1966.

CHAPTER 10

1. Weatherburn, C. E., *Differential Geometry of Three Dimensions*, Cambridge University Press, London, Vol. 1, pp. 135–139, 1927.
2. Tesar, D., and Eschenbach, P. W., "Four Multiply Separated Positions in Coplanar Motion," *Trans. ASME, Journal of Engineering for Industry*, Series B, Vol. 89, No. 2, pp. 231–234, May 1967.
3. Bottema, O., "On Instantaneous Invariants," *Proc. Int. Conf. Mechanisms*, Yale University, Shoe String Press, New Haven, Conn., pp. 159–164, 1961.
4. Sandor, G. N., and Freudenstein, F., "Higher-order Plane Motion Theories in Kinematic Synthesis," *Trans. ASME, Journal of Engineering for Industry*, Series B, Vol. 89, No. 2, pp. 223–230, May 1967.
5. Veldkamp, G. R., "Some Remarks on Higher Curvature Theory," *Trans ASME, Journal of Engineering for Industry*, Series B, Vol. 89, No. 1, 1967, pp. 84–86.
6. Roth, B., and Yang, A. T., "Application of Instantaneous Invariants to the Analysis and Synthesis of Mechanisms", *ASME paper 76-DET-12*, 1976.

CHAPTER 11

1. Pipes, L. A., *Matrix Methods for Engineering*, Prentice-Hall, Englewood Cliffs, N.J., 1963.

Problems

CHAPTER 2

1. Prove the validity of Eq. 2.6 by considering the operator $e^{i\theta}$ as a vector of unit length in the complex plane.
2. Prove Eqs. 2.8 and 2.9 using $e^{i\phi}$ as a rotation operator.
3. Verify the results given for program FOURBAR for $\theta_2 = 0.0°$ using Eqs. 2.25 and 2.28.
4. Complete the solution of Eqs. 2.29 to give analytical expressions for $\dot{\theta}_3, \ddot{\theta}_3, \dot{x}$, and $\ddot{x}$ in the offset slider-crank mechanism shown in Figure 2.3.
5. Complete the solution of Eqs. 2.30 to give analytical expressions for $\dot{\theta}_4, \ddot{\theta}_4, \dot{r}_4$, and $\ddot{r}_4$ in the crank-shaper mechanism shown in Figure 2.4.
6. Write a FORTRAN program CPLR that will compute the position, velocity, and acceleration of a specified point on the coupler of a four-bar linkage based on the algorithm outlined in Section 2.5. Check your program using the results generated for point 6 by program FOURBAR.
7. Write a FORTRAN program OFFSET that will compute the position, velocity, and acceleration of point **b** in the offset slider-crank mechanism shown in Figure 2.3. Check by a graphical solution of the position, velocity, and acceleration vector equations using the complex polar unit vectors as direction indicators.
8. Write a FORTRAN program SHAPER that will compute the angular position, velocity, and acceleration of the output link 4 in the crank-shaper mechanism shown in Figure 2.4. Check by graphical solution of the vector equations.
9. The Stephenson six-bar linkage shown in Figures 1.3a and 1.3c can be classified as kinematically complex under conditions where the vector loop equation contains more than two scalar unknowns (e.g., around the loop 1-2-3-5-6 in Figure 1.3aII). If two "independent" loops can be identified, with no more than four scalar unknowns, the resulting set of four scalar equations can be solved for four unknown kinematic parameters.

   Write two independent complex vector loop equations for the Stephenson linkage shown in Figure 1.3aII and discuss the solution of these equations and their derivatives with θ_2 as the input motion parameter.

CHAPTER 3

1. Derive

$$[\dot{W}] = \begin{bmatrix} -\dot{\theta}^2 & -\ddot{\theta} \\ \ddot{\theta} & \dot{\theta}^2 \end{bmatrix}$$

in a manner analogous to the derivation of Eq. 3.55.

2. Prove that either sequence of matrix multiplication

$$[R_{\psi, \theta, \phi}] = [R_{\phi, z''}][R_{\theta, x'}][R_{\psi, z}]$$

or

$$[R_{\psi, \theta, \phi}] = [R_{\psi, z}][R_{\theta, x}][R_{\phi, z}]$$

will lead to the Euler form of the rotation matrix as discussed in Section 3.3.

3. Derive the formulas to find the rotation pole P_0 and the rotation angle θ in terms of the elements d_{ij} of a given displacement matrix $[D]$. Use these relations to solve for $P_0 = (P_{0x}, P_{0y})$ and θ from the following numerical matrix elements.

$$[D] = \begin{bmatrix} 0 & 1 & 1 \\ -1 & 0 & 3 \\ 0 & 0 & 1 \end{bmatrix}$$

4. A reference point on a rigid body is displaced from a position $P_1 = (1., 2.)$ to a position $P_2 = (2., 3.)$ while the body rotates through an angle $\theta_{12} = 50°$. Locate the finite rotation pole $P_0 = (P_{0x}, P_{0y})$ for the displacement.

5. Using relationships from Section 3.9, calculate the screw motion parameters for a rigid body displacement described by the matrix

$$[D] = \begin{bmatrix} 0 & 0 & 1 & 1 \\ 1 & 0 & 0 & 5 \\ 0 & 1 & 0 & 5 \\ 0 & 0 & 0 & 1 \end{bmatrix}$$

6. Prove that any general screw displacement may be resolved into a series of two 180° rotations such that the rotation axes are both perpendicular to the screw axis with the common perpendicular distance between the axes equal to one half the screw linear displacement s and a twist angle between the axes equal to one half the screw angular displacement θ.

7. Prove that a screw displacement s, ϕ about an axis $\mathbf{u}$ can be resolved into a sequence of rotations about three fixed axes all perpendicular to the original screw axis $\mathbf{u}$ such that
(a) The first rotation is any angle θ.
(b) The second rotation angle equals 180°.
(c) The third rotation equals $180 - \theta$.
where a distance $s/2$ and rotation $\theta/2$ separates successive rotation axes. Let the z-axis be the original screw axis to simplify the proof without loss of generality.

8. Using the results of problem 6, resolve a screw displacement described by

$$\mathbf{u} = (0, 1, 0) \qquad P = (0, 0, 3) \qquad s = 5 \qquad \theta = 60°$$

into three rotations about axes lying in the x-y plane.

9. Resolve the screw displacement of problem 8 into a sequence of three rotations about axes all of which are perpendicular to the original screw axis. The first axis passes through the origin of the coordinates, and the first and second rotation angles are 90 and 180°, respectively.

10. Resolve the screw displacement of problem 8 into two rotations described by displacement matrices $[D_1]$ and $[D_2]$ such that

$$P_{1x} = 0. \text{ for } [D_1]$$

and

$$\mathbf{u}_2 = (1, 0, 0), P_{2x} = P_{2y} = 0., \phi_2 = 90° \text{ for } [D_2]$$

11. Complete the analysis of the offset slider-crank mechanism as indicated in Section 3.11. Solve matrix equation 3.53 for unknowns s_1, θ_2, and θ_3 in terms of specified a_2, a_3 and input angle θ_4.

CHAPTER 4

1.

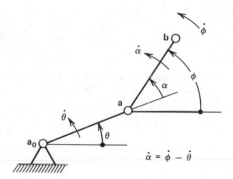

$$\dot{\alpha} = \dot{\phi} - \dot{\theta}$$

The angle α describes the *relative* angular displacement about joint **a** in the two-link dyad shown. Prove the equivalence of the two vector equations for the calculation of the velocity of point **b**.

$$\dot{\mathbf{b}} = \dot{\theta} \times (\mathbf{a} - \mathbf{a}_0) + \dot{\phi} \times (\mathbf{b} - \mathbf{a})$$

or

$$\dot{\mathbf{b}} = \dot{\theta} \times (\mathbf{b} - \mathbf{a}_0) + \dot{\alpha} \times (\mathbf{b} - \mathbf{a})$$

2. For the two-link assembly of problem 1, prove the equivalence of the two forms:

$$\ddot{\mathbf{b}} = \ddot{\theta} \times (\mathbf{a} - \mathbf{a}_0) + \dot{\theta} \times (\dot{\theta} \times (\mathbf{a} - \mathbf{a}_0))$$
$$+ \ddot{\phi} \times (\mathbf{b} - \mathbf{a}) + \dot{\phi} \times (\dot{\phi} \times (\mathbf{b} - \mathbf{a}))$$

or

$$\ddot{\mathbf{b}} = \ddot{\theta} \times (\mathbf{b} - \mathbf{a}_0) + \dot{\theta} \times (\dot{\theta} \times (\mathbf{b} - \mathbf{a}_0))$$
$$+ \ddot{\alpha} \times (\mathbf{b} - \mathbf{a}) + \dot{\alpha} \times (\dot{\alpha} \times (\mathbf{b} - \mathbf{a}))$$
$$+ 2\dot{\theta} \times (\dot{\alpha} \times (\mathbf{b} - \mathbf{a}))$$

3.

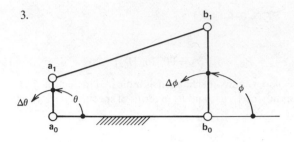

$$\mathbf{a}_0 = (0, 0)$$
$$\mathbf{a}_1 = (0, 1)$$
$$\mathbf{b}_0 = (3, 0)$$
$$\mathbf{b}_1 = (3, 2)$$

The first position of a four-bar linkage is given. Calculate the angular displacement $\Delta\phi$ of the output link for a specified input crank displacement $\Delta\theta = 30°$.

4. Calculate the relative angular velocity $\dot{\alpha}$ and relative angular acceleration $\ddot{\alpha}$ at joint $\mathbf{a}$ for the four-bar linkage of problem 3 in position 1 with $\dot{\theta} = 100$ rad/sec and $\ddot{\theta} = 0$.

5.

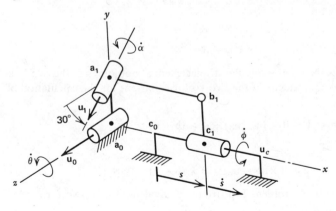

$$\mathbf{a}_0 = (0, 0, 0) \quad \mathbf{u}_0 = (0, 0, 1)$$
$$\mathbf{a}_1 = (0, 1, 0) \quad \mathbf{u}_1 = (\sin 30, 0, \cos 30)$$
$$\mathbf{b}_1 = (2, 1, 0) \quad \mathbf{c}_0 = (1, 0, 0)$$
$$\mathbf{c}_1 = (2, 0, 0) \quad \mathbf{u}_c = (1, 0, 0)$$
$$\Delta\theta = (\theta_2 - \theta_1) = 30°$$

Derive an algorithm for the *displacement* analysis of the RRSC mechanism shown. Calculate the motion parameters s_{12} and ϕ_{12} at joint c for the data given.

6. Derive an algorithm for the velocity of an RRSC mechanism similar to that shown in problem 5. Calculate motion parameters $\dot{s}_1$ and $\dot{\phi}_1$ in position 1 as shown.

7. Derive an algorithm for acceleration analysis of the RRSC mechanism of problem 5. Using the results of problem 6, calculate motion parameters $\ddot{s}_1$ and $\ddot{\phi}_1$.

8. Write a FORTRAN program RSSR that will carry out a displacement, velocity, and acceleration analysis for an RSSR mechanism specified as shown in Figure 4.5. Test using the results given for Example 4.2.

9. Write a FORTRAN program RRSS that will carry out a displacement, velocity, and acceleration analysis for an RRSS mechanism specified as shown in Figure 4.6. Test using results given for Example 4.3.

CHAPTER 6

1. Derive the equation of the circle of sliders, Eq. 6.24, from Eq. 6.23.

2. Write a FORTRAN program that will locate the moving pivot **a**, for a specified pivot $\mathbf{a}_0$ of a two-joint crank which satisfies the design equations associated with three finitely separated positions of a plane rigid body. Test your program using the data for Example 6-1.

3. Write a FORTRAN program that will locate the circle of sliders for three finitely separated positions of a plane rigid body. After the circle has been defined, the program should find the first position of a slider that moves at a specified angle alpha. Test using the data from Examples 6-2 and 6-4.

4. Write a FORTRAN program that will locate the circle of sliders for three finitely separated positions of a plane rigid body. Compute the two values of b_{1y} and α for a specified b_{1x}. Test using the data for Example 6-3. Note that the program must test for both the possibility that $\theta_{ij} = 0$ and that real solutions for b_{1y} may not exist.

5. Write a FORTRAN program that will solve the crank design equations for two finitely separated positions of a plane with velocity specified in the first position. Assume $\mathbf{a}_0$ is specified and solve for **a**, using the data from Example 6-5.

6. Write a FORTRAN program that will solve the crank design equations with velocity and acceleration of a plane specified in one position. Assume $\mathbf{a}_0$ is specified and solve for $\mathbf{a}_1$ using the data from Example 6-6.

7. Select any of problems 2 to 6 and carry out all calculations by slide rule or calculator.

8. Example 12-3 illustrates the use of PROGRAM DESIGN (listed in the appendix) for the solution of a set of nonlinear algebraic equations representing the intersection of a sphere, a plane, and a parabolic surface. The equations representing these surfaces are supplied to the Newton-Raphson equation solver SIMEQS by a *user prepared* FUNCTION YCOMP. The equations are first converted into the form YCOMP $= f_j(z_1, z_2, z_3, \ldots z_n) = 0$, then accessed in FUNCTION YCOMP(Z, J) by the value of J generated within SIMEQS. Example 12-3 illustrates access by use of a simple GO TO (1, 2, 3) J

statement. Obtain a copy of **PROGRAM DESIGN**, prepare cards for **FUNCTION YCOMP(Z, J)**, prepare the data cards as indicated either in the program listing or the output for Example 12-3, and execute the program with several different sets of initial guesses.

9. A program similar to **PROGRAM DESIGN** but written in the BASIC programming language is also given in the appendix. The equations from Example 12-3 have been inserted after statement 1000 as a test problem. Note that a linear equation solver has been included as part of the Newton-Raphson procedure. To execute the linear equation solver separately, begin execution with GO TO 6000. To solve the nonlinear equations after statement 1000, begin execution with RUN. Enter the BASIC program into the computer and test with Example 12-3 or an arbitrary set of linear equations.

10. Duplicate the results given by Example 12-4 for the data of Example 6-7. Rerun using a new variable parameter as suggested in Example 12-4.

11. Prepare a **FUNCTION YCOMP(X, J)** for the constraint equations given for Example 6-8. Test by solving Example 6-8 using Figure 6-8 as a guide for initial guesses. Develop the center point and circle point curves in small sections, changing the incremented variable as indicated by the shape of the curves.

12. Assume a plane moves through four finitely separated positions, as specified in Examples 6-1 and 6-7. Locate the first position of a unique slider pivot that will pass through four positions on a straight line. (*Hint.* Solve as the intersection of two circles of sliders associated with positions 1, 2, 3 and 2, 3, 4 with $\theta_{12} = 0$.)

13. Four positions of a point in space are specified as

$$\mathbf{p}_1 = (1.00, \ 1.00, \ 2.00)$$

$$\mathbf{p}_2 = (2.50, \ 1.75, \ 2.50)$$

$$\mathbf{p}_3 = (4.00, \ 2.00, \ 2.00)$$

$$\mathbf{p}_4 = (4.50, \ 1.00, \ 1.00)$$

Show that these points lie on the surface of a sphere described by the equation

$$(x - 2.094)^2 + (y - 2.719)^2 + (z + 0.797) = (3.460)^2$$

14. The four-bar linkage shown in Figure 6.15 has been used successfully as a prosthetic knee mechanism for above-the-knee amputees. The thigh section is guided with respect to the shank section in a manner designed to aid the amputee in the control of knee stability during weight bearing on the prosthesis.

   The stability of the knee prosthesis is influenced by the shape of the fixed shank centrode. With the shank as the fixed reference member point **a** on the fixed centrode is seen to lie on the intersection of lines $\mathbf{a}_0\,\mathbf{a}$ and $\mathbf{b}_0\,\mathbf{b}$. Assume the linkage is displaced from the initial position shown by a series of coupler displacement angles $\theta = 0, 5, 10, 20, 30, 60, 90,$ and $120°$. Calculate the displacement of points **a** and **b** and the corresponding points on the fixed centrode. Compare with the centrode given in Figure 6.15.

15. Using the results of problem 14, calculate displacement matrices $[D_{1j}]$, $j = 2, 4$ for $\theta = 10, 20$, and $30°$ and generate center-point and circle-point curves for the specified thigh section motion using the procedure outlined in Section 6-7.

16. Select a pair of guiding links $\overline{a_0\,a}$ and $\overline{b_0\,b}$ from the center-point and circle-point curves of problem 15 that would result in an equivalent but smaller linkage. Evaluate your new design by comparing its fixed centrode to the original linkage. What other considerations might be used to compare the linkages?

17. Derive expressions for constants A, B, C, and D in Eq. 6.51 in terms of specified coordinates of a point p_1, p_2, p_3, and p_4. Evaluate the numerical value of the constants for the data of problem 13.

18. As indicated in Section 6.10, a spherical linkage with all axes intersecting at a common point can be designed to guide a rigid body through four positions in space with fixed and moving axes defined by spherical center-point and circle-point curves. Assume four points on a general curve in space are specified similar to the data of problem 13. Spherical motion of a rigid body is conveniently described by two spherical angles, as shown in Figure 8.3. For assumed or specified angles β_{1j}, outline a procedure for calculating the elements of a displacement matrix $[D_{1j}]$ in terms of p_1, p_j, and β_{1j}..

19. Using the procedure developed in problem 18 to calculate displacement matrices, calculate spherical center-point and circle-point curves to be plotted on a unit sphere for the point path data of problem 13 and assumed values of $\beta_{12} = 0.$, $\beta_{13} = 30°$, and $\beta_{14} = 60°$.

20. Displacement of a rigid body through three positions in space is specified in terms of the following screw motion parameters.
    Displacement $[D_{12}]$

$$p_1 = (0.9414, 0.5214, -0.4194) \qquad s_{12} = 1.384$$
$$u_{12} = (0.3510, 0.0965, 0.9280) \qquad \phi_{12} = 133.2°$$

    Displacement $[D_{13}]$

$$p_1 = (-0.6103, 0.8371, 0.1230) \qquad s_{13} = 1.899$$
$$u_{13} = (0.1457, -0.0390, 0.9886) \qquad \phi_{13} = 70.6°$$

    Calculate numerical elements for displacement matrices $[D_{12}]$, $[D_{13}]$, and $[D_{23}]$

21. Write a function YCOMP that will locate a sphere-sphere link for guidance of a rigid body through three positions as specified in problem 20. Specify a_0 and solve for a_1.

22. Add a fourth displacement to the rigid body motion data of problem 20 described by the following parameters.
    Displacement $[D_{14}]$

$$p_1 = (0.7327, 0.6889, -0.1643) \qquad s_{14} = -1.117$$
$$u_{14} = (0.4267, -0.2464, 0.8702) \qquad \phi_{14} = 87.9°$$

    Calculate numerical elements for displacement matrix $[D_{14}]$.

23. The following problems are independent of each other, but the results can be combined in various ways to form complete mechanisms for rigid body guidance. The programs can be executed using the displacement of problems 20 or 22. Possible complete mechanisms should be tested using the kinematic analysis programs developed in Chapter 4.

24. Three finitely separated positions of a rigid body are specified by the data of problem 20.

   (a) Write a function YCOMP that, when executed with program DESIGN, will solve for unknown coordinates a_{1y} and a_{1z} of moving pivot $\mathbf{a}_1$ for a possible sphere-sphere link. Fixed pivot $\mathbf{a}_0$ and coordinate a_{1x} are to be specified.

   (b) Write a function YCOMP that will solve for link parameters $\mathbf{u}_0, \mathbf{u}_1, \mathbf{a}_0$, and $\mathbf{a}_1$ of a possible revolute-revolute guiding link with no free choice of variable.

   (c) Write a function YCOMP that will solve for link parameters $\mathbf{u}_0, \mathbf{a}_0, \mathbf{u}_1, \mathbf{a}_1, s_{12}$, and s_{13} for a possible revolute-cylinder guiding link. Any two scalar variables can be assumed.

   (d) Write a function YCOMP that will solve for link parameters $\mathbf{u}_0, \mathbf{a}_0, s'_{12}, s'_{13}, \mathbf{u}_1, \mathbf{a}_1$, s_{12}, and s_{13} for a possible cylinder-cylinder guiding link.
   How many scalar variables can be assumed in this case?

25. Four finitely separated positions of a rigid body are specified by the data of problems 20 and 22.

   (a) Write a function YCOMP that will solve for the coordinates of a moving joint $\mathbf{a}_1$ on a sphere-sphere link with specified fixed joint $\mathbf{a}_0$.

   (b) Write a function YCOMP that will solve for link parameters $\mathbf{u}_0, \mathbf{a}_0$, and $\mathbf{a}_1$ for a possible revolute-sphere guiding link with fixed revolute axis $\mathbf{u}_0$. One scalar variable may be specified.

   (c) Write a function YCOMP that will solve for link parameters $\mathbf{a}_0, \mathbf{u}_1$, and $\mathbf{a}_1$ for a possible sphere-revolute guiding link with moving revolute axis $\mathbf{u}_1$. One scalar variable may be specified.

   (d) Write a function YCOMP that will solve for link parameters $\mathbf{u}_0, \mathbf{a}_0, s'_{12}, s'_{13}, s'_{14}$, $\mathbf{u}_1, \mathbf{a}_1, s_{12}, s_{13}$, and s_{14} for a possible cylinder-cylinder guiding link. How many scalar variables can be specified?

CHAPTER 7

1. A function $y = x^2 + 1$ is to be approximated by a four-bar linkage function generator over the range $1 \leq x \leq 5$. The range in input angle θ (proportional to x) equals $60°$. The range in output angle θ (proportional to y) equals $-90°$.

   (a) Calculate k_θ and k_ϕ.

   (b) Calculate $\theta_i, \phi_i, i = 1, 4$ for Chebyshev spacing of precision points.

2. Assume a schedule of input and output angles to be approximated by a four-bar linkage function generator as listed below. The length of the fixed link $r_1 = 1.0$. Calculate the elements of the relative displacement matrix $[D_r]$ that describe the motion of the input crank relative to position 1 of the output crank for the specified position 3.

Position	θ, deg	ϕ, deg
1	10.0	30.0
2	20.0	40.0
3	30.0	55.0
4	40.0	75.0

3. Assume a schedule of input crank rotation angles θ and output slider positions x to be approximated by a slider-crank mechanism as listed below. Calculate elements of the relative displacement matrix $[D_r]$ that describe motion of the input crank relative to position 1 of the slider for specified position 2.

Position	θ	x
1	0.0	0.0
2	− 10.0	1.0
3	− 20.0	2.5
4	− 30.0	4.5

4. Assume a schedule of input slider positions x and output slider positions y as given. Calculate elements of the relative displacement matrix $[D_r]$ that describe the motion of the input slider in position 4 relative to the first position of the output slider.

Position	x, Input	y, Output
1	1.0	1.0
2	2.0	2.0
3	3.0	4.0
4	4.0	7.0

5. Write a FORTRAN program that will solve for the first position of the input crank moving pivot a_1 for a specified first position of the output crank moving pivot b_1 for combined velocity-acceleration synthesis of a four-bar function generator to generate output motion $\dot{\phi}$, $\ddot{\phi}$ for specified input motion $\dot{\theta}$, $\ddot{\theta}$. Test your program using the results of Example 7-4.

6. Write a FORTRAN program that will solve for the first position of the input crank moving pivot a_1 for a specified first position of the slider pivot b_1 and slider direction α for combined velocity-acceleration synthesis of a slider-crank function generator. Test your program using the results of Example 7-5.

7. Write a FORTRAN program that will synthesize the first position $\mathbf{b}_1$ and slope β for the double-slider function generator with velocity-acceleration motion specifications. The initial position $\mathbf{a}_1$ and slope α for the input slider are specified. Test your program using the results for Example 7-6.

8. Write a function YCOMP that will solve for points on the relative center-point and relative circle-point curves for the four-bar function generator with four finitely separated precision points. Note that the relative displacement matrices describe motion of the input crank with respect to the first position of the output crank; hence a circle point becomes point $\mathbf{a}_1$ and a center point is located at $\mathbf{b}_1$ in the constraint equations.

$$(\mathbf{a}_j - \mathbf{b}_1)^T(\mathbf{a}_j - \mathbf{b}_1) - (\mathbf{a}_1 - \mathbf{b}_1)^T(\mathbf{a}_1 - \mathbf{b}_1) = 0 \qquad j = 2, 3, 4$$

Test your function YCOMP using program DESIGN for at least five points on the center-point and circle-point curves given in Figure 7-6 for the data of Example 7-7.

9. Write a function YCOMP to solve for points on the spatial center-point and circle-point curves for a four precision point RSSR spatial function generator. Test your program using the data and results given for Example 7-8.

10. Write a function YCOMP to solve the six precision point RSSR function generator problem as outlined in Example 7-9. Test using the data and results given for Example 7-9.

11. Example 7-10 illustrates the synthesis of a spatial function generator with velocity-acceleration motion specifications. Write a function YCOMP to solve Example 7-10.

CHAPTER 8

1. In the synthesis of a plane four-bar linkage path generation linkage the unknowns are the fixed pivots $\mathbf{a}_0$ and $\mathbf{b}_0$, the first position of the moving pivots $\mathbf{a}_1$ and $\mathbf{b}_1$ plus unspecified coupler rotation angles θ_{1j}.
   (a) Write a function YCOMP that will solve the constraint Eqs. 8.1 for five specified path points $\mathbf{p}_j$, $j = 1, 5$ and specified fixed pivots $\mathbf{a}_0$ and $\mathbf{b}_0$. θ_{1j}, $j = 2, 5$ are to be left as free variables. Test using the results given for Example 8-1.
   (b) Add constant length constraint as given in Example 8-2 and solve for the conditions specified for Example 8-2.

2. Write a function YCOMP and solve for the first position of an RRSS mechanism that would guide a point through the four positions given in Example 8-3. The rigid body rotation angles for the coupler are free variables.

CHAPTER 9

1. Example 12-7 illustrates the solution of Example 9-2 as a problem in constrained minimization with the addition of inequality constraints

$$0.5 \le a_{1x} \le 1.0$$

$$0.75 \le a_{1y} \le 1.0$$

and the equality constraint

$$b_{1x} = b_{1y}$$

(a) Write a function FFUN for use with program PCON that will solve problem 9-1, optimal synthesis of a plane four-bar function generator. Test by solution of the 5 and 21 design point problems outlined in Table 9.1.

(b) Add inequality constraints to limit the location of points a_1 and b_1 to circles of radius $r = 0.25$ about $a_c = (-.12, 0.25)$ and $b_c = (0.75, -0.25)$, respectively. Solve for a_1 and b_1.

(c) Add an equality constraint $a_{1y} = -b_{1y}$. Solve for a_1, b_1 with both inequality and equally constraints active.

(d) Solve for a_1 and b_1 with only the equality constraint on the solution.

2. Duplicate the results of Example 9-3 as
(a) A seven-variable problem.
(b) A ten-variable problem.

3. Add inequality constraints to Example 9-3 that limit the solution such that a_0 and b_0 must lie in a region where

$$7.5 \leq a_{0y} \leq 10.0$$

$$7.5 \leq b_{0y} \leq 10.0$$

Solve for a_0, b_0, a_1, b_1, and d_1 with specified a_{0x}, a_{0y}, and d_{1x}.

4. Add an equality constraint to Example 9-3 that requires that the length of link $\overline{b_0 b_1} = 5$. Solve for a_0, b_0, a_1, b_1, and d_1 with specified u_{0x}, u_{0y}, and d_{1x}.

5. Solve Example 9-3 with both inequality constraints from problem 3 and equality constraints from problem 4 active.

CHAPTER 10

1. Write a function YCOMP that will solve Eqs. 10.20 for unknown first-order screw motion parameters p_{0x}, p_{0x}, u_1, ϕ, and $\dot{s}_1$ for the coupler of the spatial double-slider mechanism shown in Figure 10.2 with specified input motion a and $\dot{a}$. The location of points b and c are specified, and $\dot{c} = \dot{a}$ as shown. Test using the data given for Example 10-1 with results given in Table 10.1.

2. Write a function YCOMP that will solve Eqs. 10.23 for unknown screw motion-acceleration parameters $\ddot{p}_{0x}$, $\ddot{p}_{0y}$, $\dot{u}_1$, $\ddot{\phi}$, $\ddot{s}_1$ for the coupler of the spatial double-slider mechanism of Figure 10.2 with specified input motion $a, \dot{a}, \ddot{a}, \ddot{c} = \ddot{a}$. Test using the data of Example 10-1 with results given in Table 10.3.

3. Using the results from problem 2, calculate the second-order screw parameters u_2, $\dot{\phi}_2$, and $\dot{s}_2$ and the location of the central point c. Compare results with Table 10.2.

4. A plane has motion described in terms of a reference point **p** and angular position θ. Assume

$$\mathbf{p} = (1., 1.) \qquad \dot{\mathbf{p}} = (2., 1.) \qquad \ddot{\mathbf{p}} = (0, -10.)$$

$$\dot{\theta} = 10. \qquad \ddot{\theta} = 100.$$

   (a) Calculate the location of the velocity pole $\mathbf{p}_0$ and acceleration pole $\mathbf{q}_0$.
   (b) Calculate the pole velocity $\dot{\mathbf{p}}_0$.
   (c) Show that Eq. 10.43 is satisfied.

5. Using the motion data given for problem 4, calculate the center and radius of the inflection circle and the zero tangential acceleration circle. Draw the circles to scale and locate the acceleration pole $\mathbf{q}_0$ at their intersection. Compare with the results for problem 4(a).

6. A double-slider mechanism was designed in Example 7-6 to meet specifications such that

$$\mathbf{a} = (0, 0) \qquad \dot{\mathbf{a}} = (0, 1) \qquad \ddot{\mathbf{a}} = (0, 2) \qquad \dddot{\mathbf{a}} = (0, 0)$$

$$\mathbf{b} = (5, 15) \qquad \dot{\mathbf{b}} = (3, 0) \qquad \ddot{\mathbf{b}} = (4, 0)$$

   (a) Calculate $\mathbf{b}$, $\dot{\theta}$, $\ddot{\theta}$, and θ for the coupler motion.
   (b) Calculate numerical elements for the $[D]$, $[\dot{D}] = [V]$, and $[\ddot{D}] = [A]$ matrices that describe the coupler motion.
   (c) Calculate a series of points on the cubic of stationary curvature that satisfy Eq. 10.68.

7. Assume the double-slider mechanism of Example 7-6 with the origin of the coordinate system relocated at the velocity pole $\mathbf{p}_0$.
   (a) Locate the inflection circle using Eq. 10.55.
   (b) Locate the zero tangential acceleration circle using Eq. 10.61.

8. Examples 10.2 through 10.4 in Chapter 10 illustrate the motion analysis of plane mechanisms on a purely *geometric* basis.
   In problems 9 and 10 a FORTRAN program is to be coded which, in either case, will solve a more general problem of the same class. The program should carry out the following procedure.
   a. Locate the first-order pole $\mathbf{p}_0$ in the orginal coordinate system.
   b. Shift the coordinate system to relocate $\mathbf{p}_0$ at the origin.
   c. Calculate $\mathbf{p}_0''$ in the new coordinate system. $\mathbf{p}_0' = (0., 0.)$.
   d. Rotate the coordinate system such that $p_{0x}'' = 0$.
   e. Calculate p_{0y}'', p_{0x}''', and p_{0y}''' in the rotated (canonical) system.
   f. Write the equation, locate the center, and calculate the radius of the inflection circle in terms of the instantaneous invariant $b_2 = p_{0y}''$.
   g. Write the equation of the cubic of stationary curvature in terms of the instantaneous invariants. $b_2 = p_{0y}'' \, a_3 = p_{0x}''' \, b_3 = p_{0y}'''$.

9. Write a FORTRAN program for the geometric analysis of a double-slider mechanism. The slider at **a** moves along the y-axis. The slider at **b** moves in an arbitrary direction making an angle β with the x-axis. Test using the results for Example 10.3 where $\beta = 0$.

10. Write a FORTRAN program for the geometric analysis of a four-bar linkage where the position of all pivots $\mathbf{a}_0$, $\mathbf{a}$, $\mathbf{b}_0$, and $\mathbf{b}$ can be specified arbitrarily.
   a. Test using the data of Example 10.4
   b. Test using the following pivot coordinates

$$\mathbf{a}_0 = (0., 0.) \qquad \mathbf{a} = (0., -1.28)$$
$$\mathbf{b}_0 = (3.80, 0.) \qquad \mathbf{b} = (2.1153, 1.9009)$$

   In the transformed (canonical) coordinate system the inflection circle becomes

$$x^2 + y^2 - 36.9679y = 0$$

   and the cubic of stationary curvature is

$$(x^2 + y^2)[.022505x + .193591y] - xy = 0.$$

CHAPTER 11

1. Example 12.2 gives the results of a computer program that carries out the dynamic analysis of a plane four-bar linkage. Using the results of the kinematic analysis, form the numerical matrix equation of motion for $\theta_2 = 0.0°$ using the analytical form of the matrix shown in Figure 11.3 as a guide.
   Compute the joint force components by solution of the set of linear equations represented by the matrix equation of motion. Compare with the results given in Examples 11.1 and 12.2.

2.

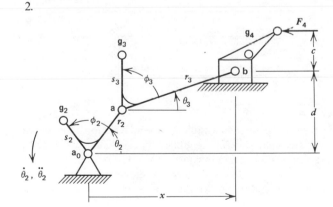

$$\mathbf{p}_2 = (\mathbf{a} - \mathbf{g}_2)$$
$$\mathbf{q}_2 = -\mathbf{s}_2 = (\mathbf{a}_0 - \mathbf{g}_2)$$
$$\mathbf{p}_3 = (\mathbf{b} - \mathbf{g}_3)$$
$$\mathbf{q}_3 = -\mathbf{s}_3 = (\mathbf{a} - \mathbf{g}_3)$$

Specified physical constants

$$m_2, I_2, m_3, I_3, m_4$$

   Draw the free body diagrams for members 2, 3, and 4 in the offset slider-crank mechanism shown using the positive leading joint force sign convention. Write the equations of motion for each member. Combine the equations of motion into a matrix equation of motion.

3. Extend the solution of problem 2.7, acceleration analysis of the offset slider-crank mechanism, to include computation of the acceleration of the mass center for links 2, 3, and 4. Compute numerical elements for the matrix equation of motion derived in

problem 2 for the data given below. Solve for unknown force components in Newtons
and the input crank torque T_2 in Newton-meters with the following data. All coordinates and dimensions are given in millimeters.

$$\mathbf{a}_0 = (0., 0.) \qquad d = 100. \qquad e_4 = 25. \qquad F_4 = 100. \text{ N}$$

$$r_2 = 100. \qquad s_2 = 50. \qquad \phi_2 = 0.°$$

$$r_3 = 250. \qquad s_3 = 100. \qquad \phi_3 = 45.°$$

$$m_2 = 1.0 \text{ kg} \qquad I_2 = 0.006 \text{ kg-m}^2$$

$$m_3 = 2.0 \qquad I_3 = 0.015$$

$$m_4 = 0.5$$

Input crank motion $\theta_2 = 60°$, $\dot{\theta}_2 = 100\pi$ rad/sec, $\ddot{\theta}_2 = 0.0$

4.

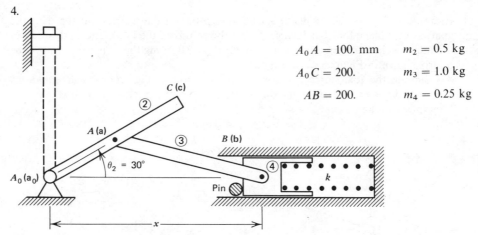

$$A_0 A = 100. \text{ mm} \qquad m_2 = 0.5 \text{ kg}$$
$$A_0 C = 200. \qquad m_3 = 1.0 \text{ kg}$$
$$AB = 200. \qquad m_4 = 0.25 \text{ kg}$$

The slider-crank mechanism shown is used as a rapid closure device for an electrical
switch. The mechanism is activated when the pin is pulled by a solenoid not shown.
Release of the pin allows a precompressed spring of constant $k = 150$ N/m to accelerate
the piston 4 in the negative x direction. Mass centers for the switch blade 2 and
connecting rod 3 are located at their midpoints. Model links 2 and 3 as long slender
rods.

The free length of the spring, $l_0 = 250$ mm. It is compressed to a length $l = 150$ at the
time of release.

(a) Compute the angular velocity of the blade 2 when $\theta_2 = 90°$ after release with
$\theta_2 = 30°$ as shown. Use this result to estimate the approximate time t for closure of
the switch.

(b) Using a time step $\Delta t = t/10$, integrate the nonlinear differential equation
(Eq. 11.36) and compute $\dot{\theta}$ and θ versus t at the end of each step Δt. What is the
final angular velocity $\dot{\theta}$?

Index

Acceleration, coriolis component, 42
  matrix, 69
  plane mechanisms, 8, 74
  pole, 252, 256
  spatial mechanisms, 79
Angular acceleration matrix, plane and
    spatial, 68
Angular velocity matrix, plane, 65
  spatial, 68
Axis rotation matrix, 49
Axodes, 240

Bennett mechanism, 109
Bresse circle, 255

Case studies in optimal synthesis, 221
Center and circle point curves, plane rigid
    body guidance, 145
  spherical path generation, 199
  spherical rigid body guidance, 151
Central point, 242
Chebyshev spacing, 206
  precision points, 166
Circle of sliders, three finitely separated
    positions, 134
  velocity, specified in first position, 135
  velocity and acceleration, specified, 135
Complex mechanisms, 3, 5
Compound links, 3
Computer programs, 305
Conjugate directions, 213
  generation of, 237
  Powell method, 213
Constant length equations — two joint
    cranks, 129

Constant slope constraint — plane sliders,
    130
Constrained minimization — penalty functions,
    219
Constraint equations, combined finite-differ-
    ential synthesis, 146
  cylindrical-cylindrical (C-C) link, 158
  four position plane rigid body guidance, 144
  revolute-cylindrical (R-C) link, 157
  revolute-revolute (R-R) link, 155
  revolute-sphere (R-S) link, 154
  slider, acceleration constraint, 131
    four-position guidance, 146
    velocity constraint, 131
  sphere-sphere (S-S) link, 152
  spherical rigid body guidance, 149
  spherical two-joint cranks, 149
  two-joint cranks, 130
Coordinate transformations, dynamic
    analysis, 284
  matrix form, 286
  point, 61
  vector, 61
Coriolis component, 42
  direction of, 44
Crank acceleration constraint equations, 130
Crank displacement constraint equations, 129
Crank input motion parameters, 136
Crank synthesis, three position, 131
Crank velocity constraint equations, 130
Cross product, vector, 15
Cubic of stationary curvature, 256
  degenerate form, 260
  roth-yang form, 257, 259
Curvature, geometric properties, 269

stationary, 256

Degrees of freedom, 103
Design equations, crank, 133
Design points, 206
Differential displacement matrix, 69
Differential geometry of motion, 238
  spatial double-slider, 244
Differential rotation matrices, 65
Differentiation of vectors, Cartesian, 16
  complex polar, 9
Directrix, 241
Displacement analysis, plane four-bar path
    generator, 204
  spatial RRSS mechanism, 201
  spatial RSSR-SS mechanism, 202
  spherical path generator, 201
Displacement matrices, numerical, 56
  plane, 52
  spatial path generation, 200
  spherical, 14
  spherical, numerical inversion, 148
  spherical path generation, 200
Dot product, vector, 14
Double slider function generator, plane,
    172, 176
  spatial, 244
  three position synthesis, 173
Dyad, oscillating slider, 25
  rotating guide, 27
  two-link, 23
Dynamic analysis, inverse dynamics, 289, 293
  oscillating slider, 280
  plane four-bar linkage, 276
  spatial mechanisms, 294
Dynamic balancing of four-bar linkage, 283
Dynamics of mechanisms, 276
Dynamics of spatial mechanisms, 294
  coordinate transformations in, 284
  in matrix form, 299
  spatial equations of motion, 297

Equality constraints, 207, 209, 221
Equivalent lower-pair mechanisms, 106
Euler equations, 298
  for a spatial rotating system, 303
Euler rotation matrix, 51

First-order screw axis, 241
Four-bar function generator, plane, 173
Four-bar linkage — kinematic analysis,

Cartesian vector notation, 17
  complex polar vector notation, 9
  relative joint rotation angles, 74
Four-bar path generation, five path precision
    points, 193
Four position plane rigid body guidance,
    crank constraint, 141
  slider constraint, 146
Function generator mechanisms, combined
    finite-differential synthesis, 173
  plane double-slider synthesis, 172, 176
  plane relative displacement matrix, 168
  plane slider-crank synthesis, 169, 175
  plane three-position guidance, 167
  scale factors for input and output, 167
  spatial RSSR four-bar, 181
  spherical four-bar, 179

Geometric analysis of plane mechanisms, 257
Geometric properties of spatial curves, 269
Global minimum, 207
Grashof criterion, 112
Grublers criterion, 103

Hartenberg-denavit notation, 62
Hessian, 211

Inequality contraints, 207
Inflection circle, 253
Inflection point, 253
Input crank motion parameters, 136
Instantaneous invariants, 257
  higher-order plane synthesis, 265
Instantaneous screw parameters, 239
Instantaneous screw calculus, 242
Instant pitch, 241
Inverse displacement matrix, 58
Inverse dynamics problem, 276, 289
  example, four-bar linkage, 293

Kinematic analysis, Cartesian vector notation,
    17
  complex polar notation, 9
  spatial numerical, 94
Kinematic analysis in closed form, crank-
    shaper mechanism, 14
  four-bar linkage, 9, 17
  offset slider-crank, 13
  oscillating slider, 19
  RCCC mechanism, 89
  rotating guide, 41

RRSS mechanism, 83
RSSR mechanism, 79
Kinematic inversion, 3
   plane function generator, 168
Kutzbach criterion, 105

Least squares minimization, 216
Local minimum, 207

Matrices, alpha-beta-gamma angles, 50
   axis rotation, 49
   Euler angles, 51
   inverse displacement, 58
   inversion, 57
   phi matrices, 248
   plane displacement, 52
   plane rotation, 46
   screw displacement, 55
   spatial displacement, 53, 55
   spatial rotation, 47
Matrix equation of motion, plane four-bar
      linkage, 279
   plane oscillating slider mechanism, 281
   spatial mechanisms, 299
Maverick mechanisms, 109
Minimum, mathematical properties of, 211
   search for, 212, 214
Mobility chart, plane four-bar linkage, 115,
      116
   RRSS mechanism, 124
   RSRC mechanism, 126
   RSSR mechanism, 119
Motion of mass center, 283

Newton-Raphson method, 141
Nonderivative search for minimum, 214
Nonlinear differential equations of motion,
      291
Nonlinear equations, damping to avoid
      instability, 144
   solution of, 141
Nonlinear programming, 207
   geometrical representation, 209
Number synthesis of plane linkages, 111

Objective function, 207
Optimal kinematic synthesis, 207
Optimal precision point spacing, 206
Optimization, case studies, 221
   conjugate directions, 213
   constrained, 219

least squares, 216
optimal kinematic synthesis, 206
Powell method, 215
scaling, 221
steepest descent, 212
unconstrained, 211
Overconstrained spatial mechanisms, 109

Parameter of distribution, 242
Partial derivatives, numerical approximation,
      218
Path curvature analysis, parametric equations,
      268
   plane, 251, 253
   spatial, 268, 269
Path generation mechanisms, displacement
      analysis, 204
   plane four-bar, 190
   RRSS, 201
   RSSR-SS, 202
   spherical four-bar, 195
Penalty functions, interior, 219
   exterior, 220
Phi matrices, 248
   plane, 252
   spatial, 248
Plane path curvature, 251
Plane path generation, five path precision
      points, 193
   with specified crank lengths, 193
Plane rigid body guidance, 129
Pole, acceleration, 252
   finite rotation, 54
   velocity, 251
Pole velocity, 252
Powell direct search method, 215
Powell least squares minimization, 217
Precision points, Chebyshev spacing, 165, 166
   optimal spacing of, 206
   specification of, 164
Predictor — corrector equations, 282
Pressure angle, 6
Programs, use of, LINKPAC, 305
   LSTCON, 335
   NONLIN, 338
   PCON, 316
   PRCR2, 341
   SIMEQ in basic, 330

Quadratic approximation of functions, 212,
      214

Quadratic form, 211

Radius of curvature, spatial, 269
Radius of torsion, 270
Range of motion analysis, 114
Reduced force or torque, 280
Reduced mass or inertia, 289
Relative motion matrices, plane four-bar
    function generator, 168
  relative acceleration, 173
  relative velocity, 173
  RSSR function generator, 182
  spherical function generator, 179
Relative spatial motion, 70
Rigid body guidance, 128
  C-C link, 158
  plane four-bar linkage, 129
  R-C link, 157
  R-R link, 155
  R-S link, 154
  spatial three-position, 159
  spherical linkage, 148
  S-S link, 152
Rotation matrices, differential, 65
  plane, 46
  spatial, 64

Screw axis surfaces, 241
Screw motion parameters, 59
  central point, 242
  directrix, 241
  displacement, 55, 59
  first-order, 245
  instant pitch, 241

  parameter of distribution, 242
  second-order, 245
Second angular acceleration matrix, 68
Slider constraint equations, 131
Slider synthesis, three position, 134
Solution of nonlinear equations, 141
Spatial curves, geometric properties, 269
Spatial rigid body guidance, 152
Spherical mechanisms, function generator,
    179
  path generation, 195
  rigid body guidance, 148, 151
Stationary curvature, 256
Steepest descent, 212
Stephensons linkage, 111
Straight line path, equation of, 130
Structural error, 165, 206

Taylors series, 212
Three position crank synthesis, 131
Transmission angle, 6
Two-link dyad, 23

Unconstrained minimization, 211

Vector product, cross, 15
  dot, 14
Veldkamp notation, 258
Velocity pole, 251

Watts linkage, 111
Work-energy method, 280

Zero tangential acceleration circle, 255